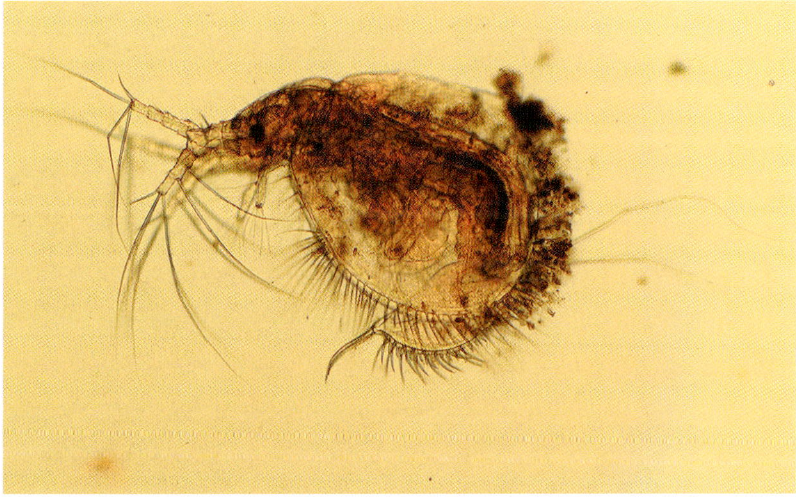

Mud-living Ilyocryptus colored red by hemoglobin. *Photographed by A.A. Kotov.*

PHYSIOLOGY OF THE CLADOCERA

PHYSIOLOGY OF THE CLADOCERA

NIKOLAI N. SMIRNOV D.SC.
Institute of Ecology, Moscow, Russia

WITH ADDITIONAL CONTRIBUTIONS

ELSEVIER

AMSTERDAM • BOSTON • HEIDELBERG • LONDON
NEW YORK • OXFORD • PARIS • SAN DIEGO
SAN FRANCISCO • SINGAPORE • SYDNEY • TOKYO
Academic Press is an imprint of Elsevier

Academic Press is an imprint of Elsevier
32 Jamestown Road, London NW1 7BY, UK
225 Wyman Street, Waltham, MA 02451, USA
525 B Street, Suite 1800, San Diego, CA 92101-4495, USA

British Library Cataloguing-in-Publication Data
A catalogue record for this book is available from the British Library

Library of Congress Cataloging-in-Publication Data
A catalog record for this book is available from the Library of Congress

ISBN: 978-0-12-396953-8

For information on all Academic Press publications
visit our website at elsevierdirect.com

Typeset by MPS Limited, Chennai, India
www.adi-mps.com

Printed and bound in United States of America

14 15 16 17 10 9 8 7 6 5 4 3 2 1

Contents

Preface

Over 700 species of Cladocera are known, and representatives of this group are often dominant in the freshwater fauna and sometimes occur in enormous quantities. They live in both small and large water bodies from arctic to tropical latitudes, in open water, on the bottom, in mud, among inshore vegetation, in acid pools on bogs, in small accumulations of water in epiphytic plants, in narrow aquatic spaces between moist sand grains. A few species even left the water and live in moist moss-like growth on tree trunks in tropical cloud forests. Some species are specialized for life in saline lakes and in the sea.

The data on organic and inorganic constituents of cladocerans are reviewed. They may accumulate physiologically important substances, those of no such importance, or toxicants. The section on nutrition of Cladocera comprises a brief review of their anatomical and environmental backgrounds, with special consideration of food quality (algae, algal lipids, bacteria, and organic debris). Feeding and digestion are discussed for both littoral and pelagic species and the digestion of protein, carbohydrate, and lipids is reviewed. Cladocera modify only a small amount of the lipids consumed with algae, and then transfer them to fish and thus further, to man. The fate of ingested chlorophyll, starvation and the impact of xenobiotics on feeding and digestion are also commented on.

Oxygen consumption and the production of CO_2 are considered in littoral and pelagic Cladocera. Littoral and bottom-dwelling species often live under conditions of hypoxia and anoxia. The dynamics of hemoglobin and the role of iron are also discussed. Studies on the impact of xenobiotics on respiration are reviewed and the blood flows and separating membranes are described. Heart rate and its controlling factors are also reviewed. Occasional heart arrest and adhesion of blood cells are reported. Studies on phagocytosis are pointed out. The data related to excretion are reviewed and the distribution of xenobiotics in the body, their bioaccumulation and the transformation of xenobiotics are described. Studies on routes of elimination of xenobiotics and on detoxification are indicated.

Ion exchange is realized via the neck organ and limb epipodites. The principal types of osmotic regulation are: hyperosmotic regulation (in freshwater species), hypoosmotic regulation (in marine *Penilia*), and amphiosmotic regulation. The effect of xenobiotics on osmotic regulation is discussed. Numerous enzymes have been determined in Cladocera and studies on their role are reviewed. The impact of xenobiotics on metabolic links is described, as well as detoxification in the cladoceran body.

Cladocera live for a week up to several months, depending on the species and the environmental conditions. Body length increments occur between molts. Mechanical damage is repaired by regeneration. Effects of environmental factors and xenobiotics on morphology and cytology are described and data on the impact of xenobiotics on molting and life duration are reviewed. The state of investigations on senescence and mortality is also commented on.

Cladocera are remarkable by their parthenogenetic reproduction, which is sometimes interrupted by bisexual (gamogenetic) reproduction. Physiological aspects of parthenogenetic and gamogenetic reproduction are discussed. Differences of movement and trajectories of littoral and pelagic species are indicated and data on the physiology of muscles, immobilization, fatigue, stress, and impact of xenobiotics are also reviewed. There is information on neurosecretion, vision, ability to discern polarized and colored light, photoperiod, chemoreception, mechanoreception, ability for orientation in space, endogenous rhythms, impact of xenobiotics.

Cladocera possess complicated behavior patterns, differing in particular species. These include migration, swarming, akinesis and escape behavior. The data on the impact of xenobiotics on behavior are also reviewed. Ecological aspects discussed comprise consideration of physiological background and the limits of physiological factors, synergism and antagonism of factors, pathway of lipids from algae via Cladocera to fish, conditioning of the environment by Cladocera, as well as the use of Cladocera in water quality testing.

Within certain limits, Cladocera can support the body's homeostasis in a dynamic environment, including the homeostatic level of their chemical constituents and of osmotic pressure. Along with special discussions, introductory remarks are made whenever it seemed to be necessary to make the matter useful both for specialists and for non-specialists.

Contributors

Stuart K. J. R. Auld School of Natural Sciences, University of Stirling, Stirling, Scotland, UK

Margaret J. Beaton Biology Department, Mount Allison University, Sackville, NB, USA

Dörthe Becker Environmental Genomics Group, School of Biosciences, University of Birmingham, Birmingham, UK

John K. Colbourne Environmental Genomics Group, School of Biosciences, University of Birmingham, Birmingham, UK

Carli M. Peters Biology Department, Mount Allison University, Sackville, NB, USA

Michael E. Pfrender Department of Biological Sciences, Environmental Change Initiative, University of Notre Dame, Notre Dame, IN, USA

Joseph R. Shaw School of Public and Environmental Affairs, Indiana University, Bloomington, IN, USA

Nikolai N. Smirnov Institute of Ecology, Leninski Prospect 33, Moscow, Russia

Elizabeth Turner School of Public and Environmental Affairs, Indiana University, Bloomington, IN, USA

Kay Van Damme Environmental Genomics Group, School of Biosciences, University of Birmingham, Birmingham, UK

Acknowledgments

The present volume is an attempt at making a summary of work of many experts throughout the world in various fields using special methods. A substantial contribution to physiology of the Cladocera is made recently by toxicologists.

Special thanks are due to Dr M.J. Burgis (UK) who generously used her time and experience to improve the manuscript.

The present review is motivated by the author's observations on living, mostly littoral, cladocerans. Some recent observations are made at the Hydrobiological Station "Glubokoe Lake" (Russia). The author is grateful to his immediate colleagues from the 'Cladocera team' for help and discussions: O.S. Boikova, N.M. Korovchinsky, A.A. Kotov, E.I. Bekker, A. Yu. Sinev, as well as to Dr Yu.B. Manteifel and Dr E.P. Zinkevich for advice.

Dr A.A. Kotov critically read the draft manuscript, suggested numerous useful additions, and used his skill for preparation of the manuscript and figures.

Many librarians, mostly personally unknown to the author, retrieved numerous publications from different countries and time periods. Their care and labor are appreciated, including those of the Biological Department Library of Russian Academy of Sciences, including Ms N.I. Gotovskaya and Ms E.V. Morozova.

The present study is supported in part by grants on Cladocera projects from the Russian Foundation for Basic Rssearch (12-04-00207, etc.) and from the Program "Living Nature".

My wife L.A. Smirnova, Ph.D. (cited here as L.A. Luferova) is tolerant (mostly) towards using a big part of my time for such ventures as this.

Formulation of ideas included in this book and its composition was much favored by creative environment at the Institute of Ecology and Evolution of the Russian Academy of Sciences, and by personal attention of academician D.S. Pavlov, Director, and of academician Yu.Yu. Dgebuadze, Laboratory Head.

The subject is made much more complete by the authors of chapters 16–18.

CHAPTER

1

General

N.N. Smirnov

1.1 SYSTEMATIC POSITION

It is now thought that there are over 700 species of the order Cladocera in the world fauna, many of which develop populations in enormous quantities and thus play a big role in the life of the biosphere. New species are still being described.

The Cladocera belong to the subclass Phyllopoda of the class Crustacea. Most Cladocera belong to the orders Anomopoda and Ctenopoda. Anomopoda principally comprise the families Daphniidae (e.g. the genera *Daphnia*, *Ceriodaphnia*, *Simocephalus*, and *Scapholeberis*), Moinidae (e.g. *Moina*), Ilyocryptidae (*Ilyocryptus*), Macrothricidae (e.g. *Macrothrix* and *Streblocerus*), Acantholeberidae, Ophryoxidae, Eurycercidae (*Eurycercus*), Chydoridae (e.g. *Chydorus* and *Pleuroxus*), Bosminidae (*Bosmina*, *Bosminopsis*); and Ctenopoda comprise the families Sididae (e.g. *Sida*, *Pseudosida*, and *Diaphanosoma*) and Holopedidae (*Holopedium*). Others belong to the order Onychopoda (*Polyphemus*, as well as marine and brackish water species) and the order Haplopoda with the family Leptodoridae (*Leptodora*).

As physiological studies should be accompanied by the reliable identification of the subjects being investigated, keys to the worldwide fauna of Cladocera are indicated: "Guides to the identification of the macroinvertebrates of the continental waters of the World," issues 1, Macrothricidae (Smirnov, 1992); 3, Ctenopoda (Korovchinsky, 1992), 11, Chydorinae (Smirnov, 1996); 17, *Simocephalus* (Orlova-Bienkowskaja, 2001), 13, The predatory Cladocera (Rivier, 1998); 21, *Daphnia* (Benzie, 2005); and 22, Ilyocryptidae (Kotov and Štifter, 2006). There are also newer, general worldwide resources for Ctenopods, created by Korovchinsky (2004); *Leydigia* (Chydoridae), by Kotov (2009); and *Eurycercus* (Bekker, et al., 2012); as well as recent regional keys.

As investigations into Cladocera are actively developing, the aforementioned summaries are rapidly becoming incomplete, and more recent literature should also be taken into consideration.

1.2 GENERAL MORPHOLOGICAL BACKGROUND

As animal functions are linked to their form, some comments on the body structure and organs of Cladocera are provided here. Most of

Physiology of the Cladocera.
DOI: http://dx.doi.org/10.1016/B978-0-12-396953-8.00001-X

the animals attributed to the order Cladocera have the same principal structure, with various modifications present in different species. Investigations into comparative and functional morphology (such as, e.g. those by Fryer, 1968, 1974, 1991, etc.) have revealed exciting data on particular species, permitting a better understanding of their lifestyles.

Cladocerans have inherited from their ancestors a weakly segmented body covered with a chitinous, mostly bivalved, shell and bearing few pairs of appendages—antennules, antennae (biramous, with the single exception of female *Holopedium*), mandibles, maxillulae, maxillae (may be completely reduced), mandibles, and five or six pairs of thoracic limbs (Figs. 1.1–1.4).

Cladocerans are mostly oval in shape, compressed from the sides, but many are spherical. In the case of *Graptoleberis*, there is a curious and unique combination of lateral and dorsoventral compression (Fig. 1.3). The appendages have numerous, but organized, setae. The posterior

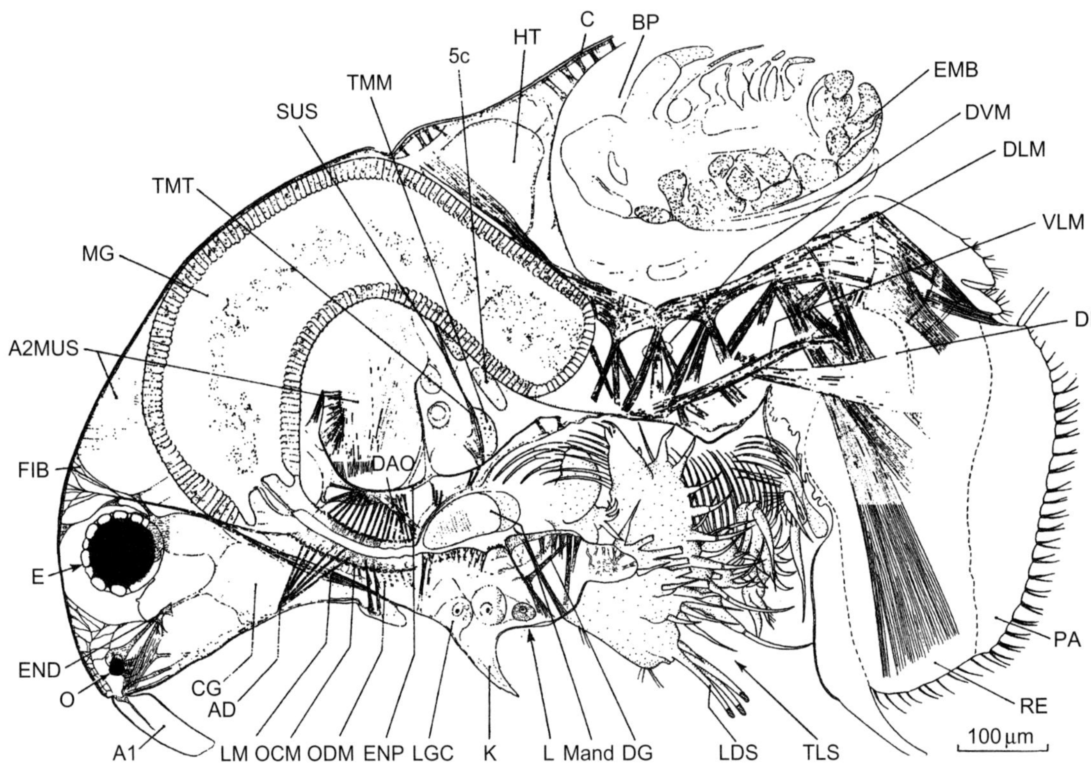

FIGURE 1.1 General anatomy of *Acantholeberis curvirostris*.
A1, antennule; A2MUS, antennary muscles; AD, apodeme; BP, brood pouch; C, carapace; CG, cerebral ganglion; DAO, dilator muscle of atrium oris; DG, duct of labral glands; DIV, diverticulum; DLM, dorsal longitudinal muscles; DVM, dorsoventral trunk muscles; E, compound eye; EMB, embryo; END, endoskeleton; ENP, endoskeletal plate; FIB, fibrils; HT, heart; K, keel of labrum; L, labrum; LDS, long distal setae of outer distal lobe of trunk limb 1; LGC, labral gland cells; LM, levator muscle of labrum; Mand, mandible; MG, mid-gut; O, ocellus; OCM, esophageal constrictor muscles; ODM, esophageal dilator muscles; PA, postabdominal lamella; RE, rectum; SUS, suspensory ligament; TLS, trunk limbs; TMM, 5c, transverse muscle of mandible; TMT, transverse mandibular tendon; VLM, ventral longitudinal trunk muscles. *Source: Fryer (1974).*

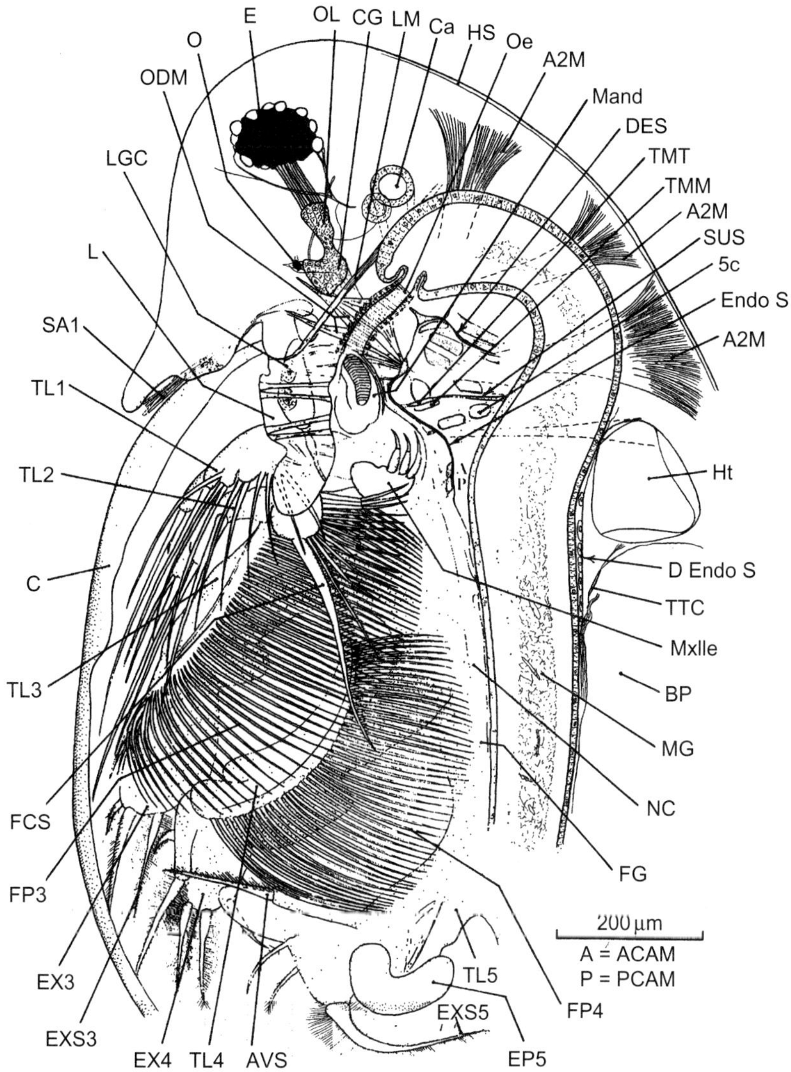

FIGURE 1.2 General anatomy of *Daphnia longispina*.
A, anterior carapax adductor muscle; A2M, antennal muscles; AVS, anterior vertical seta of trunk limb 5; Ca, cecum; D Endo S, dorsal endoskeletal sheet; DES, dorsal extension of ventral endoskeletal sheet; Endo S, endoskeletal sheet; EP5, epipodite of trunk limb 5; EX3, 4, exopod of trunk limbs 3, 4; EXS5, exopod seta 5; FCS, filter-cleaning spine of trunk limb 2; FG, food grove; FP3, gnathobasic filter plate of trunk limb 3; FP4, gnathobasic filter plate of trunk limb 4; Ht, heart; HS, head shield; LGC, labral gland cells; Mxlle, maxillule; NC, nerve cord; Oe, esophagus; OL, optic lobe of cerebral ganglion; P, posterior carapax adductor muscle; SA1, sensory seta of antennule; TL1, 2, 3, 4, 5, trunk limbs 1, 2, 3, 4, 5; TTC, thickened trunk cuticle. For other abbreviations, see Figure 1.1. *Source: Fryer (1991).*

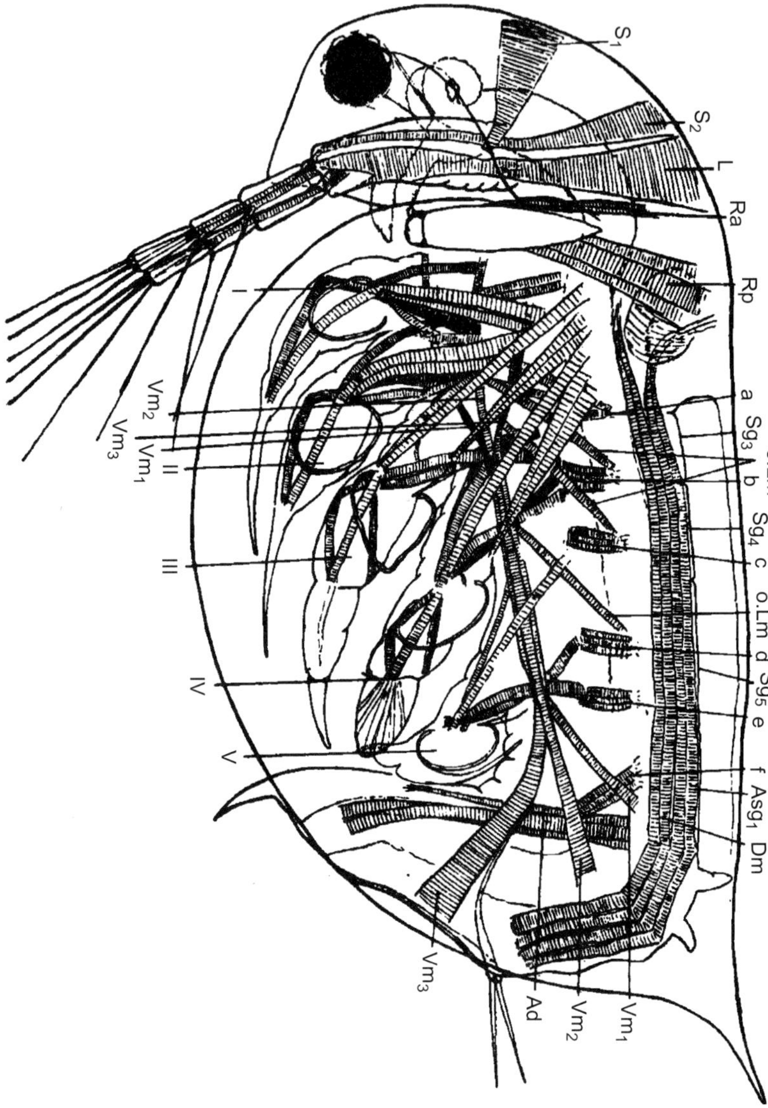

FIGURE 1.3 Muscles of *Daphnia magna*. *Source: Binder (1931).*

end of the body (postabdomen) is bent at a right angle to the abdomen, or may even be reversed. All structures tend to undergo morphological radiation, and homologous structures may occur in different species in various forms, from the ancestral state to their complete disappearance or, in contrast, enlargement and specialization.

The outer surface of the chitinous shell may be smooth, reticulated, ciliated, setose, or variously honeycombed. The dorsal space under the shell is the brood chamber into which eggs are laid and develop.

The inner organs are situated within the body rather loosely. The intestine may be

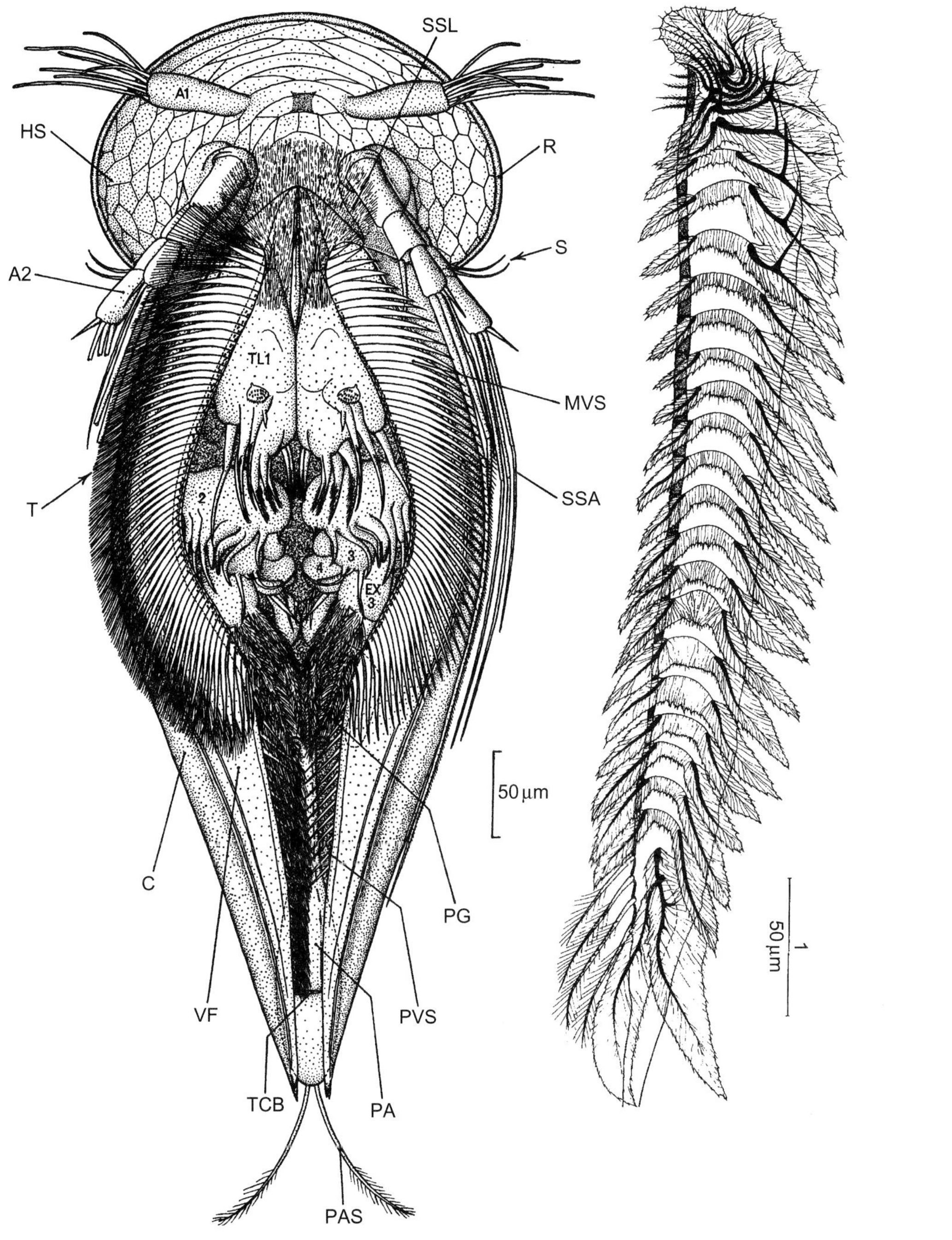

FIGURE 1.4 Modification of the ventral side for movement over flat surfaces.
Right, edge of valve of *Scapholeberis mucronata*. Left, *Graptoleberis testudinaria*. A1, antennule; A2, antenna; C, carapace; EX 3, exopod of trunk limb 3; HS, cuticle of head; MVS, medium ventral setae; PA, postabdomen; PAS, Postabdominal seta; PG, posterior gap; PVS, posterior ventral setae; R, thickened rim of head shield; S, Sensory setae of AII; SSA, Swimming setae of AII; SSL, setules of sealing seta of trunk limb 1; TCB, transverse chitinous bar; TL1, trunk limb 1; VF, ventral flange. *Sources: right, Dumont and Pensaert (1983); left, Fryer (1968).*

straight or convoluted. Muscles do not form compact masses and most of them can be seen individually (Fig. 1.3). The largest muscles are longitudinal bands stretching along the gut. Groups of muscles allow motion of the thoracic limbs and antennae. Small muscles rotate the eye and move the labrum and antennules. The intestine is supplied with circular muscles. The ovary (or testis) is paired and situated ventrally along the gut; this is also where the fat body is situated, in contact with the ovaries (Jaeger, 1935).

There are two paired remains of the coelom: the antennal gland and the maxillary gland (shell gland; Fig. 1.5). The latter is the organ of excretion, while the antennal gland has no duct and no outer orifice.

On the head of most species there are head pores, leading to an organ whose function is probably ion exchange.

The nervous system comprises a double chain of ganglia, with the brain located in the cephalic region. Sense organs comprise the unpaired eye, the unpaired ocellus (Figs. 1.1, 1.2, and 1.6), sensory papillae situated on the antennule and on some thoracic limbs, and numerous tactile setae. The eye or the ocellus may be absent in some species.

On the basis of this general structural scheme, three kinds of specialization are formed: one used for collecting food from substrata (Fig. 1.1), another for filter feeding in open water (Fig. 1.2), and the last, predatory.

Littoral cladocerans live among organic and mineral particles and require protection, supplied by thick chitinous shells, most with sculpturing; this increases the durability of their shells (Fryer, 1968). According to the same author, in littoral species the cuticle is up to 12 μm thick (in *Pseudochydorus*). In pelagic species, the integuments are many times thinner.

Further information on Cladocera may be obtained from www.cladocera-collection.cz and Lampert (2011).

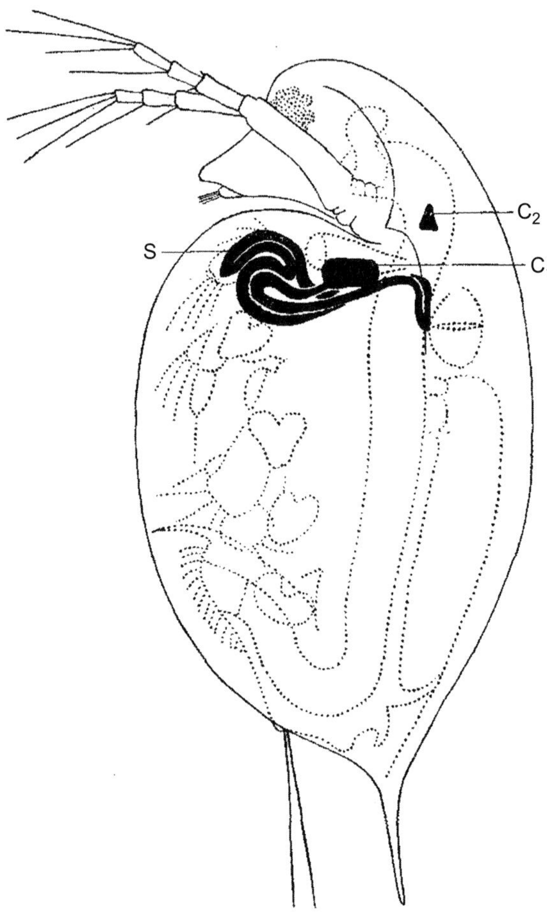

FIGURE 1.5 Position of nephridia in the body of *Daphnia*.
C_1, maxillary coelomic sac; C_2, antennal coelomic sac; S, ducts of maxillary coelomic sac. *Source: Gicklhorn (1931a).*

1.3 SIZE AND WEIGHT CHARACTERISTICS

Most Cladocera species are small: from about 0.3 mm to c. 6 mm. One of the largest, *Leptodora*, is c. 10 mm in length. Due to their small size, their surface:volume ratio is high.

Several authors have investigated their length:weight ratios [e.g. Ivanova and Klekowskii, 1972 (*Simocephalus*); Kawabata and

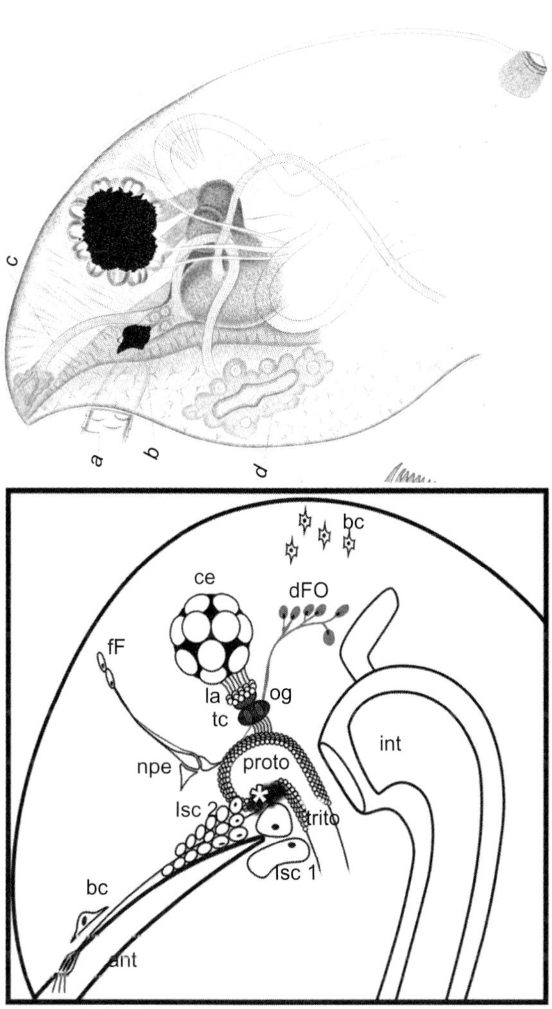

FIGURE 1.6 Cephalic region of nervous system. Above, *Eurycercus*. a, antennule; b, ocellus; d, lateral organ. Below, *Daphnia*. ant, antennule; bc, bulhed cells; ce, compound eye; dFO, dorsal frontal organ; fF, frontal filament; la, lamina; lsc 1, labral secretory cells; npe, nauplius eye; og, optic ganglion; proto, protocerebrum; te, tectum; trito, tritocerebrum; *, deuterocerebrum. *Source: above, Leydig (1860); below, modified from Weis et al. (2012).*

Urabe, 1998]. The relationship between length (L) and wet weight (W) may be described by Eq. 1.1 (Kurashov, 2007):

$$W(mg) = qL^{b}(mm), \qquad (1.1)$$

where q is 0.133 for Scapholeberis; 0.075 for other Daphniidae, 0.127 for *Eurycercus*; 0.091 for *Alona* and *Alonella*; 0.203 for *Chydorus*; 0.083 for *Macrothrix*; and 0.140 for other Chydoridae and Macrothricidae. b is 2.630 for *Scapholeberis*; 2.925 for other Daphniidae; 3.076 for *Eurycercus*; 2.646 for *Alona* and *Alonella*; 2.771 for *Chydorus*; 2.331 for *Macrothrix*; and 2.723 for other Chydoridae and Macrothricidae.

Similar equations were determined by Lynch et al. (1986):

for *Daphnia ambigua*, $W = 5.740L^{2.370}$;
for *D. galeata mendotae*, $W = 5.480L^{2.200}$;
for *D. parvula*, $W = 4.740L^{2.190}$; and
for *D. pulex*, $W = 10.674L^{2.093}$.

Dry weight (DW) increases with length, as shown for *D. hyalina* (Baudouin and Ravera, 1972). The relationship between DW (in mg) and length (L, in mm) was determined for *D. pulex* by Richman (1958) to be:

$$DW = 0.028L^{-0.022} \qquad (1.2)$$

and (by Wen et al., 1994), for *Daphnia*, *Simocephalus*, *Ceriodaphnia*, and *Bosmina* combined, to be :

$$DW - 0.013L^{2.14} \qquad (1.3)$$

Regression equations for DW and length for many planktonic and littoral species were supplied by Dumont et al. (1975), for five Amazonian species by Maia-Barbosa and Bozelli (2005), and for three species from Mexico by Nandini and Sarrma (2005).

Weight at a certain length of Cladocera was also determined and presented in tabulated form by Mordukhai-Boltovskoi (1954), Kosova (1961), and Sokolova (1974). In addition, Vasama and Kankaala reported length-carbon regressions in Cladocera (1990).

CHAPTER

2

Methods

Methods of investigating the physiology of Cladocera range from direct observation of living specimens to recording the physical and chemical manifestations of particular physiological processes. Detailed examination of preserved or living specimens supplies useful information on their anatomical background and function.

Cladocera may be collected in the littoral and pelagic zones of large and small water bodies. They should be looked for in ponds, pools, puddles, temporary pools, roadside ditches, acid bogs, fountains, all kinds of artificial basins, moist moss, and even in moist, moss-like growth on tree trunks.

Cultures may be started from living specimens. Cultures derived from a single female are called *clonal cultures* and provide relatively uniform material. Culture methods have been described by various authors either alone (e.g. *Biotechnics of Daphnia culture at fish farms*, 1958; Dewey and Parker, 1964; Ivleva, 1969; Parker and Dewey, 1969; Bogatova, 1980; Goulden et al., 1982; Dodson and Frey, 1991) or in descriptions of particular experiments. The combined culture of *Daphnia* with other cladocerans was suggested by Bogatova (1963, 1980).

Gajewskaja (1940, 1948) suggested a method of separating large-scale cultures of Cladocera and algae because combined cultivation of Cladocera and algae requires conflicting conditions. Since then, methods of large-scale culture have been developed further (Yalynskaya, 1961; Ivleva, 1969; Lampert, 1975; Bogatova, 1992).

In culture, littoral Cladocera should be fed with organic debris (detritus) collected at shore bottoms and screened to remove foreign animals (Smirnov, 1971) or with artificial detritus prepared from plants (see e.g. Rodina, 1963, Esipova, 1969, 1971; Dekker et al., 2006). Fresh detritus should be given at least every other day (e.g. a specimen may be transferred to a dish containing fresh detritus).

Cultured pelagic Cladocera should be fed with algae and adequate algal cultures should be maintained for this purpose. Some species (*Daphnia magna* and *D. pulex*) are more easily cultivated than others.

Sterilization was used to demonstrate the role of bacteria in the alimentation of cladocera (Gajewskaja, 1938; Rodina, 1965; Esipova, 1969, 1971). The latter author used Lugol's solution at a $1/22^8$ dilution to reduce the quantity of bacteria in food offered to Cladocera.

Physiology of the Cladocera.
DOI: http://dx.doi.org/10.1016/B978-0-12-396953-8.00002-1

Suspensions of latex beads have also been used in investigations of feeding behavior (Burns 1968b, 1969; Hessen, 1985).

Various kinds of intravital staining of the organ systems of Cladocera were originally investigated by Fischel (1908). The salivary gland can be stained using neutral red and Bismarck brown (Cannon, 1922). In thoracic limb IV of *Eurycercus*, there is a slime gland whose secretions are stained bright blue with Mallory's stain (Fryer, 1962, 1963). Gut contents have been stained with eosin, Congo red, methyl red, neutral red, and uranine for the determination of pH (Lavrentjeva and Beim, 1978).

Histochemical techniques have been used in the analysis of *Holopedium* slime (Brown, 1970). External slime may be distinguished by placing a cladoceran in diluted Indian ink.

Preferences for environmental factors, and for food, may be ascertained by experiments on living specimens. Examination of food composition is made easier by dissolving the soft tissues of cladocerans with 3% sodium hypochlorite (Infante, 1978). This treatment may also be useful for other purposes.

Cladocerans are very agile; therefore high-speed photography has been used (e.g. Storch, 1929; Zaret and Kerfoot, 1980) to study them. Alternatively, methods to inhibit their movement have been devised. Immobilization by narcotization is discussed in more detail in Chapter 12, section 12.5. Early techniques included attaching specimens to a substrate on a glass slide using wax dissolved in alcohol (Scourfield, 1900b; Peñalva-Arana et al., 2007). More recently, Porter and Orcutt (1980) used silicone grease to immobilize *D. magna* by its head shield to observe its feeding. For the purposes of his study, Jacobs (1980) attached *Daphnia* by the caudal spine to a plasticine bed. Specimens thus immobilized could be observed and the next instar was released into free water from the attached exuvium.

Using cyanoacrylate glue, Onbé (2002) attached *Pseudevadne* and *Evadne* to the tip of a glass capillary held in place by a stand to facilitate video recording. Peñalva-Arana et al. (2007) reported unbiased observations using computer recording combined with their immobilization system.

Methods of attachment were recently described by Seidl et al. (2002) and used by Pirow et al. (2004) for investigating oxygen transport processes in *D. magna*. The latter authors immobilized fasting animals by gluing their posterior apical spine with histoacryl to a bristle, which was then fixed to a coverslip with plastilin. Dye was then microinjected into the circulatory system from the dorsal side into the space "directly downstream of the heart." The coverslip then formed the base of a thermostatted perfusion chamber in which an immobilized daphnid was able to freely move its antennae.

Ivanova and Klekowski, (1972) achieved immobilization of *Simocephalus* by placing it into the bulb of a Cartesian diver (used for determining oxygen consumption) and leaving no free space around the animal, i.e. it was too small to allow movement.

Various kinds of microscopy, including scanning electron microscopy (SEM), can be used in investigations of Cladocera. A variation of cladoceran preparation for SEM was suggested by Laforsch and Tollrian (2000).

Modern video microscopy and digital image processing methods take advantage of their transparency (Colmorgen and Paul, 1995). Fluorescence analysis was originally used by Pravda (1950).

A special set-up for optophysiological recording was devised by Paul et al. (1997): this comprised a video recorder and PC, as well as computerized respirometry. For this procedure, *Daphnia* were placed in a glass cell of 0.45 mL volume. Pirow et al. (2001) measured tissue oxygenation using an apparatus consisting of a reversed microscope equipped with systems for measuring hemoglobin (Hb) oxygenation, reduced nicotinamide adenine dinucleotide (NADH) fluorescence (as an indicator of tissue oxygenation state), and the movements of organs. Remarkable results were obtained by Pirow et al. (2004) using injection of the oxygen-sensitive phosphorescence probe Oxyphor R2 into the circulatory system followed by phosphorescence imaging.

Surgical methods can be applied in investigations of regeneration, vision, and neurosecretion. These are described in Angel (1967) and in Chapter 10, section 10.2.1 and Chapter 13, section 13.4.3.

In biochemical and metabolic investigations, special procedures, such as homogenization (e.g. Guan and Wang, 2004b) and radiotracer or chemical methods, have been used. Studies on homogenates using modern sensitive methods have opened up the possibility of investigating metabolic pathways.

Spectrophotometry has been used for the identification of particular compounds, including Hb. Initially, Fox (1948) suggested that a quantitative estimation of Hb content in *Daphnia* in arbitrary units could be obtained against a wedge-shaped standard prepared from the worker's blood. Following this, cladoceran Hb was investigated using spectral (Hildemann and Keighley, 1955; Hoshi et al., 1968) and chemical (Hoshi, 1963a, 1963b; Smirnov, 1970) methods.

Recent toxicological studies have assessed the effects of xenobiotics on particular physiological processes.

Photographic recording of the heart rate and movement of thoracic limbs has been used (e.g. Kolupaev, 1988). Moreover, computer recording of behavior is now used (Peñalva-Arana et al., 2007). Special techniques may be found in descriptions of the original investigations.

CHAPTER

3

Chemical Composition

3.1 MOISTURE CONTENT AND CALORIFIC VALUE

The moisture content and calorific value of some Cladocera spp. are shown in Table 3.1. Moisture content is usually within 80–90% and calorific value within 3–6 kcal/g dry weight (DW). The calorific value (kcal/g) is: *Daphnia hyalina*, 6.3; *Bosmina coregoni*, 6.3; *Chydorus sphaericus*, 6.1; and *Leptodora kindtii*, 5.8 (Vijvberger and Frank, 1976). Variations in the calorific value obviously depend mostly on the fat content. The calorific value of "lean" *Daphnia magna* is only 60% of that of *Daphnia* containing more fat (Chalikov, 1951).

3.2 PRINCIPAL ORGANIC CONSTITUENTS

There have been various determinations of the chemical composition of some Cladocera (daphnids, *Moina*, *Bosmina*, and *C. sphaericus*); the general chemical composition of some cladocera is shown in Table 3.2.

Their protein content ranges from 30% upwards, their fat content is 1–20%, and their carbohydrate content is 10–30%. The protein content of *Daphnia* and *Ceriodaphnia* increases with increased protein in their natural food (Guisande et al., 1991). The content of nitrogen (N) was determined by Hessen (1990) as 8.18% DW, with little variation; the lowest values were determined in starved specimens. The content of nitrogen in five species of Cladocera was c. 8%, and of phosphorus (P) was c. 1.4% (Main et al., 1997).

It is notable that the fat content is especially variable. Some other data on the lipid content (in % DW) have been reported: *D. magna*, 17–22; *D. pulex*, 10–40; and *Bythotrephes cederstroemi*, 10–19 (Bilkovic and Lehman, 1997).

Far more informative are reports on the dynamics of general chemical composition, which depends on seasonal changes in a number of factors, on starvation, or on other particular factors. Often, scattering of the data indicates dependence on specific factors. Indeed, if arranged by season, chemical constituents demonstrate clear composition changes, as shown, for example, for *D. pulicaria* (in Fig. 3.1 of Heisig-Gunkel and Gunkel, 1982). The chemical composition of *D. pulex* generally confirms this data but depends on the period of starvation, with relative quantities of carbohydrate and fat decreasing, and those of

Physiology of the Cladocera.
DOI: http://dx.doi.org/10.1016/B978-0-12-396953-8.00003-3

TABLE 3.1 Moisture Contents and Calorific Values of Some Cladocera SPP

Species	Moisture Content (%)	Calorific value (kcal/g DW)	Ash (% DW)	Reference(s)
Bosmina longirostris	89	6.5	–	Sherstyuk (1971)
B. longispina	–	9.9	4.8	Romanova & Bondarenko (1984)
Bythotrephes longimanus	84.7	8.6	5.2	Romanova & Bondarenko (1984)
Ceriodaphnia affinis, adults	89.2	–	–	Stepanova (1967)
C. affinis, juveniles	90.4	–	–	Stepanova (1967)
C. pulchella	70–82	4–4.7	–	Sherstyuk (1971)
C. quadrangula	–	4.9	4.6	Riccardi & Mangoni (1999)
C. reticulata	87–96	5.2–5.5	–	Bogatova et al. (1971)
Daphnia cucullata	–	5.1–5.4	10.6–14.0	Riccardi & Mangoni (1999)
D. hyalina	–	4.9–5.0	12.0–12.5	Riccardi & Mangoni (1999)
D. longispina	82.6	4	22	Romanova & Bondarenko (1984)
D. magna	86.4–95.6	2.4–5.6	–	Karzinkin (1951), Ostapenya et al. (1968), Schindler (1968), Bogatova et al. (1971), Stepanova (1968), Stepanova et al. (1971), Mityanina (1980)
D. pulex	89–95	2.8–4.9	7.6–25–8	Birge & Juday (1922), Karzinkin (1951), Malikova (1953), Ostapenya et al. (1968), Stepanova (1974)
Bythotrephes longimanus	–	5.2	5.6	Riccardi & Mangoni (1999)
Leptodora kindtii	97.2	4.6	–	Sherstyuk (1971)
Moina macrocopa	95	5	–	Bogatova et al. (1971)
Moina spp.	87–88	4–4.3	–	Ostapenya et al. (1968), Stepanova (1974), Zhao (2006)
Polyphemus pediculus	80–86	4.4–4.9	–	Sherstyuk (1971)
Sida crystallina	84	5.3	–	Sherstyuk (1971)
Simocephalus vetulus	82–92.4	3.65–4	–	Sherstyuk (1971), Stepanona (1968), Stepanova et al. (1971)
Eurycercus lamellatus	80–88	4.1–4.3	–	Sherstyuk (1971)

protein and ash increasing (Fig. 3.1) (Lemke and Lampert, 1975). Seasonal variations in the biochemical composition of *D. magna* in the Luxembourg were studied by Cauchie et al. (1999). The variation (in mg/g DW) was: protein from c. 140 to 400; lipids, 120–180; chitin, 40–70; carotenoids, 70–230; and ash, 100–340.

To understand these variations, data on the dynamics of the chemical constituents in

TABLE 3.2 Composition of Some Cladocera SPP., in % DW

Species	Carbon	Hydrogen	Nitrogen	Carbohydrate	Protein	Lipid	Author
Ceriodaphnia quadrangula	49.1	7	9.8	–	54.4	5.3	Riccardi & Mangoni (1999)
Daphnia cucullata	53.3–54.5	7.8–8	10.1	14.6	57.0–62.8	19.8–22.7	
D. hyalina	50.2–52.3	7.6	11	21.3–22.9	62.2–64.0	13.1–16.5	
D. pulex	–	–	–	3.3–10.9	36.4–61.7	2.8–27.9	Birge & Juday (1922)
D. pulex	–	–	–	6.4	63.4	8.6	Stepanova (1968)
Moina brachiata	–	–	–	8.2	63.4	17.5	
Simocephalus vetulus	–	–	–	16.7	52.3	11.5	
Bythotrephes longimanus	51.2	7.6	11.1	22.0	63.9	14.1	Riccardi & Mangoni (1999)

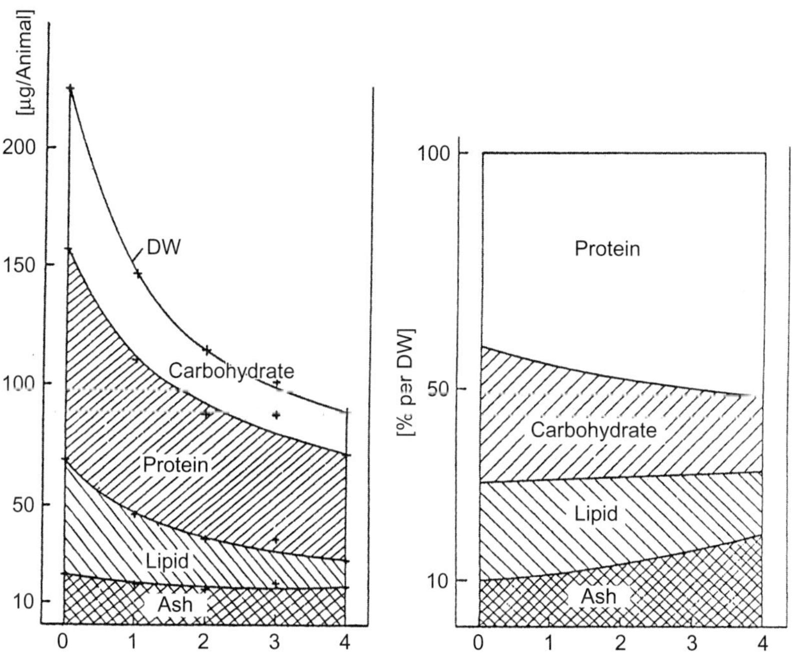

FIGURE 3.1 Changes in *Daphnia pulex* chemical composition as a result of starvation (left, per animal; right, % DW). Horizontal axis, number of days of starvation. *Source: Lemcke and Lampert (1975).*

relation to growth are also useful. McKee and Knowles (1987) studied the levels of protein, glycogen, lipid, RNA, and DNA during the growth of *D. magna*. They found that percentage DW during the first to the twenty-first day of life the protein content varied within the range of 48–62% (maximal day 8); glycogen content steadily increased from 2.4% to 7.5%; lipid content varied from c. 18% to c. 15%, decreasing to 6.4% on day 21; RNA content varied from 8.6% to 6.6%, decreasing to 4% on day 21; and DNA content generally decreased from 0.20 to 0.14%.

It has also been shown that during culture the protein and lipid contents of *Moina macrocopa* decreased in comparison with those of the initial culture, from about 25 and 0.56 mg/g wet weight (WW) to about 18 and 0.26–0.33 mg/g WW, respectively (Romanenko et al., 2004).

3.2.1 Amino Acids

The amino acid content of various cladocera is shown in Table 3.3. The amino acid composition of *D. pulex* was determined by Malikova (1953, 1956) and that of *D. magna*, *Ceriodaphnia reticulata*, *C. sphaercus* by Sadykhov et al. (1975). In the carotenoprotein complexes, the following predominant amino acids were found: in *D. magna*, alanine, glutamine, glycine, and leucine (Czeczuga, 1984); in *Moina micrura*, asparagine, glutamine, and glycine (Velu et al., 2003).

The presence of free intracellular amino acids in *D. magna* was shown by Gardner and Miller (1981).

TABLE 3.3 The Content of Different Amino Acids in Some Cladocera SPP

Amino acid	*Bosmina longirostris* (%protein) (Verbitsky, 1990)	*D. magna* (mg WW/dL) (Stepanova & Naberezhniy, 1972)	*D. pulex* (% total amino acids per DW) (Dabrowski & Rusiecki, 1983)	*Daphniopsis tibetana*, (g/100 g protein) (Zhao et al., 2006)	*Ceriodaphnia* spp. (% total amino acids per DW) (Dabrowski & Rusiecki, 1983)	*Simocephalus vetulus* (mg WW/dL) (Stepanova & Naberezhniy, 1972)	*Moina mongolica* (g/100 g protein) (Zhao et al., 2006)
Phenylalanine	1.3	228–833	3.76	1.79	3.82	107–157	3.40
Tyrosine	3.1	–	4.05	3.55	3.81	60	2.80
Leucine	3.3	222–725	3.74	5.76	4.61	425–625	5.30
Isoleucine	2.1	–	2.24	3.68	2.43	–	3.40
Methionine	–	–	1.44	3.64	1.40	–	1.50
Valine	2.9	–	3.71	3.78	3.49	–	3.90
Alanine	2.8	157–438	3.95	6.72	4.57	283–342	4.30
Glycine	2.6	194–239	2.85	3.91	2.69	236–296	3.20
Proline	2.2	211–358	2.53	4.53	2.76	100–250	2.70
Glutamic acid	5.0	317–458	5.46	7.40	6.29	522–619	8.00
Serine	1.8	205–357	2.57	3.27	3.07	255–285	3.00
Threonine	2.6	121–279	2.93	4.00	2.99	121–164	3.20
Aspartic acid	3.2	365–474	5.22	7.46	5.94	318–371	6.40
Arginine	1.9	200–401	3.40	2.85	3.45	244–348	4.30
Histidine	1.5	–	1.25	1.64	1.46	–	1.20
Lysine	3.8	–	3.78	4.78	4.36	–	3.40
Cystine	0.7	275–423	0.71	0.49	–	100–127	0.80
Tryptophan	–	–	–	0.43	–	–	1.20

It was determined that in *Daphnia*, the amino acid composition during ontogeny is rather constant (Brucet et al., 2005). However, it may vary rather widely depending on the culture conditions, as has been shown for *D. magna* (Stepanova and Naberezhnyi, 1970, 1972) and *Moina* spp. (Kokova, 1982). In well-fed *Daphnia*, the content of amino acids is 62.93 g WW/L, whereas in "lean" *Daphnia* it is 31.21–33.52 mg WW/L.

3.2.2 Mucopolysaccharide (Slime)

The chemical composition of slime was determined for the jelly capsule of *Holopedium* (Brown, 1970). Sulfated mucopolysaccharide was found to be present, as well as mucopolysaccharide modified by carboxyl groups. It has been suggested that the jelly capsule is produced by the mechanism that produces the exoskeleton at each molt.

Slime may surround the animal (as is the case with common forms of *Holopedium*). Slime is also produced by salivary glands (Fig. 3.2) and by glands of the thoracic limbs (Fig. 3.3), as has been shown for *Eurycercus* (Fryer, 1962, 1963) and for *Alonopsis elongata* (Fryer 1968), but not yet for other chydorids

The salivary glands may be shown by intravital staining with neutral red or Bismarck brown (Cannon, 1922).

3.2.3 Phosphorus-Containing Substances (Nucleic Acids: DNA and RNA)

Nucleic acids are high molecular weight compounds (polynucleotides). The nucleic acid content comprises 4.7–5.2% of the DW of *Moina* spp. (Kokova, 1982). The RNA content was determined to be c. 2–6% DW in *Daphnia* spp. and the phosphorus content to be c. 0.7–1.5% DW (Kyle et al., 2006).

Ribosomal DNA contains a significant fraction (c. 49%) of total body phosphorus (Acharya et al., 2004). The growth of Cladocera involves the requirement for a greater amount of ribosomal RNA, and especially of phosphorus (Main et al., 1997).

In juvenile *D. pulex*, an elevated DNA content was measured in the postmolt period, followed by an increase in RNA during the intermolt and premolt periods (Gorokhova and Kyle, 2002). It was found that the ratio of RNA:DNA in *Daphnia galeata* increased in response to an increase of P:C ratio in its food, and within 5 h (Vrede et al., 2002). Use of this ratio has been suggested (Markowska et al., 2011) for evaluating the condition of *D. pulex*, assuming that higher values correspond to a "good" condition. Specimens with a lower RNA:DNA ratio have lower metabolic rates and greater longevity.

3.2.4 Carbohydrates

Carbohydrates comprise glycogen and chitin.

Glycogen

Glycogen is a polysaccharide consisting of glucose bound to protein. The content of glycogen was determined by Blazka (1966): 23% in *Bosmina longirostris*, 53% in *C. reticulata*, 1–36% in *Daphnia* spp., and 33% in *Simocephalus* sp. (% of total composition, WW). Glycogen is present in developing embryos. The glycogen content in *Simocephalus vetulus* was determined to be 0.7% in the gastrula, 0.91% in the nauplius, 0.71% in released young, 1.2% in the third instar, and 2.23% in the fifth instar (Hoshi, 1953).

Chitin

The polysaccharide chitin is a polyacetylglucosamine (β-(1-4)-linked homopolymer of *N*-acetyl-D-glucosamine). The acetamide group CH_3CONH is present in the chitin molecule.

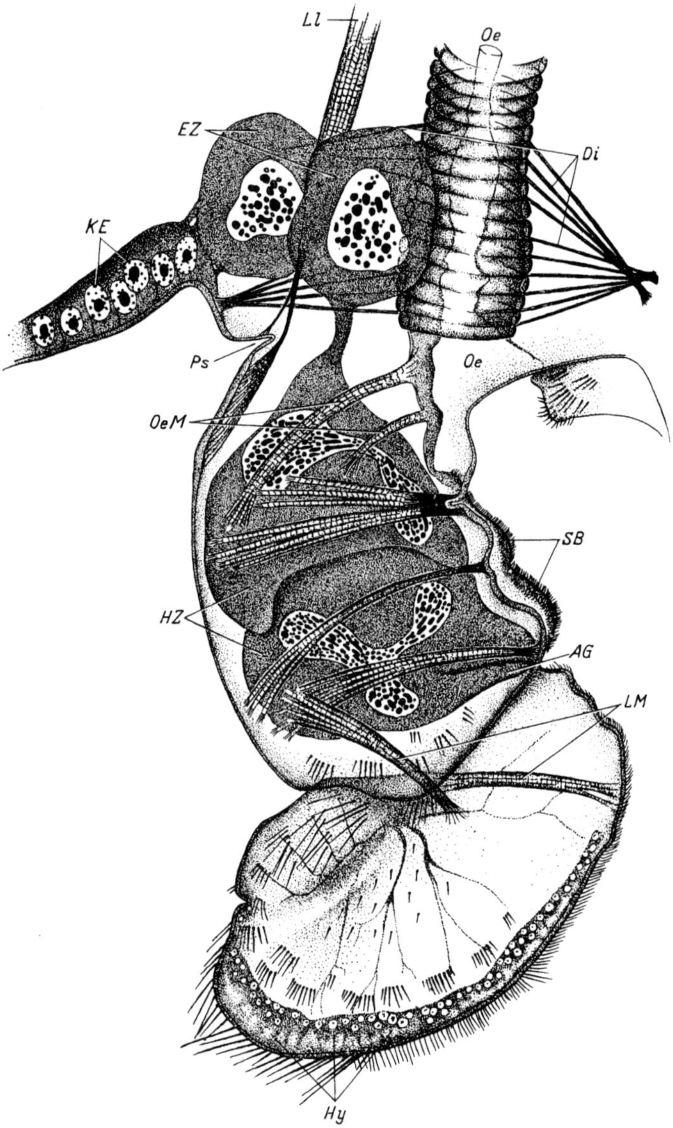

FIGURE 3.2 Labrum of *Daphnia*, showing salivary glands. AG, outlet duct; Di, dilators of esophagus; EZ, substituting cells; Hy, hypoderm of the lobe; HZ, principal cells; KE, epithelium of the head bottom; Ll, levator labri; LM, muscles of the lobe; Oe, esophagus; OeM, muscles of the esophagus; Ps, pseudosegment; SB, sensory setae. *Source: Sterba (1957a).*

Chitin is a major structural component of arthropods. The chitinous covers of cladocerans are strong and nonwettable. Cladocera, being abundant in nature, produce enormous quantities of chitin. The content of chitin in the body of *Daphnia* is approximately 15% (Chalikov, 1951) or 7% DW (Andersen and Hessen, 1991). In *D. magna*, it is 2.9–7% DW (Cauchie et al., 1995) or c. 30–70 mg/g DW (Cauchie et al., 1999).

The total annual chitin production by various cladocerans in Europe is: *D. magna*, 11.5 g/m^2 (4.6 g/m^3); *D. galeata*, 3.2 g/m^2 (0.16 g/m^3), *D. hyalina* and *D. cucullata* combined, 0.14–0.30 g/m^2 (0.09–0.2 g/m^3) (Cauchie et al., 1995).

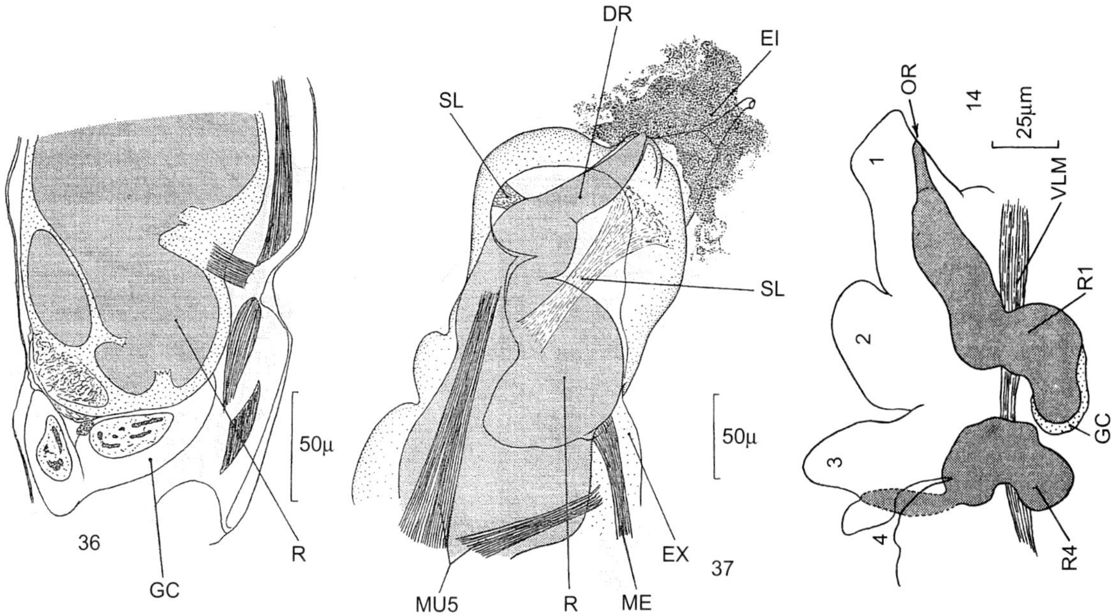

FIGURE 3.3 **Slime glands in thoracic limb 4 of *Eurycercus* and in thoracic limbs 1–4 of *Alonopsis elongata* (right). Right, *Alonopsis elongata*, left and middle, *Eurycercus*. DR, duct of reservoir; ET, entangling secretion; EX, exopodite; GC, gland cells; ME, muscle to exopodite; MUS, muscles; OR, opening of reservoir; R, reservoir; SL, suspensory ligament; VLM, ventral longitudinal muscle.** *Source: left and middle, Fryer (1963); right, Fryer (1968).*

In contrast to chitin from copepods, chitin from dead Cladocera is not decomposed in bottom sediments (or at least is not fully decomposed). After their death, it accumulates at the bottom of water bodies, sometimes forming a high proportion of the total mass of the sediment. Bottom sediment with dominant chitinous remains was termed *chitin gyttja* by Lundqvist (1927).

Chitin is not derived directly from food: it must be formed by the process of metabolism. Although chitin is specific to arthropods and highly accumulated by cladocerans, the metabolic pathways of chitin formation were not clearly described until recently. For insects, Kuznetsov (1948, p. 336) noted that "[t]here are no actual data on metabolism yielding chitin from the cycle of transformations; even more, there are no data on the metabolism intermediate as to chitin." It was noted by Hackman (1964, 499) that "[t]he biological synthesis of chitin in insects (or other animals and plants) has received little attention."

For other crustaceans, chitin is formed from glucosamine chiefly derived from glycogen, with the acetyl group furnished by oxidation of fatty acids (Vonk, 1960; Hohnke and Scheer, 1970). The latter authors suggested a general scheme of crustacean carbohydrate metabolism yielding, *inter alia*, chitin (Fig. 3.4). In Crustacea, chitin is formed and secreted by the hypodermis underlying the shell (Hohnke and Scheer, 1970).

3.2.5 Lipids

Recent investigations have supplied abundant new information on lipids. Variations in the content of lipids in Cladocera are shown in Figs. 3.1 and 3.5. Up to 17.5% DW was

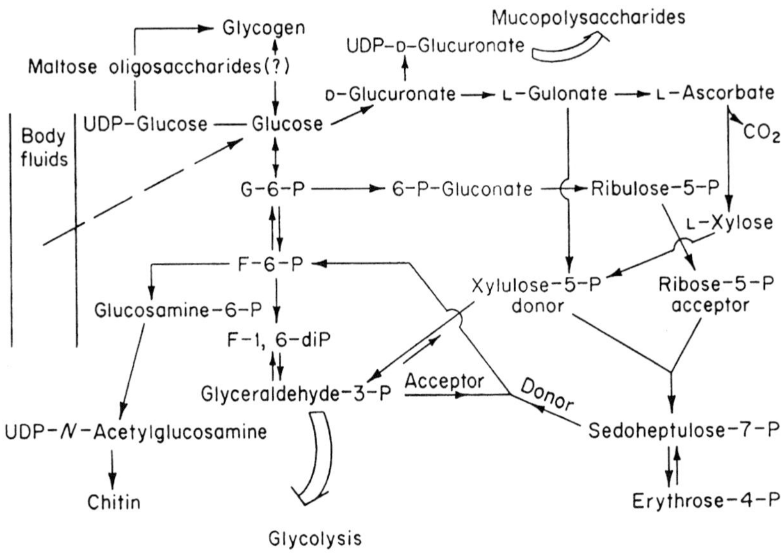

FIGURE 3.4 **Carbohydrate metabolism in Crustacea.** *Source: Hohnke and Scheer (1970).*

measured in *Moina rectirostris* (syn. *Moina brachiata*) (Stepanova and Vinogradova, 1970). The fatty acid composition of Cladocera spp. is indicated in Tables 3.4–3.8; that of *D. pulex* grown under different conditions was determined by Mims et al. (1991).

Oil drops are often clearly seen in the bodies of cladocerans. Among the algae, diatoms are a prominent group that synthesize and store lipids: after photosynthetic production of a carbohydrate, they transform it into lipids and the stored lipid is seen as oil drops. The diatoms are one of the dominant groups on the bottom substrata and in phytoplankton. They are not the only lipid-producing group, but the metabolism of other algae is less well known.

In temperate latitudes of the Northern Hemisphere, the spring peak of diatoms producing lipids as their reserve substance is followed by an accumulation of abundant oil drops by Cladocera. Due to their high-energy value, lipids are a prominent storage substance in Cladocera. Lipids are also used in the construction of biological membranes and hormones.

In addition to their basic trophic role, some lipid compounds may have exceptional properties. Pérez Gutierrez and Lule (2005) dried large numbers of *D. pulex* at room temperature, ground them, and produced 3 kg of fine powder. By extracting and fractionating it they obtained four glyceroglycolipids, all of which were found to be cytotoxic.

Lipids are discussed in more detail in Chapter 4, sections 4.1.2 and 4.4.

3.2.6 Vitamins

The vitamin content of *D. pulex* was assessed as vitamin A, 5.19 mg WW/L; vitamin B_1, 2.55 mg WW/L; vitamin B_2, 5.69 mg WW/L (Malikova, 1956); and vitamin B_{12}, μg/g DW (Stepanova and Borshch, 1970).

3.2.7 Pigments

Pigments are produced during the course of metabolism. In Cladocera, they comprise

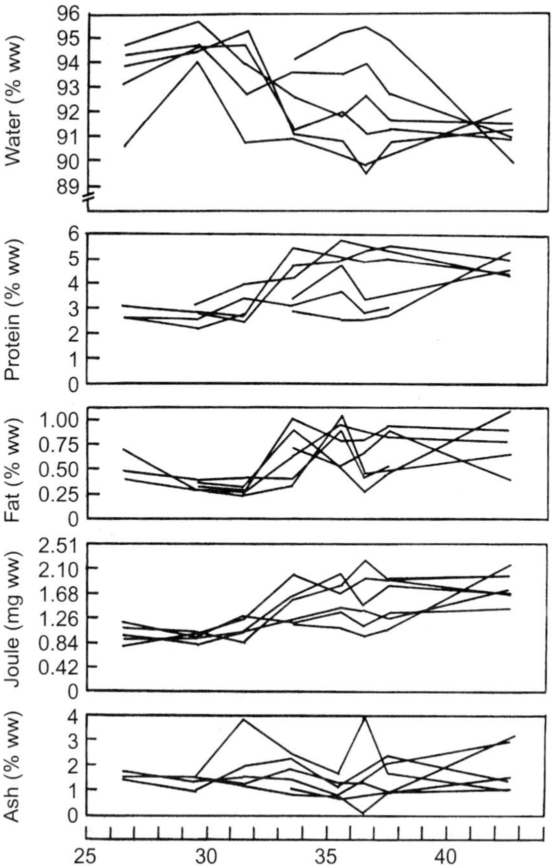

FIGURE 3.5 Seasonal changes in chemical composition of Daphnia pulicaria from six ponds, beginning from June. *Source: Heisig-Gunkel and Gunkel (1982).*

orange (carotenoid) pigments, red (hemoglobin (Hb), or bacterial carotenoids in cases of infestation by *Spirobacillus cienkowski*), green (carotenoprotein in hemolymph), and dark (e.g. ommochromes of eyes, or tanned protein formed at high pH, i.e. melanins) (Green, 1966b, 1971).

Generally, littoral cladocera are brownish, whereas planktonic species are colorless. Orange or red coloration is also observed in some species. Rarely is a species brightly colored (Weismann, 1878). Blue spots occur in *Eurycercus lamellatus* on the postabdomen, on the dorsal side of the trunk, at the base of the mandibles, and on the esophagus (Weismann, 1878; Behning, 1941; Smirnov, 1971, 1974). *Pseudochydorus* has large brown spots on its valves. Newly molted *P. globosus* are colorless, but during the intermolt period a brown spot appears and increases in intensity. With excessive solar irradiation, Cladocera are blackish (melanistic), as are their ephippia containing latent eggs.

Leydig (1860, p. 56) originally noted that the blood of Cladocera may be colorless, yellowish, reddish, bluish, or greenish. Green (1957) reported his observations of *Daphnia* with pale green blood, *Simocephalus* with green blood, and *Megafenestra aurita* with blue blood. These colors are caused by carotenoid proteins, as they produce an

TABLE 3.4 The Total Content of Lipid and Percentage Lipid Components in Some Cladocera

Species	Total lipid (% DW)	Phospholipid (%)	Triglyceride (%)	Cholesterol (%)	Cholesterol esters (%)
Bosmina obtusirostris	30.6	70	13	3	12.8
Holopedium gibberum	52.4	57	8.7	18.8	17
Bythotrephes cederstroemi	16.4	55	11.3	5	11.3

DW, dry weight.
Source: Lizenko et al. (1977).

TABLE 3.5 Content of Fatty Acids (AS % Total Fatty Acid Content)

Fatty acids	*Daphnia* spp.	*Daphniopsis*	*Bosmina*	*Holopedium*	*Leptodora*	*Bythotrephes*
SATURATED						
12:0	0.7–2.3		–	0.8	0.1–2.1	0.4
14:0	4–10	2.92	8.9	12.0	2.2–7.4	4.4
15:0	0.6–3	0.99	1.1	0.9	0.6–2.8	0.9
16:0	21–36	11.4			24–36	22–26
17:0	20.6	0.64	17.3	15.4		19.1
18:0	0.3–9.9	2.57	–	–	6.5–17.6	6.7–9
MONOUNSATURATED						
16:1ω7	6.7	5.81	4.1	3.5		2.9
18:1ω6 + 18:1ω9	9.2		12.3	8.1		10.3
18:1ω7	4.1		5.0	8.1		6.0
20:1ω9	–		1.2	2.3		–
POLYUNSATURATED						
18:3ω3	6.8	26.5	7.6	6.7		5.3
18:4ω3	11.8		10.4	11.0		4.1
20:5ω3	11.8	1.41	14.4	17.5		23.0
22:6ω3	0.9		2.6	2.1		2.1
18:2ω6	4.6		5.3	4.1		3.6
18:3ω6	1.6		0.6	0.8		0.6
20:4ω6	3.5	0.63	4.7	6.8		9.3

–, not detected.
Sources: Daphnia *spp.,* Bythotrephes longimanus, *Bychek and Gushchina (2001), Bychek and Gushchina (2001);* Daphniopsis tibetana, *Zhao et al. (2006);* Leptodora kindtii, *Bychek and Gushchina (2001);* Bosmina coregoni *and* Holopedium gibberum, *Bychek and Gushchina (2001).*

orange color when treated with desaturating agents.

The coloration of cladocerans (especially of parthenogenetic eggs) may depend on the coloration of the food consumed. Information on the pigments of cladocera, as related to their different ecology, and the transformation of ingested pigments is rather scarce (but may be influenced by seasonal changes and the actual composition of the food). This is an open field for further useful investigations.

Carotenoids

Carotenoids are lipid derivatives. Cladocerans receive carotenoids with algal food (Green 1966a), which contains significant quantities: green algae, 70–500 mg DW/L; blue-green algae, 140–530 mg DW/L (Lavrovskaya, 1965).

With reference to *Simocephalus*, Green (1955b) found carotenoids (orange) in a free state in fat globules, in the gut wall, in fat cells, linked to proteins in the cytoplasm, ovary, and eggs, and as carotenoprotein (green) in blood.

TABLE 3.6 Content of Fatty Acids (IN % DW)

Fatty acids	*Daphnia cucullata*	*D. longispina*	*Bosmina longirostris*	*Simocephalus vetulus*
14:0	7.0	2.5	2.5	3.5
14:1	0.8	3.0	2.3	3.8
14:2	0.1	0.8	2.8	1.4
16:0	16.4	11.7	14.0	12.8
16:1	5.7	14.6	10.3	12.1
16:2		0.8		
18:0	6.8	3.7	7.4	5.1
18:1	8.3	10.1	19.3	11.8
18:2	15.5	4.2	4.2	6.0
18:3	8.2	11.3	5.7	6.8
18:4	6.3	15.8	2.8	7.8
20:2	1.1	0.7	1.7	2.0
20:4	6.7	1.3	5.4	6.6
22:1		0.6		
20:5	7.6	17.4	21.7	18.9
22:5	5.4			0.4
22:6	4.1	1.5		1.0

DW, dry weight.
Source: Herodek and Farkas (1967).

TABLE 3.8 The Content of Fatty Acids (as % Total Fatty Acids) in Three Age Groups of *Daphnia magna*

Fatty Acids	Newborn	3 Days Old	Mature
C14:0	0.9	2.3	3.5
C14:1	0.5	1.5	0.9
C14:2	0.6	1.7	2.2
C16:0	13.2	21.2	24.3
C16:1	4.7	4.7	4.5
C16:2	1.5	5.5	0.3
C16:3	1.2	3.4	7.7
C16:4	1.7	1.8	3.3
C17:0	1.2	Non det.	2.8
C18:0	3.4	7.7	5.8
C18:1	53.9	9.9	9.8
C18:2	6.7	22.0	18.9
C18:3	4.6	7.7	7.7
C20:4	2.8	6.1	3.8
C20:5	0.5	1.2	1.4

Source: Bychek and Gushchina (1999).

TABLE 3.7 The Content of Phospholipid and Neutral Lipids in Three Age Stages of *Daphnia magna*

Types	Lipids	Newborn	3 Days Old	Mature
Phospholipids	Phosphatidylcholine	14.40	8.75	4.06
	Phosphatidylethanolamine	9.15	3.94	3.21
	Sphingomyelin	1.84	0.66	0.65
Neutral Lipids	Triacylglycerols	566.2	283.2	452.1
	Diacylglycerols	Traces	Traces	74.1
	Free sterols	183.6	86.2	35.5
	Free fatty acids	146.9	128.0	60.8
	Wax esters	Traces	Traces	63.4

All values in μg/100 mg WW.
Source: Bychek and Gushchina (1999).

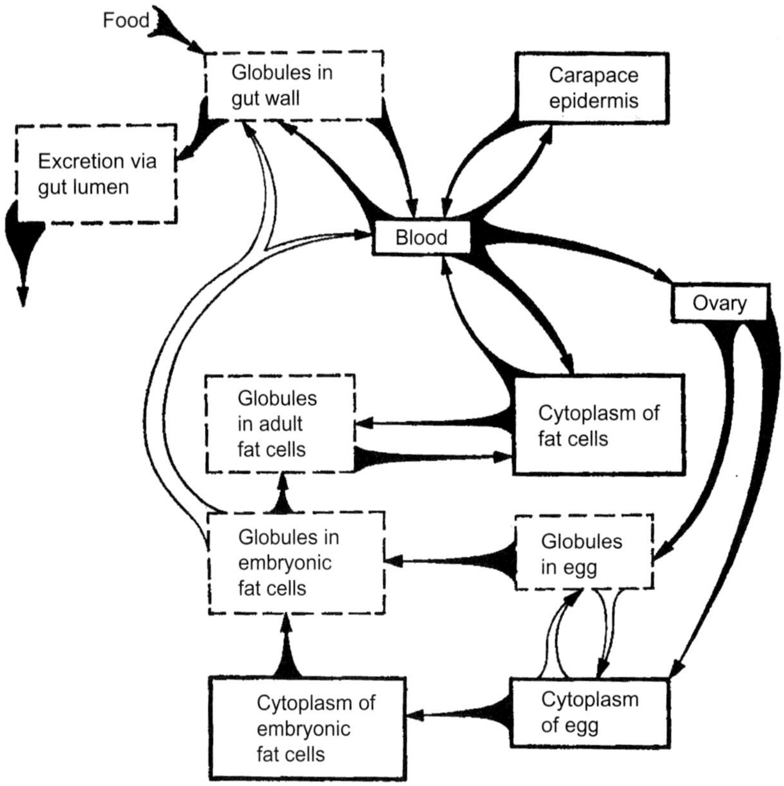

FIGURE 3.6 Pathways of carotenoid transfer in *Simocephalus*. Quadrangles of broken lines indicate stages in which carotenoid is free. *Source: Green (1966b).*

Green (1966b) reported the principal pathways of carotenoid transfer, with reference to *Simocephalus* (Fig. 3.6). Carotenoids obtained from food are passed into the blood, and from there to fat cells, to the carapace epidermis, to the ovary, and then to the eggs. A female may pass half of its total carotenoids to her eggs. In *Simocephalus*, carotenoids are either present in the cells of the gut wall and fat body or associated with proteins of fat cells, epidermis cells, the ovary, and eggs. A green carotenoprotein is found in the blood. An intermolt cycle (Fig. 3.7) and a seasonal cycle (Fig. 3.8) of carotenoids were demonstrated by Green (1966b).

The following carotenoid pigments found in the gut wall, fat cells and the ovary of *Daphnia* were reported by Green (1957): astaxanthin, β-carotene, γ-carotene, and lutein. *Daphnia* grown in the light contain much more carotenoid than those grown in the dark. Carotenoid pigments were found to be dissolved in oil drops or bound to cytoplasmic proteins in *Daphnia* (Green, 1957) and *Simocephalus* (Green, 1966b).

Later, Herring (1968) found the following carotenoid pigments in *D. magna* and other cladocerans: astaxanthin, canthaxanthin, β-carotene, echinenone, and an unidentified ketocarotenoid. Lutein was present only in the gut wall. Herring also fed *D. magna* pure carotenoids with yeast and found that the ingested β-carotene is transformed into echinenone, canthaxanthin, and astaxanthin.

In *C. reticulata*, Czeczuga (1976) found astaxanthin (quantitatively dominant), astacene, canthaxanthin, cryptoxanthin, 4-keto-4-hydroxy-β-carotene, lutein-5,6-epoxide isozeaxanthin, and astaxanthin ester. In *D. magna*, Czeczuga (1984) also found β-cryptoxanthin, isocryptoxanthin, isozeaxanthin, lutein epoxide, and zeaxanthin.

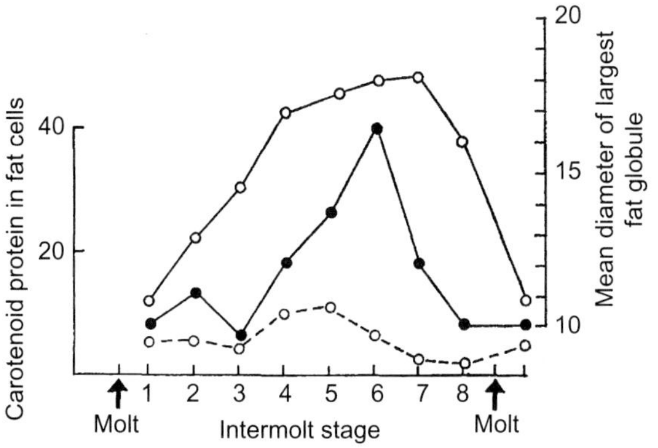

FIGURE 3.7 Intermolt cycle of carotenoid pigmentation in *Simocephalus* fat cells. Lower and middle lines, in cytoplasm (different years); upper line, diameter of largest fat globules. *Source: Green (1966b).*

According to his determinations, the total concentration of carotenoids is c. 4 g/g DW, of which astaxanthin forms 27% and β-cryptoxanthin 25%. In *M. micrura*, astaxanthin and canthaxanthin predominate (Velu et al., 2003).

In *Holopedium*, the carotenoid astacin was found by Sørensen (1936) and Goodwin (1960, p. 127) noted that *Holopedium* accumulates a lot of astaxanthin and that its "de novo synthesis is probably ruled out." As it is hardly possible that much astaxanthin is present in the food, Goodwin suggested that it is produced by oxidation of the β-carotene and its derivatives (including zeaxanthin) ingested with diatoms.

Foss et al. (1986) investigated in detail the composition of carotenoids of *D. magna* and found two new ketocarotenoids.

In various cladocerans, particular carotenoids were found (indicated in Table 3.9)

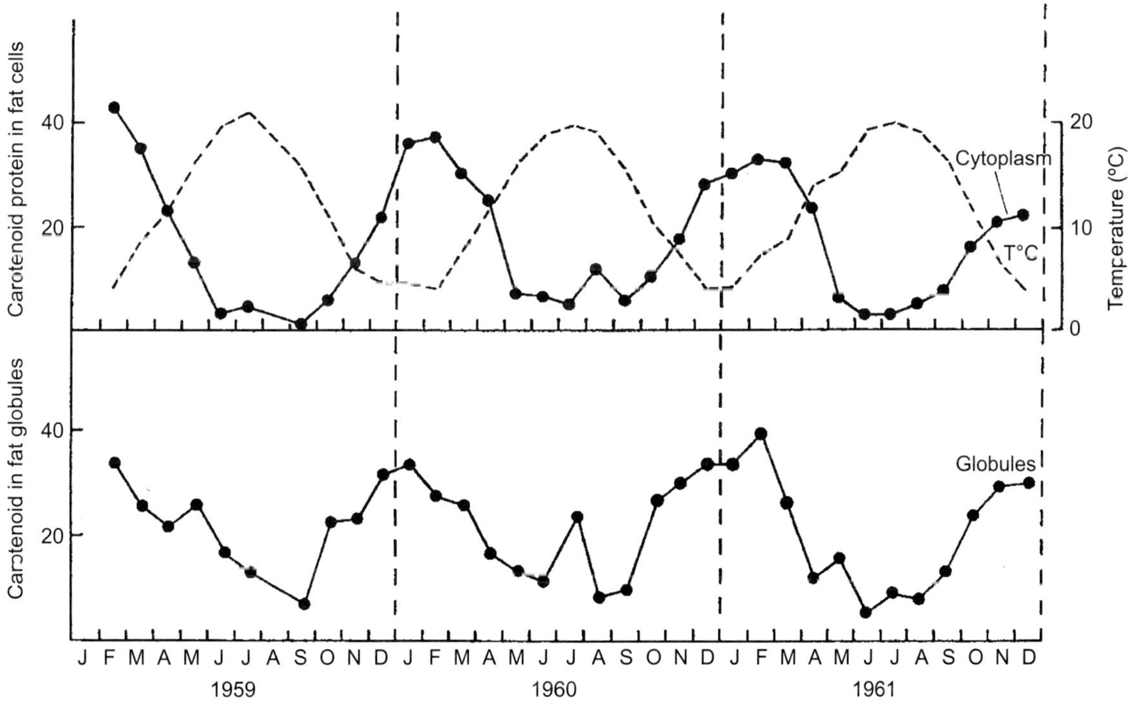

FIGURE 3.8 Seasonal variation in *Simocephalus* carotenoids. Upper, in fat cells. Lower, in fat globules. Letters indicate months. *Source: Green (1966b).*

(Herring, 1968). Carotenoids are not found in *Leptodora* (Farkas, 1958).

It seems that the excess of carotenoids ingested with food serves no physiological purpose. Having received carotenoids with their food, cladocerans just transfer them over to their eggs and progeny. About half of the mother's carotenoids are transferred to the eggs of each brood (Green, 1957, 1966a) but the developing embryos do not utilize this pigment. Moreover, the presence of free or conjugated carotenoids does not favor growth or the viability of eggs or adults.

Melanins

Dark cuticular pigmentation, including that of ephippia (Gerrish and Cáceres, 2003), is caused by melanins (Hebert and Emery, 1990). Melanins (a term relating to both brown and black pigments) are also present in the eyes. They are thought to be products of the metabolism of amino acids (tyrosine): "Chemically, the melanins are probably polymerized indole quinones formed by the action of the enzyme tyrosinase on the aromatic amino acids" (Goodwin, 1960, p. 133).

The ommochromes of eyes are rapidly dissolved with 10% alkali solution in both fresh and preserved specimens.

Arctic and high mountain populations of *Daphnia* are blackish (Hebert and Emery, 1990). In addition, chydorids collected in shallow rock pools in the hot climate of Western Australia are dark brown, as reported by B.V. Timms and M. Jocqué (personal communication). Such dark pigmentation is thought to be of protective value. *Scapholeberis* and *Dadaya* living under the surface of water are also black. Hobaeck and Wolf (1991) observed melanic populations of *Daphnia* at altitudes of 1200–1600 m above sea level inhabiting clearwater lakes and ponds, whereas transparent populations came from ponds with slightly humic water. Melanin was deposited in the dorsal area of the carapace, in the head shield, and the antennae. Its absorption maximum was c. 249 nm.

Hemoglobin

There is an extensive literature on Hb in Cladocera. Special investigations into Hb in Cladocera were made by Fox (during 1945–1955), Hoshi (during 1949–1974), and Green (1955, 1956b). Hb may be measured chemically or spectroscopically.

In contrast to Malacostraca (Prosser and Brown, 1967 [1962]), Cladocera synthesize Hb, which may be present in their blood, muscles,

TABLE 3.9 The Percentage Composition of the Carotenoid Pigments in Wild Cladocera SPP

Species	β-Carotene	Echinenone	Canthaxanthin	Ketocarotenoid	Astaxanthin	Lutein	2nd Ketocarotenoid
Ceriodaphnia megalops	5	6	5	3	72	5	2
Daphnia pulex	4–12	7–11	5–19	5–20	24–66	4–20	Trace
D. longispina	6	5	2	0	56	13	3
Eurycercus	7	0	16	0	47	19	0
Moina micrura	0.43	–	4.60	–	39.82	21.21	–
Simocephalus vetulus	6	7	8	0	38	21	0

Note: 18.65% β-cryptoxanthin and 15.29% violaxanthin have also been found in *Moina micrura* (Velu et al., 2003).
Sources: Herring (1968); Moina micrura, *after Velu et al. (2003).*

central nervous system, fat cells, ovaries, or dissolved in hemolymph of parthenogenetic eggs (Fox, 1955; Green, 1955).

The molecular weight of *Daphnia* Hb is 400,000, with an admixture of a small amount of 34,500 (Goodwin, 1960). In *M. macrocopa*, molecular weight of Hb was estimated to be 670,000, the iron (Fe) content to be 0.317%, and the number of Fe atoms per Hb molecule to be 38 (Hoshi et al., 1967). These authors conclude that each Hb subunit with one Fe atom corresponds approximately to human Hb.

The role of Hb is considered further in Chapter 5, "Respiration."

Myoglobin

Some of the Hb is present in muscles as myoglobin. Goodwin (1960) warned against the use of the latter term for Crustacea, as its constitution was then unknown. In some of my long-standing samples of filamentous algae and Cladocera preserved in formalin, the muscles of the cladocerans are stained blackish, thus indicating the presence of myoglobin. This may be attributed to tannic substances, which are known to occur in filamentous algae and produce black ink with Hb iron. However, attempts to stain similar samples with tannin failed, although further attempts may be worthwhile.

Later, the presence of myoglobin in the muscles of *Simocephalus* was directly confirmed by microspectrophotometry (Fig. 3.9) (Karnaukhov, Del Rio De Valdivia, and Yashin, 1986). Myoglobin was completely absent from oil drops.

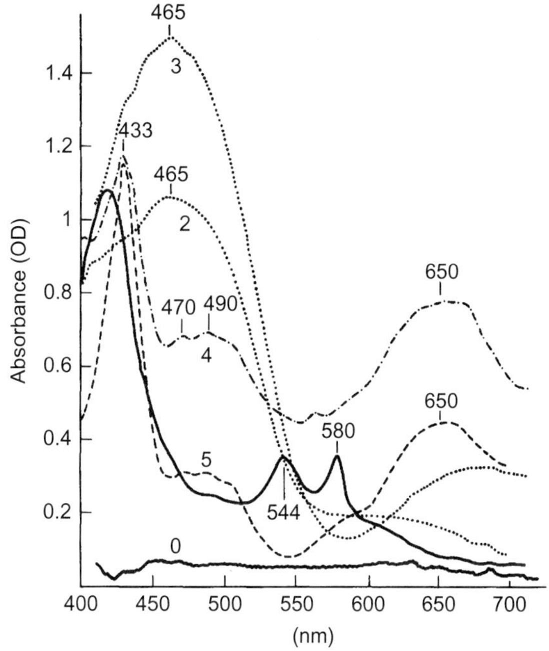

FIGURE 3.9 **Levels of different components in *Simocephalus*.** 1, myoglobin in muscles; 2,3, lipid in tissues; 4, green granules in eggs; 5, yellow-green granules in eggs. OD, optical density. *Source: Karnaukhov et al., 1986.*

Other Pigments

Cytochrome was shown by Fox (1955) to be present in cladoceran muscles (in *Daphnia*, absorption maxima at 566 and 500 nm) and by Hoshy and Akizawa (1979) (in *Simocephalus*, at 603, 563, 550, and 532–522 nm).

Coloration of *Simocephalus* and *Chydorus* results from a combination of myoglobin, carotenoids, and carotene protein, as shown microspectrophotometrically by their absorption bands (Karnaukhov, Del Rio De Valdivia, and Yashin, 1986).

Biliverdin has been discovered only in the eye of *Polyphemus*, and not in the eye of *Daphnia* (Green, 1961a).

3.2.8 Content of Particular Elements or Compounds

The content of some elements in Cladocera spp. is shown in Table 3.10. The content of mineral substances in the bodies of cladocera depends on mineralization of the environment; in *D. magna*, it ranges from 12 to 37.6% DW; and in *S. vetulus*, from 10 to 19.4% DW (Stepanova et al., 1971).

TABLE 3.10 Content of Some Elements in Cladocera

	Calcium (% DW)	Phosphorus (% DW)	Iron (mg DW/L)	Reference(s)
Daphnia pulex	9.6	1.48	950	Malikova (1953, 1956)
D. hyalina newborn	1.75	1.56	–	Baudouin & Ravera (1972)
D. hyalina mature	5.4–6.9	1–1.56	–	Baudouin & Ravera (1972), Lin & Lui (1985)

DW, dry weight.

Essential and Nonessential Elements

Cladocera need some elements, while others may be physiologically nonessential or even toxic. In addition, essential elements (e.g. copper (Cu) or zinc (Zn)) may be toxic at concentrations higher than are normally necessary. Cladocera may also accumulate, without any physiological use, substances that just happen to be around.

Generally, tissues of Cladocera consist of biogenic elements (C, H, O, and N), macroelements (Ca, Na, K, P, S, Mg, Mn, Fe, Cl, and Cu), and microelements (Zn, Co, I, Se, Mo, Li, and some others).

CARBON

Carbon (C) is one of the principal macroelements. Although the content of carbon in most Cladocera is generally about 47–48% DW, there are variations (e.g. c. 53% in *Scapholeberis mucronata)* (Hessen and Lyche, 1991). Data on the content of carbon in representatives of particular genera are available. For example, the content of carbon in the bodies of *Diaphanosoma*, *Holopedium*, *Bosmina*, and *Daphnia* was found to be 48% DW, with little seasonal variation (Andersen and Hessen, 1991). It increases with increased body length in *D. hyalina*, from 42.8% total C in DW in newborns to 44% in adults (Baudouin and Ravera, 1972).

Direct determination of the total carbon content (using the dry-combustion method of Pregl-Roth) yielded the following results (Metz, 1973). By measuring the uptake and release of radiocarbon (γC), *D. pulex* fell into two groups: 1.5–10 γC/individual (ind.) and 2–22 γC/ind. The radiocarbon content in *Daphnia* with eggs was twice that of *Daphnia* without eggs. About 52% of carbon was in the eggs, irrespective of the number of eggs. The minimum carbon content per egg was 1–1.5 γC.

The content of organic carbon was determined in *Polyphemus pediculus* by Butorina (1973): parthenogenetic females, 3–7% WW, 40–50% DW; gamogenetic females, 3.4–7.5% WW, 41–48% DW; males 4.4–5.8% WW, 42–47% DW. Moreover, it was 44.6% DW in *D. hyalina* and 49.05% DW in *Leptodora kindtii* (Lin and Lui, 1985).

NITROGEN

In natural water bodies there is always a deficiency of this element. In the bodies of Cladocera, the content of nitrogen (in % DW) was found to be 9.7 in newborn *D. hyalina* (Baudouin and Ravera, 1972); c. 9 in mature *D. hyalina* (Baudouin and Ravera, 1972; Lin and Lui, 1985); c. 9-10 in *Diaphanosoma*, *Holopedium*, *Bosmina*, *Daphnia*, and *Leptodora*, with little seasonal variation (Andersen and Hessen, 1991; Hessen and Lyche, 1991); 12.7 in *Leptodora kindtii* (Lin and Lui, 1985); and c. 7.9 in *S. mucronata* (Hessen and Lyche, 1991).

PHOSPHORUS

In aquatic environments, there are only traces of free phosphorus because, similar to N, it is avidly assimilated by algae.

A major pool of phosphorus in *Daphnia* is present in the exoskeleton (Hessen and Rukke, 2000a). Phosphorus content (as P_2O_5) in the bodies of Cladocera was found to be 3.4–3.6% DW in *D. pulex*; 3.5% DW in *Leptodora* (Birge and Juday, 1922); c. 1% DW in *Diaphanosoma, Holopedium, Bosmina*; c. 1.4% DW in *Daphnia* (Andersen and Hessen, 1991; Vrede et al., 1999), with little seasonal variation (Andersen and Hessen, 1991, 1999); and 1.1% DW in *D. magna* (Sterner and Schwalbach, 2001). Having summarized the available data, Brett et al. (2000) indicated 0.8–1.8% P DW for various species of *Daphnia*, the highest being 1.8 for *D. pulicaria* and the lowest 0.81 for *D. pulicaria*. The phosphorus content was especially high in juvenile *Daphnia*, which were thus more dependent on sources of phosphorus in their food (Main et al., 1997).

P content was found to be higher in filtrators (>1.5%) and lower in carnivorous cladocerans (c. 0.5% P DW) (Hessen and Lyche, 1991). The highest phosphorus content, c. 2%, was found in *S. mucronata* and *Ceriodaphnia quadrangula*.

CALCIUM AND STRONTIUM

Although cladocera do not possess a strongly calcified carapace, some, e.g. *Daphnia*, still retain some ability to accumulate calcium (Ca) and strontium (Sr) in their shells (Porcella et al., 1967, 1969). In both littoral and pelagic species (*Alona, Chydorus, Pleuroxus, Ceriodaphnia, Simocephalus*, and *Daphnia*), the integuments may sometimes be reinforced by $CaCO_3$ (Leydig, 1860, pl. II; Gicklhorn, 1925; Schmidt, 1943; Taub and Dollar, 1968; Porcella et al., 1969). This is similar to the deposition of calcium in the exoskeletons of ostracods or of large crustaceans. In *D. magna*, calcification was found to occur by the deposition of fine grains beginning just prior to molting (Porcella et al., 1969).

The content of calcium ranges within 2–9.9% DW in *D. pulex*, 3.2% DW in *Leptodora* (as CaO in ash) (Birge and Juday, 1922), and within 0.8–4.4% DW in *Daphnia* spp. in Norwegian lakes (Wærvågen et al., 2002). The calcium content of *D. magna* ranges from 4.2% to 1% DW, decreasing when there is a lower calcium concentration in the medium (Alstad et al., 1999). A large part of the calcium is lost with the exuvium at molting.

IRON

Iron is discussed in Chapter 5, section 5.4. See also Table 3.11.

LEAD

According to Holm-Jensen (1948), lead (Pb) is accumulated "mainly passively" in the shells of *D. magna* "probably by exchange" of Ca^{2+} for Pb^{2+}.

MAGNESIUM

In *D. pulex*, 0.5–0.9% DW magnesium (Mg) was found, and in *Leptodora*, 0.95% DW (as MgO in ash) (Birge and Juday, 1922).

COPPER

An optimum copper concentration range for *D. magna* was determined to be 1–35 μg Cu/L (Bossuyt and Janssen, 2004), and active copper regulation between 0.5–35 μg Cu/L was observed in the body. For the metabolic role of Cu, see also Fig. 3.10.

SILICA (SIO_2)

A total of 2.8% DW of silicon (Si) was reported in a sample of *D. pulex* (Birge and Juday, 1922).

SODIUM

The normal concentration of sodium (Na) in *D. magna* is 26.3 mM/kg WW (Stobbart et al., 1977). Excessive Na is not accumulated. Na uptake and release is considered further in Chapter 8, "Osmotic regulation."

3.3 XENOBIOTICS IN THE CLADOCERAN BODY

Xenobiotics affect aquatic invertebrates either by direct toxicity of dissolved toxicants or as contaminated food. The effects may be

TABLE 3.11 Bioconcentration Factors (i.e. the Accumulation of Various Substances Over Time)

Substance	Species	Factor (× Original Concentration)	Time	Reference(s)
Arsenic trioxide	*Daphnia magna*	219	21 days	Spehar et al. (1980)
Cesium-137	*D. magna*	84	5 days	Nilov (1983)
$HgCl_2$	*Ceriodaphnia affinis*	2000	In 20th generation	Gremiachikh & Tomilina (2010)
Manganese	*D. magna*	65	8 h	Kwasnik et al. (1974)
Neptunium	*D. magna*	32	48 h	Poston et al. (1990)
Nickel	*D. magna*	25	48 h	Pane et al. (2003)
Anthracene	*D. pulex*	760	3 h	Herbes & Risi (1978)
Benz(a)anthracene	*D. pulex*	10,000	24 h	Southward et al. (1978)
Benzo(a)pyrene	*D. magna*	1,000–8,000	24 h	McCarthy (1983), Oikari & Kukkonen (1990)
Chlordanes	*D. pulex*	24,000	24 h	Moore et al. (1977)
DDT	*D. magna*	23,000	24 h	Crosby & Tucker (1971)
Estrone	*D. magna*	228	16 h	Gomes et al. (2004)
Naphthalene	*D. pulex*	100	24 h	Southward et al. (1978)
Pirimicarb[a]	*D. magna*	50	48 h, per DW	Kusk, 1996
Triphenyltin chloride (0.1 mg/L in water)	*D. magna*	290	3 h	Filenko & Isakova (1979)
Water-soluble oil fraction	*Daphnia*	>500	19 h	Mikhailova et al. (1986)

DDT, dichlorodiphenyltrichloroethane.
[a]*2-(diethylamino)-5,6-dimethyl-4-pyrimidinyl dimethyl carbamate.*

immediate or delayed, be manifested as damage to metabolic links in the organism, be teratogenic, or disrupt links in the natural biocenological network. At low concentrations of xenobiotics, Cladocera continue to live and reproduce, but are less prolific. High concentrations of xenobiotics either cause death or decrease the life span. They also disturb natural adaptation processes. Generally, they have far-reaching effects at the species, population, and biocenotic levels.

Numerous experiments have shown a decrease in life span and fecundity in the presence of toxic substances. Beklemishev (1924) noted the necessity of taking into consideration the rate of penetration of toxicants through integuments. He observed that when cladocerans are placed in a toxic solution, at first they behave in their usual way ("as if nothing had happened"), then they manifest signs of anxiety, make desperate jerks, sink to the bottom, and make occasional movements there, then all movement stops, and finally the heart stops.

Excess pollution may kill all animal life or, at lower concentrations, kill animal species selectively or inhibit certain vital functions or interrelations.

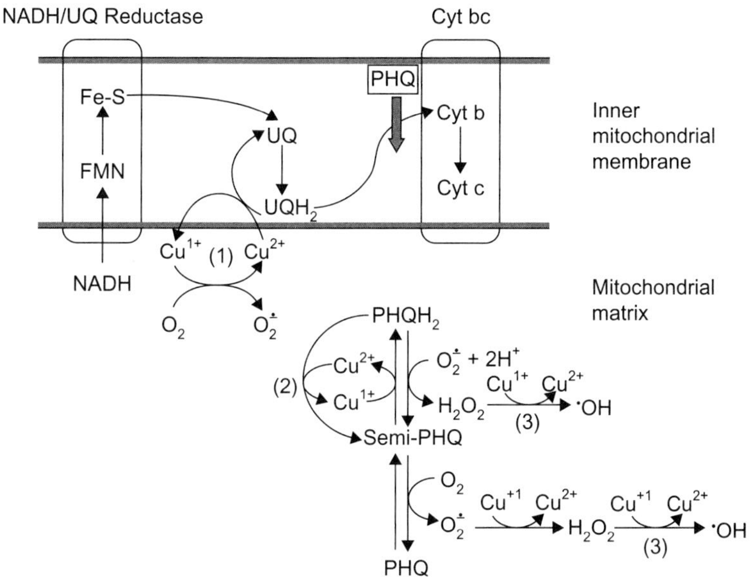

FIGURE 3.10 **Proposed roles of copper in the transformation of phenathrenequinone.** *Source: Xie et al. (2006).*

In this chapter, the principal focus is on studies demonstrating how foreign chemicals become involved in their metabolism or are transformed by cladocera, and how the chemicals or excessive factors impair particular functions or metabolic links. Accordingly, measurements of the lethal dose, quantitative characteristics of fecundity, or life span are mostly not cited here.

Particular substances may not accumulate homogeneously throughout the cladoceran body but may preferentially accumulate in particular organs or tissues. Thus, cadmium (Cd) taken from solution is concentrated in the exoskeleton of *D. magna* (Carney et al., 1986). The internal distribution of ^{109}Cd (cadmium-109) has been determined by whole-animal autoradiography (Munger et al., 1999); it mainly accumulated in the diverticula of the gut, which are sites of uptake of nutrients and Ca. Cadmium accumulated by *Moina* (Yamamura et al., 1983) was mostly bound to low molecular weight proteins (a mixture of two isoproteins). It was also shown to be absorbed onto the surface of the carapaces of *D. magna* and *Ceriodaphnia dubia* (Robinson et al., 2003).

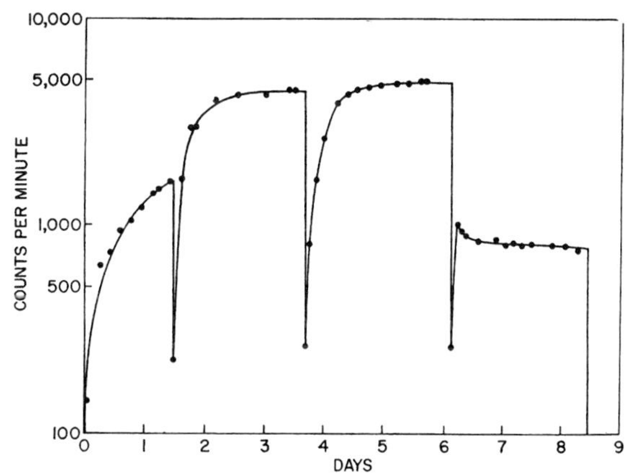

FIGURE 3.11 **Accumulation and removal of radiolabelled strontium by *Daphnia magna*.** Strontium removal occurs over the course of three complete molt cycles (followed by unlabeled medium). *Source: Marshall et al. (1964).*

Selenium (Se) is localized primarily in the cytoplasm of *D. pulex* cells. In mitochondrial matrices, excessive dense deposits were accumulated and, with time, the mitochondria degenerated (Schultz et al., 1980).

D. magna accumulates 95% of Sr (similar to Ca) in the exoskeleton; it is eliminated at each molting, thus falling to a minimum, and then the Sr content gradually increases (Fig. 3.11) (Marshall et al., 1964). Pb is also thought to be accumulated in the shells, probably by replacing calcium (Holm-Jensen, 1948). In *D. magna*, nickel (Ni) is mainly accumulated in the carapace, gut, and hemolymph (Hall, 1982).

Zinc-65 (^{65}Zn) is partly accumulated in the exoskeleton and is removed at molting (Winner and Gauss, 1986).

Anthracene accumulated by *D. pulex* is removed at a rate of 1.6 μg/ind./h. However, within *Daphnia* there are three "compartments": a rapidly eliminated compartment (with c. 30% of accumulated anthracene); a more slowly eliminated compartment (60%); and a tightly bound one (8%) (Herbes and Risi, 1978). It is thought that metabolization of anthracene by *D. pulex* is much slower than its accumulation. Exposure of *D. magna* to anthracene plus ultraviolet radiation (UVR) reduced survival and fecundity more than exposure to anthracene alone (Holst and Giesy, 1989). Exposure to UVR in the absence of anthracene produced no significant effect.

There is also a dependence on temperature. It was shown by analyses of *D. pulicaria* homogenate fractions (Heisig-Gunkel and Gunkel, 1982) that at a temperature of about 8°C the accumulation of atrazine is correlated with protein content, whereas at higher temperatures (12–29°C), it correlates with fat, and protein becomes less important.

CHAPTER

4

Nutrition

4.1 FEEDING

4.1.1 Anatomical Background

Cladocera obtain their food by means of their thoracic limbs, but there is a profound difference in the methods of food collection between littoral and planktonic Cladocera spp.

Feeding in cladocera is inseparably linked to respiration. With reference to *Daphnia magna*, it has been shown that the principal part of their oxygen supply is extracted from the feeding current (Pirow et al., 1999a).

Although generally bilaterally symmetrical, Cladocera have asymmetrical mandibles that are controlled by asymmetrical muscles. Fryer (1963, p. 347) observed that "the most important mandibular movements are those of rotation." As mandibular muscles are asymmetrical, "[t]his enables the two masticatory surfaces to move relative to each other and to exploit to the full their skeletal asymmetry" (Fryer, 1963, p. 351).

To an extent, Cladocera crush their food clumps and algal cells with their mandibles but Hebert (1978a) noted that physical damage to food particles by the mandibles during ingestion is slight. For example, in the gut of *Picripleuroxus striatus* none of the scraped off and ingested frustules of the *Cocconeis* diatom were broken (Smirnov, 1971, Fig. 233). Nevertheless, in the process of ingestion some diatoms are damaged, especially *Asterionella* (Infante, 1973; Lampert, 1978; Fryer, 1991), and *Eurycercus* spp. have been shown to crush large food clumps with their mandibles (Kotov, 1998).

Cladocera that Collect Food from Substrata

Cladocera that collect their food from substrata have evolved various specializations of their food-collecting apparatus, as discovered and described for chydorids and macrothricids by Fryer (1963, 1968, 1974, 1991). In general, bottom-living cladocera (e.g. Figs. 1.1 and 1.4, left), collect food from the surfaces of various substrata using strong, second thoracic limbs, push the collected mass forward toward the mouth, and remove excess water by undulating the exopodites of their posterior limbs. There are further specializations of this feeding activity in different genera and species. The basic (least modified) food-collecting system of the thoracic limbs in chydorids seems to be that present in *Eurycercus* and *Pleuroxus truncatus* (Fryer, 1963, 1968).

The methods of food collection used by bottom-living cladocerans are diverse (Fryer, 1963, 1968, 1974); however, more detailed information on the physiological aspects of

Physiology of the Cladocera.
DOI: http://dx.doi.org/10.1016/B978-0-12-396953-8.00004-5

feeding is scarce, in contrast to the planktonic filter-feeders.

The food particles collected by littoral cladocera are glued together by the secretions of their salivary gland (Fig. 3.2) and of glands in the thoracic limbs (the latter are known to be present in *Eurycercus* and *Alonopsis*; Fig. 3.3) and then forwarded to the intestine (Fryer, 1962, 1963, 1968). The salivary gland cells are supplied with nerve endings (Zeni and Zaffagnini, 1988). No attempts have yet been made to check whether the glands in the thoracic limbs of these genera are present in various other chydorids and macrothricids; nor is any information available as to whether these secretions contain digestive enzymes.

Littoral Cladocera, especially those which live on substrata or on the bottom of water bodies, exist within a seemingly unlimited food supply of organic debris at all stages of decomposition, as well of algae and bacteria. Fecal pellets containing variously digested organic matter also settle to the bottom and become a part of the detritus. A special place in the littoral belongs to some diatoms, mostly epiphytic and attached to substrata, which transfer their photosynthesis products to lipids and accumulate fat stocks.

Undoubtedly, therefore, among the bottom-living cladocera there are some types that are physiologically adapted to feeding on organic detritus at various stages of decomposition, from living algae to profoundly decomposed organic matter. Such a sequence of adaptations to different decomposition stages has never been specifically studied. It is likely that different decomposition levels of organic matter supply nutrition for different species of Cladocera that are physiologically specialized for their consumption. Unfortunately, there are no relevant investigations, except for observations that in an assembly of collected littoral Cladocera kept in a jar with detritus its constituent species die off gradually and in a certain sequence, which might be related to decomposition of the food material. It has occasionally been observed that *Alonella* spp. remain alive in vessels with detritus for several months when all other companion species have perished.

Van de Bund (2000) supplied what seems to be the only direct observation that a chydorid, *Chydorus piger*, reproduces especially well when feeding on fecal pellets of *Chironomus riparius* ("a highly significant effect" was observed), as well as on natural detritus; it is thought that in general chironomid fecal pellets (coprophagy) support the reproduction of chydorids.

Littoral *Scapholeberis* and *Dadaya* attach to the surface film of water and collect pollen, minor particles, and, in all probability, the unimolecular film of organic matter. Filter feeding is performed by sedentary *Sida* and *Simocephalus* spp.

Littoral *Lathonura* and *Drepanothrix* spp. may be observed to remain motionless for hours, lying on their backs and making no other movements than filter beating their thoracic limbs.

Chydorids and macrothricids can thrive for several months (e.g. Smirnov, 1964, 1965a) when fed periodically with fresh organic debris (or "detritus"). Dekker et al. (2006) cultured *C. sphaericus* under similar conditions and obtained the best growth using artificial detritus prepared from *Urtica* and *Nitzschia* diatoms.

Cladocera that Collect Food Suspended in Open Water

In contrast to those living in the littoral zone, pelagic Cladocera periodically suffer food shortages or drastic seasonal changes in food quality. In filter-feeding (mostly planktonic) anomopods and in ctenopods, food particles suspended in the outer medium are filtered by filtering fans consisting of setae on thoracic limbs (e.g. Fig. 1.2). In anomopods, the filtering fans are formed by gnathobasic setae; in ctenopods, this is done by setae on endites in combination with a smaller area of gnathobasic setae.

In daphnids, slime glands in thoracic limbs are not reported (Fryer, 1991) and are probably absent.

For pelagic cladocera, there is obviously a similar gradation of food, from fresh or recently dead algae, as well as fecal pellets, to all stages of decomposition, potentially with specialized consumers. Coprophagy in pelagic Cladocera was studied by Pilati et al. (2004). These authors obtained ^{14}C-labeled feces by feeding zooplankton with ^{14}C-labeled algae. A considerable clearance rate was found for such fecal matter: 0.084–0.089 mL/mg/h in *Holopedium gibberum* and 0.026 mL/mg/h in daphnids.

Pelagic Cladocera spp. reproduce successfully when feeding on algae (less efficiently when feeding on blue-green algae) (*Daphnia*), bacteria (*Daphnia, Ceriodaphnia, Bosmina coregoni, C. sphaericus,* and *Polyphemus*), and yeast (*Daphnia*) (Rodina, 1950). However, Lampert (1987) demonstrated that *Daphnia* grow but do not reproduce when fed on bacteria.

Planktonic cladocera consume any algae that are present in the water, as well as organic particles and bacteria. They collect their food by filtration using their thoracic limbs. The collected food is directed into the mouth opening, where it is formed into a conveniently narrow flow by rotation of the mandibles.

Carnivorous Species

In contrast to most cladocerans, the littoral chydorid *Pseudochydorus globosus* is carnivorous and feeds on the dead bodies of cladocerans, whereas *Anchistropus* is specialized for feeding on living hydra (as recently observed by Van Damme and Dumont, 2007). *Bythotrephes* ingests the soft tissues of *Daphnia* spp. (Lehman (1993). These species successfully complete their life cycles when feeding on this food alone, but nothing is known of any special traits of their nutritional physiology.

Predatory (carnivorous) Cladocera are also present in the pelagic zone. These include *Leptodora, Bythotrephes,* and *Polyphemus* spp., the latter being common in patches of open water in the littoral zone.

4.1.2 Environmental Background and Food Resources

Algal Food

In almost all waters, algae are available in quantities but the composition of their populations depends on season, latitude, and the type of water body. There are various sizes and types of algae, from those comprising small cells to those forming clumps or filaments. Frequently, algae are covered with slime. It has been shown that algal cells may be ingested repeatedly by *D. magna* before being completely digested (Schindler, 1968; Kersting, 1978).

Generally, sugar is produced by photosynthesis and is then transformed, *inter alia,* into glycerol and saturated and unsaturated fatty acids. Glycerin combined with fatty acids yields fat. Kretovich (1986, p. 278) notes that "the process of formation of fat from carbohydrates may occur at a very high rate." The schematic pathway from initial photosynthetic saccharides to lipids and amino acids in algae (Arts et al., 1997) is shown in Fig. 4.1.

The chemical composition of algal populations depends on the specific physiology of particular groups of algae. While green algae may be considered similar to green vascular plants in this respect, algae from other divisions may be highly specific in their physiology and their photosynthetic products. For example, green algae accumulate starch and cellulose, with a small quantity of fat, as the principal products of photosynthesis; diatoms (Bacillariophyceae) mainly produce lipids.

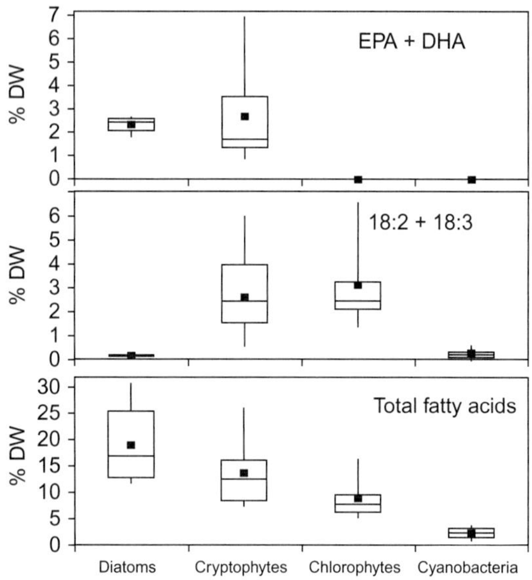

FIGURE 4.1 **Content of highly unsaturated fatty acids in dominant groups of planktonic algae.** DHA, docosahexaenoic acid; DW, dry weight; EPA, eicosapentaenoic acid. *Source: Brett and Müller-Navarra (1997).*

TABLE 4.1 Percentage Fatty Acid Content of Freshwater Seston, Diatoms, *Daphnia* spp., and Fish

Fatty Acids	Seston (Persson and Vrede, 2006)	Diatoms: *Cyclotella* (Desvilettes et al., 1997)	*Daphnia* (Persson and Vrede, 2006)	Fish: *Coregonus* (Sushchik, 2008)
Total SAFAs	51.8	17.7	37.1	–
Total MUFAs	20.2	18.6	19.9	–
Total ω3 (PUFAs)	16.8	62.0	31.2	17.0
Total ω6 (PUFAs)	6.1	4.8	9.8	5.4
Total PUFAs	22.9	71.6	41.0	22.4

Values represent % of total fatty acid content.
MUFAs, monounsaturated fatty acids; PUFAs, polyunsaturated fatty acids; SAFAs, saturated fatty acids.

The polyunsaturated fatty acid (PUFA) compositions of different organisms are shown in Tables 4.1–4.3.

Blue-green algae (cyanobacteria) also accumulate lipids, which, however, contain toxic compounds. Cyanobacteria are characterized by a low eicosapentaenoic acid (EPA; 20:5ω3) content (Müller-Navarra et al., 2000) and a low sterol content (von Elert et al., 2002) which constrains the growth and reproduction of Cladocera.

Bacterial Food

Along with other food items, Cladocera consume bacteria; they are also internally populated by bacteria, including in their intestines. To solve the question as to whether bacteria are an obligatory food component, it was necessary to develop methods for sterilizing Cladocera. To achieve this aim, Gajewskaja (1938) maintained *Daphnia* and *Bosmina* in a 0.1% solution of Rivanol (ethoxydiaminoacridine lactate) and obtained a high percentage of sterile cladocerans. It was shown that feeding on bacteria, in combination with algae, supports a long series of parthenogenetic generations of *D. magna*; in contrast, cultures failed if fed either on bacteria or on sterile algae (Gajewskaja, 1940, 1948). The same conclusion was made by Taub and Dollar (1968) for *Daphnia pulex*, and Tezuka (1971) concluded that *Daphnia* could not to survive when fed on bacteria alone.

Daphnia thrive when fed on organic debris prepared from algae or vascular plants. Good growth and reproduction of *D. magna* and *D. longispina* was obtained by Esipova (1969) using detritus prepared from phytoplankton treated with Lugol's solution to reduce the quantity of bacteria.

It has also been shown that *D. magna* feed efficiently on suspended bacteria of size 0.1–1 μm, but that for *D. galeata*, *D. hyalina*, and *D. pulicaria* feeding is less efficient (Gophen and Geller, 1984). Heterotrophic bacteria weakly support *D. magna*, but a culture developed well when supplemented with

TABLE 4.2 The Fatty Acid Composition of Some Freshwater Algae, as Percentage of Total Fatty Acids

Fatty Acids	Green Algae	Diatom *Cyclotella*	Blue-Green Algae	Chrysophyceae *Chromulina*	Cryptophyceae	Dinoflagellates *Peridinium*
14:0	0.7–2.3	2.0	2.1–17.0	11.1	1.7–2	7.6
16:0	9.8–20.4	11.6	18.5–29.1	5.0	11.1–13	29.8
16:1ω7	0.3–1.1	5.2	1.8–22.6	2.1	0.3–0.6	2.1
18:0	0.6–4.2	3.3	1.6–1.9	1.5	0.7–1.3	0.4
18:1ω9	5.0–36.5	–	1.9–3.2	1.3	1–1.1	30.0
18:1ω7	0.2–1.8	–	0.5–2.5	2.3	0.6–2.1	0.4
18:2ω6	6.7–22.1	–	2.9–11.4	7.1	0.9–1.0	0.5
18:3ω6	02–0.4	3.7	0.3	3.4	–	–
18:3ω3	14.9–37.2	–	22.8–24.6	6.6	9.7–21.3	0.2
18:4ω3	2.5–3.4	–	2.5	26.6	21.8–24	4.3
20:0	0.1–0.2	–	–	–	–	–
20:1ω9	0.3	–	–	2.4	0.7	–
20:3ω6	–	=	–	0.4	–	–
20:4ω6	–	0.5	–	0.6	–	–
20:5ω3	–	35.1	0.6	0.6	15.8–20.5	6.9
22:0	–	–	–	–	–	–
22:4ω6	–	–	–	0.4	–	–
22:5ω3	–	3.1	–	–	–	–
22:5ω6	–	–	–	9.9	2.7–4.7	–
22:6ω3	–	6.5	–	2.5	4.3–7.2	12.2

–, none or <0.1%.
Sources: Ahlgren et al. (1900); Cyclotella, *from Desvillettes et al. (1997).*

cholesterol or PUFAs (Martin-Creuzburg and Heilke, 2011); *Hydrogenophaga* and *Pseudomonas* are toxic for *D. magna*.

Daphnia consumes methanotrophic bacteria using methane-derived carbon (C) (Kankaala et al., 2006), as well and those containing a large part of carbon from this source (Taipale et al., 2007). Methanotrophic bacteria and methane are highly significant sources of carbon for *Daphnia* (Deines and Fink, 2011).

Enteropathogenic non-agglutinating vibrios are consumed by *D. magna* but do not cause pathologies and are digested as food (Avtsyn and Petrova, 1986).

Nevertheless, overall it has been determined that bacteria may make up to 87% of the total carbon ingestion of *D. galeata* (Kamjunke et al., 1999). Under conditions of mixed feeding (bacteria and algae), *D. magna* obtained 80% of its carbon from bacteria and in this situation they

TABLE 4.3 General Fatty Acid Composition of Some Algae, as a Percentage of Total Lipids

Fatty Acids	Diatoms: *Cyclotella*	Green Algae: *Pediastrum*	Dinoflagellates: *Peridinium*	Chrysophyceae: *Chromulina*	Cryptophytes: *Cryptomonas*	Blue-Green Algae: *Anabaena*
Saturated	17.7	21.5	36.4	17.6	15.1	40.5
Monoenes	10.7	12.0	30.6	8.1	3.0	18.6
Total PUFA	71.6	66.0	24.9	57.7	64.9	51.3
Total ω3	62.0	53.5	24.3	36.3	59.2	38.3
Total ω6	4.8	9.5	0.6	21.4	5.7	9.8

PUFA, polyunsaturated fatty acid.
Source: Desvilettes et al. (1997).

TABLE 4.4 The Chemical Composition of the Bottom Silt in Two Russian Lakes

Lake	Sugars, Hemicelluloses, Pentosans	Cellulose	Waxes and Resins	Fatty Acids	Organic Carbon	Total Nitrogen	Ash
Glubokoe	1.0	2.2	2.3	0.2	11.9	1.1	74.2
Beloe	7.1	6.4	6.5	1.0	27.0	2.5	46.4

Values represent % absolutely dry matter.
Source: Shcherbakov, 1967.

assimilated c. 46% of their fatty acids from bacteria (Taipale et al., 2012). *D. magna* fed solely on bacterial diets die after 5–12 days.

Organic Debris (Detritus)

Organic debris suspended in water and settling on the bottom of water bodies consist of decomposing algae, decomposing feces, and the bodies of dying animals, as well as all kinds of allochthonous material. Detritus also harbors bacteria. Organic debris forms a major component of the food of littoral and pelagic Cladocera, as noted originally by Naumann (1918).

The chemical composition of lake silt is shown in Table 4.4. Proteins and carbohydrates in the detritus (debris) are decomposed by enzymes in the feces of various animals, as has been convincingly shown for fish by Kuzmina et al. (2010), as well as by bacteria. "Old," i.e. profoundly decomposed, debris does not support *Daphnia* in culture (Lampert, 1987).

It is likely that different decomposition levels of organic matter supply nourishment for different species of Cladocera that are physiologically specialized for the consumption of material at different decomposition levels. The sequence of decomposition in sediments of lakes was described by Rodina (1963, p. 388): "The processes of decay of protoplasm of animal and vegetative tissue and the processes of oxidation and reduction are caused by different physiologic groups of bacteria: putrefactive, fat and carbohydrate decomposers, amylolytic, pectinous, cellulose decomposers, denitrifying, nitrogen fixers, phosphate liberators, sulfur bacteria, and yeasts."

Although little is known on the subject, differences in food choice were made use of in combined cultures of *Daphnia* and *Chydorus*;

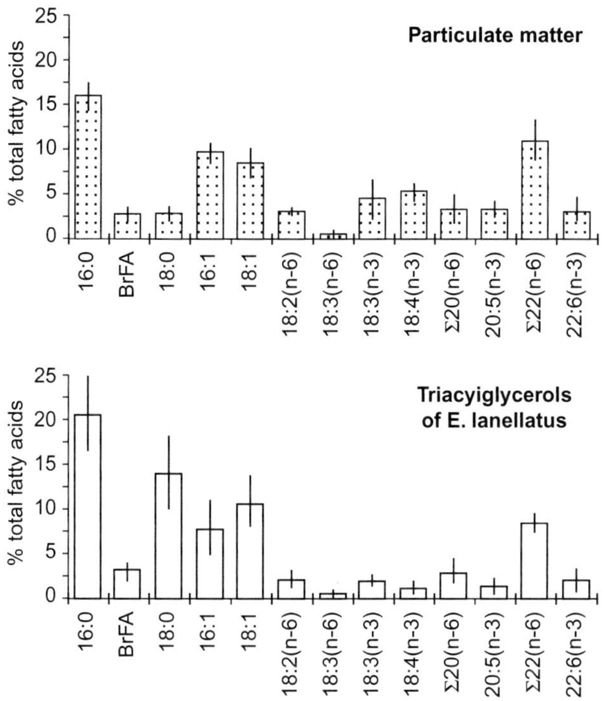

FIGURE 4.2 **Fatty acid composition in *Eurycercus* and in its particulate food.** n-, represents ω numbers. *Source: Desvillettes et al., 1997.*

production was improved by *Chydorus* feeding on the fecal remains of *Daphnia* (Bogatova, 1963, 1980).

The seasonal periodicity of amino acid composition of organic bottom material may be a contributory factor in the periodicity of littoral cladocera. Although not explored further, preliminary determinations by this author have demonstrated that various amino acids are present in organic bottom material, some of which are abundant and some present in low concentrations. The amino acid composition of seston in small water bodies of the Yenissey Basin was also studied by Kolmakova (2010).

Some authors have attempted to investigate the food quality of littoral organic debris and its changes over time. Robertson (1990) estimated the content of organic matter, chlorophyll a, and phaeopigment in River Thames mud in relation to the population dynamics of various cladocerans. The content of organic matter was higher (up to 16%) in February–April and lower in May–September (4–8%); chlorophyll a levels increased throughout the season, from c. 100 mg/L/m^2 in February to over 300 mg/L/m^2 in October. The phaeopigment content, a measure of algal decomposition, demonstrated no definite trend, fluctuating between 300 and 50–150 mg/L/m^2.

With reference to the eutrophic Lake Nesjøvatn (in Norway), it was shown that *D. galeata* obtain carbon principally from algae and detritus and phosphorus principally from algae (only a minor part of total phosphorus is obtained from bacteria) (Vadstein et al., 1995). Desvilettes et al. (1994) determined the fatty acid composition of littoral particulate matter: the prevailing fatty acids were also dominant in *Eurycercus* and *Simocephalus* (Fig. 4.2).

Even when there is a considerable inflow of terrestrial organic matter, *Daphnia* derive most of their carbon from phytoplankton (Brett et al., 2009). However, this source of food may be important for littoral species.

It has also been shown experimentally that *Daphnia* assimilate some dissolved organic matter (e.g. amino acids), but this alone is insufficient for their normal lifestyle (Rodina, 1946, 1948, 1950).

4.1.3 Introductory Remarks about Lipids

The lipid pathway in the aquatic environment starts with lipid formation by plants from initial products of photosynthesis. According to Harwood and Jones (1989, p. 12), "[d]e novo synthesis [requires] the concerted action of acetyl-CoA carboxylase and a type II (dissociable) fatty acid synthase." The chain length of the product may then be elongated and unsaturated bonds may be introduced. Fig. 4.3 shows the schematic pathway from initial photosynthetic saccharides to lipids and amino acids in algae.

Lipids are present in water in both a particulate (i.e. within algae) and a dissolved state (Arts et al., 1997). In Cladocera, they are used in energy metabolism, in the construction of cell membranes, and in metabolism; further, derivatives of lipids act as sex hormones. To understand the next sections, it is first necessary to state some basic ideas and definitions.

There are several classifications of lipids (e.g. Kucherenko and Vasilyev, 1985). In one of these, the following four groups are distinguished:

1. *Simple lipids*, which are esters of fatty acids with glycerol (fats) or aliphatic alcohols

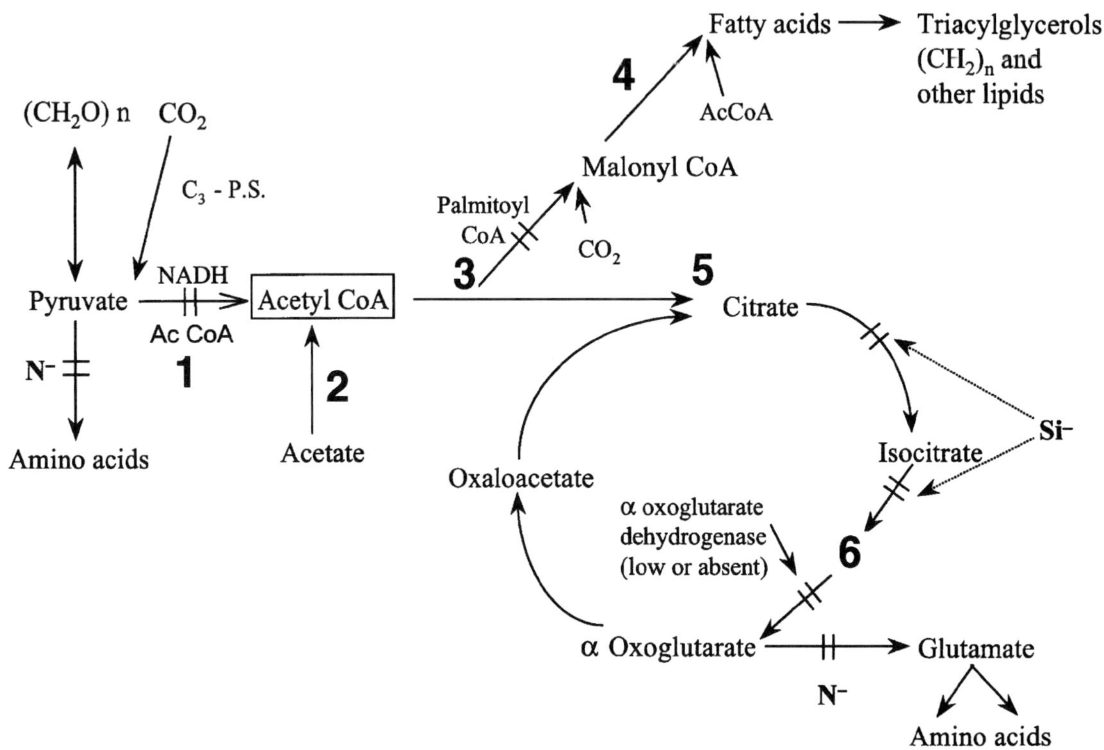

FIGURE 4.3 **Schematic pathway of photosynthetic saccharides to lipids and amino acids in algae.** C_3-P.S., photosynthesis. *Source: Arts et al. (1997).*

(waxes). Waxes comprise true waxes and esters of cholesterol, vitamin A, or vitamin D.

2. *Complex lipids*, which are esters of fatty acids with other alcohols, e.g. phospholipids, glycolipids, sulfolipids, lipoproteins, or lipopolysaccharides.
3. *Lipid derivatives* (*lipoids*), which include fatty acids (saturated and unsaturated), monoglycerides, diglycerides, sterins, steroids (vitamin D group), alcohols with β ionic ring (vitamin A group), phosphatides, and carotenoids.
   a. *Steroids* are derivatives of cyclopentano-perhydro-phenatrene. They comprise sterols and their derivatives. Ecdysteroid (ecd) concentration in whole *D. magna* is c. 200 pg ecd equivalent/mg dry weight (DW) (Bodar et al., 1990b). In diatoms, cholesterol and C_{28} sterols are the major sterols (Soma et al., 2005). Derivatives of steroids are sex hormones: male hormones, testosterone and androsterone; and female hormones, estron (progesterone) and estradiol (see Section 11.3.5).
   b. *Phosphatides* are esters of polyatomic alcohols, fatty acids, and phosphoric acid. They comprise lecithins, which consist of radicals of glycerol, phosphoric acid, choline, and higher fatty acids (saturated or unsaturated).
4. *Various others*, including vitamins E and K, and aliphatic carbohydrates.

Generally, fat is represented as a triglyceride connected to various fatty acids (R):

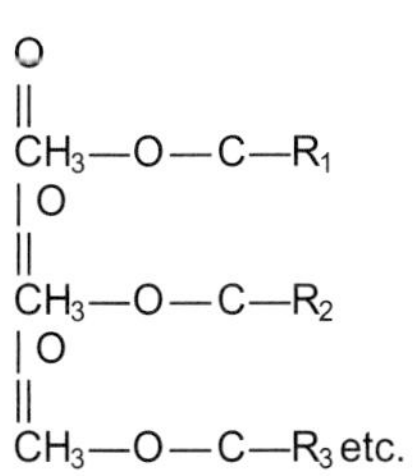

[The fatty acids may be saturated, i.e. containing no double bonds between the carbon atoms, and unsaturated, i.e. containing double bonds. The latter are described in the format *x: ywz* (e.g. Ahlgren et al., 1990), where *x* is the number of atoms, *y* is the number of double bonds (ω), and z is the position of the first double bond from the methyl end of the molecule. For example, 20:0, where 20 is the number of carbon atoms and 0 (or a certain digit) is the number of double bonds in the molecule of a fatty acid.

Some names of lipids corresponding to these designations are listed:

C14:0 myristic fatty acid
C16:0 palmitic acid
C16:1 palmitoleic acid
C18:0 stearic acid
C18:1 oleic acid
C18:2ω6 linoleic acid
C18:3ω3 α-linolenic acid
C18:3ω6 γ-linolenic acid
C18:4 octadecatetraenoic acid
C18:4ω3 stearidonic acid
C20:1 eicosenoic acid
C20:3 eicosatrienoic acid
C20:4ω6 arachidonic acid
C20:5ω3 eicosapentaenoic acid
C22:5 docosapentaenoic acid
C22:6ω3 docosahexaenoic acid

Essential Fatty Acids

It is thought that almost all PUFAs are obtained by animals from plants and are not synthesized by animals. Thus, linoleic acid (C18.2ω6) and α-linolenic acid (C18:3ω3) are essential fatty acids, whereas EPA is not strictly essential. Becker and Boersma (2007, p. 463) arrived at the conclusion that "although dietary fatty acids can be used for energy purposes, specific fatty acids [namely, PUFAs] are required to build new biomass."

Lipids in Cladocera

Oil drops are easily observed in Cladocera and are often mentioned in the literature (e.g. Flückiger, 1951). They are distributed throughout the body, although this has not been sufficiently described for representatives of various genera living in different environments. They are mostly orange in color and their size ranges from quite small to very large.

The following microchemical components are found in the fat of *D. magna* (Jaeger, 1935): butyric acid, sodium oleate, triolein, cholesterol oleate, cholesterol stearate, stearic acid, sodium stearate, tristearin, lecithin, linseed oil, and linoleic acid.

Tessier et al. (1983) found that the major lipid types in *Daphnia* are triacylglycerols and wax esters. Arts et al. (1993) confirmed this prevalence and indicated that the next most dominant lipid class is phospholipids and sterols.

It is likely that Goulden and Place were among the first to discuss the quantitative distribution of lipids in Cladocera (1993). Taking into consideration that the fatty acids are either derived from food or synthesized de novo in the body and that acetate (derived from the breakdown of carbohydrates or amino acids) is necessary for their synthesis, they used [^{14}C] acetate or $^{3}H_2O$ precursors and determined that the lipids synthesized by well-fed *Daphnia* (following incubation of up to 4 h) make up no more than 1.6% of the accumulated fatty acids. Food-restricted *Daphnia* synthesize 0.16% of their accumulated lipids; the overwhelming remainder is derived from algal food.

Thus, according to Goulden and Place (1993), daphnids selectively assimilate lipids from food and synthesize de novo not more than 1.6% of their accumulated fatty acids. Goulden and Place also indicated that lipids in daphnids consist of storage (triglycerides and wax esters) and structural (phospholipids and sterols) lipids, and that most of the lipids are transferred to the ovaries and then into eggs and used in the development of embryos.

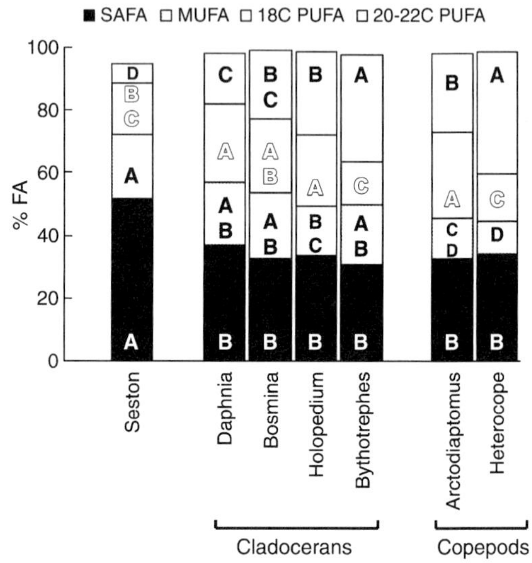

FIGURE 4.4 **Fatty acid composition in planktonic cladocerans and in their seston food.** A-D denote taxa that differ significantly in the content of the respective fatty acid group. MUFA, monounsaturated fatty acids; PUFA, polyunsaturated fatty acids; SAFA, saturated fatty acids. *Source: Persson and Vrede (2006).*

It has been determined that Cladocera predominantly contain (12–23%) EPA (Persson and Vrede, 2006) (Fig. 4.4), in contrast to copepods. This difference is assumed to be a result of their phylogenetic origin.

Bychek et al. (2005) found that *D. magna* is capable of high rates of de novo lipid radiolabeling; *D. magna* also makes direct use of dietary components (such as the PUFAs linoleate and α-linolenate). In addition, *D. magna* tolerates 24-h fasting with little change in lipoid metabolism.

Tessier and Goulden (1982) recorded an abundance of oil globules in *Daphnia* by the visually estimated lipid index (LI). Dodson (1989) observed that in presence of predators the fat content in the body (LI) decreases, which might be related to defense from predators. Hoenicke and Goldman (1987) estimated the lipid-ovary index in scores (based on visual scoring) in *Daphnia* and *Holopedium* and found that this index varies depending on the composition of their natural food.

It has also been shown that *Daphnia* (specifically with reference to *D. magna*) cannot synthesize linoleic acid (C18:2ω6) or α-linolenic acid de novo (Persson and Vrede, 2006) (Fig. 4.4). Thus, they depend on fatty acids produced by plants (including essential fatty acids, e.g. C20:5ω3, EPA).

Wacker and von Elert (2001) found that EPA (C20:5ω3) and α-linolenic acid (C18:3ω3) are not mutually substitutable resources for *D. galeata*, that their physiological functions are likely to be different, and that the former is not limiting for the growth of *D. galeata* cultivated on seston.

Having accumulated their fat reserves, *Daphnia* propagate, consume this resource, and then decline (Goulden and Hornig, 1980). Normally, the fat present in the body may be used by starving daphnias, as was observed in *D. magna* by Flückiger (1951), and short-term (24-h) starvation does not lead to profound changes in lipid metabolism (Bychek et al., 2005).

Further data are presented in Section 4.3.

4.1.4 Algal Lipids

Lipids are derived by Cladocera from various groups of algae, and different algae vary in their value as lipid sources for Cladocera (Fig. 4.3). Although diatoms supply the bulk of lipids for their consumers, their lipids are not necessarily the best types for Cladocera.

Diatoms are not alone in having the peculiar trait of not producing and accumulating starch. Fogg (1953, p. 105) noted that "starch evidently does not occur in the Xanthophyceae, Chrysophyceae or Bacillariophyceae, and the same three classes have a general tendency to store fats as reserve materials rather than polysaccharides." Sedova (1977) added that diatoms and dinoflagellates are especially rich in lipids. Particular groups of algae are characterized by specific spectra of fatty acids (Sushchik, 2008).

Of the planktonic algae, the best food items for *D. longispina*, *Bosmina*, and *C. sphaericus* are cryptomonads, with *Cryptomonas* and *Rhodomonas* containing high percentages of the PUFAs 20:5ω3 and 22:6ω3 (docosahexaenoic acid) (Ahlgren et al., 1990). These PUFA were also found to be common in fish.

It was found that in *D. obtusa* fed with the green alga *Scenedesmus* the content of lipid in the body does not positively correlate with growth rate (Sterner et al., 1992).

D. pulex fed with *Aphanizomenon* did not accumulate oil drops, although this blue-green alga is not directly toxic for them; on the other hand, when fed with the green protococcous alga *Ankistrodesmus*, *D. pulex* did accumulate oil drops (LI: 0 and 3, respectively) (Holm and Shapiro, 1984). Brett et al. (2006) studied the fatty acid composition of *D. pulex* fed on cryptomonads (*Cryptomonas* and *Rhodomonas*), Chlorophyta, and blue-green algae and found that it generally matched that of their algal food. However, there were some differences: in comparison with their food, the content of saturated fatty acids in *Daphnia* was substantially less and that of arachidonic acid was higher.

According to these data, *Daphnia* on the whole transfers somewhat modified fatty acids up the food chain (Fig. 4.5).

The fatty acid composition of some freshwater algae is shown in Table 4.1. The generalized composition of fatty acids in different algae is shown in Tables 4.1 and 4.3 and in Fig. 4.1. A high content of lipids in freshwater diatoms, i.e. 13.6% DW, has been determined by Birge and Juday (1922, p. 164): "The diatom sample contained a larger percentage of ether extract than any other sample of plant material."

Blue-green algae (or cyanobacteria) are also a source of lipids for Cladocera. Based on extensive observations and analyses, Sushchik et al. (2002) concluded that the development of planktonic *D. longispina* and *D. cucullata* mostly depends on 18:3ω3 produced by

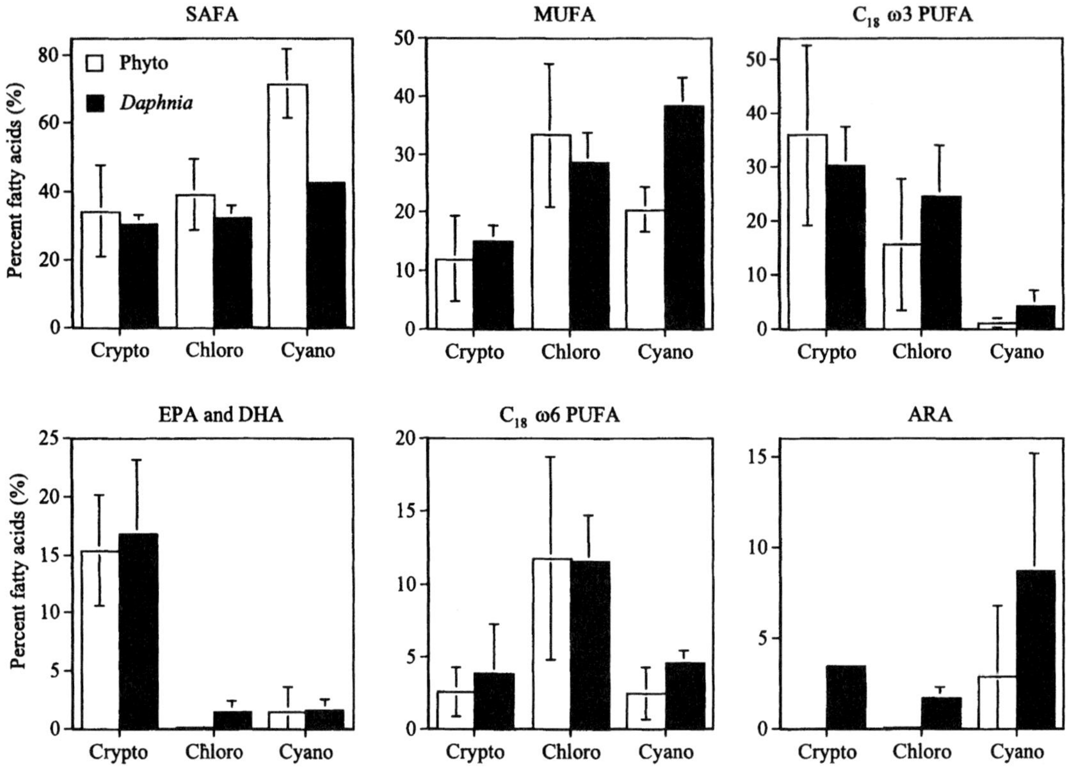

FIGURE 4.5 **Fatty acid composition of *Daphnia* after consuming various kinds of phytoplankton and of phytoplankton groups.** Chloro, Chlorophyceae; Crypto, Cryptophyceae; Cyano, blue-green algae. ARA, arachidonic acid (20:4ω6); DHA, docosahexaenoic acid (22:6ω3); EPA, eicosapentaenoic acid (20:5ω3). *Source: Brett et al. (2006).*

blue-green algae. However, at the same time, blue-green algae contain compounds that are toxic for Cladocera (see Section 4.2.4).

Not all PUFAs have a positive role: γ-linolenic acid (a PUFA obtained commercially for experiments) is acutely toxic (!) for *D. magna* at a concentration of 9 μg/mL (Reinikainen et al., 2001).

The Physiology of Diatom Photosynthesis

Of all the algal groups, diatoms (Bacillariophyceae) are the most prominent producers of lipids. However, they also have a very peculiar metabolism: they transform the bulk of the initial monosaccharide produced in photosynthesis into lipids (with the addition of some protein, pigments, and vitamins); they live in siliceous (glass) frustules; and they do not synthesize starch and cellulose from the primary carbohydrate products of photosynthesis. In contrast, they can exploit the reaction of fat formation that is generally characteristic of plants: their metabolism is largely confined to this process.

Writing specifically on diatoms, Rabinovich (1945, 1951) commented that "the oil deposits of the diatoms .... must be considered as products of comparatively slow secondary transformations" (1945, p. 43) of primary photosynthesis products. The metabolism of diatoms is better understood than that of other types of algae. The lipid content of diatoms

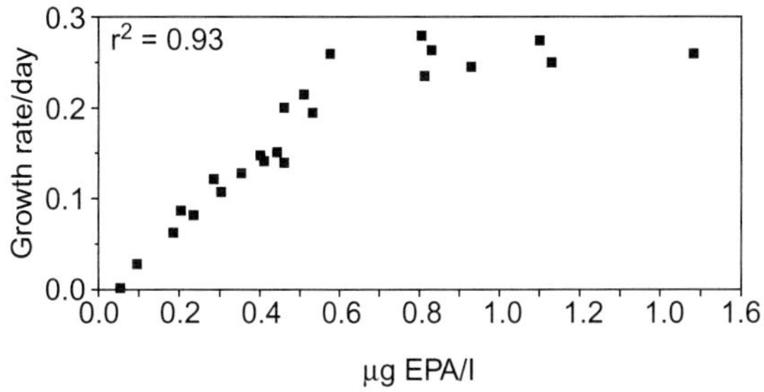

FIGURE 4.6 Relationship between available eicosapentaenoic acid and *Daphnia* growth (Lake Schösee, Germany). EPA, eicosapentaenoic acid. *Source: Brett and Müller-Navarra (1997).*

reaches about 35% of dry matter (Barashkov, 1972), mostly consisting of unsaturated fat: according to Farkas (1970), these lipids are rich in 14:0 and relatively poor in 18:2 and 18:3 fatty acids. Data on the lipid composition of diatoms are shown in Tables 3.4 and 4.1: *Daphnia* growth (in the Schösee, Germany) was shown to be proportional to the available EPA (Fig. 4.6) (Brett and Müller-Navarra, 1997).

In the freshwater diatom *Navicula pelliculosa*, the total lipids content includes the following major components: sulfoquinosyl diglyceride, digalactosyl diglyceride, monogalactosyl diglyceride, phosphatidyl glycerol, lecithin, and phosphatidyl inositol (similar to green algae). The major fatty acid constituents are palmitoleic acid, palmitic acid, EPA, and eicosatetraenoic acid (Kates and Volcani, 1966). The concentration of linolenic acid (octadecatrienoic acid), on the other hand, is low.

Opute (1974) confirmed that (1) the lipids of freshwater diatoms (*Nitzschia* and *Navicula*) consist mainly of triglycerides, monogalactosyl diglyceride, digalactosyl diglyceride, sulfoquinovosyl diglyceride, phosphatidyl glycerol, lecithin (phosphatidylcholine), and phosphatidylethanolamine; and (2) the major fatty acids are palmitoleic, palmitic, EPA, and eicosatetraenoic acids, with 16:0 and 16:1 prevailing.

Volkova (1980) also studied fatty acids in the freshwater diatoms *Stephanodiscus* and *Melosira* and discovered a prevalence of palmitic acid (16:0; 12.4–31.8% of total fatty acids) and palmitoleic acid (16:1; 23.3–31.7%). The unsaturated acid content is 53.6% and 36%, respectively, and the saturated acid content is 21.2% and 35.1%, respectively.

Accumulation of the essential, unsaturated fatty acids 20:5ω3 and 22:6ω3 in benthic (periphytonic) diatoms from March to January was assessed by Gladyshev et al. (2005). They found that the maximum of both types, c. 40 $\mu g/m^3$, occurs in August. Sushchik (2006) noted that some diatoms (*Cyclotella*) are not rich in essential PUFAs and that some blue-green algae (*Anabaena*) have a high content of these fatty acids, contrary to the generally accepted viewpoint, and thus are an important source. Sushchik (2008) also concluded that diatoms are characterized by an increased content of 14:0, 16:1ω7, and specific C16 PUFAs, and by a low content of C18 PUFAs.

Some other groups of algae also synthesize lipids, but information on their metabolism and accumulation of stock products is insufficient. There have been no reports on the controlling mechanism that channels photosynthesis to lipid formation in diatoms.

In some lakes (e.g. Myvatn and Kronotskoe), diatoms are highly dominant: as evidence, the ground surrounding the lakes consists almost entirely of diatom frustules (diatomite). Thus, the trophic chain in such lakes is based principally on thee lipids produced by diatoms.

4.2 FEEDING CHARACTERISTICS

4.2.1 Quantitative

Rate of Mandible Rolling

Cladocera eat almost continuously.

The rate of movement of the mandibles is c. 250 beats/min in *Leydigia leydigii* (Fryer, 1968), 177 beats/min in *Streblocerus serricaudata*

(Fryer, 1974), and 110–190 beats/min in *Daphnia rosea* (in direct proportion to yeast concentration) (Burns, 1968a). In *P. globosus*, "rolling and swinging take place too rapidly to permit visual determination of the rate" (Fryer, 1968, p. 335).

In *Daphnia*, Starkweather (1978) found well defined daily patterns in the rate of mandible rolling, with distinct nocturnal minima. In the daytime, the rate of mandibular beating in *D. pulex* was c. 140 beats/min, whereas at night (01:00–07:00) it was c. 80 beats/min. This rate decreased in larger specimens, from 180 beats/min for a body length of 2.0 mm to 80 beats/min for a body length of 3.4 mm. According to Murtaugh (1985), the rate of mandible movements may be taken as a measure of feeding intensity. In *D. pulicaria*, it ranges from a mean of 1.3 Hz (beats/sec) at a lower food concentration to 2.1 Hz at a higher food concentration.

Broken diatom frustules have been found in the intestine of some Cladocera. Therefore, mandibles may contribute to the breaking up of food items.

Rate of Thoracic Limb Strokes

In Cladocera, water flows created by the thoracic limbs are important for both feeding and respiration. In Anomopoda and Ctenopoda the thoracic limbs move in a metachronal rhythm, i.e. with a certain phase lag for the movement of every subsequent pair (Cannon, 1933).

The stroke rate of the exopodites of thoracic limbs, which creates a feeding and respiratory current, is 290–300 cycles/min in chydorids (Smirnov, 1971), 170 cycles/min in *C. sphaericus*, over 300 cycles/min in *D. longispina*, 60–244 cycles/min in *Ceriodaphnia* (Harnisch, 1949), and 150–470 cycles/min *D. magna* (McMahon, 1968).

The rate of strokes varies both individually and under different environmental conditions (e.g. in *Daphnia*) (Peñalva-Arana et al., 2007). In *D. pulex*, the rate of strokes of the thoracic limbs increases at decreased oxygen concentrations (Wolvecamp and Waterman, 1960). In *Daphnia carinata*, the beat frequency of the thoracic limbs is 327 beats/min for a body length of 0.6 mm; this increases slightly to 360 beats/min for a body length of 2.15 mm, and then decreases to 310 beats/min for a body length of 2.6 mm. The beat frequency decreases by c. 41% in response to natural water toxicity caused by algae such as *Cyclotella* or *Anabaena* (Forsyth et al., 1992). At too high a food concentration, *D. magna* decreases the rate of strokes of its thoracic limbs and discards the food collected in the food groove if the intestine is full.

According to Lampert and Bredeberger (1996), at low concentrations of food particles, *Daphnia* adapts to low food levels by slightly decreasing the appendage beat rate (from c. 300–550 to 300–450 beats/min in *D. pulicaria*); simultaneously, phenotypical enlargement of the area of its filtering screens occurs.

Water Clearance Rate

LITTORAL CLADOCERA

The water clearance rate in *Eurycercus* was determined to be 20.5 mL in 24 h (Marcolini, 1953) or up to 30.6 mL in 24 h for each specimen (Krychkova, 1969).

PELAGIC CLADOCERA

Determination of filtering rates differs according to different authors. For Cladocera, filtering rate generally ranges from 0.9 to 135 mL in 24 h, per specimen, depending on food abundance and other factors (Monakov, 1998). The filtration rate of *D. magna* ranges from 0.3 (and lower) to 1.8 mL/h, per specimen (Gulati, 1978). Darchambeau and Thys (2005) observed clearance rates of 2.5–13.5 mL in 24 h, per specimen for *D. galeata* in Esch-sur–Sûre Reservoir (Luxembourg). In the latter study, the clearance rate always correlated negatively with phosphorus concentration in the food.

As summarized by Sushchenya (1975), the filtering rates of different species [in mL per individual (ind.) per 24 h/mL per mg DW per

24 h at 20°C] are 0.25–115/18–7780 for *Daphnia*, 1.2–15.6/500–3800 for *Diaphanosoma*, 1–40/125–5000 for *Bosmina*, 1/560 for *Ceriodaphnia*, 8.6–135/538–860 for 26/2760 for *Moina*, and 3–133/470–3021 for Sida, *Simocephalus*. Filtering rates (in μL/μg/DW/h) are 28.3 in *D. hyalina* and 49.5 in *D. cucullata* (Steiner and Kasprzak, 2000).

The filtering rate (in mL/24 h, per specimen) was determined to be 0.9–5.15 for *D. pulex* 0.7–1.8 mm long, increasing with size in a curvilinear fashion (or 300–180 mL/mg DW/day) (Richman, 1958); for a length of 2 mm, it was about 5 mL in the daytime, increasing at night up to about 20 mL (Haney, 1985).

In ctenopods, as summarized by Korovchinsky (2004), it is 1–248 mL/ind./24 h in *Diaphanosoma brachyurum* s.l., 10–58 mL/ind./24 h in *Holopedium*, 2–657 mL/ind./24 h in *Sida*, and 32–252 mL/ind./24 h in *Penilia*.

While feeding on Chlamydomonas, filtering rates are 364 μL/ind./h in *Bosmina longirostris*, 399 μL/ind./h in *D. longispina*, 408 μL/ind./h in *Ceriodaphnia quadrangula*, and 403 μL/ind./h in *C. sphaericus* (Lair, 1991).

A very wide scattering of the experimentally obtained values is observed. This may depend on environmental factors, particular experimental conditions, or the specific clone tested. Nevertheless, Knoechel and Holtby (1986) estimated that filtering rate are mainly dependent on body length and comply with the following equation (Eq. 4.1)

$$F = 11.695L^{2.480}, \tag{4.1}$$

where F is individual filtering rate in mL/day, and L is body length in mm.

The rate complies with this equation in morphologically different forms such as *Bosmina*, *Ceriodaphnia*, *Chydorus*, and *Daphnia* spp. In *Daphnia* species, the exponent is 2.48.

In *Daphnia*, *Diaphanosoma*, and *Simocephalus*, the filtering rate is inversely proportional to the concentration of algae in the water (Sushchenya, 1958a, 1958b). However, at a certain upper limit of food particle concentration (over 8 g/m^3 for *D. longispina*), regulation of the filtration rate is no longer possible (Monakov and Sorokin, 1960). In *D. magna*, food consumption is not increased further when the food concentration reaches 2×10^6 cell/mL of *Chlorella* or 1×10^6 cell/mL of yeast (Rigler, 1961b; McMahon and Rigler, 1963, 1965); the feeding rate increases with increasing *Chlorella* concentration up to a maximum and then does not increase further at higher concentrations (oxygen uptake behaves in a similar way) (Kersting, 1978). This concentration was termed the *incipient limiting level* (McMahon and Rigler, 1965).

In hungry *D. magna*, filtering rates (in mL/ind./h) and feeding rates (in μm^3/ind./unit time) are higher than in well-fed specimens (Ringelberg and Royackers, 1985). *D. magna* quickly increase their respiratory rate with increasing concentration of food algae (Lampert, 1986).

Obtaining food particles from suspension was shown, at least partly, to occur not only through mechanical screening. Gerritsen and Porter (1982) believe that collecting food particles also depends on viscosity and the electric charge of both the filtering limbs and the food particles. The filtering rate also depends on oxygen consumption, and this relationship seems to be species dependent. In contrast to *D. galeata*, oxygen consumption in *D. magna* rapidly declines at concentrations below c. 3 mg/L (Heisey and Porter, 1977).

In their analysis of food filtration by daphnids, Gerritsen and Porter (1982) measured the Reynolds number (*Re*): for *D. magna*, the *Re* for filtering combs of limbs III and IV was 0.4 – 2.0, for a single seta it was 10^{-2}–10^{-4}, and for a setule it was 10^{-3}–10^{-4}. The region of reduced flow around a setule extends far beyond the next setule. Thus, there is little flow between setules and, according to these authors, the appendage functions as a solid wall. As *Daphnia* collect smaller particles (i.e. smaller than the mesh size of the limbs) at a greater rate than larger particles, it was concluded that the mechanism of food collection does not involve direct sieving. It was assumed to be food capture by electrostatic attraction. In support of this, neutralization of the negative surface charge of small particles further increases the retention efficiency.

However, subsequent authors, first of all Fryer (1991), confirmed the fundamental significance of direct filtration. It was shown in *Daphnia* that the strokes of the thoracic limbs are repetitive, with short delays between pumping actions; in the presence of food the delays between pumping sessions become longer (Peñalva-Arana et al., 2007).

Abrusán (2004) increased the viscosity of the medium using a sucrose polymer produced by *Leuconostoc* or with Ficoll. Increasing the viscosity of the medium at a constant temperature decreased the growth of three *Daphnia* species (except for *D. galeata*) (Abrusán, 2004). It is likely that the decrease was due to interference with filtration. The latter is thought to occur at an *Re* of approximately 10^{-1}–10^{-4}.

Quantities of Food Consumed

LITTORAL CLADOCERA

There is little information on littoral Cladocera. In chydorids, the daily ration (in μg DW/specimen/24 h) was determined to be 80 in *C. sphaericus* (Aksenova et al., 1969), 250 in *Acroperus* (Smirnov, 1971), and 2500 in *Eurycercus lamellatus* (Smirnov, 1962). *E. lamellatus* ingest periphyton within the range of 0.5 – 462 μg DW/specimen/ 24 h (Balayla and Moss, 2004).

PELAGIC CLADOCERA

The absolute quantity of food ingested (I) in the case of *Daphnia pulicaria* fed on *Scenedesmus*, depends on the length (L) and is described by Eq. 4.2 (Pott, 1982):

$$I(\mu g\,C/h\times \text{animal}) = 0.214\times L^{2.105}[C(\mu g\times \text{animal}) = 2.957\times L^{2.962}] \quad (4.2)$$

The daily ration of pelagic species (in % WW) was determined to be 17–140 in *Daphnia*, 16–101 in *Ceriodaphnia*, 129–285 in *Simocephalus*, 30–74 in *Moina brachiata* (Ostapenya et al., 1968; summarized in Smirnov, 1969, 1975), up to 750 in juvenile *Moina macrocopa* fed on *Chlorella* (Vorozhun, 2001), 5.7–21 in *Sida*, 29–49 in *D. cucullata*, and 3.5–24 in *D. longispina* (Kryuchkova, 1989). Sushchenya (1975) reported the range of daily rations to be 67–129% WW, but also described the data of various authors as ranging from 12.5 to 246% WW. He further commented that his own previous data (Sushchenya, 1958a, 1958b) was strongly underestimated.

The daily ration (i.e. ingested food; in % WW) of those feeding on detritus has been determined to be 69 in *D. magna*, 70 in *D. longispina*, 179 in *Simocephalus vetulus*, 23 in *C. quadrangula*, and 51 in *Moina rectirostris* (syn. *M. brachiata*) (Esipova, 1971). Generally, the daily ration for pelagic Cladocera is estimated to range from 0.8% to 247% WW, depending on the food, and probably also on the methods used (Monakov, 1998). The average daily ration for *Leptodora* (a predator) is estimated to be 30% of its WW (at 17°C) (Hillbricht-Ilkowska and Karabin, 1979).

Food requirements (or rations) may also be estimated using the oxycalorific coefficient.

Investigations into the diurnal rhythm of feeding rates have indicated that the feeding rate in *Daphnia* and *Holopedium* spp. is higher and less variable at night than during the day (Haney, 1985).

Ingestion rates of algae for *D. longispina* are different for different groups of algae (Schindler, 1971): highest values (μg/h/10) were determined for *Gloeocystis* (green alga),

11.0; *Coelastrum* (green), 16.8; *Asterionella* (diatom), 17.7; and *Oscillatoria* (blue-green), 22.7.

Movements of the thoracic limbs are reduced by nicotine, epinephrine, cyanide, strychnine, metrazol, and carbon dioxide (i.e. the corresponding striated muscles are affected) in *D. magna* (Sollman and Webb, 1941).

At increasing concentrations of available food, the rate of oxygen consumption by *D. magna* increases: this phenomenon is known as *specific dynamic action of food* (Barber et al., 1994).

It should be noted that the values provided above vary greatly depending on the dynamics of environmental factors. See also below on the redundancy of food intake (Section 4.5.3).

4.2.2 Excessive and Insufficient Diets

Luxury Consumption (Superfluous Feeding)

High concentrations of available food (above a certain level) may lead to excessive consumption of food particles (overcollection) (Porter et al., 1982). The term *luxury uptake* was initially used to describe the acquisition of a nutrient in excess of current demands (see Sterner and Schwalbach, 2001). It is now understood to mean the consumption of food in quantities exceeding physiological requirements. In this situation, the quantities of food material passing through the intestine increase but less of the food is digested, i.e. some food is discarded unused.

Redundancy of food consumed under conditions of excess available food (*Oocystis*) has been noted, e.g. in *D. magna* (Myrand and de la Noüe, 1983); it is thought that overcollection may lead to reduced assimilation rates (Schindler, 1968; Porter et al., 1982).

Sterner and Schwalbach (2001) applied the term *luxury consumption* to the storage of phosphorus during favorable periods to be used during times of less food availability. A significant fact discovered by van Donk and Hessen (1993) is that P-saturated cells of green algae are efficiently assimilated by *Daphnia*, whereas P-limited cells passed through the gut largely intact.

In Cladocera, food rations (i.e. ingested food) may reach high values. In juvenile *M. macrocopa* fed on *Chlorella*, the ration reached 750% body weight (Vorozhun, 2001) in adult *E. lamellatus* fed on detritus, 2500% body weight (Smirnov, 1962).

If data on oxygen consumption are available, then the extent of redundant food consumption may be estimated using the oxycalorific coefficient (described, e.g. in Dediu, 1989): its value is 4.86 cal/mL O_2 or 3.38 cal/mg O_2, or 14.2 J/mg O_2 (1 mL O_2 = 1.6 mg O_2). Multiplying the values of oxygen consumption by the oxycalorific coefficient yields the minimum physiological energy demand (assuming that the food is physiologically balanced). The ratio of the ration value obtained by direct determination to that determined respirometrically is termed the *excess index* (Gajewskaja, 1948).

For *E. lamellatus*, the daily relative ration determined directly was 2500, and 220 when determined respirometrically (Smirnov, 1962). Thus, the excess index is c. 11 for adults; it is higher for juveniles.

As a result of staying within the intestine for only a short period, especially under conditions of "excessive" feeding, algal cells are digested incompletely or may even remain alive. The latter was demonstrated for algae ingested by *Daphnia* (Porter, 1975), especially for algae covered with gelatinous sheaths.

Insufficient Food

Daphnias that are undernourished throughout their life live about 40% longer than well-fed daphnias (Skadovskiy, 1955). If starving daphnias are given abundant food, they catch up with well-fed daphnias and finally produce the same quantity of eggs. However, if well-fed daphnias are given limited food, then their fecundity decreases by 2.5 times, their longevity decreases, and they manifest signs of senescence. Jacobs (1961) supplied cultures of *Daphnia* with a low (1/100 dilution) concentration of *Chlorella* vs. standard cultures, and observed that under such conditions growth was slower, a much longer time was needed to reach the first instar, maturity was not reached at the fifth instar, and no eggs were deposited.

Accumulated saturated fatty acids (20:0, EPA) are used by *D. magna* during periods of inadequate food (Becker and Boersma, 2005). *D. magna* receiving a low quantity of food and low-P algae produce neonates with an increased lipid content (Boersma and Kreutzer, 2002).

It has been shown experimentally that under low-food conditions, *Daphnia* species develop larger filtering screens on their thoracic limbs (Lampert, 1994).

4.2.3 Qualitative Characteristics: Selectivity

The presence of selectivity in filter-feeding daphnids was shown experimentally by Gajewskaja, (1949). Selectivity was quantitatively investigated and used to estimate the ecological advantages of particular species (DeMott, 1982).

Morphological Background

The distance between filtering setae on the thoracic limbs may be one cause of selectivity. This distance was determined to be from 0.2 μm in *Diaphanosoma* to 4.7 μm in *Sida crystallina* (Geller and Mueller, 1981). In accordance, *Diaphanosoma*, *C. quadrangula*, *D. cucullata*, *D. magna*, and juvenile *D. hyalina* were found to be efficient consumers of bacteria, and *D. hyalina* (adult), *D. galeata*, *D. pulicaria*, *B. coregoni* to be low efficiency consumers of bacteria. *S. crystallina* and *H. gibberum* were found not to be able to consume suspended bacteria by filtration.

In a suspension of latex beads, *Holopedium* selected the largest particles and *Diaphanosoma* was almost nonselective (Hessen, 1985).

Postabdominal Rejection

Cleaning of the thoracic limbs by the postabdomen was observed in chydorids and *Daphnia rosea* (Burns, 1968a, 1968b). Rejection of too large or too abundant algae by *D. magna* was observed by Porter and Orcutt (1980). Suspended clay increases the frequency of postabdominal rejection and decreases the rate of thoracic limb beating (Kirk, 1991). In *Daphnia*, Burns (1968a, 1968b) also noted that the action of the thoracic apparatus was interrupted during postabdominal rejection.

Labral Rejection

Labral rejection is the rejection of food intake by deflection of the labrum, thus covering the mouth. This has been shown in *D. rosea* (Burns, 1968a, 1968b).

Regurgitation

Regurgitation has been observed by Rankin (1929) only in *Simocephalus* fed with palmella algae stained with methyl violet.

4.2.4 Feeding is Influenced by Environmental Factors

In *D. pulex* and *D. galeata mendotae*, about 85% of the daily filtering occurs during the dusk-dawn period; filtering rates are lowest in deep water during the daytime, increased with reduced depth (ascent), decreased at midnight, and increased again before their descent in the morning (Hancy and Hall, 1975). Other authors (Starkweather, 1975; Steiner and Kasprzak, 2000) noted that in daphnids individual filtering rates are higher at night.

With reference to *C. quadrangula*, it has been shown (Gladyshev et al., 1997) that feeding discontinues in the dark and that rhythmical changes of feeding intensity occur with a periodicity of 2–3 h. *C. quadrangula* grown under conditions of 12 h light:12 h darkness and transferred to constant light manifested two maxima in the circadian rhythm of grazing; this does not occur when they are maintained under constant light (Kolmakov et al., 2002).

Influence of Suspended Minerals

Mineral suspensions, both natural and industrial, interfere with the filtration process (Dubovskaya et al., 2003; Gladyshev et al., 2003) by decreasing the feeding efficiency (Rylov, 1940), which leads to decreased fecundity and other biological parameters of *Daphnia, Moina*, and *Ceriodaphnia* (Gorbunova, 1988, 1993).

Turbidity becomes lethal at 6400 mg/L (at particle sizes $<5\,\mu m$) (Gorbunova, 1988, 1993). Al_2O_3 nanoparticles cause partial mortality of *Ceriodaphnia dubia* at concentrations >200 mg/L, which is aggravated by the presence of other toxicants (Wang et al., 2011).

Suspended clay decreases the rate of beating of thoracic limbs, decreases the algal ingestion rate by 60–70% at low algal concentrations, and increases the frequency of postabdominal rejection in *Daphnia* (Kirk, 1991). According to Kirk, the rejected boluses contained both clay and algae. At 50 mg/L suspended clay, the rate of beating of the thoracic appendages decreased from 246 to 93 beats/min, the rate of mandible movements from 83 to 34 beats/min. In addition, a negative impact on *D. hyalina* was observed by the presence of suspended particles at a concentration of 25 mg/L (Rellstab and Spaak, 2007).

Suspended ground material decreases the sensitivity of the investigated Cladocera species to phenol (Gorbunova, 1993); in contrast, the toxicity of copper was reduced by turbidity (if the concentration of copper ions was not above 0.1 mg/L).

4.2.5 Impact of Xenobiotics on Feeding

The filtration rate of *D. magna* is decreased by 52% (of the control) in the presence of 0.15 mg Cd/L (Domal-Kwiatkowska et al., 1994); that of *D. pulex* (feeding on a green alga *Scenedesmus*) is decreased by 40% following preexposure to glyphosate and by 30% following preexposure to *p,p'*-dichlorodiphenyldichloroethylene (Bengtsson et al., 2004). The decreased filtration rate in *D. magna* consuming fluorescing microbeads was used to estimate the acute toxicity of metals (Ginatullina and Kamaya, 2012).

The rate of filtration and of ingestion in *Daphnia* decreased in the presence of the insecticide methylparathion (Fernández-Casalderrey et al., 1993) and the acaricide tetradifon (4-chlorophenyl 2,4,5-trichlorophenyl sulfone) (Villarroel et al., 1998). In the case of tetradifon, the reduction was 50% in comparison with the control, which occurred at 0.02 mg/L for filtration and 0.24 mg/L for ingestion.

In *D. magna* exposed to the insecticide cypermethrin (at 0.1 μg/L) or the fungicide azoxystrobin (at 0.5 mg/L), the limb filtering rate, the mandibles rolling rate, and the heart rate decreased (Friberg-Jensen and Nachman, 2010); in contrast, all of these parameters increased in the presence of 1 μg/L cypermethrin.

4.3 DIGESTION

4.3.1 Anatomical Background

The cladoceran intestine consists of a stomodaeum (foregut or esophagus) passing into the mesenteron (midgut) and proctodaeum (hindgut) (Hardy and McDougall, 1984). The inner epithelium of the foregut and the hindgut is of ectodermal origin (embryologically), whereas that of the midgut is of endodermal origin (Chatton, 1920). The stomodeum and hindgut (proctodeum) have a cuticular lining. The intestine may be convoluted or have no convolutions.

A blind cecum may be present on the ventral side of the midgut, before the beginning of the hindgut, as in chydorids and *Acantholeberis* (Fig. 4.7) (Fryer, 1970). Food is transferred from the midgut to the blind cecum and accumulates there before being finally evacuated (Smirnov, 1969, 1971). The ventral blind cecum and its function were first described by Claus (1876). In many species of chydorids, the blind cecum on the ventral side of the intestine is rather long, as in *Alona intermedia* (Tollinger, 1909) and species of the genera *Pleuroxus*, *Chydorus*, *Graptoleberis*, *Acroperus*, and *Camptocercus* (Smirnov, 1971). The equivalent cecum is short and wide in *Eurycercus* and *Leydigia*.

On the dorsal side of the midgut, a pair of blind diverticula (hepatic ceca) is present in Daphniidae, *Ophryoxus* (Macrothricidae), and in some *Eurycercus* species. As reviewed by Korovchinsky (2004), in ctenopods a pair of short dorsal ceca is present in *Holopedium* and *Latona parviremis*; *Sida* and *Limnosida* have one dorsal cecum; *Diaphanosoma* has no dorsal ceca. Chydorids also have no dorsal ceca.

Glands

The labrum comprises paired salivary glands. In *Daphnia*, the salivary gland on each side consists of a defined number of cells: 2 main cells, 2 substitutional cells, and 32 basal cells (Fig. 3.2) (Sterba, 1957a). The labral gland of *Simocephalus* was described by Cannon (1922).

Slime glands are present in the thoracic limbs: in limb IV of *Eurycercus*, in which the slime gland secretions are stained bright blue with Mallory's stain (Fryer (1962, 1963), and in

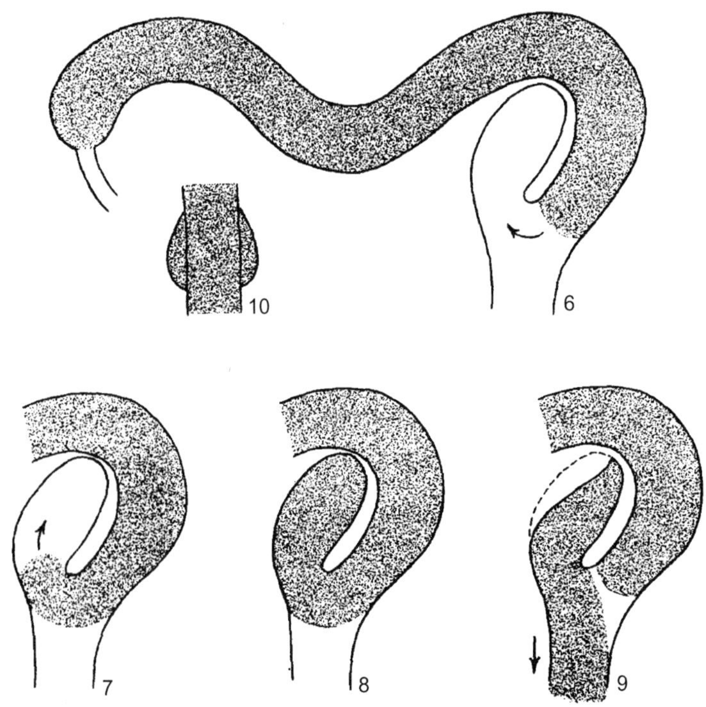

FIGURE 4.7 **The ventral blind cecum of *Acantholeberis curvirostris*, showing its filling and evacuation.** *Source: Fryer (1970).*

Alonopsis elongata (Fig. 2.1) (Fryer, 1968). The secretions of this gland aid the accumulation of food particles. No other species of chydorids or other cladocerans have yet been investigated for the presence of slime glands in thoracic limbs.

The inner surface of the gut is folded (Fryer, 1969) and densely covered with numerous microvilli (Fig. 4.8) (Quaglia et al., 1976; Avtsyn and Petrova, 1986). The microvilli are long, occasionally branching, and situated close to each other, as shown in *Daphnia* (Fig. 4.9) (Quaglia et al., 1976). The fine structure of the *Daphnia* gut is described by Schultz and Kennedy (1976a). They conclude that enzyme secretion is holocrinous. Microvilli also line the inside of the digestive diverticula.

Within the gut lumen of *Daphnia*, the pH is 5.6–6.2 (Lavrentjeva and Beim, 1978).

4.3.2 Peritrophic Membrane

The food collected in the intestine is surrounded by a peritrophic membrane; this membrane has been shown in *Daphnia* (Chatton, 1920; Fox, 1952; Quaglia et al., 1976) and *Acantholeberis* (Fryer, 1970). The tubular peritrophic membrane surrounds the food and extends through the midgut and hindgut.

Chatton (1920) investigated the peritrophic membrane in *Daphnia* in detail and demonstrated that two peritrophic membranes are present. The posterior peritrophic membrane is formed, according to Chatton (1920), by delamination of the proctodeal cuticle and is attached to the posterior circular furrow between the midgut and the hindgut. It surrounds the posterior part of the anterior peritrophic membrane (Fig. 4.10).

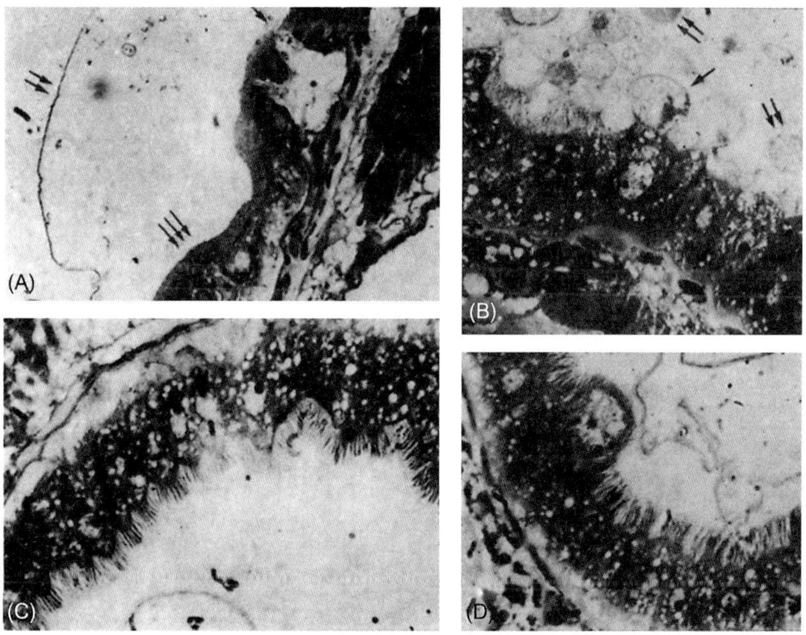

FIGURE 4.8 **Cross section of the gut of *Daphnia* showing plications and villi.** A, anterior part; one arrow, secreting epitheliocytes (holocrine type of secretion); two arrows, peritrophic membrane; three arrows, chitinous membrane. B, middle part; secreting epitheliocyte; one arrow, macrolemmocrine secretion; two arrows, secretion granules. C, middle part; epitheliocytes with villi. D, protruding secreting epitheliocyte (macroapocrine type of secretion). *Source: Avtsyn and Petrova (1986).*

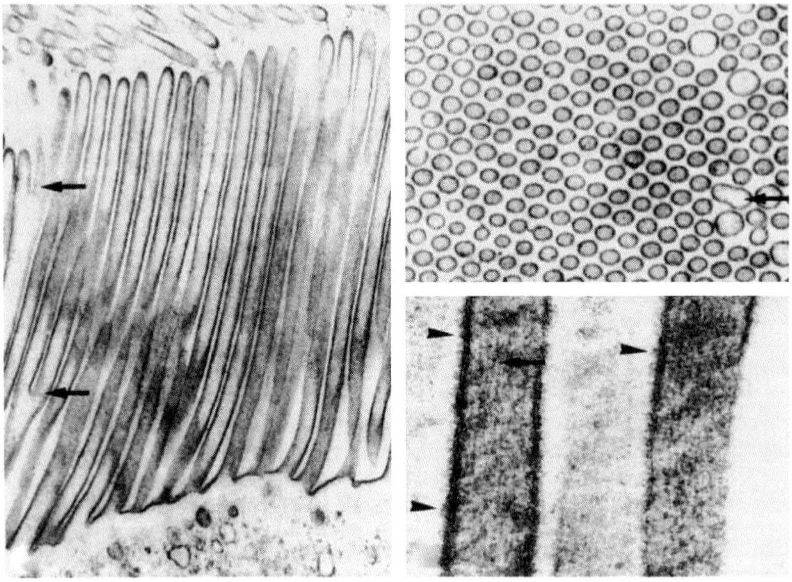

FIGURE 4.9 **Villi in the gut of *Daphnia* in longitudinal and transverse section.** *Source: Quaglia et al. (1976).*

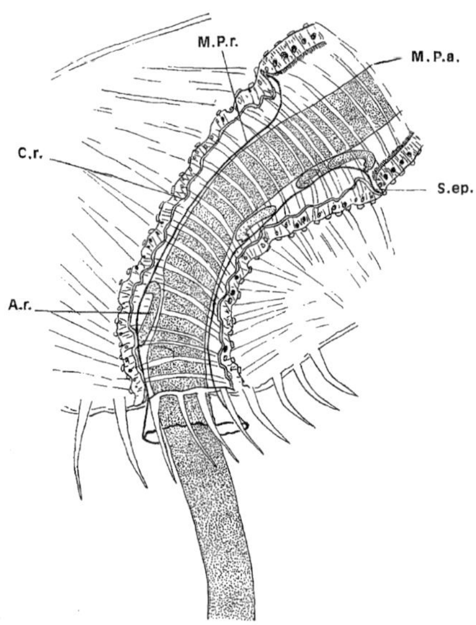

FIGURE 4.10 **The posterior peritrophic membrane of *Daphnia magna* surrounding the anterior peritrophic membrane.** A.r., Amoebidium on the inner side of the rectal peritrophic membrane; C.r., cuticle; M.P.a., part away from anus; M.P.r., rectal peritrophic membrane; S.ep., emargination in the gut tissue. *Source: Chatton (1920).*

The inner ectodermal lining of the esophagus and the entodermal lining of the midgut are also separated by a circular furrow. Both linings are in contact but are not fused. The circular furrow seems to be the site of formation of the anterior peritrophic membrane (Fig. 4.11). Chatton, however, believes that the anterior peritrophic membrane is a lining of the esophagus that is disconnected at the mouth but connected in the area of the furrow between the esophagus and the midgut, turned inside out, and extended by secretions of the midgut. This membrane fully surrounds the flow of food and feces to the anus. It isolates the midgut mucosa from direct contact with ingested food and acts as a kind of dialyzer.

The food within the peritrophic membrane of *Daphnia* is moved to and fro by antiperistalsis of the intestinal wall (Chatton, 1920, Fox, 1952).

It is not quite clear whether peritrophic membranes are discarded with each act of defecation or at each molting. This could easily be observed in representatives of various genera,

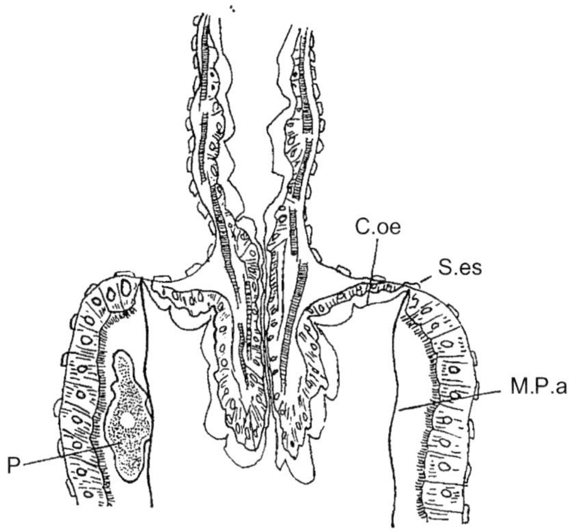

FIGURE 4.11 **The anterior peritrophic membrane (p. a.) of *Daphnia magna* in the place of transition of the esophagus to the midgut, in cross section.** C.oe., separating ectodermal cuticle of esophagus; M.P.a., peritrophic membrane; P, *Pansporella* in space between the peritrophic membrane and the gut wall; S.es., entero-stomodeal emargination. *Source: Chatton (1920).*

but has not yet been done. Chatton (1920) noted that at strong contractions of the rectal opening, the posterior part of the anterior peritrophic membrane may be torn off.

According to Hansen and Peters (1997/1998), the peritrophic membranes are formed by secretion of the midgut epithelium. They consist of a network of chitin-containing microfibrils embedded in a matrix of proteins, glycoproteins, and proteoglycans. The thickness of the peritrophic envelope is 280 nm. The lumen of a peritrophic membrane is termed the *endoperitrophic space* and the space outside the peritrophic membrane is called the *ectoperitrophic space*. Hansen and Peters (1997/1998) experimentally determined that the peritrophic membrane of *D. magna* is permeable to dextrans with a molecular weight of up to 2000 kDa (a Stokes radius of 31 nm) and to latex beads with a diameter 139 nm.

4.3.3 The Fat Body

The fat body is a special organ involved in fat metabolism, as well as having other functions. It has a variable structure. In addition to their role in fat metabolism, fat cells play a role in glycogen metabolism and are the sites of hemoglobin (Hb) formation (Goldmann et al., 1999).

The fat body, formed from special cells containing oil drops, was first observed in the Cladocera by Leydig (1860, p. 51–52). It may be slightly yellowish or reddish. Leydig also noted that the appearance of these cells depends on the season and the developmental stage. In addition, Hardy (1892) believed that blood cells, which absorb fat, may attach to the inner substrata and form fat tissue.

Special investigations into the fat body of Cladocera were initiated by Jaeger (1935) and continued by Sterba (1956a). Jaeger noted that oil drops in Cladocera are present not only inside the body but they are also included in special cells. According to Jaeger (1935), the fat body is connected to the ovaries and is situated ventrally along the gut (Fig. 4.12). The sequence of development of the fat body and the ovaries is shown in Fig. 4.13.

The structure of the fat body is rather irregular and complex. Some of its cells are scattered and attached to various organs and membranes. According to Jaeger (1935), the cells of the fat body are situated near to muscles at sites where they contact the membranes that direct blood flow, on intestinal muscles, and in the abdomen. In the abdomen, they comprise the plate between adductors of the abdomen and two plates on both sides of the gut, joining at the abdominal membrane. In thoracic limbs, the fat body extends to the base of epipodites and into endopodites, but not into exopodites. Plates of the fat body are present in valves.

Fat cells are present on both sides of the membranes that separate blood flows, except for the dorsal membrane (Jaeger, 1935). They are clearly separated from the membranes. The head does not contain fat cells.

In Cladocera, fat is concentrated in the cells of the fat body. As long as germinal groups are present in the ovary, the fat content of the fat body increases. After the start of yolk formation in the ovary, the fat body decreases in size (Jaeger, 1935; Sterba, 1956a). The cells of the fat body are basophilic; their only inclusions are fat drops; and there is no intercellular substance. Due to the accumulation and expenditure of fat, the cells of the fat body are in either a fat-free or a fat phase.

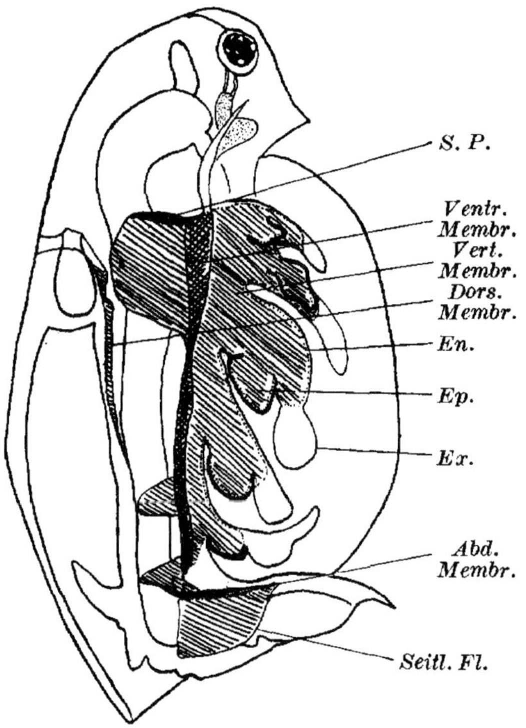

FIGURE 4.12 **Position of fat body in *Daphnia*.** Fat body is shaded. En., endopodite; Ep., epipodite; Ex., exopodite; Seitl.Fl., lateral lobe. *Source: Jager (1935).*

With fat expenditure, the fat cells are completely reduced. Jaeger (1935) observed that during this process part of the cell may be reduced. At the same time, substituting cells are formed. Restoration of the fat body occurs through the action of blood cells (Jaeger, 1935): blood cells loaded with fat may settle onto tissues and turn into cells of the fat body.

Jaeger (1935) did not observe any division stages in the substituting cells or the fat cells, but assumed that division stages are initially present in both. Sterba (1956a) presented data on "the life cycles" of fat cells in response to feeding on material rich in fat or rich in carbohydrate (Fig. 4.14).

In *Daphnia*, fat globules and other particles are carried from the gut lumen through the gut wall by blood corpuscles (Hardy, 1892) and via cells of the anterior region of the midgut (Hardy and McDougall, 1894). According to Jaeger (1935), osmic acid stains both fat in *Daphnia* blood cells and minute fat particles (hemokonia) in hemolymph, thus

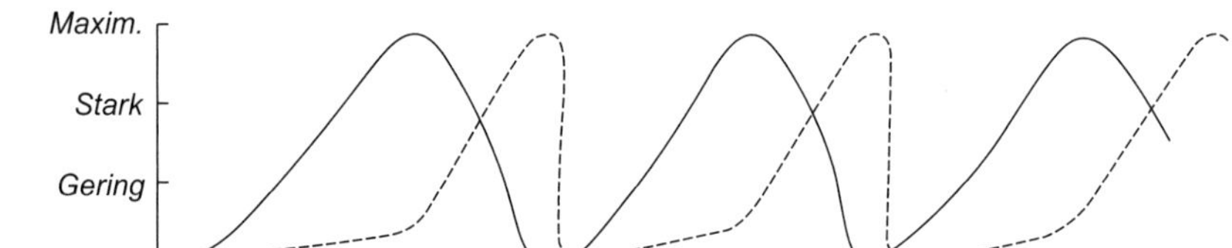

FIGURE 4.13 **Sequence of fat body (continuous line) and ovaries (interrupted line) development.** Abscissa, days. *Source: Sterba (1956a).*

making the hemolymph turbid. This indicates a distribution of fat from the alimentary canal throughout the body both in blood cells and as minute drops.

Despite much information on lipids in Cladocera now being available, no further investigations have been made into the fat distribution.

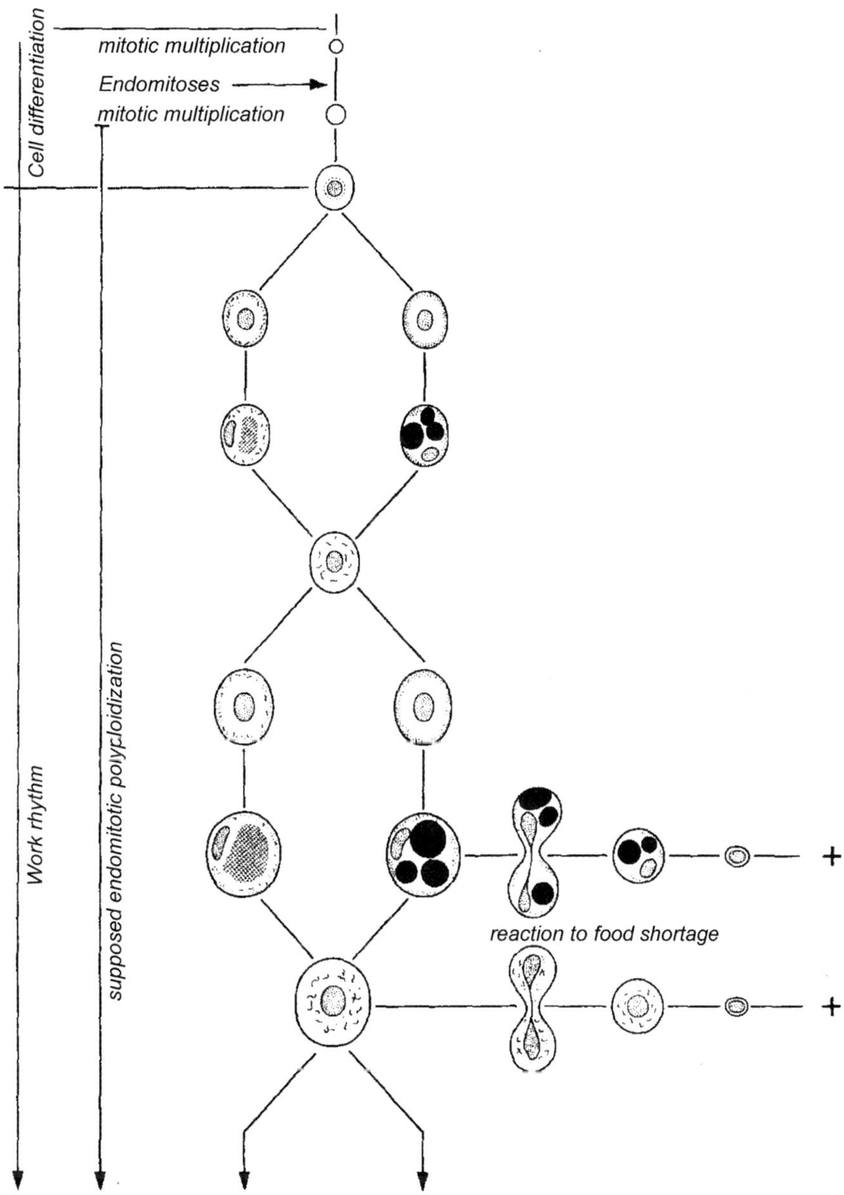

FIGURE 4.14 **Fat cell cycle in *Daphnia*.** Left, with food rich in carbohydrate; right, with food rich in fat. Black drops indicate fat. Lower arrow to the right indicates the response to food shortage. *Source: Sterba (1956a).*

4.3.4 Histo-Hemolymphatic Intestinal Barrier

Beim and Lavrentieva (1978; 1981) investigated the pH in the intestinal canal of *Daphnia* by staining it with eosin, Congo red, methyl red, neutral red, and uranine. The stained material passed through the gut in 20 min. During this time, these fluorochromes did not stain hemolymph, oil drops, or the brood pouch. Therefore, the authors suggested the presence of a histo-hemolymphatic intestinal barrier to these stains. They also stated the possibility that organic toxicants may stay within the intestine and not penetrate the gut wall. See also the data of Fonviller and Itkin (1938) on the permeability of integuments in Section 8.3.

4.3.5 Peristalsis

Peristalsis and antiperistalsis of the gut can be easily observed and were reported for *Daphnia* as early as 1894 by Hardy and McDougall. Hardy and McDougall note especially (1894, p. 45) that peristaltic movements of the proctodaeum "undoubtedly lead to the entrance and exit of water."

Rankin (1929) observed strong reverse peristalsis in *D. magna* and *Simocephalus serrulatus* fed on stained food.

The caeca situated on the dorsal side of the anterior gut also make rhythmic contractions, 6–10 times/min in *D. magna* and much less frequently in *S. serrulatus* (Rankin, 1929).

Peristalsis is stimulated by pilocarpine hydrochlorate and some other purgatives in *Sida* (McCallum, 1905) and by pilocarpine, mecholyl, physostigmine, and guanidine in *D. magna* (Sollman and Webb, 1941). Strong contractions of the intestine are caused by acetylcholine or physostigmine (eserine) in *D. magna* (Obreshkove, 1941a) and by acetylcholine, carbaminoylcholine, and prostigmine in *Simocephalus* (Mooney and Obreshkove, 1948).

Atropine blocks the actions of acetylcholine and physostigmine (Obreshkove, 1941a). Inhibition of peristalsis by atropine has also been shown in *S. crystallina* (McCallum, 1905), and by epinephrine, quinine, quinidine, metrazol, and cocaine in *D. magna* (Sollman and Webb, 1941).

In *Daphnia*, calcium ions (Ca^{2+}) cause contracture of the intestine, whereas potassium ions (K^{+}) cause relaxation of intestinal muscles (Ermakov, 1936).

4.3.6 Defecation

The formation of feces is confined to the posterior region of the midgut and to the proctodaeum (Hardy and McDougall, 1894). These authors noted that peristalsis in the proctodaeum is independent of the central nervous system as it takes place in the isolated proctodaeum.

Defecation occurs very frequently in Cladocera. About 10–20 min after the intestine is filled, it is evacuated. Defecation was initially described in *Daphnia* by Hardy and McDougall (1894), who noted the presence of a sphincter muscle at the junction of the midgut and the hindgut.

In chydorids and *Acantholeberis*, food is passed through midgut, transferred to the blind cecum, and evacuated when the cecum is filled and dilated (Smirnov, 1969, 1971; Fryer, 1970). The discharge of feces is followed by anal water intake. With regard to littoral species, the intestine is evacuated every 7 min in *E. lamellatus* (Smirnov, 1962) and every 11–19 min in *Acantholeberis* (Fryer, 1970).

Fryer (1969, p. 371) reported that *L. leydigii* "was watched for 2 min 5 s as it fed, during which time it discharged 15 fecal ribbons. Rough calculations indicate that at such a rate of ejection the entire alimentary canal could be evacuated in less than 6 min, and possibly considerably less."

For daphnids (mainly planktonic species), the intestine is evacuated every 31–35 min in *D. magna*, *D. longispina*, and *S. vetulus*, and 20–24 min in *C. quadrangula* and *M. brachiata* (Esipova, 1971). In *Bosmina*, it takes 10 min for the intestine to fill with food; the average daily ration is about 100% of its weight; and a portion of food is digested in 30–35 min (Semenova, 1974). In *M. macrocopa*, the intestine is filled in about 20 min and fully evacuated in less than 50 min (Morales Ventura et al., 2011). In periods of food scarcity, the intestine remains full for longer periods (Fryer, 1968). Such a short retention of food in the intestine is somewhat compensated for in chydorids and in many macrothricids by elongation of the intestine, through making convolutions.

Feces form cylindrical columns that do not immediately dissolve in water; they settle to the bottom.

Further comparative studies of living cladocera are desirable.

Intestinal evacuation in *D. magna* is accelerated by incubation in 100–250 μM Na_2CO_3, 100 μM NaNO3, or 1 μM KCl (Gophen and Gold, 1981). Gut clearance in *D. magna* is more efficient when green algae are available than in clear water.

Despite food being retained in the intestine only for short periods, cladocerans grow normally and produce young. Thus, digestion for a short period provides sufficient material to sustain normal life.

In the anterior part of the intestine, holocrine secretion takes place; in the middle part, macrolemmocrine and macroapocrine secretion occur (Avtsyn and Petrova, 1986). We must remember that the holocrine secretion is transformation of the whole cell for secretion and apocrine secretion involves breaking off the ends of gland cells for secretion, while the functioning part of the cell with the nucleus is retained and continues the same kind of secretion.

4.4 DIGESTION OF PARTICULAR SUBSTANCES

Cladocera feed on various algae, bacteria, and organic detritus at various stages of decomposition. These food sources contain protein, carbohydrates, lipids, and indigestible materials. Protein, lipids, and soluble carbohydrates are definitely digested. A diagram of the intake and transformation of food substances in an adult daphnid is shown in Fig. 4.15 (based on Hallam et al., 1990).

Digestion and assimilation of food occurs in the mesenteron. This tissue is not differentiated, but Hardy and McDougall (1894) drew attention to the unique and remarkable fact that digestion takes place in the middle region and absorption in the anterior region. Antiperistalsis and peristalsis are involved in this process. Feces formation occurs in the posterior region.

The chemical composition of *D. magna* is much influenced by the chemical composition of its food (Stepanova et al., 1971). Thus, the highest content of protein, fat, and carbohydrate was found in *Daphnia* fed on yeast. Cladocera also may accumulate various substances, including xenobiotics.

The enzymes that take part in cladoceran digestion have been investigated since 1929 (Rankin, 1929). An amylolytic enzyme has been identified in *D. magna* and *S. serrulatus* (Rankin, 1929), and proteinase and polypeptidases were reported in *D. pulex* (Hasler, 1937). In *Polyphemus* (a mixed herbivore and carnivore), proteinases, amylases, and lipases were identified (Dehn, 1930; Hasler, 1937).

Much later (Schoenberg et al., 1984), endogenous cellulase was detected in the gut of *Acantholeberis*, *Daphnia*, and *Pseudosida*; as no microbial gut flora was found, this enzyme enables cellulose to be digested. Cellulose digestion has also been reported in *Daphnia* by Lampert (1987) and De Coen and Janssen (1997).

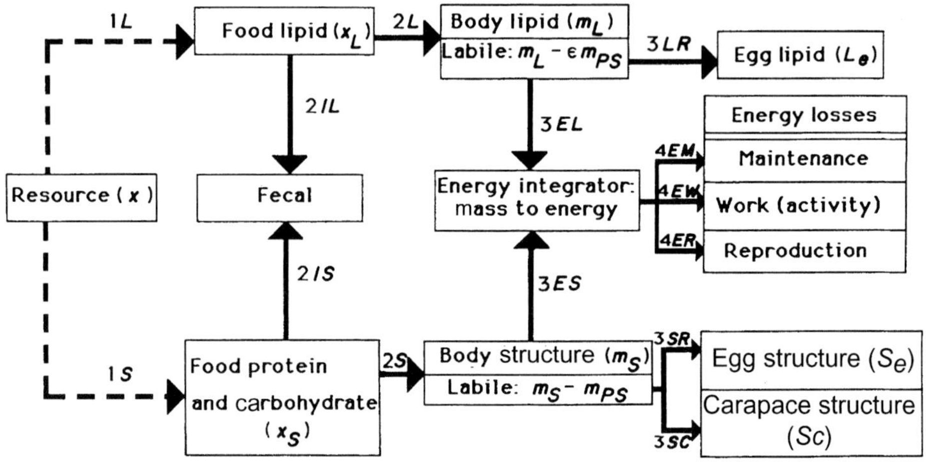

FIGURE 4.15 **Metabolic flow diagram for an individual adult daphnid.** *Source: Hallam et al. (1990).*

4.4.1 Protein Metabolism

Protein is obtained by Cladocera from algae and seston. Indeed, it was shown with reference to *D. obtusa* that the amounts of released nitrogen and phosphorus are generally proportional to the amounts ingested with food algae (Sterner and Smith, 1993).

The principal final product of protein metabolism in Cladocera is ammonia (Parry, 1960); of the secondary products, the most abundant is urea.

Proteolytic enzymes (trypsin, β-galactogenase, and esterase) were found and measured in homogenates of *D. magna* by De Coen et al. (1998). The activities of these enzymes decreased during 90-min incubations with various inorganic and organic toxicants. Von Elert et al. (2004) also identified proteases in homogenates of *D. magna*: two major ones (trypsin and chymotrypsin) and nine others. The two major proteases account for up to 83% of the proteolytic activity of the gut contents. Trypsin is strongly inhibited by *N-p*-tosyl-lysine chloroketone and 4-amidinophenylmethanesulfonyl fluoride; chymotrypsin is strongly inhibited by chymostatin. Both activities have alkaline optima.

Proteases (trypsin, chymotrypsin, elastase, and cysteine protease) were also found in whole body homogenates of *M. macrocopa* (Agrawal et al., 2005a). Their activity was significantly inhibited by an extract of the blue-green alga *Microcystis aeruginosa*. In *D. carinata*, proteinase activity was on average, 0.42 U/mg protein/min, that of chymotrypsin was 0.49 U/mg protein/min, and that of trypsin was 0.21 U/mg protein/min (Kumar et al., 2005b). Peptide metabolites of *Microcystis* that inhibit the trypsin-like activity in *Daphnia* were identified by Czarnecki et al. (2006).

For *D. magna*, algae are a better source of nitrogen than bacteria (Schmidt, 1968). According to this author, an adult *D. magna* consumes 0.37–0.47 μg N from green algae per 24 h per μg body N and excretes 0.17–0.19 μg/24 h of N per μg body N. Release rates in *Daphnia* were measured: nitrogen was 3–3.42 and phosphorus was 1–1.24 μg/mg C/h. No inter-species differences were found (Pérez-Martínez and Gulati, 1998/1999).

It has been suggested that Hb functions as both a respiratory protein and a protein store (Rudiger and Zeis, 2011).

4.4.2 Phosphorus Metabolism

In Cladocera, phosphorus forms part of nucleic acids and phospholipids. Other potential pools of phosphorus in Cladocera are minor metabolites (adenosine triphosphate and adenosine diphosphate), reduced nicotinamide adenine dinucleotide, reduced nicotinamide adenine dinucleotide phosphate, and calcium-associated phosphorus (hydroxyapatite) in integuments (Vrede et al., 1999). In *D. galeata*, 35–69% of the total phosphorus content is associated with nucleic acids (Andersen and Hessen, 1999; Vrede et al., 1999).

Cladocera are commonly deficient in phosphorus. Inorganic phosphate is taken from solution by *D. schødleri*, mostly through the epipodites (Parker and Olson, 1966). With a longer exposure, phosphorus accumulates in muscles, a cerebral ganglion, and the ovaries. The addition of 3 μM $AgNO_3$ to the surrounding water reduced the uptake of phosphorus as a result of silver impregnating the surface of epipodites.

Having determined that *D. magna* can extract phosphorus from food both rich and scarce in phosphorus, Sterner and Schwalbach (2001, p. 415) asked the following further questions: "In what chemical form is phosphorus stored? Where is it stored? How does it relate to the entire phosphorus budget? Are some species better able to exploit temporal variation due to enhanced storage capabilities? And, perhaps most fundamentally, how should we relate growth studies on simplified, constant foods to in situ conditions where these parameters vary in time and space?"

According to Vrede et al. (1998, 1999), about 67% of phosphorus is located in the body of *Daphnia*, 24% in eggs, and 14% in the carapace. The latter percentage was confirmed by Vrede et al. (1999) for the carapaces of *D. magna* and *D. galeata*. Phosphorus present in the carapace of *Daphnia* is discarded during molting and is thought to be inorganic (Vrede et al., 1998).

Macroergic phosphorus compounds (adenine nucleotides: ATP, ADP, and AMP) were found (Romanenko et al., 2004) to decrease considerably (by two to three times) during the growth of an enrichment culture of *M. macrocopa*.

With P-deficient food (C:P of c. 700), *D. magna* had a higher alkaline phosphatase activity compared with the amount present following phosphorus-rich food (C:P of c. 100); poor phosphorus nutrition also lowers the activity of alkaline phosphatase in released materials (containing dissolved enzymes).

In *D. pulex*, the daily renewal rate of phosphorus in the body ranges from 35% to 60% (i.e. 35–60% of phosphorus is turned over each day) (Lehman, 1980). Most of the liberated phosphorus is from freshly assimilated material, although a fraction is from phosphorus incorporated into tissues. Phosphorus content in *D. pulex* was found to be higher when it is fed with green algae rich in phosphorus (Lehman and Naumoski, 1985).

The phosphorus release rate in *D. galeata* varies from 0.16–1.18 μg P/mg C/h in juveniles and 0.06–0.96 μg P/mg C/h in adults (Vadstein et al., 1995).

4.4.3 Lipid Metabolism

There is an extensive record of investigations into the role of lipids in Cladocera digestion. Some publications report the chemical composition of lipids in Cladocera, whereas other studies investigate the sources of lipids in Cladocera and try to answer the question of whether Cladocera transform the ingested lipids.

The lipid composition of Cladocera is shown in Tables 3.4–3.8. Total fatty acid concentrations in herbivorous cladocerans is 117–147 mg/g C, and 104 mg/g C in predatory Bythotrephes (Persson and Vrede, 2006).

Cladocera obtain lipids from their food. Goulden and Place (1993) determined that de novo synthesis of fatty acids by daphnids is less than 2%. According to D'Abramo (1979), Cladocera, in particular *M. macrocopa*, are entirely dependent upon the sources of fatty acids in their diet. Farkas et al. (1981) found that *D. magna* cannot form docosapolyenoic acid.

Hardy (1892) observed that in Cladocera fat globules and other particles are carried from the gut lumen through the gut wall by blood cells. However, these observations do not seem to be followed up. Hardy and McDougall (1894) also noted that fat globules are ingested by columnar cells of the anterior region of the midgut, including the liver diverticula.

Farkas et al. (1984) suggested that *D. magna* cannot overwinter in an active state owing to a failure to adapt the phospholipid composition of its membrane and its physical state to the temperature (see Section 15.1).

Pelagic Cladocera show seasonal changes in the lipid content that are clearly dependent on the available algal food. In Humboldt Lake (Canada), triacylglycerol was found to be the principal component in *D. pulex*, especially during September–May, after the decline of blue-green algae (Arts et al., 1992). An abundance of bacteria following the collapse of a blue-green algal bloom resulted in increased lipid content in *C. sphaericus* and *Diaphanosoma leuchtenbergiana*. The authors concluded that there is an outstanding role for cryptophytes (*Cryptomonas* and *Rhodomonas*) as lipid sources, while *Daphnia* were at near starvation during the blue-green algal bloom.

Significant seasonal changes in the relative content of fatty acids were measured in *D. galeata*, *Leptodora*, and *Bythotrephes* in the Middle Volga (Fig. 4.16) (Bychek and Guschina, 2001). For example, the content of linolenic acid in *Daphnia* was especially high in July. Further, the content of C16:0 (palmitic acid) was especially high from June to August, similar to that in seston, in both herbivorous *Daphnia* and the predators *Leptodora* and *Bythotrephes*.

As shown in *Daphnia*, PUFAs are components of cell membranes and precursors of eicosanoids (hormone-like mediators) (Schlotz and Martin-Creuzburg, 2011).

Lipid composition has been determined in total lipid extracts from whole dried Cladocera (Arts et al., 1992; 1993; Sushchik et al., 2002).

According to von Elert et al. (2002), *D. galeata* can convert docosahexaenoic acid and C18 PUFAs into EPA (20:5ω3). α-Linolenic acid is the limiting PUFA, as EPA cannot be converted into α-linolenic acid. Thus, these algal sources (i.e. either α-linoleic or EPA) are usable and nonsubstitutable for *Daphnia*, and α-linolenic acid may be a source of EPA.

Following feeding with EPA-free or EPA-enriched *Scenedesmus* (green alga), 18:3ω3, 18:4ω3, 20:3ω3, and 20:4ω3 tissue concentrations were higher in *D. galeata* than in *D. hyalina*, indicating that assimilation and biosynthesis of PUFAs is higher in *D galeata* (von Elert, 2004).

As Schlechtriem et al. (2006) summarized: *Daphnia* feeding on highly unsaturated fatty acids [HUFAs; e.g. EPA (20:5n3) or arachidonic acid (20:4n6)] become enriched with these fatty acids; docosahexaenoic acid (22:6n3) is not accumulated but is converted to EPA; and α-linolenic acid (18:3n3) and linoleic acid (18:2n6) are used as precursors of EPA and arachidonic acid (ARA).

Littoral Cladocera

The sources of lipids in littoral *Eurycercus* were determined by Desvilettes et al. (1994, 1997). These authors found a striking similarity between the lipid composition of *Eurycercus* and its seston food (Fig. 4.16). The food of *Eurycercus* consisted of *Cryptomonas*, diatoms, dinoflagellates, ciliated strombiids, and bacteria. Cladoceran triacylglycerols comprise odd-branched fatty acids, a high percentage of ω6 PUFAs, and significant levels of 18:4ω3, 20:5ω3, and 22:6ω3. The latter three were thought to be derived from *Cryptomonas*, whereas the

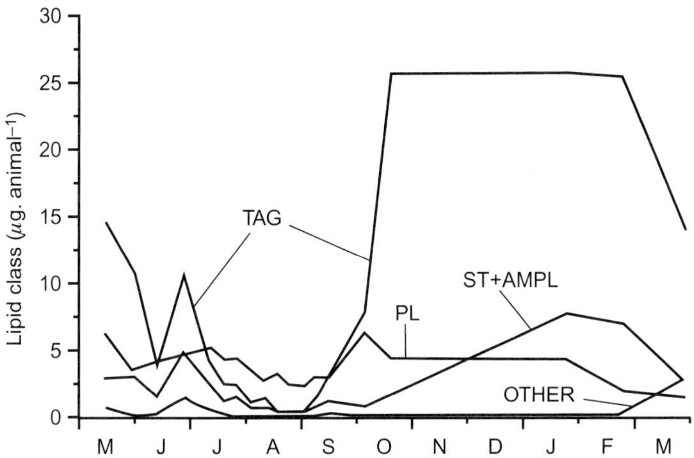

FIGURE 4.16 Seasonal changes in the content of lipid classes in *Daphnia pulex*. AMPL, acetone-mobile polar lipids; PL, phospholipids; ST, sterols;. TAG, triacylglycerols. Abscissa, months. *Source: Arts et al. (1992).*

odd-branched fatty acids came from consumed bacteria.

No similar investigations have been done in any other littoral Cladocera.

Pelagic Cladocera

There is good evidence that PUFAs are important components of Cladocera food. Similarities between phytoplankton fatty acids and the neutral lipids of *Daphnia* were noted by Bourdier and Bauchart (1986/1987), namely an abundance of 14:0, 16:0, 16.1, and 18:0 PUFAs. The growth of *Daphnia* is favored by ω3 HUFAs from lipid reserves or the diet (DeMott et al., 1997).

A very high correlation was found between *D. galeata* growth and the seston content of EPA (20:5ω3), a HUFA (Müller-Navarra, 1995a). Sixty-nine percent of the variation in *D. magna* growth was explained by the algal ω3 PUFA content (Park et al., 2002).

The lipid content of some planktonic cladocerans, and its fractional components, was determined by Lizenko et al. (1977) (Table 3.7), who showed that fatty acid percentages change with age (Table 3.8).

The total lipid content of *D. magna* depends on the composition of its food (Cowgill et al., 1984). The quantity of fat in *D. magna* was found to be directly proportional to the fat in their food algae (green algae, *Ankistrodesmus* and *Selenastrum*; Fig. 4.17) (Cowgill et al., 1984); this effect lasted for five generations. Only C14:0, C14:1, and C18:0 were more concentrated in *Daphnia* than in their algal food (Cowgill et al., 1984). In *Daphnia* fed with yeast, the monounsaturated fatty acid (MUFA) concentration was the highest, compared with those fed with other foods; the MUFAs were mainly C16:1ω7 and C18:1ω9. If the food was *Chlorella* (from seawater), the PUFA concentration was comparatively high (Huang et al., 2001). For *Daphnia rosea*, HUFAs comprise a highly favorable food component (Park et al., 2003).

In *D. cucullata*, the level of PUFAs 22:5ω6 and 22:6ω3 was determined to be 1.76% by weight of total fat (lipids), 23.9% of monosaturated fatty acids (18:0, 9.3%), and 28.4% of saturated fatty acids (16:0, 12.6%) (Farkas, 1970). In *D. pulex*, triacylglycerols are the dominant lipid class, followed by phospholipids and sterols (Arts et al., 1993).

Weers et al. (1997) found that C16:4ω3 and C22:5 were in lower abundance in *D. galeata* than in their algal food. For various different amounts in their food, *Daphnia* tends to contain high levels of EPA (C20:5ω3). *Daphnia* demonstrates a low rate of [^{14}C]linoleic acid

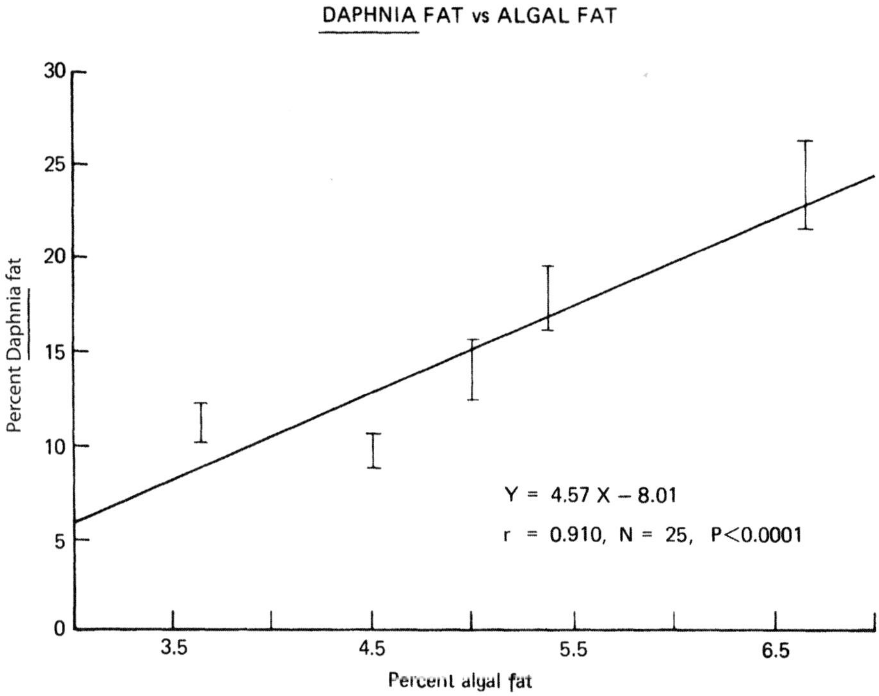

FIGURE 4.17 **Fat content of *Daphnia* vs. fat content of the consumed algae.** *Source: Cowgill et al. (1984).*

(C18:2ω6) and [^{14}C]linolenic acid (C18:3) conversion into C20 PUFAs. Thus, the latter are trophically important.

Brett and Müller-Navarra (1997) and Demott and Muller-Mavarra (1997) found that the efficiency of cladoceran growth is proportional, first of all, to the availability of fatty acids, and, in second place, to HUFA, and much less so to phosphorus, nitrogen, or carbon (Fig. 4.5); EPA is the most significant food component. Zooplankton growth and egg production strongly correlates with the 20:5ω3 to carbon ratio (Müller-Navarra et al., 2000).

Daphnia growth is favored by HUFAs from lipid reserves or the diet (DeMott et al., 1997). PUFAs present in phytoplankton, EPA especially, are important food ingredients for somatic growth and egg formation in *Daphnia* (Ravet et al., 2003). *Daphnia* growth is also favored by EPA at a threshold concentration of 13 mg/L (Makhutova et al., 2009); EPA is an ω3-type PUFA.

There is a differential response of *D. magna* to different fatty acids consumed in food, as shown by Becker and Boersma (2005, 2007). PUFAs are predominantly stored, whereas saturated fatty acids (20:0; EPA) are metabolized. In eggs, EPA is present in higher concentrations than arachidonic acid. The accumulated saturated fatty acids are used during periods of inadequate food supply, and are also passed into the eggs. Egg production is a major drain on *Daphnia* fatty acids. Changes in fatty acid concentrations are smaller compared with phosphorus changes.

In newborn *D. magna*, there is a high content of free sterols and phospholipids (phosphatidylcholine, phosphatidylethanolamine, and sphingomyelin) (Bychek and Guschina, 1999). These authors also found that the content of triacylglycerols decreases with the growth of *D. magna* fed on *Chlorella*. Fatty acid desaturation is lower in newborns, and wax esters are only detectable in adults. Fatty acid content has been determined by Lizenko et al. (1977) (Table 3.4) and Bychek and Guschina (1999).

In *Daphnia* spp., the EPA content varies least among the ω3 PUFAs; less so than in the food available to them (Müller-Navarra, 2006). Müller-Navarra fed *Daphnia* with a *Nitzschia* culture and published the fatty acid profiles of this alga and of *Daphnia*: in *Daphnia*, they reported a higher content of the fatty acids that are present in high concentrations in the diatom.

On a PUFA-rich diet, *D. magna* accumulate more ω3 PUFA at 15°C than at higher temperatures (Sperfeld and Wacker, 2011).

Copper may induce lipid accumulation, especially at high concentrations such as 12 μg/L Cu in later generations (e.g. the fourteenth) of *D. magna* (Bossuyt and Janssen, 2004).

Abnormal Fat Metabolism

Fat droplets are commonly present in the body of cladocerans and are used as a stock material. However, *Daphnia* that consume food deficient in a vitally necessary material may accumulate pink fat droplets and die soon thereafter. Flückiger (1951) described a case in which in a culture of *D. magna* fed with green algae, the daphnias became filled with pink fat droplets grouped abundantly around the intestine (from the first third section to the anus), in thoracic limbs, and around nephridia. In such animals, reproduction stopped, the ovaries could not be discerned, molting happened very rarely, and they died within 2 days. The addition of yeast or egg yolk normalized the situation and the animals continued with parthenogenetic reproduction. However, it was not possible to conclude which substance deficiency caused the metabolic disturbance.

4.4.4 Carbohydrate Metabolism

The general scheme of carbohydrate metabolism of Cladocera conforms to that shown in Fig. 3.3 (Hohnke and Scheer, 1970). The diets of Cladocera mostly contain an excess of carbon.

In *D. magna* and *S. serrulatus*, the presence of an amylolytic enzyme was demonstrated by Rankin (1929). The ingested starch, stained blue with iodine, changed color to pinkish as it passed down the gut, thus indicating a change from starch to dextrin.

Littoral Cladocera

The only available data on carbon assimilation by littoral Cladocera are likely to be those obtained by Infante (1973). *E. lamellatus* fed with *Scenedesmus* incorporated 12 μg C/100 μg C animal/24 h; much less (2.5 μg C/100 μg C animal/24 h) was incorporated when it was fed with *Staurastrum*.

Pelagic Cladocera

Carbon assimilation in filter-feeders (*Daphnia* and *Sida*) is 1.7–4 μg C/100 μg C animal/24 h, depending on the type of algal food (Infante, 1973). Stable isotope analysis showed that *D. hyalina* derives almost all of its body carbon from algal sources (Grey and Jones, 2007).

Urabe (1991) determined the distribution of carbon use in subsequent instars of *B. longirostris* and showed that the proportion of assimilated carbon used for respiration increases in subsequent instars and that used for growth decreases (Fig. 4.18).

Cellulose is also digested and assimilated with an efficiency of < 11.5%, as found for *Acantholeberis curvirostris*, *D. magna*, and *Pseudosida bidentata* (Schoenberg et al., 1984). Carbon assimilation in *D. magna* was estimated to be approximately 1.8 μg C/ind./h (Myrand and de la Noüe, 1983). Carbon release per single *Daphnia* containing 1 γC was calculated to

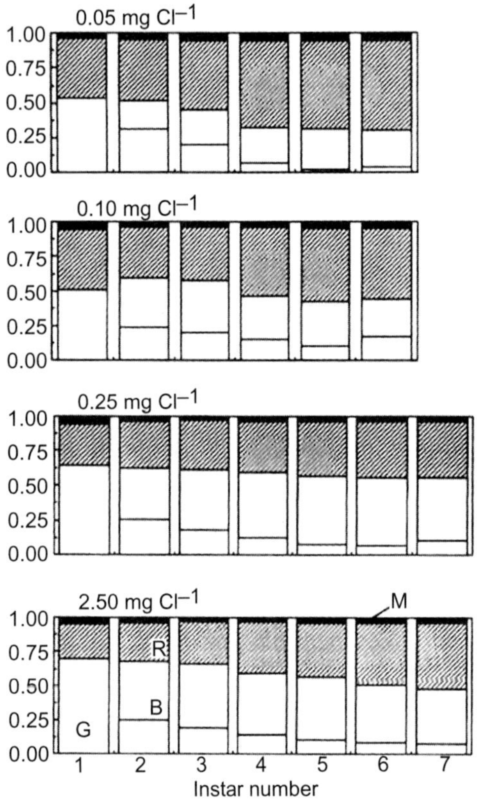

FIGURE 4.18 Carbon allocation in *Bosmina longirostris*. Proportion of carbon allocated to body growth (G), reproduction (B), respiration (R), and molting (M) at four food concentrations. *Source: Urabe (1991).*

be c. 0.26 γC/day (Metz, 1973). Blue-green algae are characterized by a low sterol content, which leads to a low carbon transfer efficiency into *Daphnia*; this constrains cholesterol synthesis and thus growth and reproduction (von Elert et al., 2002). The obvious excess of carbon in the food of *D. magna* and its fate during digestion were investigated by Darchambeau et al. (2003). *Daphnia* were fed cultures of the green alga *Selenastrum* that differed with respect to their C:P ratios (400 and 80). The respiration rate of *Daphnia* fed with high C:P algae was significantly higher than that of *Daphnia* fed with low C:P algae. The rate of excretion of dissolved organic carbon was higher in *Daphnia* fed on the high C:P algae (13.4% of body C/day), in comparison with *Daphnia* fed on low C:P algae (5.7% of body C/day).

Lampert (1978) determined that 10–17% of the carbon contained in algae ingested by *D. pulex* is transformed and liberated as dissolved organic carbon. Most of the carbon is liberated into the environment from algal cells: 4% from algal cells swallowed whole; and the rest resulting from *Daphnia* secretions and leaching from their feces.

Thus, *D. magna* must dispose of excess ingested carbon by means of respiration and a higher excretion rate of dissolved organic carbon in order to maintain their carbon balance and support their homeostatic carbon content. Juvenile *D. magna* liberate 55–72% of their carbon as dissolved organic carbon and 9–37% of the total carbon loss as carbon dioxide (CO_2). Carbon lost by *D. magna* as dissolved organic carbon consists mainly of the high molecular weight organic fraction (He and Wang, 2006a). For adults, Total loss of organic carbon and high-molecular organic compounds were 44–64% and 20–47%, respectively. The release of some excess carbon as CO_2 by *D. magna* was also noted by Jensen and Hessen (2007).

4.4.5 Metabolism of Various Elements

In Cladocera, the metabolic significance of some elements has been investigated. It has been shown for *D. magna* that Na, Ca, Fe, Se, Zn, and Cu are the essential elements, while cadmium and U are not (Barata et al., 1998).

Calcium

Cladocera derive calcium from food and water. The lowest level of calcium in water in which *D. magna* has been recorded is 0.1–0.5 mg/L (Hessen and Rukke, 2000a,

2000b), and 0–2 mg/L for *D. galeata* (Rukke, 2002). There are also interpopulational differences in tolerance to a low ambient calcium concentration, as has been determined for *D. galeata* (Rukke, 2002).

The calcium content of *D. magna* decreased from 4.2% to 1% DW over a range of 0.25–0.013 mM Ca in the culture medium; saturated calcification was reached at a calcium concentration of >0.13 mM; about 40% of calcium is lost with an old exuvium (Alstad et al., 1999).

Strontium

Strontium is concentrated by cladocera to a much lower degree than is calcium (Marshall et al., 1964). Similar to calcium, 95% of strontium is accumulated in the exoskeleton and is eliminated at each molting, thus dropping to minimum; thereafter, the content of Sr gradually increases, as shown in *D. magna* by Marshall et al. (1964) (Fig. 3.10).

Selenium

The minimum essential concentration of Se to support a culture of *D. pulex* and *D. magna* was determined to be 0.1 ppb (Keating and Dagbusan, 1984). Below this concentration in the environment, *Daphnia* progressively lost distal antennal segments at molting, manifested cuticle deterioration, and had a shortened life span. Elendt (1990) demonstrated that in *Daphnia* necrotic bands are formed in basal segments of antennal rami in the absence of selenium, and that distal parts of antennal branches are rejected and not restored in subsequent molting. Selenium deprivation causes alterations in mitochondria and the sarcoplasmic reticulum, and complete lysis of muscle fibrils in antennal muscle cells. This damage periodically occurs in the absence of Se, a constituent of glutathione peroxidase, which protects cells and their membrane polyunsaturated lipids against peroxidation. Uptake of selenium by unfed *D. pulex* reaches a maximum after 12 h, and depuration is slow (Schultz et al., 1980).

4.5 ASSIMILATION

By averaging data from various authors, Sushchenya (1975) calculated the average assimilation of food consumed by freshwater planktonic cladocera to be 58.4% of that ingested. In addition, the percentage assimilation of consumed food (green algae, bacteria, pink yeast, and bacteria) was determined radiometrically to be 40–56.7% in *Daphnia* and *Simocephalus* (Fedorov and Sorokin, 1967). In *Daphnia*, assimilation is highest for yeast and assimilation is highest in *Simocephalus* fed bacteria. The assimilation efficiency of *D. pulex* is 35% for *Ankistrodesmus falcatus* (a green alga) and 11% for *Aphanizomenon flos-aquae* (a cyanobacterium) (Holm et al., 1983). *Bythotrephes* ingests the soft tissues of consumed *Daphnia* and the assimilation efficiency is estimated to be 85% (Lehman, 1993).

Anderson et al. (2005) highlighted the fact that the food available for herbivorous Cladocera is frequently nutritionally unbalanced and its composition fluctuates. Therefore, to support overall homeostasis it is necessary to release excess substances, especially carbon, which may be done either as selective absorption in the gut or the excretion of excess substances in an organic form. With reference to carbon, regulation of homeostasis of the chemical composition has been investigated in *D. magna* by Darchambeau et al. (2003). However, the assimilation values also depend on environmental conditions. According to experimental data obtained by Buikema (1975), the percentage assimilation [i.e. the total energy accumulated during growth (for young specimens) and used in respiration] in *Daphnia* is highest at illumination above 7 foot-candles (fc; approximately

75.3 lx), but is higher with polarized light than with nonpolarized light of the same intensity.

The feeding ratio (i.e. food units spent per unit of weight increment) of algal food for *D. magna* is about 4 but has been shown experimentally to be 6 (Gajewskaja, 1945). In *Daphnia* and *Moina* fed on green algae, the percentage assimilation was 48–60% of the consumed food; the weight gain was 16% in *Daphnia* and 7.6% in *Moina* (Ostapenya et al., 1968). $K_1 = P/R$ and $K_2 = P/R + T$ were found to be 22.5% and 49.9% in *Daphnia*, respectively, and 11% and 23% in *Moina*, respectively (Ostapenya et al., 1968), where P is weight gain, R is the ration, and T is the energy expenditure for metabolism.

Food items are of varying composition and assimilation by cladocera is dependent on the composition of the consumed food. In experiments on feeding of *D. magna* with green algae (*Scenedesmus*), the phosphorus content per DW of *Daphnia* decreased when the algal content was low (DeMott et al., 1998). When the ratio of C:P in the food increased from 120 to 900, carbon assimilation declined and the phosphorus content in *D. magna* decreased from 1.47% to 1.08%. The assimilation of algal food by Cladocera differs for different groups of algae (Schindler, 1971): high values (in μg/h/10) were determined for *Oscillatoria* (blue-green alga; 5.8), *Asterionella* (diatom; 6.8), *Ankistrodesmus* (green alga; 7.2), and *Cryptomonas* (Cryptophyceae; 7.7).

At higher food concentrations, the assimilation efficiency decreases, as has been shown for *D. magna* (Schindler, 1968) and *D. galeata* (Urabe and Watanabe, 1991), but the assimilation efficiency by *D. galeata* becomes higher with decreasing food concentration, which affects the carbon balance (Urabe and Watanabe, 1991). In addition, the clearance rate decreases with increasing food concentration. Assimilation efficiencies reach c. 50% in *B. longirostris*, *D. longispina*, *C. quadrangula*, and *C. sphaericus* (Lair, 1991). In *Bythotrephes* feeding on the soft tissues of *Daphnia*, assimilation efficiency was estimated to be 85% (Lehman (1993).

Hallam et al. (1990) proposed a model summarizing the flow of assimilated resources in the adult daphnid, with special reference to lipids as a labile energy source (Fig. 4.15). This model has greatly contributed to the organization of various relevant data available in the literature.

4.5.1 Digestion Efficiency

Gut Passage Time

Food stays in the cladoceran intestine for only a few minutes. The intestine has been shown to be evacuated every 7 min in *E. lamellatus* (Smirnov, 1962); every 11–19 min in *Acantholeberis* (Fryer, 1970); every 31–35 min in *D. magna*, *D. longispina*, and *S. vetulus*; and every 20–24 min in *C. quadrangula* and *M. brachiata* (fed on detritus) (Esipova, 1971). Cauchie et al. (2000) summarized the data of various authors on gut passage time in *Daphnia* and *Bosmina* species to be 4–6 min in *D. galeata*, 15–25 min in *D. longispina*, 2–55 min in *D. magna*, <2–60 min in *D. pulex*, 19–40 min in *D. pulicaria*, 5–6 min in *D. rosea*, and 5–10 min in *B. longirostris*. The only report of a longer retention time was for *D. schoedleri*, which has a gut passage time of 135 min.

In chydorids, food passes through the intestine in about 5–7 min (Smirnov, 1969, 1971, 1974), while food passes through the intestines of planktonic *Daphnia* spp. in 25–54 min (Geller, 1975; Pavlutin, 1983; Haney et al., 1986; Christofersen, 1988); Murtaugh (1985) reported a retention time of 4–106 min in *D. pulicaria*. Pavlutin (1983) observed the rate of food evacuation to increase in larger animals and at higher food concentrations. Stained material passes through the intestine of *Daphnia* in 20 min (Beim and Lavrentieva, 1981).

Digestion in cladocera is surprisingly efficient. With this in mind, Fryer (1970, p. 266) noted: "A puzzling feature of the Anomopoda is the rapidity with which the food passes through the gut. A further, related, problem ... is the effect of the tremendous dilution of the gut contents brought about by the great intake of water."

This high efficiency may depend on the following facts:

1. The dense cover of microvilli in the intestinal lumen (Figs. 4.1 and 4.6) greatly increases the gut surface. Quaglia et al. (1976) propose that food is absorbed from the lumen of the midgut in a digested form and liquids are reabsorbed in the proctodeum (hindgut);
2. The inner surface of the intestine is plicate (or folded), which further increases the inner surface;
3. In addition to the huge surface area created by microvilli, the efficiency of digestion is increased through intensive mixing of the food by peristalsis;
4. Food in the intestine is intensively mixed by anal water intake;
5. Powerful enzymes are present in the intestine; and
6. Dialysis occurs through the peritrophic membrane.

In *Daphnia*, the collected food particles are glued together by salivary gland secretions and then sucked into the gut by periodic dilations of the esophagus (Sterba, 1957a). In *Daphnia* and *Simocephalus*, the food also receives secretions from the hepatic caeca. The principal function of the latter is thought to be secretion of digestive enzymes. However, hepatic caeca are completely absent in many cladocerans (e.g. in chydorids). The hepatic caeca contract 6–10 times/min in *D. magna*, but is much less frequent in *S. serrulatus* (Rankin, 1929). It is unknown whether enzymes are present in these secretions.

The tonus (or contraction) of the circular muscles of the intestine of *Daphnia* was found by Flückiger (1952) to be increased by the sympathomimetics L-adrenaline bitartrate, D-adrenaline-bitartrate, L-noradrenaline–bitartrate, L-ephedrine hydrochloride, and oxyphenylethanomethylamine tartrate.

As for digestive enzymes, amylases and lipases have been identified in homogenates of *Daphnia* peptidases (Dehn, 1930; Hasler, 1937; Rodina, 1950). The major proteinases in *D. magna* are trypsins and chymotrypsins (Agrawal et al., 2005b); these were identified in a gut homogenate. Agrawal et al. (2005b) also demonstrated that the blue-green alga *Microcystis* produces several inhibitors that differ in their specificity for these enzymes.

In *Simocephalus*, the pH of the anterior part of the midgut is acidic and it changes to alkaline nearer to the anus (Rankin, 1929). In *Daphnia*, the pH of the anterior part of the midgut is 6.0–6.2 (Rodina, 1950; Beim and Lavrentieva, 1981; Beim et al., 1994), i.e. lower than in the external medium, and increases to 7.4 in the hindgut (Rodina, 1950). Secretions of the hepatic ceca of *Daphnia* and *Simocephalus* are acidic (Rankin, 1929), and enzyme secretion in the middle gut of *Daphnia* is holocrinous (Schultz and Kennedy, 1976a).

A schematic representation of digestion and food transformation in Cladocera (Hallam et al., 1990) is shown in Fig. 4.15.

4.5.2 Incomplete Digestion

It is clear that food is incompletely digested by Cladocera. Some of the substances expelled into the environment by *Daphnia* have been identified as unsaturated fatty acids (von Elert, 2000) and are considered to be infochemicals, i.e. biologically active substances that influence the companion species of algae and other animals. The release of such substances may be a result of incomplete resorption.

In a culture of *D. magna*, the following substances were identified as influencing cenobium formation in *Scenedesmus* (von Elert, 2000): linoleic acid, α-linolenic acid, myristic acid, palmitic acid, palmitoleic acid, stearic acid, oleic acid, and stearidonic acid. Considering the great abundance of Cladocera, such large amounts of released infochemicals may have a considerable influence on the metabolism and behavior of hydrobionts.

4.5.3 Unbalanced Diet

The available food always lacks some components: nitrogen or phosphorus deficiencies are especially common. This fact partly explains excessive food consumption by cladocera. Characteristics used to assess food quality for planktonic Cladocera (e.g. *Daphnia*) are the C:P ratio (Schulz and Sterner, 2000) and the C:N ratio. Factors influencing the former ratio have been better investigated than those affecting the latter. It was noted that *Daphnia* species grow fastest under conditions of phosphorus-rich food (Acharya et al., 2004) and that the phosphorus content of *D. magna* directly depends on the phosphorus level of their food (Sterner, 1993; Urabe et al., 1997; Becker and Boersma, 2005). According to the latter authors, the phosphorus content in *D. magna* decreased from c. 16.7 to c. 10 mg/g DW as the C:P ratio increased, and enrichment of food with inorganic phosphorus (PO_4) significantly stimulated *D. dentifera* growth (Elser et al., 2001).

The phosphorus content of *Daphnia* species is directly proportional to the phosphorus content of their food (i.e. it decreases with an increasing C:P ratio in the seston) (DeMott et al., 2004) in both natural and experimental conditions. Interestingly, phenotypic variation prevails over genetic differences among species. Müller-Navarra (1995b) fed *D. galeata* on algae saturated with phosphorus or on P-limited diet: *Cyclotella* (a diatom) or *Scenedesmus* (a green alga). *Cyclotella* was a better food, as estimated by *Daphnia* growth, owing to its substantially higher content of 20:5ω3 (both phosphorus saturated and phosphorus limited) than *Scenedesmus* under both conditions. *D. galeata* growth, survival, and reproduction were all higher when fed on non P-limited *Scenedesmus*, compared with P-limited algae (Sundbom and Vrede, 1997). The impact of phosphorus on these biological outcomes was higher than that of ω3 HUFA.

In *Bosmina*, phosphorus content is thought to be lower, at c. 0.7–1% DW. Growth and fecundity were unaffected in *Bosmina* fed on low-phosphorus algae, while in *Daphnia* these parameters declined (Schulz and Sterner, 1999). Thus, under natural conditions *Bosmina* are better able to survive a low-phosphorus diet.

The phosphorus content in Cladocera has been studied when C:P ratios ranging from 140 to 1000 are present in their food (Ferrão-Filho et al., 2007). The phosphorus content changed substantially over this C:P range in food in *Daphnia ambigua* and a hybrid clone of *D. pulex-pulicaria*, whereas *Ceriodaphnia richardi* and *Moina micrura* showed tight phosphorus homeostasis. The critical algal food quality required for the propagation of *Daphnia* spp. was determined to consist of a C:P ratio of 225–375 (Brett et al., 2000); Anderson and Hessen (2005) indicate that for *Daphnia* carbon is limiting at low levels of available food and phosphorus becomes limiting with abundant food. At a C:P ratio above 230 in scarce seston, phosphorus becomes limiting.

The consumption of unbalanced food may lead to the failure of Cladocera cultures provided with abundant food. When fed exclusively on green protococcaceous algae, *D. magna* accumulated oil drops and stopped reproducing (Flückiger, 1951). The daphnias became filled with pink fat droplets, which were grouped abundantly around the intestine (from the first third to the anus), in thoracic limbs, and around nephridia. In such animals, the absence of a certain vitally necessary food ingredient caused reproduction to stop; further, ovaries could not be discerned, molting occurred very rarely, and they died within 2 days. These daphnias did not use their accumulated fat; in contrast, those fed normally use their fat reserves in the absence of food. The addition of egg yolk or yeast improved conditions but the essential component remains unknown.

Sterner et al. (1993) fed *D. obtusa* with *Scenedesmus* containing different levels of nitrogen and phosphorus (moderately N limited, severely N limited, and severely P limited) and found that "no amount of low quality food would support rapid *Daphnia* growth." The nitrogen content and N:P ratio in *Daphnia* were essentially constant despite varying in the *Scenedesmus*. Feeding rates were lower with P-limited than with N-limited *Scenedesmus*. These authors concluded that the mineral nutritional value of algae may influence the demographics of herbivorous Cladocera more than is commonly estimated. In *D. pulicaria* fed on N- or P-limited green algae, fecundity was reduced compared with controls fed on non-limited algae (Kilham et al., 1997).

If fed solely with starch the functioning of muscles of the antennae and of thoracic limbs is depressed, work of different heart fibers is uncoordinated, a fat body is not seen, and *Daphnia* die in about six days (Flückiger and Flück, 1952). Thus, *Daphnia* is not able to produce fat directly from starch. Addition of vitamin B_1 (aneurin) to the culture medium of *Daphnia* fed solely on starch restored a normal heart rate (Flückiger and Flück, 1952), and the addition of pantothenic acid to a culture of *Daphnia* fed solely on *Chlamydomonas* increased the duration of their life by three times (Fritsch, 1953).

In Cladocera, HUFA is necessary to support the normal condition of cell membranes. Schlechtriem et al. (2006) cultivated *D. pulex* at 11°C and 22°C for 1 month on a HUFA-free diet. The conversion of C18 fatty acid precursors to EPA (20:5ω3) and ARA (20:4ω6) was observed, and it was concluded that "HUFA such as ARA and EPA are highly conserved during starvation" (Schlechtriem et al., 2006, p. 397).

4.5.4 The Fate of Ingested Chlorophyll

This question of the fate of ingested chlorophyll arose rather a long time ago, but it remains unanswered. Hardy and McDougall (1894, p. 5) noted that "[i]n the chlorophyll of the algae which form so large a portion of the diet of Daphniidae, we have a substance whose fate we can to a certain extent trace, and we can find that as digestion proceeds the food mass in the middle region loses its green tint, whereas the fluid contents of the anterior become colored a vivid green. Further there is evidence that this dissolved chlorophyll is absorbed, for the striated border of the epithelium becomes colored an intense green and the cells charge themselves with yellow pigment masses." This is, however, the exterior view.

Indeed, Cladocera consume great quantities of chlorophyll with algae. Their algal food contains a lot of chlorophyll: green algae contain 1.24–3.00 g DW/L and blue-green algae contain 0.68–4.83 g DW/L (Lavrovskaya, 1965). The chlorophyll content in lakes may average 5–14 mg/m^3 and reach even higher concentrations (Westlake, 1980; Schulz and Sterner, 2000; Brandl et al., 2010a; Brandl et al., 2010b).

Surprisingly, information on the fate of chlorophyll during digestion and its further metabolism is scarce, both in the Cladocera literature and from experts on photosynthesis and on animal physiology (including ruminants). For insects, Kuznetsov (1948, p. 111) noted: "Very interesting theoretically processes of digestion of chlorophyll and of other food pigments are almost not traced in insects"; "The question on the origin of chlorophyll in hemolymph of insects is very interesting," (p. 12); and "information is yet insufficient for final judgment on so important biological issue" (p. 13)

Daphnids fed on the algae *Scenedesmus* and *Rhodomonas* release most of the consumed chlorophyll in their feces; some of it is degraded to chlorophyllide a, pheophytin a, and pheophorbide a. It is assumed that either the pigment or the chloroplasts themselves are poorly assimilated. It has also been shown that Daphniidae convert ingested algal chlorophyll into pheophorbide (Fundel et al., 1998; Pandolfini et al., 2000). At low concentrations of food algae, less pheophorbide a and more colorless digestion products were liberated, thus indicating that more pheophorbide a was digested (Fundel et al., 1998).

The levels of pheopigments in the gut of filter feeding Cladocera were used by Gorbunova (pers. comm.) to indicate decreased consumption of algae in the presence of mineral turbidity. It may be added that the decreased chlorophyll content in the gut of *D. magna*, determined chromatographically, has also been used to indicate reduced feeding in the presence of cypermethrin (Christensen et al., 2006).

Fox et al. (1949) believed that Hb formation in Cladocera is not influenced by the presence of chlorophyll in their food. As a protein, chlorophyll (tetrapyrrole) should undergo the general process of digestion. No answer was found to questions such as: Into which components is chlorophyll degraded during digestion? and Are these components used in the construction of Hb? It would be wasteful if chlorophyll, or partly decomposed chlorophyll, is simply discarded from the organism: yet, in Cladocera this is probably so. Kuznetsov (1948) reviewed the opinions of various authors that in herbivorous insects, which consume a lot of chlorophyll, the hemolymph becomes green due to the permeation of chlorophyll or to the creation of green pigment formed from chlorophyll derivatives.

It seems that little is actually known, for both invertebrates and vertebrates, about the process of chlorophyll digestion. At best, handbooks on animal husbandry note that ruminates need magnesium and obtain it from plants, where a minor fraction of magnesium is bound to chlorophyll. Further studies on the fate of chlorophyll in the process of Cladocera digestion are therefore highly desirable.

4.5.5 Chemical Composition of Feces

Little is known about the chemical composition of cladoceran feces. Steryl chlorin esters (SCEs) are products of the transformation of chlorophyll. The composition of sterols in SCEs within *D. magna* fecal pellets was studied by Soma et al. (2005), who showed that they contain both sterols formed by metabolism and unaltered sterols from dietary algae. C_{27} sterols (except for cholesterol and C_{28} sterols, major sterols in diatoms) were scarce in these SCEs. Cholesterol, probably a product of crustacean metabolism, was relatively abundant in these SCEs.

4.6 STARVATION

The life span under starvation conditions was summarized by Threlkeld (1976): it was 4–6 days for *D. magna*, 11 days for *D. pulex*, 5 days for *D. pulex* var. *pulicaria* Forbes, and 4 days for *M. macrocopa*. The mean survival time of starved *D. magna* was 7.6 days (Elendt, 1989), with a maximum of 10 days (Porter and Orcutt 1980). Fully starving adult *D. magna* survived for 160–246 h, *D. galeata* for 105–140 h, neonates of *D. magna* for 66–134 h (longer with a higher maternal lipid content), and neonates of *D. galeata* for 60 h (Tessier et al., 1983).

It is clear that these organisms use their intrinsic resources in the complete absence of food. When maintained in autoclaved tap water filtered through a 0.22 μm pore filter, *Ceriodaphnia cornuta*, *M. macrocopa*, *Diaphanosoma sarsi*, and *Scapholeberis kingi* could reproduce for three generations, although they matured later than did females fed with *Chlorella* (Kumar et al., 2005). In contrast, *Macrothrix* spp. were barely able to undergo a single generation.

Over 3 days of starvation, *D. pulicaria* weight reduced from c. 8.5 to c. 5.5 μg (DeMott et al., 1997). In starving *D. magna*, the greatest losses were those of DW, proteins, and lipids (including those due to the release of young formed before starvation), and reductions in total carbohydrate and glycogen occurred during the first day of starvation (Elendt, 1989). The triglyceride:total lipid ratio decreased during 5–6 days of starvation from 0.52 to c. 0.15.

The changes in the chemical composition of *D. pulex* caused by starvation are shown in Fig. 3.1 (Lemcke and Lampert, 1975). The chemical composition of starved *D. magna* was analyzed in detail by Elendt (1989) and Hessen (1990). Hessen (1990) found that the mean phosphorus content of starved *Daphnia* was 1.38% DW; this was unaffected by the quantity of available food. With reference to *Simocephalus* (Green, 1966b), starvation resulted in carotenoids being depleted from all organs. In *D. magna* starved for 1–3 days, the beating rate of the thoracic appendages decreased; it increased immediately (from c. 6 Hz to c. 8 Hz) when food was added (Plath, 1998). In these organisms, the respiration rate was low and the addition of food was followed by a three- to fourfold increase in the respiration rate and increased swimming activity (Schmoker and Hernández-León, 2003). By the fourth day of starvation, the oxygen consumption of *D. obtusa* had decreased to 50% of the level of well-nourished specimens (Vollenweider, 1958).

In *D. pulex*, the respiratory quotient (RQ) decreased from 1.13 to 0.71 over 5 days of starvation (Richman, 1958). This decrease in RQ demonstrates a change from predominant carbohydrate utilization in the metabolism to protein and fat metabolism.

In starved female *D. magna*, the following events occur in midgut enterocytes (Elendt and Storch, 1990): depletion of lipid and glycogen reserves, swelling of mitochondria, reduction of the endoplasmic reticulum and of dictyosomes, and a decrease in cell height. Prolonged starvation of *D. magna* also results in a loss of lactate dehydrogenase activity (anaerobic metabolic activity) in a *Daphnia* homogenates (Hebert, 1973). In starving *Daphnia*, there is also a drop in blood concentration (Fritsche, 1917). Belayev (1950) interpreted the decrease in blood hypertony relative to the external water as an indication that food is the source of substances that support hypertony. Carotenoids in all tissues are depleted during starvation in *Simocephalus* (Green, 1966a) and the heart rate in starving *D. longispina* decreases (Ingle et al., 1937).

Hungry daphnias actively filter their food particles; it was noted that a period of about 30 min is sufficient to overcome the starvation effect (Lampert, 1987). At starvation, daphnid growth continues for a time as a result of internal stores, and refeeding results in "catch-up" growth (Bradley, Baird, et al., 1991; Bradley, Perrin, et al., 1991).

Ingle et al. (1937) demonstrated that under conditions of minimum food *D. longispina* produce fewer young; when they are again fed abundantly, they "promptly produce many more young in each brood." In *D. magna*, temporary starvation has been demonstrated experimentally to be followed by immediate cessation of energy allocation to reproduction (by ceasing egg production) (Bradley et al., 1991a; Bradley et al., 1991b). This energy allocation is confined to the first half of the instar. Refeeding with *Chlorella* results in a recovery of fecundity. Trubetskova and Lampert (1995) also found that with starvation, the eggs of *D. magna* are fewer and smaller. Of course, the reaction to food shortage (i.e. a decrease in egg number) is delayed. Makrushin (1966) demonstrated that in *D. pulex* and *D. longispina* ovicells decompose if starvation occurs at stage I or early stage II of their development, but not if starvation occurs later. Stage I was characterized by Makrushin as the state of the ovary when either no fat is present in the cytoplasm of ovicells or it is present as tiny droplets; the brood pouch may contain eggs that have not started cleavage. Such a time lag, dependent on the energy reserves of the body, was also reported by Goulden and Hornig (1980), who noted that smaller *Bosmina*, with insufficient energy reserves, die comparatively sooner. As indicated for *Daphnia*, the critical time point at which starvation determines the number of eggs produced is about halfway through (0.5) the intermolt period; variation in the abundance of food at later stages does not change the number of eggs (Bradley et al., 1991a; Bradley et al., 1991b). Ebert and Yampolsky (1992) investigated how food shortage leads to the production of fewer eggs in *Daphnia*. They found that the number of eggs decreased when the "females were starved at least for 0.6 of the adult instar duration before egg laying." In addition, the number of eggs increased when abundant food was given to the 0.6 instar, before the eggs were deposited into the brood chamber.

4.7 NATURAL TOXICITY

Cladocera encounter many toxic agents under natural conditions. Blue-green algae periodically develop in great quantities and either unfavorably modify the environment or are directly toxic. They may be directly toxic or decrease the feeding rate in *Daphnia* (e.g. Rohrlack et al., 2001). Extracts from *Microcystis* can inhibit protease activity in *M. macrocopa* (Agrawal et al., 2001). Further, Rohrlack et al. (2004) found that *Daphnia* feeding on *Microcystis* are unable to shed their old integument, despite a new integument being produced. These authors also found that blue-green algae contain the protease inhibitor microviridin J. From the gut of *Daphnia*, it penetrates into the blood and disrupts normal metabolism, which leads to disturbances in molting and interference with normal swimming and filtration, which ends fatally.

Paralytic shellfish toxins are obtained by *Moina mongolica* following ingestion of the dinoflagellate *Alexandrium* and then are transferred to fish that predate on the *Moina* (Jiang et al., 2007).

4.8 IMPACT OF XENOBIOTICS ON DIGESTION

Large fat cells (i.e. storage cells) in *D. magna*, which are especially numerous at the posterior curve of the digestive tract, became smaller, contained less glycogen, and their mitochondria became more spherical following at an exposure to 20 μg/L Cd (Bodar et al., 1990b).

In solutions of phenol and aluminum sulfate, pH in the intestine of *Daphnia* increases from the normal value of 6 up to 8 (Beim et al., 1994).

Cadmium is accumulated by *C. dubia* from their diet (from green algae loaded with cadmium) and from water, but uptake from the diet is slower (Sofyan et al., 2007). Exposure of *D. magna* for 48 h to sublethal solutions of $CdCl_2$ or $HgCl_2$ inhibits cellulase, amylase, galactosidase, trypsin, and esterase; in contrast, exposure to $K_2Cr_2O_7$ increases activity of these enzymes (De Coen and Janssen, 1997). Assimilation of the green alga *Selenastrum* by *D. magna* decreased several fold upon exposure to 100 μM Cd^{2+} (Baillieul and Blust, 1999).

CHAPTER

5

Respiration

5.1 ANATOMICAL BACKGROUND

Most gas exchange in cladocera occurs through the body surface (Peters, 1987; Pirow et al., 1999a, 1999b). The gills do not have this function, since it has been shown that most of the oxygen necessary is extracted from the feeding current (Pirow et al., 1999a; Seidl et al., 2002). Oxygen is then distributed throughout the body and a branch of hemolymph reaches the brood chamber (Fig. 5.1) (Seidl et al., 2002; Pirow and Buchen, 2004). Another area of active respiration is the rectum, during the frequent process of anal water intake.

As has already been mentioned, in Cladocera respiration and feeding are closely interconnected, with both being performed with the participation of the thoracic limbs. It was previously thought that the sites of active gas exchange were the gills (epipodites) because it is easily observed that the gill surface is the area of active silver reduction in a $AgNO_3$ solution. Due to this dual function of movement of the appendages, the rate of their movement is labile and depends on both the concentration of food particles and the oxygen concentration (Pirow and Buchen, 2004). According to these authors, hypoxic exposure results in tachycardia under food-free conditions and in ventilatory compensation under food-rich conditions.

5.2 ENVIRONMENTAL BACKGROUND

Littoral Cladocera frequently live under conditions of oxygen deficiency, especially those that bury themselves in the bottom sediment. In general, pelagic Cladocera suffer less often from oxygen deficiency, except under conditions such as the mass decomposition of algae or the morning peak of algal respiration. They are also exposed to diurnal and seasonal changes in the distribution of oxygen concentration, as well as the decreasing vertical oxygen gradient, which has been well described in the context of limnology (see, e.g. Hutchinson, 1957, Ch. 9; Macan, 1963; Dodson, 2005; Kitaev, 2007). It should also be remembered that the oxygen concentration in water decreases with increasing temperature.

5.3 OXYGEN CONSUMPTION

The oxygen affinity of blood is numerically expressed by the partial pressure of oxygen

DOI: http://dx.doi.org/10.1016/B978-0-12-396953-8.00005-7

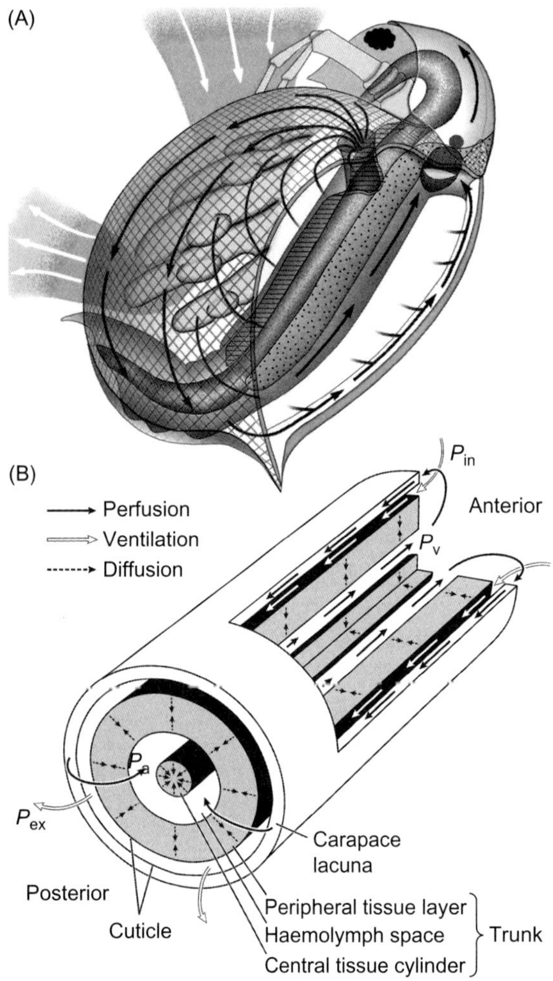

FIGURE 5.1 **Blood flow in *Daphnia magna*.** P_a, oxygen partial pressure of the hemolymph entering the trunk; P_{ex}, expiratory oxygen partial pressure; P_{in}, inspiratory oxygen partial pressure; P_v, oxygen partial pressure of the hemolymph leaving the trunk. *Source: Pirow and Buchen (2004).*

that induces half saturation [semisaturation (P_{50}), i.e. the formation of 50% of oxyhemoglobin at a certain partial pressure of oxygen (mm Hg)]. The ability of cladocerans to exist at higher or lower concentrations of oxygen gas (O_2) is characterized by the property of their blood to reach semisaturation at a certain oxygen pressure. For example, at 17°C in *Ceriodaphnia laticaudata*, semisaturation occurs at 0.8 mm Hg of O_2, i.e. much lower than that in *Daphnia magna* (3.1 mm Hg) (Fox, 1945). Different species of *Ceriodaphnia* can withstand different oxygen limits; *C. laticaudata* is an example of a species able to live at lower oxygen concentrations (Burgis, 1967).

The hemolymph of cladocerans becomes oxygenated while it flows through the ventral part of the carapace and the rostral region via direct diffusion (Pirow, Wollinger, and Paul, 1999b).

Winberg (1950) summarized all quantitative data on oxygen consumption by crustaceans available at that time. These data included, *inter alia*, freshwater planktonic cladocerans and copepods. Winberg concluded that oxygen consumption for all crustaceans complies with the equation (Eq. 5.1):

$$R = 0.105W^{0.81} \tag{5.1}$$

where R is the rate of oxygen consumption in mg/h at 15°C, and W is the weight in g.

As summarized later by Threlkeld (1976), the rate of oxygen consumption in cladocerans is:

$R = 0.2935w^{-0.184}$ in *D. magna*;
$R = 0.207w^{-0.124}$ (R in μg/day, w in μg) in *D. pulex* var. *pulicaria*; and
$R = 0.133w^{-0.107}$ in *Simocephalus vetulus*.

The rate of oxygen consumption is used as a measure of intensity of metabolism and, accordingly, of food requirements.

Oxygen consumption increases with increasing temperature, up to a maximum. For *Daphnia middendorffiana*, the maximum was found to be 26°C, above which inactivation of respiratory enzymes started; for *Bythotrephes cederstroemi*, the upper limit is 23°C—thus, it is more stenothermic (Yurista, 1999). Oxygen consumption in *S. vetulus*, measured using Cartesian divers or 100–130 mL bottles, did not change within a pH range of 4.5–9.5 in immobile specimens, although their general oxygen consumption (including movement and filtration) did increase with increasing pH (Ivanova and Klekowski, 1972). *S. vetulus* immobilized in the small space in the diver's bulb consumed 29.5% less oxygen than did moving animals (Ivanova and Klekowski, 1972). Further, O'Connor (1950) calculated that muscular activity in *Daphnia* requires 32% of their total metabolic energy.

A model of oxygen transport from the environment into tissues has been suggested with reference to *D. magna* by Moenickes et al. (2010). Oxygen consumption correlates with the respiratory electron transport system (ETS); Simčič and Brancelj (1997) reported that the ETS is localized in the inner mitochondrial membrane and comprises a multienzyme complex containing flavoproteins, metallic proteins, and cytochromes. This redox system transports electrons from nicotinamide adenine dinucleotide (NADH), nicotinamide adenine dinucleotide phosphate (NADPH), and succinate to O_2. Formazan produced in this process closely correlates with O_2 consumption; thus, a method of measuring formazan production spectrophotometrically in homogenates has been found to be extremely sensitive and applied to five species of *Daphnia*.

5.3.1 Respiration in Littoral and Bottom-Dwelling Cladocera

Benthic cladocera, especially those that bury themselves in the bottom sediment, live under conditions of frequent oxygen deficiency and an unlimited supply of food material. Littoral chydorids may survive at c. 1.9 mg/L O_2 or even as low as 0.4–1.7 mg/L O_2 (Pacaud, 1939; Bogatova, 1962). For specific types of chydorids, the minimum oxygen concentrations in their environment at 18–22°C is 1–1.7 mg/L O_2 for littoral cladocerans (*Acroperus*, *Alona*, *Camptocercus*, *Eurycercus*, and *Pleuroxus*) and 0.29–2.43 mg/L O_2 for pelagic cladocerans (*Daphnia* and *Simocephalus*) (as summarized by Smirnov, 1975). *Chydorus ovalis* consumes 19–21 mg O_2/24 h/g wet weight (WW) at pH 9 and 23°C (Yatsenko, 1928). Information on the respiration of bottom-dwelling Cladocera is very scarce.

Reduced oxygen concentrations induces upward swimming in *Chydorus sphaericus* and *Pseudochydorus globosus* (Meyers, 1980) at decreased light intensities (if not attached to a substratum). *C. sphaericus* reacts to higher oxygen levels (2.5 ppm) than does *Pseudochydorus* (1.15 ppm).

Tissue oxygenation has been estimated by NADH fluorescence in the limb muscles of *Simocephalus* (Forasacco and Fontvieille, 2008), using the assumption that NADH content positively correlates with oxygen consumption.

5.3.2 Respiration in Pelagic Cladocera

Pelagic species usually enjoy a relatively good oxygen environment (by living at normoxia). Water flow created by the beating of their thoracic appendages supplies both oxygen and food to pelagic species. Beating rates of the appendages depend on both the oxygen concentration and the concentration of available food particles (Pirow and Buchen, 2004).

Oxygen tension (pO_2) within the *Daphnia* body is about 4–5 times lower than of the surrounding water (Fox, 1945; Wolvecamp and Waterman, 1960): in *D. magna*, it is 3.1 mm Hg, but in *C. laticaudata* it is only 0.8 mm Hg.

There have been numerous measurements of oxygen consumption rates in cladocera, mainly for *Daphnia* species. At 20°C, *D. longispina* consume 19.1 mg O_2/24 h/g WW (Shcherbakov, 1935), i.e. 0.8 mg/h. Under an average "normal" temperature, *Daphnia* consume c. 1 μL O_2 per 1 mg dry weight (DW) per hour (as summarized by Peters, 1987). Oxygen consumption is about 0.2–1.3 mL O_2/g WW/h at 20°C in *Daphnia* and 0.29 mL O2/g WW/h in *Chydorus* (according to Wolvecamp and Waterman, 1960), and 0.14–1.18 μL O_2 per individual (ind.) per day at 16°C (according to Hillbricht-Ilkowska and Karabin, 1970).

Oxygen consumption by *D. magna* is maintained at approximately the same level when the oxygen concentration in the water is 4–1.5 mg/L, but it abruptly declines with a further decrease in the external concentration (Skadovskiy, 1955). The respiration of *D. magna* is higher in feeding animals, lower in animals provided with low-food levels, and still lower in fasting specimens (Jensen and Hessen, 2007). Kersting (1978) also demonstrated that *D. magna* respiration reaches a maximum at a concentration of food algae that is close to the critical level (i.e. where the food uptake does not increase further and becomes constant). With an excess of food (*Chlorella*), *D. magna* respiration decreases somewhat (Kersting and Leew-Leegwater, 1976).

From a quantitative aspect, the relationship between the body weight (W) and oxygen consumption (Q) in planktonic cladocera can be described by the following equation (Eq. 5.2), and is somewhat below the average level for all crustaceans (Sushchenya, 1972):

$$Q = 0.143W^{0.803} \tag{5.2}$$

For *D. pulex*, Richman (1958) determined the following relationship between weight and oxygen consumption (Eq. 5.3):

$$\text{Log } O_2 = \log 0.0014 + 0.881 \log W \tag{5.3}$$

The lethal minimum oxygen concentration was found to be 0.29 mg/L in *D. obtusa* and 0.99 mg/L O_2 in *D. hyalina* (Herbert, 1954).

Heisey and Porter (1977) investigated oxygen consumption and filtering rates in oxygen concentrations ranging from air saturation levels to almost zero. The oxygen consumption in *Daphnia galeata mendotae* turned out to be directly proportional to the oxygen concentration. In *D. magna*, it was slightly increased at O_2 concentrations above c. 3 mg/L; below this concentration, the rate of oxygen consumption declined rapidly.

Paul et al. (1997) and Pirow et al. (2001, 2004) devised special computerized recording techniques and have achieved remarkable results in the field of *Daphnia* respiration. A special set-up for optophysiological recording with a video recorder and PC, as well as computerized respirometry, was devised by Paul et al. (1997). For this, *Daphnia* were placed in a glass vessel of 0.45 mL volume. The apparatus used by Pirow et al. (2001) consisted of an inverted microscope equipped with systems for measuring hemoglobin (Hb) oxygenation, NADH fluorescence (as an indicator of the oxygenation status of tissues), and the movements of organs.

Pirow et al. (2004) also investigated oxygen transport processes in *D. magna* using an oxygen-sensitive phosphorescence probe, Oxyphor R2, injected into the circulatory system. Fasting animals were immobilized by gluing their posterior apical spine with histoacryl to a bristle that was then fixed to a coverslip with plastilin. The dye was then microinjected into the circulatory system from the dorsal side into the space "directly downstream of the heart." A single daphnid was fixed onto a coverslip and then placed onto the bottom of a thermostatted perfusion chamber; in this apparatus, the individual daphnid was able to move its antennae freely. Measurements of the distribution of oxygen partial pressure (P) in the hemolymph were followed by phosphorescence imaging of the phosphorus range (0–6 kPa). The resulting three-dimensional profiles were different for Hb-rich and Hb-poor specimens. A steep gradient was discovered in Hb-poor specimens, whereas Hb-rich individuals showed flat gradients (Fig. 5.1).

With increasing temperature, oxygen consumption increases but the oxygen content of water tends to decrease. Oxygen consumption also depends on various other factors, such as age (as shown for *Simocephalus* by Obreshkove and Banta, 1930), illumination (Buikema, 1972), and vessel size (Zeiss, 1963). The respiratory rate rapidly increased in previously starved *D. magna* provided with increasing concentration of algae until the "incipient limiting level" of food concentration was reached (Lampert, 1986).

According to Flückiger (1953), sympathomimetics (i.e. L-adrenaline, D-adrenaline, and L-noradrenaline) increase oxygen consumption by *Daphnia*. The presence of carbaminoylcholine chloride (a parasympathomimetic) also increases oxygen consumption.

5.4 HEMOGLOBIN AND IRON

Hb has been reported in various littoral and pelagic Cladocera by many authors. With oxygen deficiency, Hb appears in the blood of cladocera and its concentration increases over time (Chandler, 1954; Smaridge, 1956; Kobayashi, 1970).

5.4.1 Hemoglobin in Littoral Cladocera

Although Hb is more important in littoral and bottom-living Cladocera than in pelagic Cladocera, Hb has mainly been studied in pelagic daphnids and in *Moina*. In littoral chydorids, such as *Acroperus harpae, Alona affinis, Eurycercus lamellatus, Picripleuroxus striatus, Pleuroxus trigonellus*, and *P. truncatus*, Smirnov (1970) demonstrated the presence of Hb using three different methods:

1. By producing crystals of pyridine hemochromogen in the bodies of the chydorids, (according to the method of Hoshi, 1963a, p. 88);
2. By producing crystals of hemin (according to the method of Hoshi (1963b, p. 94); and
3. By staining Hb blue with benzidine base (according to the method of Pearse, 1969).

The blue staining was clearly observed to be localized and the crystals produced by methods (1) and (2) were characteristic of each substance. Hb was found in both reddish specimens and specimens with no visible reddish coloration.

The red color of mud-living *Ilyocryptus sordidus* is caused by Hb, as shown by Fox et al. (1951). If its young are cultivated in well-aerated water, they grow into pale adults. It was also demonstrated that bottom-living and littoral cladocera, such as *Ilyocryptus* and *C. sphaericus*, gain or lose Hb, respectively (Fox, 1955). The presence of Hb in bottom-living cladocera may be related to an oxygen deficit in such conditions or even to anoxia. However, some species living on the mud surface are not colored red, as was observed e.g. in *Drepanothrix dentata* (in Lake Glubokoe, Moscow region). However, the presence and levels of Hb in most bottom-living and inshore cladocerans remain unexplored.

5.4.2 Hemoglobin in Pelagic Cladocera

Hb has been found in planktonic *Daphnia* and *Moina*, and also in *Simocephalus* (Fox, 1948; Hoshi, 1957; Hoshi and Nagumo, 1964). The maximum Hb absorption in an absorption spectrum is different in representatives of different species, and even in related species; for example, it is at 314 nm in *D. magna* and 418 nm in *Ceriodaphnia quadrangula* (Czeczuga, 1965).

Generally, Hb is formed when there is an oxygen deficit. The Hb content of *Daphnia* decreases before a molt, but increases in the ovaries at the same time (Fox et al., 1949). During an instar (the intermolt period), Hb passes from the blood into parthenogenetic eggs, as has been shown for *Daphnia* (Green, 1956), and this occurs a short time before molting. The Hb concentration in *Daphnia* increases when there is a low oxygen concentration in the environment [shown by Fox, (1955), Green (1956), and Landon and Stasiak, (1983)]. The latter authors also demonstrated that the Hb concentration in *Daphnia* is directly proportional to depth in Arco Lake (Minnesota, USA). In aerated water, the Hb content in *Daphnia* decreases. The Hb content of pink *Moina macrocopa* was shown to be 1.7 g Hb/100 mL blood and 1/15 of this value in pale specimens (Kobayashi, 1981b). Studies on variations of Hb content in response to environmental factors, including oxygen concentration, have been reviewed by Kobayashi and Hoshi (1984). The Hb concentration in *D. magna*, as determined by Kobayashi (1981a), ranges from 0.1 to 1.7 g Hb per 100 mL blood. According to later measurements, the Hb concentration of *D. magna* ranges from 0.24 to 241 mg Hb/g DW (Kobayashi and Nezu, 1986). It increased at low oxygen concentrations: rapidly in immature specimens and more slowly in larger animals, although animals longer than 3 mm did not become red. The lethal oxygen concentration for pale *D. magna* is c. 0.22 mL O_2/L at 15°C and c. 0.5 mL O2/L at 30°C; for red specimens, it is c. 0.1 mL O2/L at 15°C and c. 0.12 mL O2/L at 30°C. Under the same conditions, *D. magna* males accumulate more Hb than do females at low oxygen concentrations (Kobayashi, 1970)). Hb is not just a passive chemical, however. In Hb-rich *D. magna* its contribution to the circulatory transport of oxygen to tissues is much greater than in Hb-poor specimens (Bäumer et al., 2002). Extra Hb ensures an adequate oxygen supply to limb muscle tissue in *D. magna* at moderate oxygen concentrations in the ambient water (Pirow et al., 2001). The tissue oxygenation state can be estimated by NADH fluorescence in muscles.

Daphnia manage well without Hb; however, at low oxygen concentrations pale *Daphnia* make up for the absence of Hb by increasing the rate of movement of their thoracic limbs (Hoshi and Inada, 1973). Curiously, although Hb appears in animals at reduced oxygen concentrations, inactivation of Hb by carbon monoxide (CO) did not make daphnias much less vigorous or decrease their survival in short-term experiments. According to Fox (1948), *Daphnia* containing deoxygenated Hb survive for a long time: it was therefore thought that Hb takes no part in their respiration. It seems to have only a supplementary role. Long-term cultivation, however, revealed that red daphnias do live longer, feed more successfully, and produce more eggs (Fox, 1948; Fox et al., 1951).

According to Kobayashi (1981b), in poorly aerated water the respiration rates of pale *M. macrocopa* and of CO-treated red specimens are identical. The lethal oxygen concentration becomes higher in *Daphnia* at increasing CO concentrations and is more obvious in pink *Daphnia* (Kobayashi and Hoshi, 1980). Nevertheless, *Daphnia* containing Hb live longer, swim more actively, consume more food, and produce more parthenogenetic eggs (Fox et al., 1951). Kobayashi and Hoshi (1982) believe that Hb functions only at low O_2 concentrations. In their experiments, all specimens of pale *D. magna* died at c. 0.5 mL O_2/L, whereas specimens containing 1–1.5 g Hb/100 mL blood died at c. 0.1 mL O_2/L. The critical values for *C. quadrangula* are: 0.9 mL O_2/L for pale specimens and 0.3 mL O_2/L for pink specimens (Kobayashi, 1983a). Above this level, the rate of oxygen consumption by *C. quadrangula* is constant: c. 12 μL O_2/ind./h for pale specimens and 11 μL O_2/ind./h for pink specimens.

Survival may also depend on the former environment of the tested specimens. For example, *S. vetulus* taken from a pond survive much longer in deoxygenated water than do specimens taken from a river (Shkorbatov, 1953).

Rudiger and Zeis (2011) found an inverse relationship between temperature and Hb concentration in *D. longispina* from oxygen-rich water, with the seasonal minima of Hb concentration coinciding with low food availability and the peak of reproduction. These authors suggested that in *Daphnia* Hb functions as both a respiratory protein and a protein store.

Generally, the role of Hb in cladocera is facultative and these organisms may successfully live and reproduce without it (e.g. when it is inactivated).

The presence of Hb in fat cells was noted by Green (1955); it was later shown that Hb is synthesized in the fat cells and epithelial cells of epipodites (Goldmann et al., 1999). Hb can also be decomposed in fat cells without yielding bile pigments (Green, 1971). Using radioactive iron, Hoshi and Sato (1974) found that there is a continuous process of Hb synthesis and breakdown in the *Daphnia* body. Hb breakdown is now known to occur in fat cells (Smaridge, 1954; Green, 1955), whereas it was previously thought that Hb is normally excreted through the shell glands (Fox, 1948).

Populations of *Daphnia* have also been observed to be colored brown by Hb in combination with hematin, which contains three-valent iron in the heme moiety (Fox, 1047–1948; Goodwin, 1955).

The Hb values indicated by these various authors therefore depend on various factors (e.g. environmental factors and species).

5.4.3 Iron

The iron component of Hb is present in water and more abundantly on the bottom of water bodies in both ferrous and ferric forms. Obviously, it can therefore be ingested by cladocerans directly or in algae. The iron content of Hb has been measured as 0.033%, in *D. magna* and 0.317–0.353% in *M. macrocopa* (Hoshi et al., 1967; Hoshi and Kobayashi, 1971). Hb formation in *Daphnia* is favored by the presence of iron in the environment (Fox and Phear, 1953; Goodwin, 1960); Dave (1984) found that the optimum concentration for Hb formation is 2 μg Fe/L ($FeCl_3.6\,H_2O$). Hb formation is also favored by higher temperatures and is not influenced by carbon dioxide levels (Fox and Phear, 1953). In the body of *D. magna*, its distribution was described by Smaridge (1956), who found three forms of iron in *Daphnia* tissues: "loosely bound" iron, "firmly bound" iron, and iron within heme compounds (Hb and cytochromes). Loosely bound iron can be stained by, for example, Prussian blue and firmly bound iron can be revealed by treatment with acid alcohol. The iron in Hb and cytochromes may be revealed by treatment with, for example, hydrogen peroxide or by microincineration. Radioactive iron–59 (^{59}Fe) is also incorporated by *Daphnia* (Smaridge, 1956).

Loosely bound iron is found in the gut wall, gut caeca walls, and fat cells (both as ferrous and ferric iron), as well as in ovaries, blood plasma, and appendage walls (ferrous iron). It is assumed that iron is incorporated into newly synthesized Hb in the fat cells and ovaries. Some iron may be accumulated in fat cells and then excreted. Iron is also incorporated into heme and is present in all tissues, even in the lenses of the eyes. Within the bodies of *D. magna* specimens with increasing Hb, iron is absorbed by the gut wall and is present in newly synthesized Hb in fat cells and the ovaries (Smaridge, 1956). In specimens that are losing Hb, it is especially present in walls of the gut ceca and fat cells, and in the excretory shell glands; Hb is probably broken down in fat cells. Tazima et al. (1975) also reported that in *Daphnia* ferruginous compounds accumulate in the gut ceca, and that during Hb breakdown they can be observed in fat cells and the shell glands.

Iron used for Hb synthesis is absorbed from the midgut and used by fat cells for Hb synthesis; Hb destruction is aided by gut and fat cells, and Hb is excreted via the maxillary glands (as loosely bound ferric iron).

5.5 EVOLUTION OF CARBON DIOXIDE AND THE RESPIRATORY QUOTIENT

Carbon dioxide (CO_2) production by *D. magna* is estimated to be 0.03 mg/g air-dried weight by females at 22°C and 22% higher in males (MacArthur and Baillie, 1929). In *D. pulex*, as determined by Richman (1958), its values are close to those of O_2 consumption. Later, CO_2 production by *D. magna* was measured by Kolupaev (1989) to be 0.38 mg/g WW/h in January–April and 0.51 mg/g WW/h in May–August; oxygen consumption was 0.43 mg/g WW/h and 0.62 mg/g WW/h, respectively; and the frequency of thoracic limb beating 299 per min and 346 per min, respectively. According to data published by Kolupaev (1989), the RQ in *D. magna* is <1.

RQ is the ratio of the carbon dioxide released to the oxygen consumed. The RQ is 1.00 for carbohydrate metabolism, 0.711 for fat metabolism, and 0.781 for protein metabolism (Koshtoyants, 1951). RQ values of <1 may indicate the transformation of carbohydrate into fat accompanied by oxygen binding (e.g. for deposition into eggs) or anoxybiosis (CO_2 formation during fermentation in an oxygen-free medium). The RQ in well-fed *D pulex* is 0.92–1.11 for those of 1.6 mm in length and 0.95–1.24 in larger specimens (Richman, 1958). The RQ in starved *D. pulex* decreases to 0.7, thus indicating a shift from carbohydrate metabolism to the utilization of fat and protein (Richman, 1958).

5.6 ENERGY BUDGET

Richman (1958) determined the energy budget of *D. pulex* (Table 5.1) to comprise expenditures for the growth and production of young, with the latter using the major part of the stored energy.

An energy-channeling scheme for *Daphnia* was proposed by McCauley et al. (1990), as shown in Fig. 4.11. Basal metabolism (i.e. metabolism in *Daphnia* immobilized by D-tubocurarine) was determined by means of a Cartesian diver (O'Connor, 1950). It was found that the expenditure for muscular activities is 32% or more of the total metabolism. In *D. magna* immobilized with uretane, basal metabolism takes 1/4–1/6 of the metabolism of an active *Daphnia* (Postnov and Philippova, 1988). The expenditure for respiration was determined by Galkovskaya (1970) to 21–44% in *D. longispina* and 32% in *Bosmina*. Male cladocerans have a higher rate of metabolism than females (Banta and Brown, 1924).

5.7 HYPOXIA

Hypoxia (low oxygen concentrations) is commonly experienced by Cladocera. Oxygen deficiency may occur in the early morning following the prevalence of algal respiration over photosynthesis, or when there is a complete covering of floating *Lemna*. In the absence of oxygen, *Daphnia*, *Simocephalus*, *Scapholeberis*, *Bosmina*, and *Alona* spp. have been shown to remain immobilized for 1–6 h and *C. sphaericus* for 18 h (Nikitinsky and Murezowa-Wyss, 1930).

There have also been investigations of cladocera respiration under conditions of low oxygen content, specifically for planktonic Cladocera. It has been shown that during an oxygen shortage *Daphnia* swim toward regions of higher oxygen concentration (Pardi and Papi, 1961).

Rivier (1986) reported the presence of *Daphnia* with numerous embryos during winter in the bottom of ice-covered Lake Siverskoe, at a low oxygen content (c. 2 mg/L O_2). She also found *D. longiremis*, *D. galeata*, *D. cristata*, and *Bosmina longirostris* reproducing under the ice (Rivier, 2012).

TABLE 5.1 Energy Budget of *Daphnia pulex* at a Moderate Food Concentration

Age	Energy Consumed (Cal)	Energy for Growth (Cal)	Energy for Growth and Young (Cal)	Energy of Respiration (Cal)	Energy of Egestion (Cal)
Pre-adult	0.469	0.062	0.050	0.050	0.357
Adults after 34 days of growth (young not included)	5.671	0.071	–	0.791	3.872
Adults after 40 days of growth	6.140	–	1.070	0.841	4.229

The low relative value of energy for growth is to be noted in slow-growing adults.
Source: Richman (1958).

In *D. magna* at 1.8 mg/L O_2, the following parameters decrease in comparison with controls at higher oxygen concentrations (Homer and Waller, 1983): DW, number of days to first brood, number of young in the first brood, total number of young produced over the 26 days of the experiment. The median lethal dose (or LD_{50}; after 4 h) is 0.2–0.7 mL O_2/L for *Daphnia* and *Simocephalus*, and 1–1.6 mL O_2/L for *Leptodora* and *Bythotrephes* (Herbert, 1954).

Pink *D. magna* have been shown to survive for at least 24 h at 5% air saturation while pale ones perish within 1 h at 7% air saturation (Usuki and Yamaguchi, 1979). At low oxygen concentrations, the lactic acid content in *Daphnia* increases (Fig. 5.2). In *D. magna* with embryos, under hypoxic conditions there is a higher rate of blood flow through the brood chamber (Seidl et al., 2002).

In contrast to *D. galeata*, oxygen consumption in *D. magna* rapidly declines at O_2 concentrations below c. 3 mg/L (Heisey and Porter, 1977). An oxygen concentration c. 0.2 mg/L O_2 suppresses the formation of large filtering screens in *D. magna* at a low food concentration, and thus reduces the filtering rate (Hanazato, 1996). Growth was also retarded.

Hypoxia also frequently induces the appearance of Hb in the body (Gorr et al., 2004; Zeiss et al., 2004). In hypoxic *Daphnia*, the Hb content is increased in the blood, muscles, and nerve ganglia, and the content of cytochromes in the tissues also increases (Fox, 1945).

In addition, oxygen-dependent Hb induction takes place in *D. magna* under hypoxic conditions, thus favoring enhanced oxygen transport. At oxygen with a partial pressure of 3 kPa, newly synthesized Hb can be detected within 18 h and a steady state concentration is

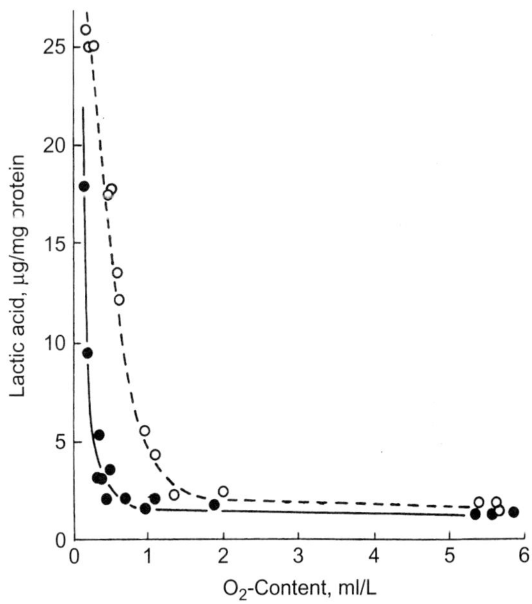

FIGURE 5.2 Lactic acid content in pale (open circles) and pink (solid circles) *Daphnia* at different oxygen concentrations over a period of 1 h. *Source: Usuki and Yamaguchi (1979).*

reached within 11 days (Zeiss et al., 2003); in the same species, hypoxia-inducible factors (HIFs) are accumulated and facilitate O_2 delivery to hypoxic tissues (Gorr, 2004).

Juvenoid hormones contribute to the elevation of Hb levels in *D. magna* (Gorr et al., 2006): at high O_2 tension (pO_2), these hormones induce the robust reactivation of juvenoid response elements, but at low pO_2 this reaction is inhibited.

In addition, in *Moina micrura* not acclimated to hypoxia, the metabolism and the stroke frequency of their antennae is decreased (Hubareva, 2000a). At 0.7–0.8 mg O_2/L, pale *M. micrura* may stop filtration but continue swimming, in contrast to red specimens (Svetlichny and Hubareva, 2002b). In *D. magna*, there is no compensatory increase in the beating rate of thoracic limbs during hypoxia: it is maintained at c. 180 per min (Colmorgen and Paul, 1995). It was later observed that the beating rate of their appendages does increase under hypoxic conditions, and that compensatory tachycardia also occurs, followed by an increase in the Hb concentration (Pirow and Buchen, 2004).

Under conditions of acute hypoxia in *M. micrura*, the O:N ratio is 8.5, but this increases to 34–40 under oxygen saturation (Hubareva and Svetlichny, 1998). Under acute hypoxia, largely anaerobic protein catabolism occurs. During longer periods of hypoxia, active *M. micrura* became acclimated and use whatever oxygen is available for lipid oxidation. *M. micrura* adaptation to a prolonged period in low-oxygen environments involves a greater role for lipids as metabolic substrates (Hubareva, 2000b). Under conditions of acute hypoxia (i.e. a decrease in O_2 concentration to c. 0.5 mg/L), the typical protein-lipid catabolism of *M. micrura* changes completely to the anaerobic protein catabolism (Hubareva, 2000a, 2000b; Svetlichny and Hubareva, 2002a, 2002b, 2004). Further, the excretion of NH_3-N drastically decreases from 0.079 μg N/L/h to 0.019 μg N/L/h, in proportion to the O_2 concentration.

Seidl et al. (2005) exposed both the parental generation and progeny of *D. magna* to 10–19% air saturation. Acclimation to hypoxia involved adjustments to the Hb level and the metabolic level: Hb concentration increased by 266% and oxygen affinity increased by 32%. Smaller offspring, combined with reduced Hb and metabolic levels, reduced the critical ambient oxygen tension by about 50%. Transgenerational effects were not observed.

Zeiss et al. (2009) studied the composition of the *D. pulex* proteome under normoxic (po_2 20 kPa) and hypoxic (po_2 3 kPa) conditions. Hypoxia caused a strong inducement of Hb and carbohydrate-degrading enzymes, i.e. glycolytic enzymes (enolase) and enzymes involved in the degradation of storage and structural carbohydrates.

In hypoxic water, Hb-rich *Daphnia carinata* preferred a temperature of 19°C (vs. 16°C for controls), whereas in normoxic water all the animals gathered at regions of 23°C (Wiggins and Frappell, 2000).

In *Daphnia*, hypoxic stress and heat (30°C) caused oscillation in reactive oxygen species, glutathione redox system activity, and HIF-1 (hypoxia-inducible factor) activity (Becker et al., 2011).

5.8 ANOXIA

There are good reasons for discussing anaerobiosis in Cladocera. Most of the known species live in the littoral zone and on the bottom of water bodies, where there is, at least periodically, a deficiency of oxygen. Many species, probably including all Ilyocryptidae, live in the surface of mud under permanent conditions of very scarce oxygen.

For the upper mud layer of lakes (0–2 cm), Martynova (2010) reported the presence of O_2 at 0.75–9 mg/L (the bottom water layer), N_2 at

42–175 mg/kg WW, CH_4 at 1–21 mg/kg, and CO_2 at 1.5–156 mg/kg. Brand (1946) listed the findings of various authors: the inhabitants of the mud surface of freshwater water bodies where the oxygen content is close to zero include the cladocera *Lathonura rectirostris*, *Daphnia pulex*, and *Simocephalus exspinosus*. The lethal minimum oxygen concentration for chydorids is 0.4–1.7 mg/L O_2 (Pacaud, 1939; Bogatova, 1962); however, in general, information on anaerobiotic Cladocera is very scarce.

Under severely hypoxic conditions, *C. sphaericus* movement in water saturated with hydrogen was discontinued for 18 h, and in *Alona* for 4 h (Mudretzowa-Wyss, 1933); and in nitrogen-saturated water movement ceased in *C. quadrangula* for c. 35 min for red specimens and 23 min for pale specimens (Kobayashi and Ichikawa, 1987). In the presence of hydrogen sulfide (H_2S), *Daphnia*, *Simocephalus*, and *Bosmina* became immobilized within c. 10–20 sec (Nikitinsky and Mudrezowa-Wyss, 1930).

It has been observed that *Chydorus sphaercus* does not feed in anoxic water layers (Lair, 1991). As studied by Colmorgen and Paul (1995) and Paul et al. (1997), during anoxia the frequency and amplitude of the movements of thoracic limbs decreased in *D. magna*; the heart continued to beat at a rate similar to normoxia but the stroke volume decreased. According to these authors, during hypoxia the frequency and amplitude of their thoracic limb movements remained at the usual rate of c. 180 min, but the heart rate increased (i.e. compensatory tachycardia).

Thus, under anoxic conditions the metabolism of Cladocera changes radically. Anaerobiosis is generally characterized by the exploitation of carbohydrates in metabolism, suppression of protein metabolism (Brand, 1946), and high RQ values due to the final stage of CO_2 production (Koshtoyants, 1951).

Lactic acid concentration in the body of *Daphnia* is generally very low, but at low oxygen concentrations *D. magna* accumulated more lactic acid prior to death (Usuki and Yamaguchi, 1979). In Hb-colored *Daphnia*, the lactic acid content is higher at lower oxygen concentrations (Fig. 5.2).

The responses of *D. magna* to anoxia were studied further by Paul et al. (1998). During the first 2 h of anoxia, their anaerobic metabolism is characterized by L-lactate accumulation in the body (0.36 μmol lactate/g WW/min) and a decrease in pH in the body (metabolic acidosis). During the first 2 h of anoxia, the heart rate does not change much, but at longer exposures the heart rate decreases. Under subsequent conditions of normoxia, the heart rate and the body pH return to normal.

In *S. vetulus* under anaerobic conditions, the glycogen content is 0.5% in the gastrula, 0.67% in the nauplius, 0.53% in the released young, and 0.64% in the third instar (Hoshi, 1953). Unlike in aerobic conditions, it does not increase during postembryonic development.

Hydrogen sulfide is another factor that limits the availability of bottom habitats for Cladocera; for example, a low H_2S concentration (0.15 mM) causes irreversible immobilization of *C. ovalis* (Bryukhatova, 1928). However, there is a report of a ctenopod-like cladoceran that lives in the anoxic depths of the Black Sea, which is saturated with hydrogen sulfide (Sergeeva, 2004; Korovchinsky and Sergeeva, 2008).

5.9 IMPACT OF XENOBIOTICS ON RESPIRATION

Information on this subject is very limited. Oxygen consumption in *Simocephalus* is inhibited by 2.5 mM 2,4-D (Andrushaitis, 1972). In *Simocephalus*, the presence of >5 mM 2,4-dichlorophenol-Na in the environment leads to decreased oxygen consumption (Klekowski and Zvirgzds, 1971). At 5–8 mg/L naphthalene, the

Hb content and oxygen consumption of *D. magna* decreases by about a 25% compared with controls (Crider et al., 1982). Phenol at a concentration of 0.1 mg/L reduces the beating rate of the thoracic limbs in *D. magna* from 305 beats/min (control) to 204 beats/min (Kolupaev, 1988). In contrast, the respiratory demand increases in *D. magna* after a 23–day exposure to 0.99 mGy/h americium (Am-241), thus indicating an increase in metabolic expenditure (Alonzo et al., 2006).

Oxygen consumption decreases in *D. magna* in presence of copper and zinc during the first hour, then (as determined after 3 h) compensatory mechanisms are activated (Sladkova and Kholodkevich, 2012). Therefore, these authors suggest that measurements of this effect should be made after the first hour of exposure.

CHAPTER

6

Circulation

6.1 ANATOMICAL BACKGROUND: BLOOD CELLS

Cladocera possess an open circulatory system and a myogenic heart. Heart beats and the flow of blood (hemolymph) can be easily observed, as Cladocera are semitransparent. In *Leptodora*, an additional propulsive organ has been described (Gerschler, 1910; Saalfeld, 1936; Maynard, 1960), which can easily be found and observed in living specimens in the first thoracic limb, at the distal end of the first segment: it takes the form of a disk supplied with a propulsive muscle.

For *Daphnia magna*, it has been determined that the hemolymph makes up 58% of its wet weight and 61% of its volume (Kobayashi, 1983b). The heart has lateral ostia for the inflow of hemolymph and a short anterior vessel from which hemolymph is squirted at each heart beat (Fig. 6.1). Venous and arterial blood flows are separated by a membrane in the area of the heart (Fig. 6.2) (Herrick, 1884). It was later found that the whole body space is subdivided by ventral and dorsal membranes into three blood spaces: the ventral lacuna, the dorsal lacuna, and the intestinal lacuna. In addition, two vertical membranes divide the ventral lacuna and the limbs into the medial lateral compartments (Hérouard, 1905; Pirow, Wollinger, and Paul, 1999b). It may be added that the inner membranes of the cladoceran body need to be better described.

The carapace covering the body is double walled, with the blood flowing into the internal carapace lacuna.

The heart is innervated by the cardiac nerve, which reacts to adrenaline (thus increasing the heart rate); however, the central nervous system does not take part in heart regulation, as has been shown by the elimination of its various parts (Ermakov, 1936). Later, Ermakov (1937) demonstrated the presence of nerve fibers that accelerate and retard the heart rate.

Cladoceran blood is light yellow, but may also be red due to the presence of hemoglobin (Hb). Artemocyanin (biliprotein) has been found in the blood of *Daphnia pulex* (Peeters et al., 1994).

Blood cells were originally noted in the blood flow by early authors (e.g. Gruithuisen, 1828; Leydig, 1860, p. 56) and can be easily observed. Gruithuisen (1828) and many subsequent authors (e.g. Hardy, 1892) believed that Cladocera possess a single type of blood cell. Cuenot (1891) saw numerous amoebocytes of size 6–9 μm in the blood of *Daphnia schefferi*, *Simocephalus vetulus*, *Chydorus sphaericus*, and

Physiology of the Cladocera.
DOI: http://dx.doi.org/10.1016/B978-0-12-396953-8.00006-9

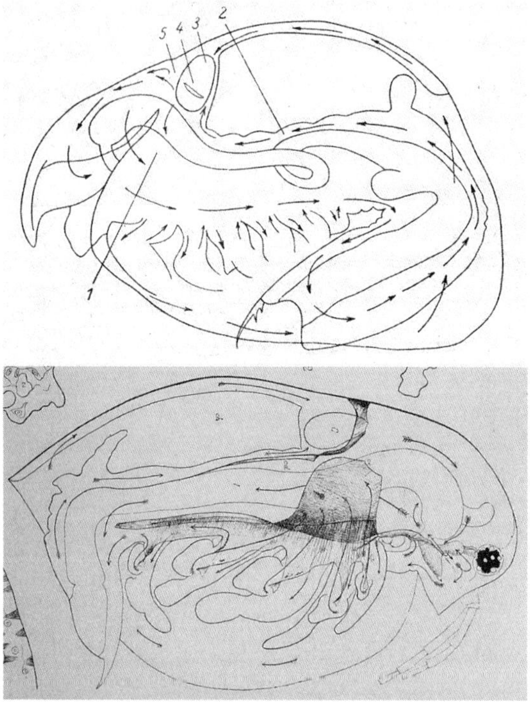

FIGURE 6.1 Upper, actual organism. Lower, schematic structure. Upper, blood flow in *Eurycercus*. 1, arterial flow; 2, venous flow; 3, heart; 4, lateral ostium; 5, eddie. Lower, blood flow in *Daphnia*. (C, heart; cd, dorsal septum; ch, brood chamber; cp, cloison separating external ventral space from the inner space; cv, reflected part of ventral septum; I, gut). *Sources: Upper, Smirnov (1971); lower, Hérouard (1905).*

Sida crystallina. According to Cuenot, amoebocytes produce pseudopods. He noted local aggregations of amoebocytes but did not locate any lymphatic glands.

Metchnikoff (1892) also observed a single type of blood cell in *Daphnia*: he termed and understood them to be leucocytes. He also observed *D. magna* hemocytes in the process of phagocytosis (Fig. 6.3) (Metchnikoff, 1882, 1884).

There are many thousands of hemocytes in *Eurycercus*, and several hundred in the smaller *Acroperus* (Smirnov, 1975). According to Jaeger (1935), the blood cells loaded with fat may settle on tissues and turn into cells of the fat body.

Where the blood cells are formed and the process of hemopiesis is unknown. Early authors observed only one type of blood cell in *Daphnia* (Hardy, 1892; Fritsche, 1917) and in *Eurycercus* (Smirnov, 1970, 1971). In *Eurycercus*, the average length of these cells is 15 μm, but in *Daphnia* they are only 7–8 μm in diameter. The blood cells of *Eurycercus* are of variable structures and have occasional pseudopods.

Closer examination of *D. magna* revealed two cell types circulating in the hemolymph (Auld et al., 2010): spherical cells (granulocytes) and irregular-shaped amoeboid cells (about 9 μm long, and identified as the cells initially described by Metchnikoff (see 1898, Fig. 17.3). The latter cells attack the bacterial parasite *Pasteuria ramosa*; in parasitized *Daphnia*, there is a large increase in the number of amoeboid cells.

6.2 BLOOD FLOW

The movement of hemolymph is powered by the heart and by body movements. Gruithuisen (1828) was probably the first to present a general scheme of circulation, within *Simocephalus*. Following the movement of blood cells he indicated arterial flows from the heart to the cephalic region, to antennae, over valves, and along the trunk, as well as reverse venous flow. According to this author, the blood flows along capillary canals.

The time for the complete transit of a blood cell through the circulation of *Daphnia* is about 10–20 sec (Dearborn, 1903; Maynard, 1960). The systolic heart volume was determined to be 2.53 nL in *D. pulex* and 1.28 nL in *Holopedium* (Maynard, 1960). Blood flow can be clearly seen by the movement of blood cells (Fig. 6.1), and the patterns of flow are generally permanent, being separated and directed by membranes within the body. Herrick (1884) reported in *Eurycercus* the presence of

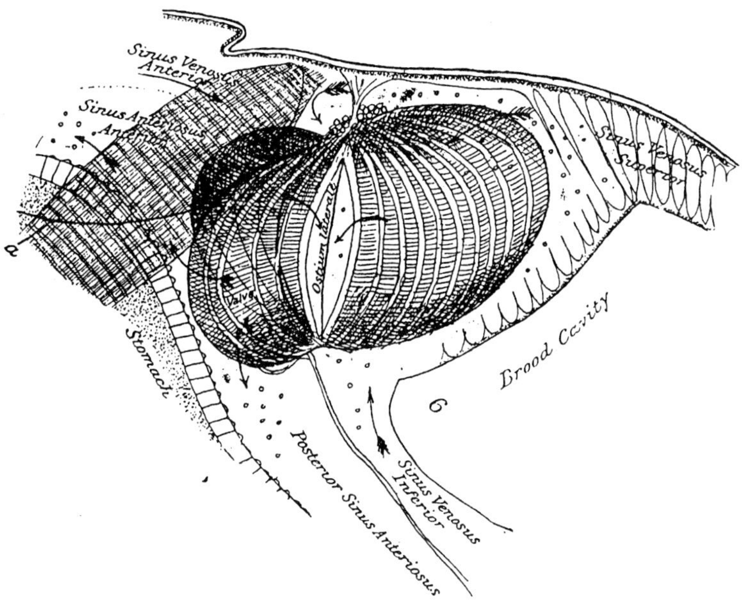

FIGURE 6.2 ***Eurycercus* heart and membranes dividing arterial and venous blood flow.** *Source: Herrick (1884).*

membranes that divide the venous and arterial blood flows (venous and arterial sinuses) in the area of the heart (Fig. 6.2). In general, blood flow is directed forward from the heart (arterial flow), and along the dorsal side of the body; in the area of the head it turns and then passes through the limbs, follows along the ventral side to the postabdomen, and then returns along the dorsal side of the body and of the shell, as observed in *Daphnia* by Hérouard (1905) (Fig. 6.1). In different groups of cladocera, the system of blood flow varies, but it has not yet been accurately described.

Blood flow in *D. magna* has been investigated in detail by Pirow and Buchen (2004) (Fig. 5.1). Pirow et al. (2004) also reported details of oxygen partial pressure throughout the body of Hb-poor and Hb-rich *D. magna*, and represented the data obtained as a three-dimensional diagram (Fig. 6.4).

In *D. pulex* and *Holopedium*, the systole is about 1.5 times longer than the diastole (Fig. 6.5) (Storch, 1931). At each contraction, about half of the blood is expelled from the heart. The systolic volume of the heart in *D. pulex* is 2.53 nL (Maynard, 1960). At its contraction in systole, the cranial region of the heart is first contracted, then the ostial region, and then the caudal region, resulting in complete contraction of the heart (Proksova, 1950). In diastole, the cranial region is dilated first, and so on, further resulting in a complete widening of the heart.

6.3 HEART RATE

The heart rate can be determined by direct counting; however, this becomes too difficult at a rate of over 400 beats/min. A special device for recording the heart rate was proposed by Présig and Véro (1983), based on an optoelectronic circuit that transforms changes in light intensity caused by heart beats into changes in potential. Numerous studies on heart rate in cladocera have been mainly confined to *Daphnia*, but they have been reported by Smirnov (1965b) for representatives of

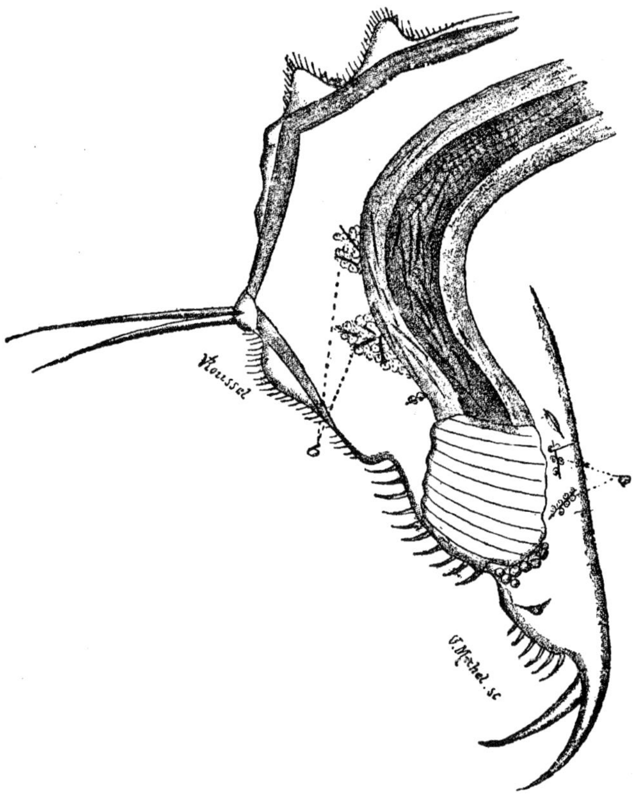

FIGURE 6.3 **Phagocytosis in *Daphnia*.** a, spores of *Monospora* surrounded by leucocytes. *Source: Metchnikoff (1892).*

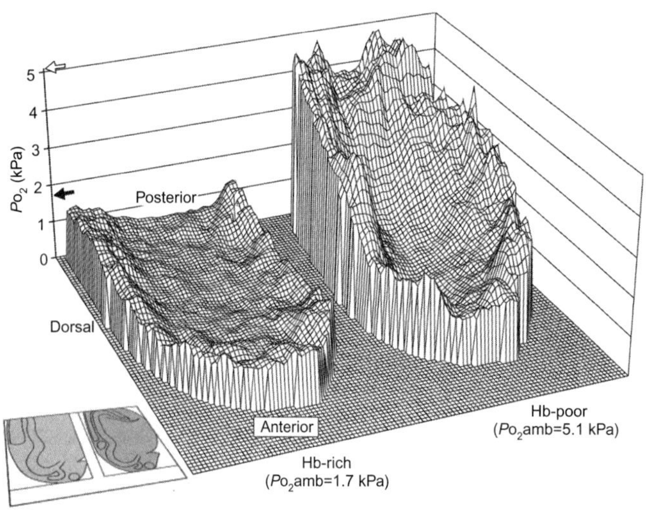

FIGURE 6.4 **Oxygen partial pressure in the hemolymph throughout the body of Hb-rich and Hb-poor *Daphnia magna*.** *Source: Pirow et al. (2004).*

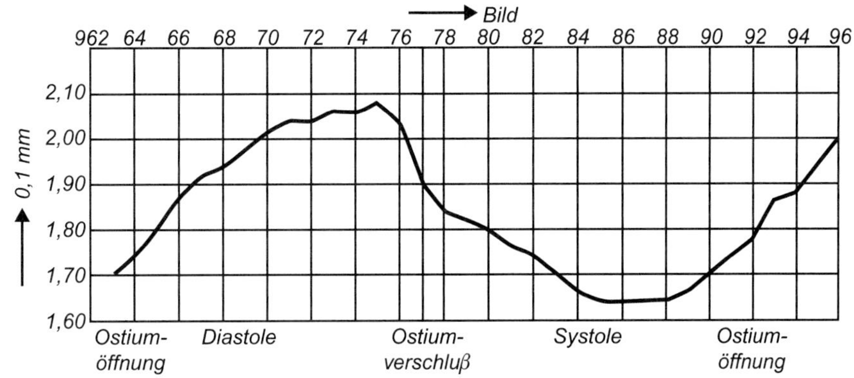

FIGURE 6.5 **Diastole and systole in the *Daphnia* heart.** Changes in heart size were measured. *Source: Storch (1931).*

various genera (for 22 species belonging to 16 genera from 6 families).The heart rate in littoral and pelagic species, as determined by direct counting, ranges from 190–320 beats/min at 17–18°C in females. The very slow-moving macrotricid, *Ilyocryptus agilis*, was a notable exception: it has a heart rate of only 120 beats/min at 20°C.

Kimographic recording of the *Moinodaphnia macleayi* heart rate was made by Tonapi et al. (1984): it was 105 beats/min in females without embryos, and 120 in females with embryos, 90 in males, and 120 in males after molting (at 22°C). This is very slow compared to most other Cladocera. In *D. magna* without embryos, the heart rate is 350 beats/min: with embryos, it is 379 beats/min (Kolupaev, 1989).

Remarkably, the heart rate increases in both females and in males by about 10–20% following every kind of disturbance, i.e. the increase is emotional, (Smirnov, 1965b). When the disturbance is discontinued, the former heart rate is rapidly restored. Otherwise, the heart rate is stable and only significantly drops just before death.

There are numerous data characterizing the level of heart rate as being dependent on various factors. Ermakov (1936) found that the heart rate of *D. magna* changes with starvation, molting, reproduction; potassium ion (K^+)-induced systolic arrest, and calcium ion (Ca^{2+})-induced diastolic arrest; the heart rate also increases with increased body size, up to a certain limit, and the heart rate decreases in starving *Daphnia longispina* (Ingle et al., 1937).

There is a diurnal cycle of heart rate: in *Daphnia*, the maximal heart rate occurs in late afternoon (Tonolli, 1947), which may be linked to the fact that the heart rate generally increases at higher temperatures (Fig. 6.6) (Maynard, 1960). The heart rate also increases after feeding (Baylor, 1942; Maynard, 1960), depending on temperature and level of illumination, whereas starvation is followed by a decreased heart rate. The heart rate also decreases upon illumination with a beam of light directed at the heart (Schulz, 1928) and by ultrasound (up to full arrest) (Nikonov et al. (1970).

Salo (1960) presented data showing that the heart rate in *D. magna* and *D. pulex* may be equal to, higher than, or lower than the beating rate of the limbs. The sum of both parameters is far from constant, but there was suggested to be a compensatory relationship between the rate of movement of the thoracic limbs and the heart rate. In view of this fact, their conclusion of a compensatory relationship between the

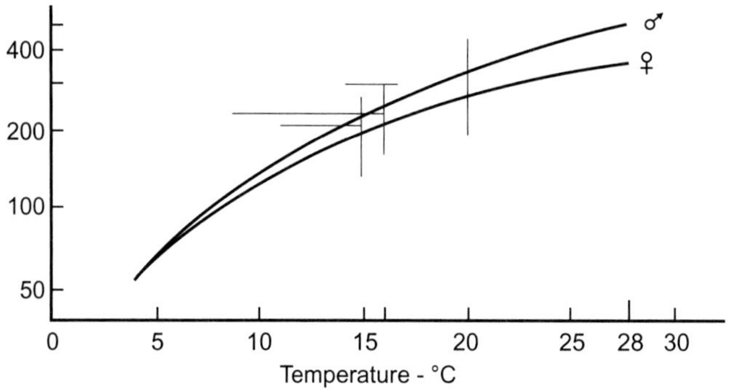

FIGURE 6.6 Heart rate of female and male *Daphnia* at different temperatures. *Modified from Maynard, 1960.*

movement rate of the thoracic limbs and the heart rate (as suggested by Salo, 1960) seems to be highly questionable and demands further investigation. However, the existence of compensatory tachycardia has been convincingly demonstrated in *D. magna* under conditions of hypoxia (Paul et al., 1997; Pirow and Buchen, 2004) and a lack of food (Pirow and Buchen, 2004). The heart rate is below 300 beats/min at 20 kPa O_2 and increases to c. 400 beats/min at 3–7 kPa O_2.

Skadovskiy (1955) thought that the heart rate in daphnias may serve as an index of metabolic intensity. It is therefore tempting to apply the heart rate as an index of metabolic intensity, assuming that they are quantitatively proportional (as was thought also, e.g. by Meijering, 1975). Indeed, the heart rate increases with increasing temperature (Fig. 6.6). However, its relationship with the metabolic rate is certainly not direct (see also Maynard, 1960, p. 206) and, in view of the emotional reactions, and the occasional retardation and arrest of the heart in active animals, it can hardly be used for estimating metabolic intensity. Our observations convince us that the heart rate is a stable parameter, which is mainly changed by emotional factors, and tends to maintain a constant average until death.

Despite its variability, the heart rate of *Daphnia* has also been suggested as a measure for the estimation of pollution by means of a recording device (Kiknadze et al., 1983).

6.4 HEART REGULATION

Fischel (1908) described the heart nerve in *Daphnia* with its ganglia; according to Ermakov (1937), the heart of *D. magna* is innervated by nerves that accelerate or retard its activity. Obreshkove (1942) also reported that the movements in the heart and intestine of *Daphnia* are accelerated or inhibited through nervous stimuli. However, the heart is myogenic and acetylcholine is the inhibitory transmitter substance (Postmes et al., 1989); further, an extract of *Daphnia* has been reported to contain acetylcholine-like substances (Artemov and Mitropolitanskaya, 1938). Bekker and Krijsman (1951) noted the existence "of a myogenic pacemaker in the heart of *Daphnia*, inhibited by extracardiac cholinergic nerves," in the absence of a neurogenic pacemaker. This view is supported by the experimental evidence described below. The influence of various substances on the heart rate has been studied in either intact cladocerans or specially prepared specimens.

For standardizing the testing of chemical effects on the heart of *D. pulex*, Lévy (1927a) placed the animal in Ringer solution and made two incisions in the dorsal integuments: one in front of the heart and another behind the heart. In most cases, the heart rate stabilized after about half an hour. Excess potassium ions in the solution resulted in a rapid intensification of the heart rate, whereas excess calcium ions inhibited the heart (Lévy, 1927b).

Other observations have been made on intact specimens. The regulatory influence of various substances on the heart rate had been investigated by the end of the nineteenth century. According to Pickering (1894), atropine sulfate (3 mg/mL in water) increases the heart rate of *Daphnia*, and diastolic stoppage occurs after about half an hour; caffeine also increases the heart rate, and large doses culminate in systolic stoppage; muscarine nitrate, veratrine, theobromine, and xanthine do not greatly influence the heart rhythm.

Solutions of pilocarpine (1:100 w/v) and ergotamine (1: 10,000) both retard the heart rate: the action of pilocarpine ends in arrhythmia and diastolic arrest, that of atropine—in arrhythmia and systolic arrest (Ermakov, 1937).

Atropine retards the heart rate at a concentration 1:50 (w/v) according to Ermakov (1936, 1937), but accelerates it at 2.5:10,000–6.25:100,000 (w/v), according to Bekker and Krijsman (1951). Obviously, the difference in these reactions depends on the concentration of the tested substance.

In *D. magna*, parasympatheticomimetics (e.g. mecholyl) do not slow the heart, whereas cardiac tonic drugs (e.g. digitalis) cause slowing of the heart and its dilatation (Sollman and Webb, 1941). According to these authors, the *Daphnia* reactions differ qualitatively in many particular ways from those of vertebrates. Baylor (1942) observed an inhibitory effect of atropine, acetylcholine, $NaHSO_3$, and KCl on the heart rate of *D. magna*, whereas adrenaline had an accelerating effect. Baylor (1942) noted a similar regulation by acetylcholine and potassium (inhibition) in the *Daphnia* heart to that of vertebrates.

The heart rate of *Daphnia* is also accelerated by thyroidine (Hykes, 1926), and by adrenaline (a 1:100,000 solution) (Hykes, 1926; Suomalainen, 1939; Ermakov, 1936, 1937). However, Bekker and Krijsman (1951) found adrenaline to accelerate the heart at higher concentrations (2:10,000) but to retard the heart rate at lower concentrations (2:1000,000).

Flückiger (1952) found that the *Daphnia* heart rate is accelerated by L-adrenaline bitartrate, D-adrenaline bitartrate, L-noradrenaline bitartrate, L-ephedrine hydrochloride, dihydroergotamine methansulfonate, and oxyphenyl ethanomethylamine tartrate. However, L-adrenaline, D-adrenaline, and L-noradrenaline retard the heart rate during the first 5–10 min. According to Beim et al. (1970), the *D. magna* heart rate is retarded ("with some exceptions") by adrenaline, aminasin, aprophen, atropine, dihydroergotoxin, ephedrine, isadrin, noradrenaline, pilocarpine, pituitrin, and strophanthin. Contradictions may be caused by differences in the concentration tested. The heart rate of *Daphnia* has also been shown to be retarded by choline (Suomalainen (1939), acetylcholine, tetraethyl pyrophosphatase, digitalin, rotenone (Bekker and Krijsman, 1951), caffeic acid, propranolol (Campbell et al., 2004), ergotamine (Ermakov, 1936, 1937), pilocarpine (Bekker and Krijsman, 1951; Ermakov, 1936, 1937; Bekker and Krijsman, 1951), pituitrin (Hykes, 1926), and an extract of the thymus gland of mammals (Hykes, 1926). Phenobarbital strongly decreases the heart rate of *D. pulex* to about 25% at a concentration as low as 2.9 mg/L (Postmeret al., 1974). Lactose at 50–200 mM also inhibits the heart rate and causes severe arrhythmia in *D. pulex* (Campbell et al., 2004). It is thought that lactose directly affects ion channels in the heart, an effect that is reversible within 3–4 h.

Eserine (physostigmine) produces a toxic effect on the heart of *D. magna* (Baylor, 1942).

The heart rate of *D. pulex* increases in the presence of serum from mammals, frog (*Rana*), and carp, and then stops in systole (Vatovec and Timet, 1952). Such stimulation is irrespective of the sex of the donor animal (Vatovec and Timet, 1955).

Postmes et al. (1989) studied the effect on *D. magna* of adrenoceptor agonists (1-epinephrine-bitartrate, 1-norepinephrine-bitartrate, 1-phenylephrine HCl, phenylterol HCl) and adrenoceptor antagonists (DL-metoprolol tartrate and DL-propranolol-HCl). It was found that epinephrine could not be blocked by propranolol (an adrenoceptor antagonist), thus suggesting that the drug action is not meditated by adrenoceptors. Postmes et al. (1989) found epinephrine to inhibit the heart rate, contrary to previous reports by Sollman and Webb (1941), Baylor (1942), Bekker and Krijsman (1951).

The consistent experimental results on the efficiency of cholinolytics, obtained using *D. magna* and rats by Tonkopiy, et al. (1994b), led these authors to the conclusion that *Daphnia* possess M-cholinoreceptors of a structure similar to those in mammals: a case of parallelism between remote groups. The cholinolytics tested (amedin, amizil, artropine, cyclodol, glycine, and spasmolytin) provide a protective action following intoxication of *D. magna* by arecoline and armine (Tonkopiy, et al., 1994a).

Dzialowski et al. (2006) investigated the effect of beta-adrenergic receptor antagonists on *D. magna*. The lowest observed effective concentration for decreasing the heart rate was 55 μg/L for propranolol and 3.1 mg/L for metoprolol.

Dopamine in a culture solution was found to produce only a slight effect on the *Daphnia* heart rate: from c. 66 beats per 10 sec in the control, it increased to c. 70 at a dopamine concentration of c. 1 mM (Peñalva Arana et al., 2007).

Results obtained by Villegas-Navarro et al. (2003) suggest the presence of Na^+-ATPase receptors to verapamil and of adrenergic receptors in *D. magna* heart.

The heart rate of *D. magna* is decreased by diphenhydramine (DPHM) and increased in the presence of curcumin (Vaidya et al., 2009). These authors suggest the following explanations: DPHM may prevent the sympathetic action of histamine, and parasympathetic acetylcholine (Ach) may bind (ACh-R) onto myocardial cells and reduce the heart rate. Curcumin may antagonize histamine *N*-methyl transferase and thus prevent histamine methylation. Vaidya et al. (2009) think that histamine may act as a primary cardiac sympathetic neurotransmitter.

An estrogenic substance (ethynylestradiol) causes a significant decrease (to 1/3 of the normal rate) in the heart rate of *Daphnia* (Walker et al., 1998).

Tasse and Camougis (1965) were the first to investigate the cardiac activity of *Daphnia* electrographically. They used three methods:

1. The application of remote electrodes revealed a correlation between muscle action potential and body movements (Fig. 6.7).

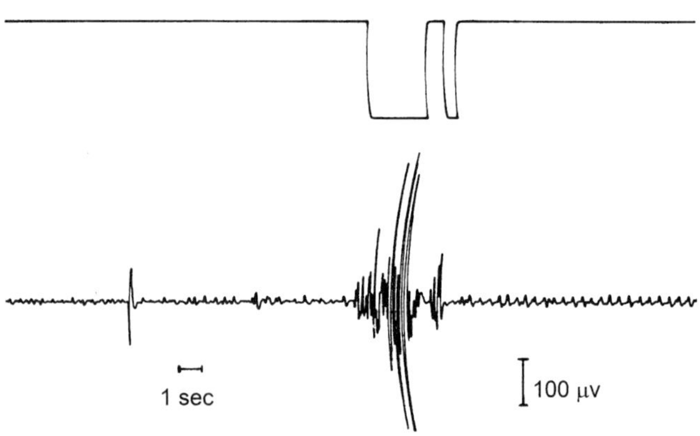

FIGURE 6.7 Electric activity of *Daphnia* recorded with remote electrodes (lower trace) correlates with the observed body movements (upper trace). *Source: Tasse and Camougis (1965).*

2. Metallic electrodes placed on the ventral abdominal region revealed slow, biphasic waves on which faster, smaller waves were superimposed. The magnitude of the biphasic waves varied from 200 to 500 μV, and their frequency from 1.5 to 4 Hz. There were periods of no electrical activity.
3. Glass capillary electrodes punctured the carapace and were placed in the dorsal region of the heart. This revealed variations in the amount of electrical activity, ranging from a slow, biphasic wave to a fast monophasic and larger biphasic wave (Fig. 6.8). Fundamental frequencies ranged from 3 to 9 Hz, averaging at 5.4 Hz, with amplitudes from 100 to 150 μV. Handling of the *Daphnia* sometimes led to cardiac arrest.

In a $1/10^6$ dilution of γ-aminobutyric acid (or GABA), the frequency of electrograms decreased Camougis, 1965).

6.5 HEART ARREST

The heart rate of Cladocera is stable and drops immediate prior to death. Despite this, the heart may stop for rather a long time (from minutes up to an hour) in response to provocation without any detrimental consequences.

For reasons that are unclear, regular activity of the heart is sometimes interrupted for long intervals in *Eurycercus*, *Camptocercus*, *Bosmina*, *Ilyocryptus*, and *Lathonura* (Smirnov, 1965b). In

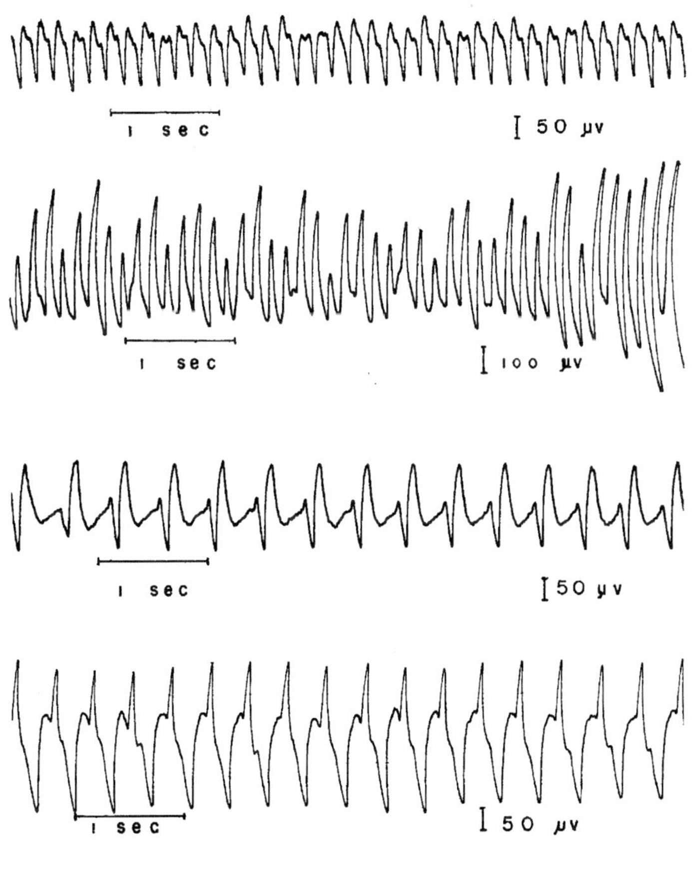

FIGURE 6.8 Electric activity of the *Daphnia* heart recorded with capillary microelectrodes in different specimens. *Source: Tasse and Camougis, 1965.*

L. rectirostris, after a prolonged interval the heart resumed its activity, without any obvious reason; the discontinuation of heart beats had no negative effect on the animal whatsoever.

In copulating *C. sphaericus*, the heart rate either remains normal or decreases markedly in both males and females; in females, the heart sometimes completely stops.

It was observed that if the bend of the digestive tube of *Daphnia* is touched with a fine glass needle, then the heart stops immediately (in systolic phase) and the posterior region of the intestine exhibits intensive peristalsis (Obreshkove, 1942). The period of heart arrest can reach 20 min. After a while, both the heart and intestine return to their normal activities. Heart arrest is also caused by touching the dorsal surface at the posterior heart area. It is thought that nerve endings convey the inhibitory impulses to the heart. After its arrest, the heart resumes beating, but for several minutes the beats are feeble and irregular. Addition of 0.01–1 μg/L acetylcholine results in immediate normalization of the heart beats. In *Daphnia* treated with 10 μg/L physostigmin prior to heart arrest, the recovery from inhibition is rapid and complete (Obreshkove, 1942). The general impression is that Cladocera manage well without their heart and do not need it much.

Events related to the mechanism of heart arrest in Cladocera deserve further investigation.

6.6 ADHESION OF BLOOD CELLS

Blood cells are adhesive and sometimes stay on the surface of organs. This fact may be observed especially if a local irritation, injury, or induction shock, is inflicted—to such an extent that none may remain in circulation (Hardy, 1892). In *Acroperus*, all hemocytes sometimes become immovable (Smirnov, 1971). Maynard (1960) indicated that in *Daphnia* blood cells adhere to tissues in the event of irritation; for example, in *Leptodora* this happens when it sticks to the surface film and the cells become freely circulating again when the *Leptodora* is liberated. In *Leptodora*, blood cell adhesion also occurs during ether narcotization (Saalfeld, 1936).

Hardy (1892) believed that blood cells that absorb fat may attach to the inner substrata and thus participate in the formation of the fat tissue. There has been no explanation of this observation in Cladocera and a mechanism for understanding this process has yet been outlined.

6.7 PHAGOCYTOSIS

When consumed by *D. magna*, *Metschnikowiella bicuspidata* (syn. *Monospora bicuspidata*, Ascomycetes) spores are liberated from their coat and penetrate the body cavity through the wall of the intestine. However, Metchnikoff (1884) observed that *D. magna* leukocytes can surround and destroy these spores by phagocytosis (Fig. 6.3); this process allows some individuals of *Daphnia* to survive. However, leukocytes do not respond to *Saprolegnia* penetrating the body of *D. magna*, and to some other bacterial parasites. Metchnikoff initially described these facts in detail in 1884, and then reported them in 1885 and in 1892 in "Lectures on comparative pathology of inflammation." He also observed (Metchnikoff 1888) phagocytosis of spores of the bacterium *Pasteuria ramosa* infesting *Daphnia*. Metchnikoff (1885) had also earlier noted *Daphnia* leucocytes surrounding wounds.

Hardy (1892) observed that fat globules and other particles are carried from the gut lumen through the gut wall by blood cells, and that blood cells may attach to inner substrata, thus contributing to the formation of the fat tissue.

Phagocytosis was also observed by Metchnikoff in some other invertebrates and warm-blooded animals, and he developed these observations into a theory of general biological and medical importance. Despite the convenience of transparent cladocerans for these observations, it seems that no one continued with these observations on daphnids or their relatives. For a long time, there was therefore no direct information on phagocytosis in Cladocera. Moreover, Vonk (1960) and Quaglia et al. (1976) did not observe phagocytosis in the midgut of crustaceans and Quaglia et al. (1976) assumed that the food is liquefied within the gut and absorbed by the epithelium.

A closer examination of *D. magna* revealed the presence of two cell types circulating within the hemolymph (Auld et al., 2010): amoeboid cells (about 9 μm long, and identified as the cells initially described by Metchnikoff) can phagocytosis. They attack *Pasteuria*, and there is a large increase in the number of amoeboid cells in parasitized *Daphnia*.

6.8 IMPACT OF XENOBIOTICS ON HEART RATE

The available data indicate that xenobiotics generally depress the heart rate of Cladocera spp. In *D. magna*, the addition of $CdCl_2$ is followed by a noticeable decrease in the heart rate and the beating rate of thoracic limbs (Smirnova, 1960); 500 mg/L Dikonirt (a herbicide containing 2,4-D) caused reduction in the heart rate from 5 beats/sec to 1 beat/sec after 6 h of exposure (Présig and Véro, 1983). In contrast, hydroquinone stimulates heart beats (Kiknadze et al., 1983).

As little as 0.1 mg/L phenol depresses the heart rate of *D. magna* from 368 beats/min (control) to 254 beats/min (Kolupaev, 1988). A 0.1 × lethal concentration, 50% (LC_{50}) of the pyrethroid pesticide, phosphatak, was shown by Saprykina (1996) to depress the heart rate in *Daphnia* (in the fourth, fifth, and sixth generations), stimulate the beating of thoracic limbs, and considerably stimulate oxygen consumption in the first, second, and fourth generations—by 39%, 49%, and 48%, respectively.

At a comparatively high concentration of $CuSO_4$, the thermoresistance of the cardiac muscle of *D. magna* is significantly decreased (Pashkova et al., 1998).

It may also be added that an alternating magnetic field with a frequency similar to that of *D. magna* heart beats strongly intensified the heart rate (Usanov et al., 2001b). In an alternating low-intensity magnetic field, the heart rate of *D. magna* decreased; at low concentrations of phenol, inhibition of the *D. magna* heart rate by a magnetic field is higher than that by this chemical factor (Usanov et al., 2001b, 2003).

CHAPTER

7

Excretion

7.1 ANATOMICAL BACKGROUND

Although Cladocera possess a functioning organ of secretion, Peters (1987, p. 219) believes that excretion occurs mostly through the body surface in Cladocera: "the soluble excreta of *Daphnia* are released through the general body surface." The organ of excretion consists of the paired maxillary glands (shell glands) comprising the nephridium and its convoluted efferent ducts (Figs. 1.5 and 7.1). The maxillary gland is situated in the hemocoel, and is therefore exposed to the blood, under the anterior area of the carapace, and opens to the outside in the anterior part of the body (Claus, 1875a). Judging by the detailed drawings of Claus, the opening to the outside is situated within the anterior part of the brood pouch. A detailed scheme of the structure of the nephridium is given by Gicklhorn (1931a) (Fig. 1.5).

Hérouard (1905) described the various stages of increasing complexity of the maxillary gland that occurs with age and stated that it is relatively simple in *Eurycercus*. He also noted that the carapace adductor muscles are attached at the inner surface of its convolutions. The lengths and arrangements of the convolutions of this gland are somewhat different in different representatives of particular families (Claus, 1875a, pl. XI; Hérouard, 1905). For example, the convolutions are numerous in *Daphnia* and *Sida*, whereas the gland is shorter in *Ceriodaphnia* and *Moina*. It is shortest in *Macrothrix* and *Acroperus*, but all of these have a convoluted efferent duct.

The maxillary gland of *Penilia* (a marine species) is very different to those of other cladocerans (Leder, 1915): its duct is short, has no convolutions, and is dilated in the distal part, producing a kind of urinary bladder (Fig. 7.2).

The nephridium of the maxillary gland and the antennal gland (the latter has no channel and is probably nonfunctional) are all that remain of the coelom in Cladocera. Although the transparent maxillary gland is discernible without any treatment, coelomic sacs may be made clearly visible by means of intravital staining (Fischel, 1908; Dejdar, 1930; Gicklhorn, 1931a, 1931c) with neutral red, methylene blue, Nile blue sulfate, or Bismarck brown.

7.2 THE PROCESS OF EXCRETION

The final products of metabolism are generally low molecular weight compounds of nitrogen, carbon dioxide (CO_2), and water. Like other Crustacea, Cladocera are mostly

Physiology of the Cladocera.
DOI: http://dx.doi.org/10.1016/B978-0-12-396953-8.00007-0

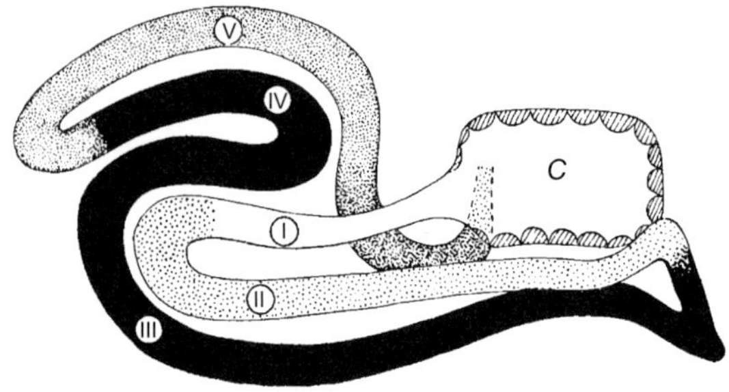

FIGURE 7.1 **Nephridium of *Daphnia*.** C, coelomic sac; I-V, its ducts; variously colored by intravital staining. *Source: Gicklhorn (1931a).*

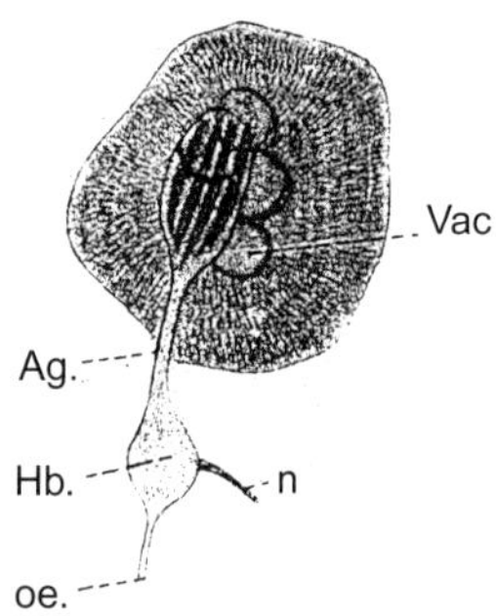

FIGURE 7.2 **Nephridium of *Penilia*.** Ag, efferent duct; Hb, urinary bladder; n, nerve; oe, opening; Vac, vacuoles with liquid. *Source: Leder (1915)*

ammonotelic animals, excreting mainly ammonia (Vonk, 1960); at least, this is what occurs under aerobic conditions. What little information is available on Cladocera excretion relates mainly to daphnids, and occasionally to *Bosmina* spp.

Aladin and Plotnikov (1985) carried out an outstanding and unique investigation on the concentration, estimated by the temperature depression point, of the liquid excreted from the maxillary gland (i.e. urine) of *Daphnia magna*, in comparison with the ambient water and the hemolymph. This is probably the only case in which a reduction in freezing temperature has been determined for these three media: the hemolymph, urine, and external water. Microcryoscopic techniques were applied: the reduction in freezing point was −0.34°C for the liquid in the coelomic sac of the maxillary gland, −0.05°C for the convoluted duct, and −0.01°C for the external water. These data indicate that the urine excreted by *Daphnia* is hypoosmotic to hemolymph and slightly hyperosmotic to the external water. Thus, in fresh water, the urine of *D. magna* is isotonic with the hemolymph but hypertonic to the ambient water. The liquid in the excretory canal is hypotonic to both urine and hemolymph; this was interpreted as providing evidence of the reabsorption of salts from the urine and of water being excreted. In the case of *D. magna* acclimated to water with a salinity of 7‰ (i.e. 7 parts per 1000), the hemolymph, urine, and liquid in the excretory canal of the *D. magna* are isotonic to the water.

7.2.1 Nitrogen Compounds

In Cladocera, the final products of nitrogen metabolism are principally ammonia derived from protein metabolism, along with a smaller percentage of urea (Fig. 7.3) (Parry, 1960; Wiltshire and Lampert, 1999).

Although I failed in my attempt to stain the coelomic sac of the maxillary gland with freshly prepared Nessler's reagent in order to demonstrate the presence of ammonia, further attempts would be worth making.

There are few quantitative estimates of the liberation of excretion products by Cladocera. According to Schmidt (1968), adult *D. magna* fed on green algae excrete 0.17–0.19 μg N/24 h/μg body N. The excretion of ammonia nitrogen by *D. pulex* fed on the green alga *Chamydomonas* was found to be 0.20 μg N/individual (ind.)/24 h or 5.11 μg N/mg dry weight (DW)/24 h (Jacobsen and Comita, 1976). The mean, steady-state release of ammonia by *D. magna* was determined to be 11 nmol/mg DW/h by Gardner and Scavia (1981). The mean release of nitrogenous

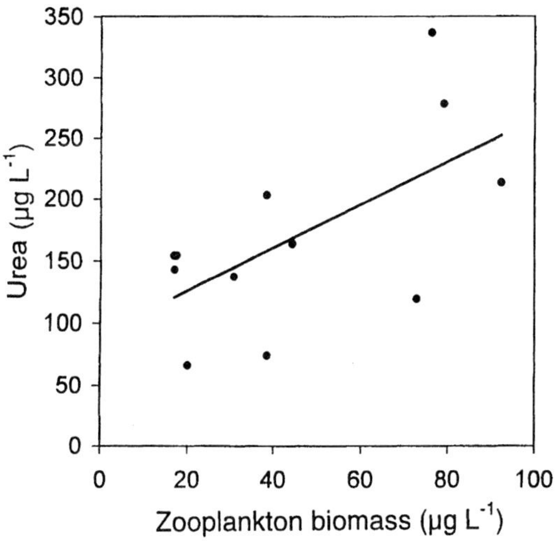

FIGURE 7.3 **Urea concentration in lake water (Lake Schösee, Germany) in relation to crustacean zooplankton biomass.** *Source*: *Wiltshire and Lampert (1999)*

products by well-fed *Daphnia* was estimated to be 0.76 µg/mg/h ammonia and 0.36 µg/mg/h (Wiltshire and Lampert, 1999). In contrast, starved *Daphnia* liberate 0.45 µg/mg/h ammonia and 0.06–0.1 µg/mg/h urea. A total of 6.5 min after feeding was discontinued, *D. magna* release ammonia nitrogen at a rate of 80 nmol/mg/h and phosphorus at 4.5 nmol/mg/h P (i.e. soluble reactive P) (Scavia and Garner, 1982).

In other Cladocera (i.e. *Bosmina*, *Ceriodaphnia*, *Daphnia*, and *Scapholeberis*), the excretion rate was determined by Ejsmont-Karabin (1984) to be 0.8–2.6 µg/mg DW/h N-NH_4 and 0.26–0.58 µg/mg DW/h phosphorus. In *Daphnia*, the excretion rate (as summarized by Peters, 1987) is c. 1 µg DW/h for nitrogen and c. 0.5 µg DW/h for phosphorus. In addition, the rate of release of ammonia nitrogen by *Ceriodaphnia reticulata* has also been shown to depend on temperature: 1.0 mg/g wet weight (WW)/24 h at 15°C, 1.9 mg/g WW/24 h at 22°C, and 2.4 mg/g WW/24 h at 27°C (Gophen, 1976).

The products of hemoglobin (Hb) decomposition are excreted by Cladocera through their maxillary glands (i.e. shell glands) (Fox, 1948) and iron appears in the shell glands of *Daphnia* that are losing Hb (Smaridge, 1954). The site of Hb breakdown is the fat body.

Amino acids are also excreted to the external environment, as shown with reference to *D. magna* by Gardner and Miller (1981). The estimated quantity is c. 0.13 nmol amino acids/mg DW/h and c. 21 nmol amino acids/mg DW/h. They also release ammonia, urea, and phosphorus compounds. In the presence of *Daphnia galeata* fed on green algae, the concentration of dissolved free amino acids in the environment increases from 1 µM by up to 3 µM over 2 h (Riemann et al., 1986). Recalculated for carbon, the release rate was c. 0.2 µg C/ind./h, i.e. 12% of that ingested.

7.2.2 Carbohydrates

In addition to the release of carbon dioxide, it has been shown that excess ingested carbon is excreted by *D. magna* as dissolved organic carbon (Darchambeau et al., 2003). In the presence of *D. magna*, the surrounding water is enriched with dissolved organic carbon, mostly by excretion following food digestion; much lower amounts are liberated from feces (He and Wang, 2006a, 2006b).

7.2.3 Phosphorus Compounds

Inorganic phosphorus (PO_4-P) is released into the ambient water by *D. magna* at a rate of 8.4 ng/ind./h (Rigler, 1961a); phosphatase is also released. The release rate of phosphorus by *Daphnia* spp. is variously estimated to be 0.91 µg P/mg DW/h (Peters and Rigler, 1973), 1.1–1.5 µg P/mg DW/h (Olsen and Østgaard, 1985), or 0.05–1.5 µg P/mg DW/h (as summarized by Yuan Hua Wen, 1994 from data from various authors obtained using radiotracer or chemical methods).

Adult *D. galeata* release 0.06–0.96 µg P/mg C/h into the surrounding water (Vadstein et al., 1995). The excretion rate of phosphorus for *D. galeata* was estimated by Pérez-Martínez et al. (1995) to be 6.3% of the total phosphorus/h for adults and 16.2% for juveniles. The specific P release rate in *D. pulex* decreases as the P:C ratio of its food decreases; there is no further release of phosphorus to the environment below a P:C ratio of 6–8 μg P/mg C (Olsen et al., 1986).

Pérez-Martínez et al. (1995) also obtained quantitative data on phosphorus metabolic rates in their three compartment model: the gut, metabolic pool, and structural pool.

Scavia and McFarland (1982) measured phosphorus release during the molting cycle of individual specimens of *D. magna* in a specially designed incubation flow cell. The rate of phosphorus release was 6.7 times higher at and after ecdysis than at other phases of the life-cycle.

Phosphorus excretion rates by natural populations in the Bay of Quinte (Lake Ontario) were estimated by Peters (1975) to be 2.2–4.8 mg P/m^2/day for *Daphnia*, 0.6–1.2 mg P/m^2/day for *Diaphanosoma*, and 0.6–1.3 mg P/m^2/day for *Leptodora*. Phosphorus release by Cladocera has seasonal fluctuations: in Swarzędzkie Lake (Poland), a peak of 10–15 μg P/L/day was reached in June (Kowalczewska-Madura et al., 2007).

7.2.4 Water

As noted above, Cladocera drink water and surplus water is rapidly excreted. According to Peters (1987), about 8% of the body's water is removed by *Daphnia* every hour.

7.3 BIOACCUMULATION OF TOXIC SUBSTANCES

It has been shown that *Daphnia* accumulate various substances that may be useful, alien, or harmful to their normal life, and some substances accumulated by Cladocera are unaffected by their metabolism.

7.3.1 Accumulation of Inorganic Substances

If not killed by their high concentrations, *D. magna* may accumulate xenobiotics, e.g. copper (Cu) or lead (Pb) (Holm-Jensen, 1948). When fed with green algae and *Euglena*, *Daphnia* selectively accumulated higher concentrations of sodium (Na), calcium (Ca), scandium (Sc), lanthanum (La), neodymium (Nd), zirconium (Zr), chlorine (Cl), bromine (Br), and nickel (Ni) than were present in *Euglena*, whereas the latter contained higher concentrations of various other elements [including mercury (Hg) and arsenic (As)] (Cowgill and Burns, 1975). The accumulation of such substances is represent by their bioconcentration factor (i.e. accumulation coefficient), which is the ratio of the concentration of a pollutant within the body to its concentration in the environment. The bioconcentration factor of *D. magna* for ^{54}Mn (manganese-54) was 65, reached after 8 h of exposure to the isotope solution, with excretion having started within 2 h of exposure (Kwassnik et al., 1978).

There are numerous studies on the accumulation of various metals and compounds, mostly restricted to daphnids. According to Yu and Wang (2002), in *D. magna* the assimilation efficiency from algal food is 30–77% for cadmium (Cd), up to 44% for chromium (Cr), 24–58% for selenium (Se), and 7–66% for zinc (Zn). Routes for metal removal into the dissolved phase were different for different metals: excretion was the most important route, molting represented 50–70% of daily Cd efflux and 20–70% of Zn; and the major routes of Cr efflux were via excretion and feces egestion. Substantial amounts of Se were released via the production of offspring. The release of metals has an important impact on the biogeochemical cycling in lakes. The following data have been obtained for particular elements.

Arsenic

Arsenic from food algae is accumulated by *Moina* predominantly as nonmethylated arsenic and dimethyl arsenic compounds (Maeda et al., 1992). Folt (2005) found that *Daphnia* accumulate As and Hg, with levels accumulated by *Daphnia* being 2–3 times higher than those in copepods.

Cadmium

The Cd content in *Holopedium* collected from natural lakes in Canada ranged from 0.9 to 31 μg/g (Yan et al., 1990). Bodar et al. (1990a) determined that *D. magna* can accumulate c. 115 ng Cd/mg DW/24 h and can develop resistance to Cd in a single generation. In addition, lethal concentrations of Cd differ for different clones of *D. magna*.

Upon exposure to Cd, *D. magna* respond by reducing their Cd assimilation efficiency and ingestion; in addition, metallothioneins are formed in its body during detoxification (Guan and Wang, 2004a). *D. magna* juveniles obtain twice as much Cd from water than from algae (*Chlorella*) (Barata et al., 2002). The gut ceca are the main sites of Cd and Ca intake.

D. magna sensitivity to Cd increases following irradiation by a millimeter-range low-intensity electromagnetic field (Gapochka et al., 2012).

D. magna may accumulate Cd over several generations and recover its biological parameters in Cd-free water over one or two generations (Guan and Wang, 2006); therefore, they have a high potential for recovery following Cd contamination.

Cesium

In a solution of ^{137}Cs (cesium-137), *D. magna* accumulate radioactive Cs: the accumulation coefficient reaches 84 at day 5, but decreases at molting and with the liberation of juveniles (Nilov, 1983).

Mercury

More Hg accumulates in *D. magna* from the *Chlorella* it consumes in the form of methylmercury CH_3HgCl than as mercuric chloride $HgCl_2$ (Boudou and Ribeyre, 1981, 1983). After 48 h exposure to methylmercury (in food), the Hg concentration in *D. magna* is c. 440 ng/g. The bioconcentration factor of $HgCl_2$ by *Ceriodaphnia affinis* reaches 2000 in the twentieth generation (7.2 μg/g DW) (Gremiachikh and Tomilina, 2010).

Manganese

The Mn concentration factor of *D. magna* is 65 and maximum uptake is reached after 8 h of exposure despite active excretion (Kwassnik et al., 1978).

Neptunium

Neptunium (Np) is concentrated by *D. magna* to 40 μg/g in 48 h when present in the medium at a concentration of 1.23 μg/mL (Poston et al., 1990).

Nickel

D. magna exposed to Ni in solution accumulate it in the F_1 generation to a concentration exceeding 18 times that of untreated controls; at a Ni concentration of up to 42 μg/L, the glycogen content in the body is noticeably decreased (Pane et al., 2004). When fed with Ni-contaminated algae, *D. magna* can accumulate up to 72 μg Ni/g DW (Evens et al., 2009). Both Ca and Mg reduce Ni toxicity, but Na has no effect (Deleebeeck et al., 2008). According to Vandenbrouck et al. (2009, p. 18), in *D. magna* Ni affects functional gene classes that are "involved in different metabolic processes (mainly protein and chitin related processes), cuticular turnover, transport and signal transduction."

Silver

Silver (Ag) is known to be highly toxic for cladocerans (Rodgers et al., 1997), although the presence of organic matter alleviates this effect (Glover et al., 2005b). Ag is incorporated into *D. magna*, and not merely adsorbed onto its surface (Bianchini et al., 2002a, Bianchini and Wood, 2003a, 2003b). Ag accumulation by neonates is higher in the presence of sulfide. Acute Ag toxicity in *D. magna* is proportional to the whole-body uptake of Na^+ (Bianchini et al., 2002b). Bianchini and Wood (2003a) determined that Ag causes ionoregulatory disturbance resulting in decreased levels of whole-body Na concentration due to the inhibition of Na^+ uptake; in contrast, whole-body Cl^- concentration is unaffected. Exposure to $AgNO_3$ also leads to the inhibition of Na^+, K^+-dependent adenosine triphosphatase (N^+, K^+-ATPase), as a direct result of Ag accumulation in the body. Inhibition of Na^+ uptake by chronic Ag exposure is explained by the inhibition of Na^+ channels at the apical "gill" membrane (Bianchini and Wood, 2003b).

Diet-borne Ag or Cu (contained in algae grown in metal-contaminated media) contributes no more to *Ceriodaphnia dubia* survival or reproduction than do the equivalent waterborne metals (Kolts et al., 2009). In a solution of $AgNO_3$, *D. magna* accumulate Ag in their gills and digestive tract, but the accumulation is twice as high in the presence of sulfide (Bianchini et al., 2005). The latter fact was explained by the presence of sulfide-bound silver in the digestive tract. Transfer to clean water for 5 h leads to a significant decrease in Ag concentration in all body compartments. In a solution of $AgNO_3$, *D. magna* can accumulate c. 11 ng Ag/mg WW in 24 h (Glover et al., 2005b).

Zinc

In laboratory cultures, in the author's experience, the green alga *Scenedesmus* flourishes in Zn-coated metal trays in a mineral culture medium. However, *Daphnia* perish when fed with these algae, which indicates high levels of Zn accumulation by the algae. Nevertheless, toxicity may depend on Zn concentration. For example, the green alga *Pseudokirchneriella* was grown at different Zn levels and accumulated different Zn concentrations suitable for *D. magna* feeding (Canli, 2005). When Zn concentrations in the algae increased from 100 to 220 ng/mg DW, the protein content of *D. magna* increased from 220 to 300 μg/mg DW.

For *D. magna*, the optimum range of Zn is 300–600 μg/L; at these concentrations, *Daphnia* contain 212–254 μg Zn/g DW (Muyssen and Janssen, 2002). Decreased or increased Zn levels within the body occur within 1 day. At 600 μg Zn/L, Zn concentrations in the body fluctuate over 2–3-day intervals, suggesting a role for molting in the elimination of Zn.

After 24 h of exposure of *Daphnia* to ^{65}Zn (zinc-65), the bioconcentration factor was 1110; after 30 days in a solution of 250 μg Zn/L, *D. magna* accumulated c. 800 μg/g DW (from both solution and particles containing Zn) (Nilov, 1980). Zn bioaccumulation is enhanced by preexposure to Cd (Guan and Wang, 2004a, 2004b).

Increased humic acid chelation of Cu, Cd, and Zn decreases their bioavailability (Winner and Gauss, 1986).

Radionuclides such as ^{110}Ag, ^{60}Co, ^{137}Cs, and ^{54}Mn ingested with food are accumulated in *D. magna* (Adam et al., 2002).

Table 3.11 shows that bioaccumulation rates by Cladocera may vary widely.

7.3.2 Accumulation of Organic Substances

Benzo(a)pyrene

D. magna mainly takes up benzo(a)pyrene (a polynuclear aromatic hydrocarbon) from solution; the presence of particles containing this absorbed chemical decreases its accumulation (McCarthy, 1983). If fed with *Chlorella* contaminated with hexachlorobenzene (HCB), *D. magna* can accumulate 1.7 μg HCB/kg DW over 6 days (Muños et al., 1996).

Bioaccumulation of benzo(a)pyrene by *D. magna* is reduced by humic substances (McCarthy, 1983; Gourlay et al., 2003); the effect is highest at pH 6.5 (Kukkonen, 1991). In addition, the accumulation of dehydroacetic acid from humic water by *D. magna* is lower than that from a nonhumic water control at a pH higher than 6; in contrast, at pH 5.5 and lower, the effect is reversed (Kukkonen, 1991). Fluoranthene and pyrene bioaccumulation by *D. magna* is reduced by the presence of humic acids in the medium (Gourlay et al., 2003).

Dichlorodiphenyltrichloroethane

Dichlorodiphenyltrichloroethane (DDT) is accumulated by *D. magna* from ambient water principally through their integuments (Table 3.11) (Crosby and Tucker, 1971). The total concentration of DDT in *Daphnia* may reach over 4.2 g/kg; a large part of the DDT is found in the carapace and is removed during molting. The 100% lethal dose (or LD_{100}) was found to be 1100 ng/mL.

Fenvalerate

Fenvalerate causes a decreases rate of filtration of *Chlamydomonas* by *D. galeata* and *Ceriodaphnia lacustris* at sublethal concentrations (i.e. in *C. lacustris* that already contains 0.01 μg/L) (Day and Kaushik, 1987). Rates of algae assimilation also decrease, especially in fenvalerate concentrations >0.1 μg/L.

Lindane

According to Hansen (1980), *D. magna* accumulates lindane (a chlorinated hydrocarbon).

Polychlorinated Biphenyls

D. magna can accumulate c. 8 μg polychlorinated biphenyls (or PCBs)/g in 24 h from ingested algae and this concentration is maintained over the subsequent 72 h, i.e. no significant elimination occurs within 72 h (Joaquim-Justo and Thomé, 1998). Concentrations reached up to 6 ng/g malathion in *Simocephalus vetulus* after 8 h of exposure (Olvera-Hernández et al., 2004) and up to c. 0.180 mg/g tetradifon in *D. magna* after 8 h of exposure (Ferrando et al., 1996).

7.4 TRANSFORMATION OF XENOBIOTICS

Consumed toxic substances may be metabolized by Cladocera. Thus, As consumed by *Moina macrocopa* fed on algae containing Na_2HAsO_4 accumulates up to 111 μg As/g DW in the form of inorganic As (75%), mono-CH_3 (8%), and di-CH_3 (16.6%), and is then mostly excreted (Maeda et al., 1992).

An example of the transformation of an organic toxicant in the body of a cladoceran is the fate of heptachlor absorbed by *D. magna* (Fig. 7.4). Heptachlor is metabolized in the daphnid body to 1-hydrochlordene, 1-ketochelordene, and 1-hydroxy-2,3-epoxychlordene, as well as derivatives such as glucosides and sulfates. (Feroz et al., 1990).

Pyrene consumed by *D. magna* is transformed into water-soluble metabolites (Ikenaka et al., 2006).

7.5 THE ROUTES OF ELIMINATION OF XENOBIOTICS

Pollutants can penetrate the body of a cladoceran via its carapace or its gut along with ingested water and food. They then undergo biodegradation and, especially those that are only slightly biodegradable, are accumulated (to different extents in different tissues), and are gradually removed. The routes of elimination are excretion, defecation, molting, and

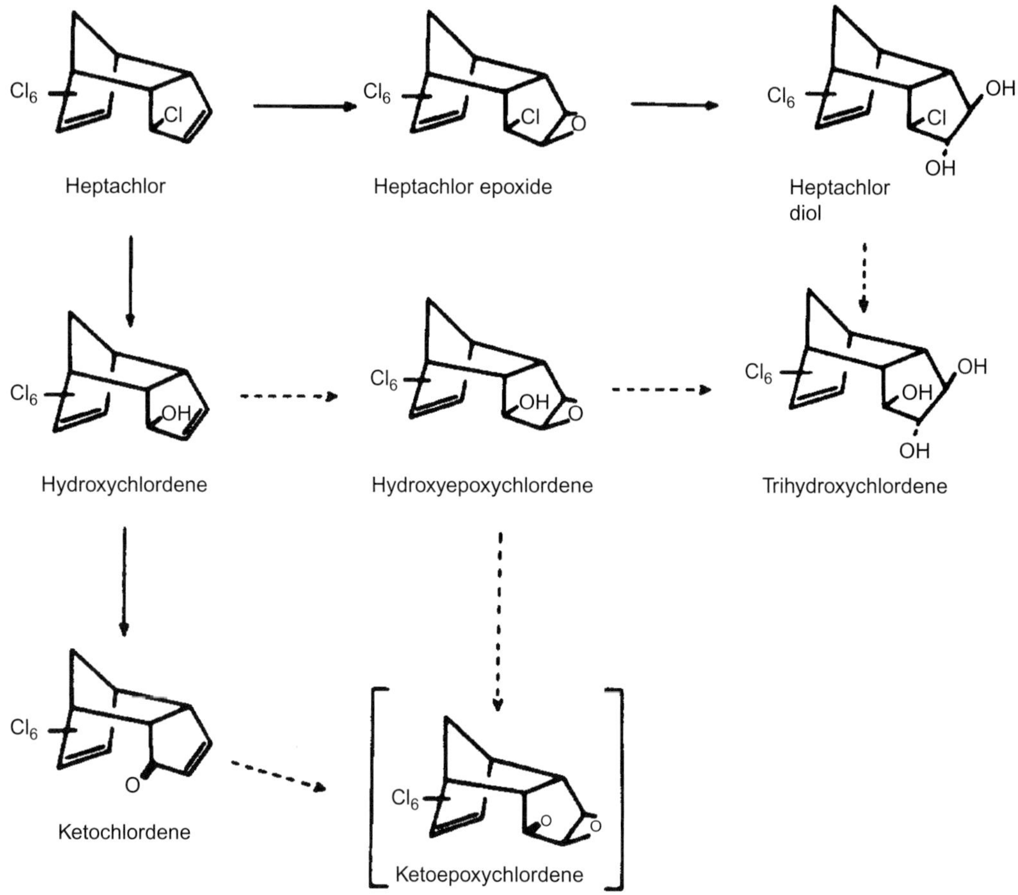

FIGURE 7.4 **Biotransformation of heptachlor within the *Daphnia* body.** *Source: Feroz et al. (1990)*

transfer to eggs. The actual accumulation is the difference between the intake level and the capacity for elimination via these routes.

According to Carney et al. (1986), Cd obtained from solution is concentrated in the exoskeleton of *D. magna*, and ecdysis frees the animal from a considerable part of the accumulated Cd. Similarly, ^{65}Zn partly accumulates in the exoskeleton and is then removed at molting (Winner and Gauss, 1986).

The elimination rate of metals via these routes is little affected by the concentration of Cd, Se, or Zn in the ingested food (Guan and Wang, 2004a). However, rapid elimination of Se and Zn might depend on the transfer of these metals from mother to offspring. The principal pathway of Cd and Zn elimination from the body of *D. magna* is excretion to the water, a secondary pathway is transfer to neonates, and the pathways used least are by molting and through feces (Guan and Wang (2004b).

Cu is principally accumulated by *D. magna* from its food; the Cu efflux rate is 0.20%/day at high food concentrations, and excretion accounts for 82–84% of the total Cu loss, while c. 6.6% is transferred to the offspring within 7 days (Zhao et al., 2009).

CHAPTER

8

Osmotic Regulation

8.1 POTENTIAL ANATOMICAL BACKGROUND

Ion exchange in Cladocera is thought to be carried out via the neck organ and the epipodites of the thoracic limbs. Possible routes of ions through the neck organ of cladocerans were illustrated by Potts and Durning (1980) (see Fig. 8.1). On the surface of the epipodite of *Daphnia*, which is reported to be slightly alkaline (pH = 8) (Lavrentjeva and Beim, 1978; Beim and Lavrentjeva, 1981; Beim et al., 1994), a special cell membrane, presumed to participate in active ion transport (Kikuchi, 1982), was identified on the side in contact with the external medium.

8.2 ENVIRONMENTAL BACKGROUND

Different species of Cladocera live in fresh water, brackish water, and saline inland waters, and a few live in seawater. A small number of polyphemid species and one ctenopod (*Penilia*) are oceanic.

Different species within a genus may tolerate different salinities (such as e.g. *Moina* sp.). Some species can endure variable salinity within rather limited ranges but the presence of salts in the external water is obviously necessary. *Chydorus ovalis* perished in triple-distilled water (Winberg, 1933), and ordinary distilled water caused high mortality levels in *Daphnia magna* within 24 h (Stobbart et al., 1977). The obvious reason for this is disturbance of their salt metabolism.

Total salinity (i.e. total dissolved solids; TDS) of fresh waters, where most Cladocera live, is within 100 200 mg/L (see e.g. Hutchinson, 1957), reaching c. 1000 mg/L in bodies of hard water (Kitaev, 2007). The salinity of some inland lakes may be c. 30‰ and may reach the level of saturated brine. As is well known, oceanic salinity is c. 32‰ and may be lower in enclosed seas (16‰ in the Black Sea—the conventional boundary between marine and brackish waters). It should be noted, however, that both the total salt content and the ionic composition of inland saline waters and marine waters differ; they may also be different in various saline lakes.

Physiology of the Cladocera.
DOI: http://dx.doi.org/10.1016/B978-0-12-396953-8.00008-2

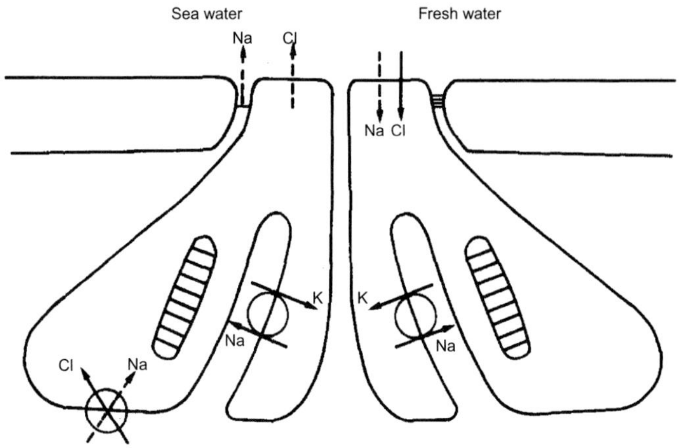

FIGURE 8.1 Possible routes of ion movements through neck organ in fresh and sea water. Solid lines, active transport; broken lines, passive diffusion. *Source: Potts and Durning (1980).*

8.3 WATER BALANCE: THE IMPERMEABILITY OF CHITIN TO WATER

The general water and salt balance of the cladoceran body is maintained by the process of osmotic regulation through their water intake, intake of minerals with food, and excretion, and through special channels for the removal of ions.

There are two methods of water intake: Cladocera take some water in through their integument and they also drink water. The latter seems to be the principal source of water intake.

8.3.1 Impermeability of Integuments

Since chitin is unwettable, cladocera are almost entirely isolated from their aquatic medium; they are connected to the external water by drinking, taking in water through the anal opening, and partly by osmotic regulation. Cladocera must excrete any excess water.

Some direct permeability of the *Daphnia* body was measured using intravital staining (trypan blue) by Fonviller and Itkin (1938). They found that a 0.01% solution of trypan blue stained first of all the esthetascs of the first antenna, and then penetrated further to the basal grains, nerves, an antennal ganglion, and the brain. The contents of the intestine were then stained and from there the stain penetrated the nephridia, epipodites, and finally the intestinal epithelium. (See also Section 4.3.1.)

8.3.2 Water Drinking

Cladocerans drink water and also take in some water with their food. Water intake was observed by Fox (1952) in *Daphnia hyalina* (24 swallowing movements/min), *Moina brachiata* (43 swallowing movements/min), *Diaphanosoma brachyurum* (29 swallowing movements/min), *Penilia avirostris* (37 swallowing movements/min), and *Bythotrephes longimanus* (95 swallowing movements/min). Water intake through the mouth serves to mix digestive enzymes with the food and pushes the food through the intestine. It is also involved in osmotic regulation.

8.3.3 Anal Water Intake

Anal water intake is a normal characteristic of Cladocera. Several antiperistaltic contractions of the hindgut follow defecation, and the evacuated volume is compensated for by anal water intake. Such water intake has been known for a long time in *Daphnia* (Hardy and McDougall, 1894; Chatton, 1920), *Sida*, *Leptodora*, *Bythotrephes* (Fox, 1952), *Ceriodaphnia*, *Moina*, and *Bosmina*, and observed later in chydorids and macrothricids (Fryer, 1970; Smirnov, 1971), but not in the marine *Evadne* and *Podon* (Fox, 1952).

Anal water intake may be conveniently observed in a specimen placed in a solution of neutral red: after defecation, red liquid fills the posterior gut. Anal intake of stained water was also observed by Chatton (1920), Rankin (1929), and Fryer (1970). It is possible to see that after defecation, there generally follow several antiperistaltic contractions of the hindgut. If the fecal particles were not expelled sufficiently far away, some are occasionally drawn back inside with the water (as observed in *Daphnia* and *Sida*).

8.4 PROCESS OF OSMOTIC REGULATION

The majority of Cladocera spp. live in fresh water. However, some species can prosper in a wide range of salinities, up to highly saline water bodies (Aladin, 1981; Frey, 1993). However, it is notable that species that can live successfully in saline lakes do not penetrate marine environments. This may be caused by differences in the salt composition of these two environments. The osmotic pressure within cladocera is thus supported by the intake and efflux of various substances and of water.

Cladocera can propagate in four different environments that differ radically in their salinity:

1. freshwater (hundreds of species of Cladocera);
2. slightly saline water (dozens of species), the critical boundary of salinity being 5–8‰ (Khlebovich, 1974);
3. highly saline continental waters (a few species); and
4. the marine environment (a few species of cladocera).

There must therefore be special mechanisms among the cladocera that enable them to exist in such different situations. Species within the same genus, e.g. *Moina*, live either in fresh waters or in saline inland waters. Among the ctenopods, *Penilia avirostris* is a marine species, whereas its morphological counterpart, *Sida crystallina*, is strictly a freshwater species. Their capacity for osmotic regulation demonstrates that the physiological background of salinity tolerance varies within particular species.

Extensive investigations into osmotic regulation, mostly in Cladocera and ostracods, were made by Aladin (1978–1996). Having applied microcryoscopic methods, Aladin (1981) described three principal types of osmotic regulation in Cladocera:

1. hyperosmotic regulation—in Sididae (except for *Penilia*), Holopedidae, Daphniidae, Moinidae (except for *Moina mongolica*), Bosminidae, Chydoridae, and Macrothricidae;
2. Hypoosmotic—in *Penilia*;
3. Amphiosmotic—in Aralo-Caspian Podonidae and *M. mongolica*.

8.4.1 Freshwater Cladocera

In freshwater cladocera, hyperosmotic regulation maintains the hemolymph hyperosmotic to the external medium. They drink water (Fox, 1952), must excrete excess water, and must supplement ions lost through excretion by obtaining them from food and by reabsorption of salts via the nuchal organ (Aladin, 1991). It has been determined that there is sodium and chlorine ion uptake in *Simocephalus* kept in a solution of sodium chloride; in this way, the hemolymph is maintained at a level hyperosmotic to the medium (Nimmo, 1966).

According to Fritsche (1917), the freezing temperature (or the freezing point) in *D. magna* can be reduced from −0.20°C to −0.61°C. It is higher in younger specimens, and there is some relationship with nutrition and egg production, since it is higher in fed specimens than in starving ones. It decreases during prolonged parthenogenesis and is, on average, high in specimens with ephippia. Belayev (1950) found there to be considerable interindividual variation in temperature depression (Δ°) of the blood in *Daphnia pulex* (0.24–0.45°C), *Eurycercus glacialis* (0.36–0.39°C), and *Bythotrephes longimanus* (0.35–0.46°C). He concluded that *D. pulex* can regulate its osmotic pressure if the external salinity is not above 5‰.

8.4.2 Marine Cladocera

Marine cladocera (podonids and *Penilia*) have a hypoosmotic type of osmotic regulation, i.e. they keep their hemolymph hypoosmotic to the external seawater (Aladin, 1979, 1994, 1996). To achieve this, they must continuously excrete ions through their epipodites and the nuchal organ (Aladin, 1991). In these cladocerans, increased activity of succinate dehydrogenase was found in the cells of the nuchal organ, indicating intensive cellular metabolism and active salt excretion against the concentration gradient.

By applying the microcryoscopic method (a version of determination of osmotic pressure by the freezing point), Aladin (1978, 1979, 1982a, 1996) found that temperature depression of hemolymph in marine podonids ranges from −0.5°C to −0.76°C (depression of seawater at water salinities above 12‰ is −0.7°C to −1.39°C), whereas in *Penilia* it is −0.72°C (that of seawater is −0.99°C). Aladin also determined that marine cladocera support a lower osmotic concentration within the brood pouch, which favors normal development of the embryos. According to Aladin, marine cladocera swallow water and secrete ions: chlorine ions in the area of the nuchal organ in podonids, and from the epipodites of the thoracic limbs in *Penilia*. At the sites of these organs a special succinate dehydrogenase activity was identified, also indicating excretion of salts against a concentration gradient.

8.4.3 Amphiosmotic Regulation

Aladin (1982b) also discovered that Caspian and Aral podonids can use hyperosmotic osmotic regulation at salinities below 8‰ and may exploit hypoosmotic osmotic regulation above 8‰. This is a special case of cladocerans living in continental hypersaline waters. *M. mongolica* can live in salinities up to 88‰ (with the salt composition of the Aral Sea). At salinities over 8‰, it exploits hypoosmotic osmotic regulation; under 8‰, hyperosmotic osmotic regulation occurs (Aladin, 1982c, 1983).

8.4.4 Sodium Exchange in Hyperosmotic Cladocerans

It has been established that sodium is retained better at pH 3 by an inhabitant of acid environments (pH 3.4–6.3), *Acantholeberis curvirostris*, than by *D. magna* (pH 6.9–10.2); its uptake is reduced in acid waters in both species, but more so in *D. magna* (Potts and Fryer, 1979). Sodium loss is lower at pH 3 than at pH 7 in *A. curvirostris*, but is four times greater in *D. magna* (Potts and Fryer, 1979).

Havas and Likens (1985) studied the effect of pH on sodium regulation in *D. magna* and *Daphnia middendorffiana*. The rate of Na loss (efflux) at pH 4 and below (in hard water) was compared with that at pH 8.0: Na uptake was the same in both. In soft water, Na uptake was inhibited by 69% at pH 4.5 (compared with the pH 6.5 control), and loss (efflux) increased to 125% of the control at pH 4.5. Thus, there are problems with Na regulation below pH 5.5 in soft water and below pH 4.5 in hard water.

Glover and Wood (2005) found that sodium metabolism in *D. magna* may be disrupted by acidification and by ionoregulatory toxicants. Acidification inhibits Na intake. At low pH, calcium (in the form of $CaSO_4$ and as calcium gluconate) inhibits Na uptake by *D. magna*, whereas at higher pH and at high Na concentrations, calcium stimulates Na uptake.

Daphnias that are depleted of sodium can restore their normal content of 26.3 mM/kg WW within 15 h (Stobbart et al., 1977). They do not accumulate excessive Na. *Daphnia galeata mendotae* survive better when they have a higher Na content (at 6–10 mg/g dry weight) (Havens, 1992). At pH 4.5, both the Na content and survival was reduced. A concentration of 200 g aluminum (Al)/L at pH 4.5 enhances Na content in the body of *Daphnia* and prolongs survival.

Experimental exposure of *D. magna* to natural organic matter promoted the Na loss from the daphnid to the water, thus resulting in reduced whole-body Na levels; this was a labile process, dependent on the period of pre-exposure (Glover et al., 2005a).

Na uptake and release in *D. magna* adults and neonates was investigated more recently by Bianchini and Wood (2008): these processes turned out to have different mechanisms in adults and in neonates (Fig. 8.2). According to these authors, in neonates, a proton-coupled Na^+ channel is important for whole-body Na^+ uptake at the apical membrane. In contrast, this membrane does not contribute to whole-body Na^+ uptake in adults: adults possess only the Na^+ channel associated with Na^+/H^+ exchange. In both cases, protons (H^+) for the

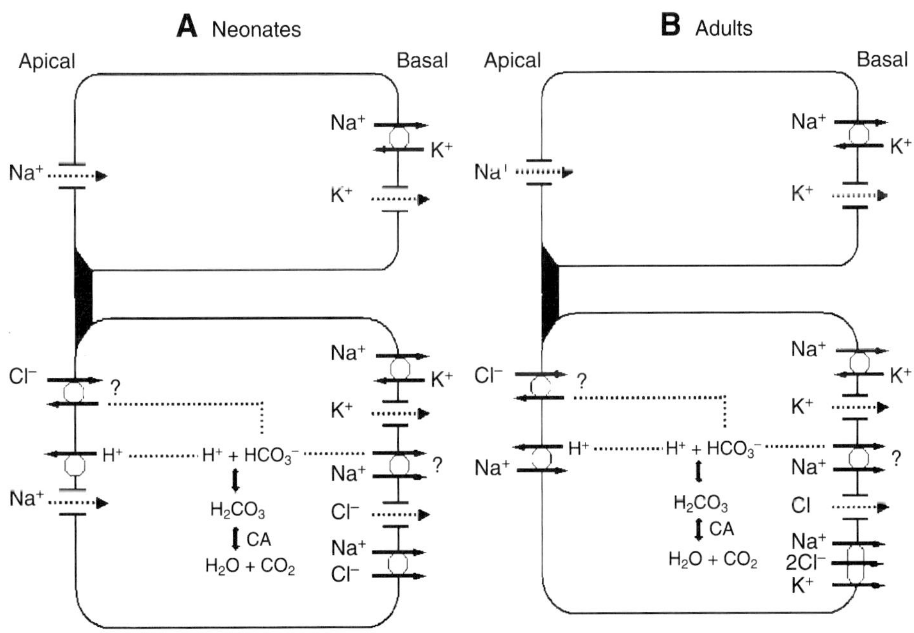

FIGURE 8.2 Ion transport in *Daphnia magna*. *Source: Bianchini and Wood (2008).*

transporters are supplied by carbonic anhydrase. In neonates, at the basolateral membrane Na^+ is pumped to the extracellular fluid by a Na^+, K^+-ATPase and a Na^+/Cl^+ exchanger, whereas K^+ and Cl^- move through specific channels. In adults, a Na^+/Cl^- exchanger is replaced by a $Na^+/K^+/2Cl^-$ cotransporter. Accordingly, the sensitivity of adults and neonates to osmoregulatory toxicants is different.

8.4.5 Turgor

Hydrostatic pressure, or turgor (termed also the hemocoel pressure by Downing, 1974), within the cladoceran body is supported in the same way as has been described for osmotic regulation. Its function is the upkeep of pressure necessary for mechanical actions. According to Krogh (1939), turgor requires only a small fraction of osmotic pressure.

Homeostasis is supported by both osmoregulation and excretion, in cooperation with other physiological processes.

8.5 EFFECT OF XENOBIOTICS ON OSMOTIC REGULATION

The following data have been reported. Ni is toxic to *D. magna* due to Mg^{2+} antagonism: exposure to Ni affects Mg^{2+} homeostasis (Pane et al., 2003). In the body of *D. magna*, Mg^{2+} is decreased by 18%, and its uptake also decreased. No noticeable effect of Ni on the Ca^{2+} or Cl^- balance was observed and there was no acute (i.e. in short-term experiments) toxic effect on oxygen consumption.

Amiloride (*N*-amidino-3,5-diamino-6-chloropyrazinecarboxamide hydrochloride) inhibits Na influx in *D. magna* (Glover and Wood, 2005).

CHAPTER

9

Cell and Tissue Metabolism

9.1 METABOLISM IN TISSUES

Steinberg et al. (2010) measured the protein content, antioxidant capacity, internal hydrogen peroxide, total ascorbic acid, and free proline levels (ascorbic acid and proline are both antioxidants) in *Daphnia magna* homogenates.

Enzymes are specific proteins that catalyze metabolic processes. For Cladocera, numerous individual enzymes have now been discovered. Those involved in food digestion are found in the digestive tract of *Daphnia*: these include proteases, lipases, amylases, and cellulase (Hasler, 1937; and as summarized by Hebert, 1978a and in Section 4.4). Enzymes involved in various metabolic processes have also been found in Cladocera homogenates (e.g. Hebert, 1973). In addition, the following enzymes have been used to explore genetic variability in *Daphnia* spp.: alkaline phosphatase, esterase, fumarase, glucose-6-phosphate dehydrogenase, lactate dehydrogenase, glutamate oxaloacetate transaminase, leucine aminopeptidase, phosphoglucose isomerase, tetrazolium oxidase (Hebert and Moran, 1980; Hebert and McWalter, 1983; Hebert et al., 1989; Hebert and Finston, 1996, 1997; Hebert and Wilson, 2000), malate dehydrogenase, xanthine dehydrogenase (Hebert and McWalter, 1983), aldehyde oxidase, amylase, phosphoglucomutase (Hebert et al., 1989), aspartate aminotransferase, fumarate hydratase, and mannose-6-phosphate isomerase (Kořínek and Hebert, 1996).

Berges and Ballantyne (1991) found the following enzymes in whole-body homogenates of *D. magna*: citrate synthase (CS), lactate dehydrogenase (LDH; for anaerobic metabolic activity), pyruvate kinase (PK; for anaerobic metabolic activity), alanine aminotransferase (ala AT), aspartate aminotransferase (asp AT), glutamate dehydrogenase (GDH), nucleoside diphosphate kinase (NDPK; an anabolic enzyme), and glucose-6-phosphate dehydrogenase (G6Pdh; an anabolic enzyme).

The acetylcholinesterase (AChE) content in *D. magna* was determined to be 12.7 mU/mg protein (Day and Scott, 1990). The AChE content in *Daphnia* is inversely proportional to the body size; this situation is probably caused by an increase in total protein that is not proportional to substrate hydrolysis (Printes and Callaham, 2003). AChE activity 2.5 μM/L/min/g protein in 1-mm long *Daphnia* and decreases to c. 0.5 μM/L/min/g protein in 3-mm long *Daphnia*. In this system, parathion (*O*,*O*-diethyl-*O*-*p*-nitrophenyl phosphorothioate) inhibits AChE activity and phenobarbital negatively affects the protein.

Physiology of the Cladocera.
DOI: http://dx.doi.org/10.1016/B978-0-12-396953-8.00009-4

Moina macrocopa AChE was extracted from homogenate supernatant by Martinez-Tabche et al. (1997, 1998), and differences in its activity relative to control has been used to estimate water quality (Martinez-Tabche et al., 1998). AChE activity is inhibited by lead (Pb), sodium dodecylbenzenesulfinate, mixtures containing these components, and, especially, crude oil (Martinez-Tabche et al., 1997). *D. magna* AChE is also inhibited in vivo by at 48-h exposure to the median lethal concentration (LC_{50}) of sodium dichromate and sodium molybdate (Diamantino et al., 2000). Both substances inhibit growth and reproduction, with sodium dichromate being the more toxic.

Choline esterase from *D. magna* hydrolyzes acetyl thiocholine iodide, propionylthiocholine iodide, and butirylthiocholine iodide; it is highly sensitive to the organophosphate dichlorvos (2,2-dichlorovinyl dimethyl phosphate; DDVP) (Menzikova, 1988).

Acid phosphatase maintains the level of inorganic phosphorus in cells. It is found in *Ceriodaphnia affinis* (Tsvetkov et al., 2012), and in *D. magna* the overall activity of the acid phosphatase complex increases following exposure to $CdCl_2$ (Zarubin et al., 1997).

Alkaline phosphatase (AP) is a potential biomarker of phosphorus limitation. The highest AP activity in *D. magna* was found at lower temperatures (10°C), and for every 10°C reduction in body temperature from 25°C, AP activity increases by a factor of 1.67 (Wojewodzic et al., 2011).

Phenoloxidase activity is increased in well-fed *D. magna*, and wounding stimulated an increase in enzyme activity. In addition, clones with higher phenoloxidase activity were more resistant to *Pasteuria* (Mucklow, 2003).

Borgeraas and Hessen (2002b) observed diurnal variation in the activity of *Daphnia* antioxidant enzymes [catalase, glutathione *S*-transferase (GST), and superoxide dismutase]. Hyaline *Daphnia* showed higher GST and superoxide dismutase activities, but lower catalase activity in comparison with a melanic type. Superoxide dismutase and catalase demonstrated diurnal variations in activity, with a maximum value at midday. In arctic melanic *D. tenebrosa*, however, diurnal fluctuation is insignificant.

Boavida and Heath (1984) confirmed that phosphatases liberated into the environment are produced by *D. magna* and not derived from its algal food.

9.2 EFFECTS OF XENOBIOTICS ON CYTOLOGY AND METABOLIC FACTORS

When investigating the lethal effect of solutions of NaCl and $HgCl_2$ on *Daphnia*, *Simocephalus*, and *Sida*, Beklemishev (1923a, 1923b) found that the effect described a curve, with either one peak or two or more peaks. The latter was observed under conditions when some specimens survived and then suffered slower intoxication. This indicates that several physiological mechanisms are affected by the toxic agent.

The lethal concentrations are different for different substances, and can sometimes be very low. It has been shown that 100% of *Chydorus sphaericus* (Smirnov, 1971, 1974) perish (i.e. 100% lethal dose, or LD_{100}) within 24 h in a 0.00001% solution of $AgNO_3$ (precipitating mainly on the epipodites), but they can survive in a 0.0000003% solution. They also died in a 0.0005% solution of potassium dichromate, but survived in a 0.0001% solution; and died in 0.004% neutralized formalin, but survived in a 0.001% solution. In solutions of most other salts tested, death occurred at concentrations of 0.5% or more.

A series of similar measurements was made with *D. magna*: Ag, Cu, and Hg caused a loss of ions in this *Daphnia* sp., e.g. of Na^+ ions (Holm-Jensen, 1948). Prior to their death, Na^+ concentrations were reduced to one-third of the normal value. It is thought that this situation is caused by inhibition of the mechanism responsible for the active uptake of ions. Toxic effects could be eliminated by the addition of glutathione or (for Ag and Cu) of cysteine.

D. magna exposure to metals led either to an increase or decrease in the protein content of the body: the maximum reduction of total protein (9%) occurred upon exposure to Ni, and the maximum increase (40%) upon exposure to Mg (Biesinger and Cristensen, 1972).

In a lake with an elevated content of Cu, Ni, and Al, *Holopedium* had a lower lipid content and lipid droplets in the body were smaller by 21%, compared with those in an unpolluted lake (Arts and Sprules, 1987). In the presence of $CuSO_4$, the excretion of amino acids to the external environment by *D. magna* was noticeably increased (Gardner and Miller, 1981).

Mercury is thought to produce soluble albuminates that may penetrate deeper into tissues. In addition, mercuric chloride ($HgCl_2$) has proven to be highly toxic to *Simocephalus*: the actual killing rates were determined at concentrations ranging from 0.5 M to 50 μM (Breukelman, 1932).

In *Daphnia*, Cu^{2+} and triphenyltin chloride cause chromosomal aberrations in cells of the intestines, and especially in cells of the embryos (Filenko and Lazareva, 1989). At chronic levels of exposure, Zn accumulates mainly in cellular organelles and in heat-stable protein fractions in *D. magna* (Wang and Guan, 2010).

Gut diverticula (hepatic ceca) are the organs that are principally attacked by xenobiotics. In *D. magna* exposed to 12–52 μg Cd^{2+}/L, the gut diverticula became shrunken and paralyzed, and Cd granules were found in the mitochondria and microvilli of the distorted ceca (Griffiths, 1980). Sublethal solutions of $CoCl_2$, $NiCl_2$, $Al_2(SO_4)_3$, and $Cd(NO_3)_2$ cause degeneration of the hepatic ceca in *D. magna* (Luzgin, 1982a). Degradation starts from the apex of the cecum, thus indicating that this is the site of the most intensive assimilation of salts. Exposure of *D. magna* to sodium selenate (Johnson, 1989) also leads to Ca deposition in the mitochondria of hepatic ceca.

The impact of 3-amino-1,2,4-triazole (a herbicide) on newborn *Daphnia* was investigated by Schultz and Kennedy (1976a). Young *Daphnia* are immobilized at a concentration of 0.1 mg/L; immobilization of all individuals occurs after about 22 h. The length of exposure required to reach immobility is decreased in animals approaching ecdysis (molting). In particular, mitochondria, especially those of muscle cells, are affected. Other effects included tissue swelling, myofilament disarrangement, and dissociation of membranes.

In *D. magna*, the toxic effect of phenanthrene and 9,10-phenanthrenequinone (PHQ) (polycyclic aromatic hydrocarbons) was tested both with and without Cu (Xie et al., 2006). Copper (Cu) and PHQ generated reactive oxygen species, and this process involved mitochondria. The proposed pathways for Cu cycling and the roles of Cu are shown in Fig. 3.9. In living *D. magna* cells, Cu+ or superoxide dismutase may reduce superoxide radicals to hydrogen peroxide and cause oxidative damage (Xie et al., 2006).

9.2.1 Disturbances to Enzyme Activity

Exposure of *D. magna* to metals (Biesinger and Cristensen, 1972) leads to either an increase or a decrease in glutamic oxaloacetic transmutase activity within the body: the maximum decrease (26%) occurred upon nickel exposure, and the maximum increase (100%) upon exposure to Mn (a 65% increase occurred following exposure to Mg).

Exposure of *D. magna* to xenobiotics inhibits enzyme activity: copper (II) chloride ($CuCl_2$) inhibition of succinate dehydrogenase activity occurs long before a lethal effect is observed (Luzgin, 1982b); parathion (at concentrations of 0.05–5 μg/L) (Dortland, 1978); and the surfactants dodecylbenzyl sulfonate and sodium dodecyl sulfonate inhibit AChE (the enzyme that hydrolyzes acetylcholine, a neurotransmitter) (Guilhermino et al., 2000).

Anticholinesterase compounds are present in waste water and, being physiologically active, represent dangers for animal and human health. They are released into the environment as final waste products from agricultural, medical, and military enterprises or as a result of accidents. Tonkopiy et al. (1993) investigated the cholinergic system of *D. magna* by applying various anticholinesterase compounds, including reversible inhibitors, organophosphates, and carbamates. Anticholinesterase compounds increased the toxic effect of the myorelaxant ditilin. These authors demonstrated that central m-cholinolytics reduce the toxicity of armine and aminostigmine.

Cholinesterases from *D. magna* are noticeably inhibited by zinc (Zn) (Diamantino et al., 2003). Sturm and Hansen (1999) studied cholinesterase (ChE) inhibition in *D. magna* by parathion, dichlorvos, and aldicarb [2-methyl-2(methylthio)propioaldehyde-*O*-(methylcarbamoyl)oxime]. The presence of pesticides decreases ChE in a dose-dependent manner: parathion is efficient at 0.1 μg/L, dichlorvos at 1 μg/L, and aldicarb at 100 μg/L.

AChE and carboxylesterase (CbE) inhibition by organophosphorus pesticides (malathion and chlorpyrifos) and carbamate pesticides (carbofuran) was assessed in *D. magna* by Barata et al. (2004). CbE is more sensitive than AChE to organophosphorus pesticides, but both are equally sensitive to carbofuran. Mortality increased at low levels of AChE inhibition by carbofuran, while upon exposure to organophosphorus pesticides mortality increased when the level of inhibition was $> 50\%$.

Glutathione is a thiol-containing tripeptide (γ-glutamyl-cysteinyl-glycine), which functions as a reducing agent in cells (Elliott and Elliott, 1997). It is found in *D. magna* cells and its concentration is reduced by treatment with acridine, 1,10-phenathroline, benzo(a)quinoline, phenantridine, and phenazine (*N*-heterocyclic polyaromatic hydrocarbons) (Feldmanová et al., 2006).

Following an initial 10-h exposure, malathion decreased both protein content and AChE activity in *Simocephalus vetulus*; lipid content was reduced after 24 h exposure, but lipid peroxidation levels increased over the whole exposure period (4–50 h) (Olvera-Hernández et al., 2004).

Methoprene (a juvenile hormone agonist) is toxic to endocrine-related processes, i.e. growth, molts, and fecundity, in *D. magna* (Olmstead and LeBlanc, 2001a, 2001b).

PK and malate dehydrogenase were inhibited after a 7-day exposure of *D. magna* to 0.05 mg/L 3,4-dichloroaniline (Morgado and Soares, 1995). However, after exposure for 14–21 days, the activities of these enzymes were stimulated relative to control. *D. magna* steroid hydroxylase is differentially modulated by exposure to the toxicants phenobarbital, β-naphthoflavone, piperonyl butoxide, and malathion (Baldwin and LeBlanc, 1994b). *D. magna* exposure to 811 μg/L di-2-ethylhexyl phthalate for 21 days resulted in decreased glycogen content (Knowles et al., 1987).

A polycyclic aromatic hydrocarbon, pyrene is transformed in *D. magna* with the participation of cytochrome P450 monooxygenase (Akkanen and Kukkonen, 2003), whereas the presence of piperonyl butoxide inhibits pyrene biotransformation.

Exposure of *D. magna* to 60 μg/L chlordecone for 7 days resulted in a reduction in RNA levels from 20 μg to c. 9 μg per individual (McKee and Knowles, 1986); the DNA concentration decreased from c. 0.6 μg to c. 0.2 μg per individual, respectively; and ADP and ATP concentrations decreased after longer exposure times.

Enzyme activation by xenobiotics may also occur under certain conditions. Carbohydrases (estimated by amylolytic activity and saccharase activity) in *D. magna* were activated by the herbicide Roundup (glyphosate) *in vitro* at concentrations up to 50 mg/L (Filippov et al., 2010). Upon exposure to Roundup, the overall proteolytic activity of *D. magna* increases and the overall amylolytic activity decreases (Papchenkova et al., 2009); this was interpreted to indicate an increased role for proteins in metabolism at sublethal concentrations of Roundup. No adaptation was observed over four generations.

The impact of cadmium (Cd) on delta-aminolevulinic acid (ALA-D) synthesis, an early step in the biosynthesis of heme (porphyrin), was tested by Berglind (1985). *D. magna* was exposed to cadmium (as $CdCl_2.2.5\ H_2O$) in concentrations of 0, 0.1, 0.2, 0.4, and 1.6 μg Cd/L: ALA-D concentration fluctuated around the control values. However, after 16 days the hemoglobin (Hb) content decreased to 31–80% of the control. Further experiments (Berglind, 1986) were made to test the effect of Cd (at concentrations of 0, 0.2, and 2.0 μg/L) and other heavy metals, both separately and in combinations. Pb (in concentrations of 0, 0.26, and 260 μg/L) was added in the form of lead acetate $Pb(CH_3COO)_2$ and Zn (0, 020, and 200 μg/L) as zinc sulfate ($ZnSO_4.7\ H_2O$). ALA D activity was enhanced by Cd and by high concentrations of phosphorus (P) + Zn, but inhibited by Pb, Pb + Cd, Pb + Zn, and Pb + Cd + Zn. Stimulation by Cd was abolished by Zn. Hb content did not decrease, even at 260 μg/L Pb.

The effect of 48-h menadione, paraquat, endosulfan, Cd, or Cu exposure on *D. magna* enzymes varied (Barata et al., 2005). A low response of antioxidant enzymes to menadione and endosulfan was observed to correlate with increasing levels of lipid peroxidation; antioxidant activities were enhanced by paraquat; there was a low antioxidant enzyme response to Cd; and high levels of both antioxidant enzyme activities and lipid peroxidation were induced by Cu. The most responsive biomarkers of oxidative stress were glutathione peroxidase, catalase, and GST (the key antioxidant enzymes).

D. magna lactate dehydrogenase activity is noticeably inhibited by Zn (Diamantino et al., 2001). *D. magna* β-galactosidase activity is decreased in the presence of fluoranthene (a phototoxic polycyclic aromatic hydrocarbon) following ultraviolet illumination (Hatch and Burton, 1999).

The activity of *D. magna* heme peroxidase is significantly increased in kerosene-contaminated groundwater (Connon et al., 2003).

In addition, the inhibition of *Daphnia* enzymes by pollutants, seen as reduced fluorescence, has been suggested as a sensitive toxicity test (Kubitz et al., 1995).

The toxicity of strophanthin for *D. magna*, determined by LD_{50}, was found to be different in different seasons, being highest in January–February and lowest in October–November (Rautenberg, 1953).

Daphnia exhibit a three-phase attraction or repulsion effect in response to coumarin or copper (II) sulfate ($CuSO_4$) (Heintz, 1964): the animals are repulsed at low and high concentrations, and attracted at intermediate concentrations. Maximum attraction occurred at 10 ng/L for coumarin and 600 mg/L for $CuSO_4$.

9.3 DETOXIFICATION

Cladocera liberation from a toxic substance may be a result of depurination in a clean environment or targeted metabolic processes.

GSTs, the enzymes that participate in chemical detoxification, have been isolated from *D. magna* (LeBlanc et al., 1988; Baldwin and LeBlanc, 1996). *D. magna* GSTs are inhibited by 1,4-benzoquinone and 2,4-dichlorophenoxyacetic acid (Dierickx, 1987a). The presence of five major GSTs has been demonstrated in *D. magna*, as has detoxification of 1-chloro-2,4-dinitrobenzene (CDNB) (Dierickx, 1987b). Agents that inhibit or increase the GST content either increase or decrease CDNB toxicity, respectively.

Detoxification of tannins, which are thought to be ingested with algal food, was studied in *Daphnia* and *Simocephalus* and found to be carried out by special enzymes (cytochromes P450, esterases, and GSTs) (Ray et al., 2000).

Pentachlorophenol is metabolized in *D. magna* by sulfate conjugation; it is then excreted at a rate of 2.65 nmol/g/h (Kukkonen and Oikari, 1988).

In benzo(a)pyrene depurated from *D. magna*, 0.1 of its initial quantity remained after 40 h (McCarthy, 1983).

Metallothioneins (special soluble proteins, the *proteins of detoxification*) with a molecular weight of c. 10 kDa are found in *Daphnia*. It was found that over 80% of the Cd burden in the body of *D. pulicaria* grown in Cd-containing water is bound by this 10-kDa protein (Gingrich et al., 1984). A similar percentage (up to 75%) of the Cd burden in the body of *D. magna* was also determined to be bound to metallothioneins; increased Cd in daphnids induces subcellular reorganization of essential metals (Cu and Zn), and a higher content of these metals in the soluble cellular fraction (Fraysse et al., 2006).

Cytochrome P450 enzymes may contribute to detoxification and acclimation of *D. magna* to chronic toxaphene exposure, as cytochrome P450 inhibition by piperonyl butoxide led to a decline in growth rate, fecundity, and survival (Kashian, 2004).

CHAPTER

10

Growth and Molting

10.1 GROWTH

Increases in the body length of Cladocera principally occur between molts. For example, within 10 sec after molting a *Daphnia magna* was observed to increase in length from 1.3 mm to 1.6 mm (Green, 1956a).

Newborn Cladocerans grow rapidly. Their growth during prereproductive instars is linear, and then it becomes slower, especially in species that bear two parthenogenetic eggs. Their growth slows down upon reaching maturity. In mature chydorids, for example, increments in length become very small; in contrast, in mature daphnids slow growth continues (Zaffagnini, 1964; Smirnov, 1971). Information on Cladocera growth was summarized by Frey and Hann (1985). In general, growth is accelerated at higher temperatures and food concentrations but is retarded at lower temperatures, low food concentrations, moderate illumination of 301 lx, and illumination with green light (Buikema, 1972). Yerba and Hernández-León (2004) identified a significant relationship between somatic growth and aminoacyl-tRNA synthetase activity, both in terms of protein and dry weight.

10.1.1 Life Span

The life span of Cladocera species can reach several months, for example, up to 182 days in planktonic cladocerans (Fritsch, 1953). In ctenopods, the maximum life duration is between 19 and 74 days, as summarized by Korovchinsky (2004).

In culture, chydorids live for about 3 months. For example, *Pleuroxus denticulatus* lives for 121 days at 15°C (Shan, 1969), and *Alona costata* lives for up to 140 days, frequently producing parthenogenetic broods (Smirnov, 1965a, 1971), and then dies under the same culture and feeding conditions. Much shorter life spans have been recorded by various authors for *Moina macrocopa*: a maximum of 14 days (Terao and Tanaka, 1928a), 11 days (Razlutskii, 1992), and an average longevity of 12.5 days (Chuah et al., 2007). The same species lived for up to 15 days in experiments carried out by Garcia et al. (2004), but in the studies of Makrushin (Makrushin, 2011; Voronin and Makrushin, 2006) its life span was only up to 9 days. For *Moina rectirostris* (syn. *Moina brachiata*) the maximum life span was determined to be 28 days in culture (Razlutskii, 1992).

Physiology of the Cladocera.
DOI: http://dx.doi.org/10.1016/B978-0-12-396953-8.00010-0

Naturally, the longevity of any species depends on its living conditions, including food and temperature. The life span of *Daphnia* spp. decreases under conditions of high temperature (Armitage and Landau, 1982) and oxygen deficit (MacArthur and Baillie, 1929; Zhukova, 1955). At comparatively low temperatures (8–10°C), which slow down metabolic processes, the life span of *D. magna* is greatly increased (as determined by MacArthur and Baillie, 1929): it is 25 days at 28°C, 42 days at 18°C, 88 days at 10°C, and 108 days at 8°C. Female *Daphnia longispina* can live for up to c. 66 days (1600 h) at 20°C, but for only 234 days at 5°C, whereas male *D. longispina* lived up to 50 and 170 days, respectively, at the same temperatures (Munro and White, 1975). It has been shown that other suboptimum conditions may also prolong the life span of *Daphnia* (Ingle, 1933). Limited food contributes to increased life span in *Daphnia* (Ingle et al., 1937; Hirosaki, 1953; Skadovskiy, 1955; Sterba, 1956b; Steinberg et al., 2010) and *Simocephalus* (Hirosaki, 1953) spp. Ingle et al. (1937) demonstrated that *D. longispina* live for 40% longer under conditions of minimum food compared to well-fed specimens. Starved *Daphnia* produced fewer young, but the total number of young produced is nearly identical in poorly fed (which live longer) and well-fed (with a shorter life span) *Daphnia*.

The life span of *D. magna* is prolonged by heparin (Schechter, 1950). Interestingly, vitamin K reduces the life span of *D. magna* at a concentration of 400 μg/L, whereas it is somewhat stimulatory at a concentration of about 200 μg/L (Schechter, 1950). The longest life span of *D. longispina* was recorded when calcium pantothenate was added to the culture medium (Ingle et al., 1937).

Meijering (1958) suggested measuring the life span in heart beats; *D. magna* reached 47 million heart beats at 30 instars (about 65 days). In another communication, 50% of those that survived were reported to have had about 40 million heart beats (Meijering and Redfern, 1962). About 8000 heart beats occur between molting and the liberation of eggs into the brood pouch, according to Meijering (1960). Fritsch (1953) recorded that *D. magna* males live for about 30 days and have 19.1 million heart beats.

Extreme environmental conditions can reduce life span. For example, the mean life span of the low-saline species *Diaphanosoma celebensis*, decreased from 24 to 5 days at a salinity of 30 PSU (practical salinity units) (Achuthankutty et al., 2000).

10.1.2 Mortality

Gainutdinov et al. (1997b) found that the presence of sugars in the surrounding water, which is controlled by algae, modifies *D. magna* mortality by a mechanism involving sensory cells, amplification of the signal, and modification of neuroendocrine units in the nervous system.

Mortality unrelated to predation and related to an unidentified infection that caused a 10% daily loss of *Daphnia hyalina* was reported for Lake Constance (Germany) (Gries and Güde, 1999).

10.2 MODIFICATION OF FORM

There is sometimes a significant difference in the body forms of adults and juveniles of the same species, e.g. in some macrothricids. Adults of the same species of Cladocera are also variable in form; striking examples are species of *Daphnia* and *Bosmina*, which show numerous phenotypes. This morphological plasticity affects the general form, the body size, and the development of dorsal and posterior spines. Previous reports have frequently investigated these morphological changes, but the physiological prerequisites for such modifications of form are currently unknown.

10.2.1 Mechanical Damage and Regeneration

Mechanical Damage

Cladocera living on substrata are well protected from mechanical damage by their comparatively thick carapaces, and their body form is modified to enable their penetration of various littoral obstacles. The thickest carapace is probably that of *Pseudochydorus globosus*, which can be up to 12 μm (Fryer, 1968). In addition, broken structures are frequently present, as well as wounds inflicted by predators.

Turbulence

With respect to modification of the general *Daphnia* form, according to Hutchinson (1953, p. 155), turbulence provides "a continual random stimulation of the sense organs." Hutchinson continues (1953, p. 156) that "there are an enormous number of interconnected internal things." However, the physiological mechanisms by which external turbulence modifies the body form are currently unknown.

With moderate turbulence, the neonates of *Daphnia cucullata* developed large cephalic helmets at their older age (Hrbacek, 1959; Laforsch and Tollrian, 2004a).

Ermakov (1927) collected a sample of *D. pulex* and *D. magna* in which about 5% of the specimens had damaged antennae or anomalous regenerated structures. In *Sida* (females), the percentage of damaged specimens can reach 25% (Ermakov, 1929) or even 30% (Korovchinsky, 2004). Wave beating kills a noticeable fraction of *Chydorus*, *Daphnia*, and *Bosmina* (Manuilova, 1955). She also recorded that the ratio of living to dead *Chydorus sphaericus* reaches 2.9:1 after storms (Manuilova, 1956). In a shaker (at 200 shakes/min in 60 mL of water), 50% of *D. pulex* perished within 6 h, and 50% of *Eurycercus lamellatus* died within 7 h (Smirnov, 1971). *Daphnia* can perish within a day in shaken or intensively aerated cultures (Harvey, 1972).

Specimens (e.g. of *Bosmina*, *Daphnia*, or *Ceriodaphnia*) with obvious signs of injury and wound healing, caused for example by predatory cyclopoids (Li and Li, 1979), are frequently found in Cladocera samples (Kerfoot, 1975; Murtaugh, 1981). The proportion of damaged prey may be used as a measure of predation pressure but these abnormalities can also stem from disturbances in embryonic development.

Regeneration

In a case of a wound inflicted on a *Daphnia* as a hole in the valve and followed by regeneration, a decreased hole size was seen in successive instars (Anderson and Brown, 1930; Anderson, 1933). Within a few hours after an injury, the edge of the wound is surrounded by a brown material, which is assumed to be clotted blood with oxidized tyrosine. Over the course of successive instars, the wound area decreases until it is completely closed. Anderson and Brown (1930) have shown that during wound healing in *D. magna*, chitin secretion starts 60% of the way through the intermolt period.

Clot Formation

Formation of a blood clot was observed by Ermakov (1927, 1929) at sites where an antenna or the shell had been dissected. Some blood initially flows from the wound, but then a clot is formed and hemorrhage stops. Most cladocerans can survive such operations.

A lost appendage is usually regenerated, although not completely. It has been found experimentally that missing (amputated) segments of an antenna may not regenerate; nevertheless, the setae do (Fig. 10.1) (Agar, 1930). Remarkably, regeneration either restores the

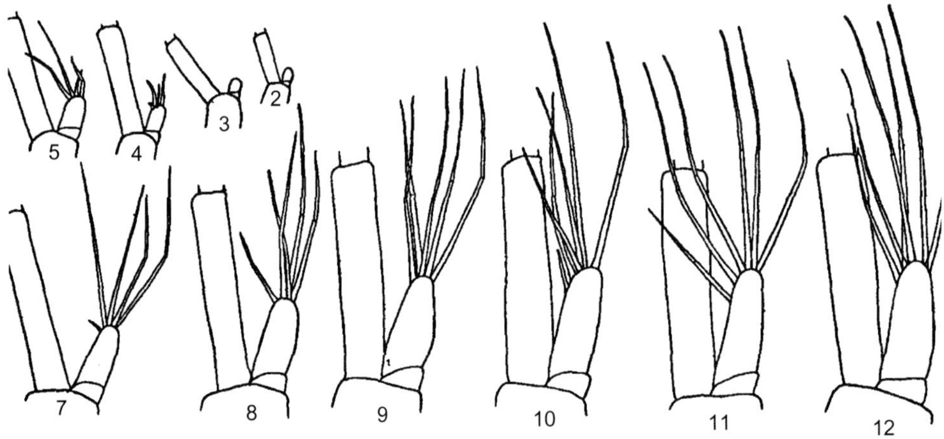

FIGURE 10.1 **Antenna regeneration in *Daphnia* over successive instars.** *Source: Agar (1930).*

normal structure or may produce hypermorphoses, hypomorphoses (submorphoses), teratomorphoses, or dichotomy (as defined by Ermakov, 1927, 1929). These facts are important for understanding the morphological (morphogenetic) potential of body structures. Strictly speaking, there is no hard boundary between variation and teratology (i.e. formation of abnormal structures).

According to Shimkevich (1909, 1925), the trends shown by anomalies and teratologies may be exploited by normal morphogenesis. Teratologies may easily be reproduced experimentally and the events that occur during subsequent moltings observed. In addition, experimental extirpation of the eye has mainly been successful (see Chapter 13, section 13.4.3).

10.2.2 Chemical Growth Factors

Somatic growth is constrained by food availability, including the availability of polyunsaturated fatty acids (Schlotz and Martin-Creuzburg, 2011). Cyanobacteria are characterized by a low content of eicosapentaenoic acid (20:5ω3) (Müller-Navarra et al., 2000) and a low sterol content (von Elert et al., 2002), which constrains the growth and reproduction of Cladocera.

The lowest calcium threshold for the growth of *D. pulex* juveniles was measured as c. 1.5 mg/L (Riessen et al., 2011). Early juveniles of *D. magna* (Hessen et al., 2000a) and *Daphnia galeata* (Rukke, 2002) have especially high calcium requirements.

Chemomorphoses

The term *chemomorphosis* was suggested to describe morphological variation caused by exogenous chemical agents (Hebert and Grewe, 1985). It has been repeatedly reported that the formation of spined morphs of *Daphnia* (i.e. those with neck teeth) is caused by the presence of a predator (*Chaoborus*) and its excreted kairomones (Havel, 1985; Hanazato, 1991; Hanazato and Ooi, 1992; Hunter and Pyle, 2004). Both females and males of *Daphnia* develop larger protective helmets or dorsal denticles on their shells in medium enriched with an extract derived from predators (*Chaoborus* larvae or *Anisops*) (Grant and Bayly, 1981; Krueger and Dodson, 1981; Hebert and Grewe, 1985; Spitze, 1992; Hanazato, 1990; Hanazato and Ooi, 1992;

Hanazato, 1995; Repka and Pihlajamaa, 1996). They also develop a cephalic spine and an elongated tail spine (Brancelj et al. 1996).

The morphology of various *Daphnia* spp. has been shown to change under the influence of chemicals (kairomones) liberated into the environment by predators (e.g. invertebrates and fish) (Black, 1993; Larsson and Dodson, 1993; Tollrian, 1990, 1993, 1994, 1995). In *D. pulex* grown in the presence of predators, characteristics of their life cycle are also changed (Black, 1993), for example, including the comparatively rapid growth of juveniles. Generally effects include elongation of both cephalic helmets and the tail spine. This reaction also causes an increased rigidity of *Daphnia* carapace (Laforsche et al., 2004).

While working with *Daphnia*, Jacobs (1980) concluded that there are specific growth determinants that preferentially act on mitotic rates within the Cladocera helmet. Helmet cells are not supplied with nerve fibers, and their growth determinants are carried in the hemolymph. Later, Beaton and Hebert (1994, 1997) identified polyploid cells in the cephalic epidermis of *Daphnia*; the DNA content in these cells is higher than in the thoracic regions, and mitotic activity is concentrated around them. Thus, these cells are assumed to be developmental control centers that govern the shape of the head.

Morphological reactions to chemicals start in embryos that have shed the third membrane; these have liberated chemosensilla that can detect chemicals (Laforsch and Tollrian, 2004b). However, morphological changes have been observed in the third instar in *D. cucullata* and *D. pulex*, but not in three other species.

A reverse morphogenetic process was discovered by Riessen (1984): loss of helmets was observed in generations of helmeted *D. retrocurva* produced in the absence of predators and with abundant food. The opposite process, i.e. a reduction in tail spine length, was also observed by Burns (2000), who cultivated nine species of *Daphnia* in water from crowded cultures of the same or another species.

A reduction in body size and changes to the carapace morphology have also been observed, including reduction in the tail spine length (in *D. lumholtzi* and *D. ambigua*). Kairomones or physical factors seem to either induce or inhibit the proliferation of tissues. Thus, identification of the chemical controlling factor(s) is the next priority.

However, identification of the chemical nature of this factor is easier than elucidating the actual pathway through which it effects changes in morphology. It is much more difficult to find out how proliferation is channeled to produce the modified structure. Barry (2002) applied substances to *D. pulex* that differentially affect neurotransmission, including those that either enhance (nicotine and physostigmine) or inhibit (atropine) cholinergic transmission; or stimulate (*cis*-4-aminocrotonic acid, diazepam, and muscimol,) or antagonize (bicuculline, picrotoxin, and SR95531) the action of γ-aminobutyric acid (or GABA). Neckteeth development was enhanced by physostigmine and picrotoxin but suppressed by atropine. It is thought that these compounds influence the cells that release hormones responsible for the development of neckteeth.

Khlebovich and Degtyarev (2005) assumed that two alternative hereditary programs, corresponding to the typical and defensive (long-spined) forms, are present in *D. pulex*, as actinomycin D inhibits the transformation between these forms.

10.3 IMPACT OF XENOBIOTICS ON MORPHOLOGY

Inorganic and organic xenobiotics cause deformities of the carapace, the formation of abnormal setae on antenna, the disappearance of antennal segments, and the abortion of eggs and embryos (Fig. 10.2) (Shcherban, 1986).

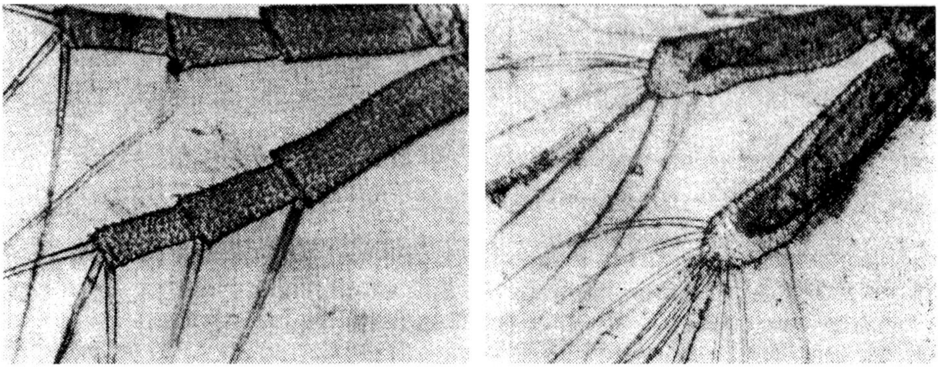

FIGURE 10.2 ***Daphnia* antenna deformation in a toxic environment.** Left, control; right, antenna deformed by a toxic environment. *Source: Shcherban (1980).*

D. magna growth is inhibited by the vertebrate antiandrogen, cyproterone acetate; this compound has no effect on molting and developmental parameters at concentrations up to 5 mM (which are nontoxic to *Daphnia*) (LeBlanc and McLachlan, 1999).

Copper (Cu) and nickel (Ni) reduced the induction of neck teeth in *D. pulex* in the presence of a *Chaoborus* kairomone (Hunter and Pyle, 2004); in contrast to Cu, 200 μg/L Ni decreased the length of neck teeth but increased their number. Later, Mirza and Pyle (2009) found that *D. pulex* neonates from mothers exposed to a kairomone + Cu had fewer and shorter neck teeth than those from mothers exposed to kairomone alone.

Zinc (Zn) insufficiency leads to "an increased demand on the animal's pool of available selenium" in *Daphnia*, resulting in cuticle deterioration (resembling senescent specimens) and depressed reproduction. The situation is improved by the addition of 5 ppb (parts per billion) Se (Keating and Caffrey, 1989).

In *D. magna*, Banlen solution (a herbicide mixture of 2-methoxi-4-chlorphenoxiacetic acid and 2-methoxi-3,6-dichlorbenzoic acid) at a concentration of 1.33 g/L produced a high percentage of juveniles with a deformed carapace, obviously due to disturbance in embryonic development (Trofimova, 1979). In addition, Laurox-9 causes teratogenic changes in *Daphnia* progeny such as deformation of the carapace, reduction in the number of segments in the branches of their antennae, and an increased number of setae on their segments (Shcherban, 1980).

The effects of the insecticide carbamyl and the *Chaoborus* kairomone on the formation of neck teeth in *D. pulex* are thought to be synergistic (Hanazato and Dodson, 1992). Carbaryl inhibits the formation of antipredator morphological features in *Bosmina longirostris* (Sakamoto et al., 2009).

In *D. magna*, embryonic abnormalities are also caused by fenarimol (an agricultural fungicide) via its antiecdysteroidal activity (Mu and LeBlanc, 2002). Malformations of the carapace and swimming setae are caused by a mixture of fluxetine and clofibric acid (Flaherty and Dodson, 2005).

Following exposure to tributyltin chloride, *D. magna* large fat cells (storage cells), which are especially numerous at the posterior curve of the digestive tract, become smaller, glycogen is lost, rough endoplasmic reticulum became less abundant and more vesicular, and their mitochondria are modified (Bodar et al., 1990b).

10.4 MOLTING

In Cladocera, molting and growth are inherently interconnected. Periodical molting is interconnected with periodic physiological processes. The integuments of freshly molted specimens are soft and thin.

At molting, all integuments of the body, including those of its minutest parts, are removed. Thus, exuvia may provide a convenient material for the investigation of morphology.

Molts may be numerous, reaching e.g. 48 in *Acroperus* (Smirnov, 1965a). The number of molts is likely to be different in different cladocerans.

In Anomopods, an embryo undergoes a molt just after leaving the brood pouch (Kotov, 1997). After this follow two molts (and accordingly two instars) prior to maturation in female chydorids and *Bosmina*, and up to six or more in daphnids and *Eurycercus* (Frey and Hann, 1985); and two to six molts in ctenopods (and even up to 10 in *Sida*) (Korovchinsky, 2004). In *D. magna*, five or more prereproductive molts have been observed (Anderson, 1932).

Juveniles are liberated from the brood chamber a few hours before molting, and eggs are released to the brood chamber a few minutes after molting (Rammner, 1929; Shan, 1969). However, molting also occurs in the absence of eggs or when eggs within the brood pouch decompose (Rammner, 1929).

In some benthic forms, such as *Ilyocryptus*, *Monospilus*, and *Oxyurella*, valves are retained during molting in parthenogenetic specimens. However, prior to bisexual reproduction, the accumulated valves are discarded. *Daphnia obtusa* grown on severely phosphorus (P)-limited green algae (*Scenedesmus*) are often observed with a postmolting exuvium attached to their posterior end (Sterner et al., 1993).

Cladocerans are attacked by various epibionts and endoparasites. Some of the epibionts are removed by movements of the postabdomen, but some stay in inaccessible places until the next molting. As molts occur every few days, they can be considered an instrument of sanitation for the animal's surface. Green (1974) counted an increasing number of attached epibionts during the intermolt period. He also discovered that some Peritricha leave *Daphnia* prior to molting, probably due to changes in the cuticula.

Molting (e.g. in *Daphnia*) is accompanied by a drain of calcium (Ca), P, and carbon (C) (Hessen and Rukke 2000a). In *D. magna* during one molt cycle (from one episode of neonate liberation and molting to the next ones) the content of potassium (K) and Zn decreased but that of Ca and iron (Fe) did not (Yukleevskikh, 2000).

Chitobiase (*N*-acetyl-β-D-glucosaminidase) takes part in exoskeleton degradation and recycling, as shown in *D. magna*. There is some flux of chitobiase from the old to the new cuticle. Chitobiase activity varies significantly over the molt cycle, with a fivefold increase 6 h or less before the next molt (Espie and Roff, 1995). During molting, chitobiase is released with the molting fluid into the aqueous environment. Duchet et al. (2011) found a positive correlation between chitobiase activity in the water and the number of *Daphnia* neonates produced during the observation period. It was therefore assumed that chitobiase activity may be used to assess the condition of the population without the destruction of specimens.

Molting is controlled by neurosecretion. Bosch de Aguilar (1969, 1972) concluded that the esophageal group (see Section 13.2) of the anterior region of the nervous system produces a factor that inhibits molting. Martin-Creuzburg et al. (2007) measured ecdysteroid levels during a complete molt cycle in *D. magna*. Free ecdysteroids predominated in whole-body extracts: their concentration increased sharply in the early premolt stage, followed by a sharp decline prior to ecdysis. It was concluded that molting is probably induced by 20-hydroxyecdysone. Only small amounts of ecdysteroids are found in newly deposited eggs.

It is notable that treatment of *D. magna* with ecdysteroids (ecdysone and 20-hydroxyecdysone) at concentrations of up to 5 μM in the culture medium led to unsuccessful exuviation (Bodar et al., 1990c). When tested on *D. magna*, estrogens (i.e. 17β-estradiol, diethylstilbestrol, and 4-nonylphenol) also inhibited molts, whereas bisphenol A did not (Baldwin et al., 1995).

In general, molting does not depend on illumination or darkness, although illumination with green light accelerates molting (Buikema, 1972).

In *D. magna*, the heart rate increases before molting, decreases after molting, and then increases again within several minutes (Meijering and von Reden, 1962). An intermolt pigmentation cycle (of carotenoid protein) of fat cells in *Simocephalus* was reported by Green (1966b); a sequence of development of the fat body and the ovaries (Sterba, 1956a) also follows the intermolt cycle (Fig. 4.13).

More experimental work on metabolic events, such as on respiration, during the molting cycle is necessary. An example of such a study is the uptake of ^{85}Sr (strontium-85) after molting, which is lost at every molting (Fig. 3.10) in *D. magna*: this was recorded over several moltings (Marshall et al., 1964).

Under natural conditions, *Daphnia* feeding on blue-green algae (*Microcystis*) were unable to shed their old integument although a new one had already been produced (Rohrlack et al., 2004). These authors found that blue-green algae contain the protease inhibitor microviridin J. From the gut of *Daphnia*, it penetrates into the blood and disrupts normal metabolism, which leads to a disturbance in molting, and thus of swimming and filtration, and is followed by death. Cladocera undergo periodic molting. Phylogenetically, Livanov (1955, p. 211) observed that "molting delimits the size of arthropods—abundant small forms are created—and in this way they adapt to exploit the smallest life spaces and, accordingly, undergo extreme adaptation to the highly restricted life possibilities of the 'empty cavities' in nature that they fill" [translated from Russian].

10.4.1 Impact of Xenobiotics on Molting

Some xenobiotics can interfere with the hormone-regulated molting process in arthropods by acting as antagonists of endogenous ecdysteroids. The exposure of *D. magna* to 20 μg Cd/L for 8 days led to a decrease in the whole-body ecdysterol concentration from c. 200 to 70 pg ecd equivalent/mg dry weight (Bodar et al., 1990b); this was probably the reason for their unsuccessful exuviation.

When exposed to PCB 29 (2,4,5-trichloribiphenyl), arochlor 1242, and diethyl phthalate, *D. magna* takes longer than the controls to complete molting (Zou and Fingerman, 1997). Exposure of *D. magna* to a solution containing 20-hydroxyecdysone or ponasterone caused incomplete ecdysis, followed by premature death (Baldwin et al., 2001).

10.5 SENESCENCE

There is only circumstantial evidence concerning senescence in Cladocera. Meijering (1958a, 1958b, 1962b) suggested measuring biotic time in heart beats, and found that mortality increases at the *age* of 18 million heart beats for *D. magna*. There are morphological signs of senescence as well as physiological signs, e.g. decreased heart rate and decreased fecundity (Dudycha, 2003). Dudycha demonstrated that such events may be different in different species of *Daphnia*. In senile females, growth is inhibited and may become *negative*, as noted by Frey and Hann (1985).

In moribund *D. magna*, the heart rate decreases (MacArthur and Baillie, 1929), and in old age fat within the fat body is depleted, the midgut epithelium is progressively degraded, the muscles degrade, and general activity decreases (Schulze-Robbecke, 1951).

As observed in *D. magna*, old males may become sterile (Meijering, 1962a). According to Skadovskiy (1955), if well-fed daphnias are given limited food, then their fecundity decreases by 2.5 times and they manifest signs of senescence (e.g. decreased heart rate).

Meijering and Redfern (1962) have named the period before death in *D. magna* the *debile phase*. It was noted that moribund *D. magna* lose sodium (Na) (Stobbart et al., 1977). Makrushin (3011) observed and confirmed histologically that natural death occurs in *M. macrocopa* before they lose the ability to reproduce, whereas senescent *Moina* sink to the bottom as their antennae stop beating.

With reference to *D. magna*, the hypothesis was proposed by Gainutdinov et al. (1997a) that death results from a program of elimination of the organism that is regulated by changes in the environment, perceived via thermo- and chemoreceptors and resulting from the integrative processing of sensory information.

In *Ceriodaphnia affinis* cultivated in a 5.5 nM solution of the antioxidant SkQ1 [10-(6′-plastoquinonyl)decyltriphenylphosphonium], longevity increased due to interruption of the senescence program involving the mitochondria (Anisimov et al., 2008). In a 55 nM solution, the opposite effect was observed, obviously due to toxicity.

10.6 IMPACT OF XENOBIOTICS ON LIFE SPAN

Cladocera live in a highly dynamic solution containing a great many inorganic and organic substances, including xenobiotics. There are numerous factual data on mortality rates at increasing concentrations of xenobiotics, and only some are discussed here. *D. magna* mortality was used as a parameter for the assessment of water quality and bottom sediments by Romanenko et al. (2011).

A 100% lethal dose (or LD_{100}) of Cd comprised a 2-day exposure to 2 mg/L Cd^{2+} (Mardarevich et al., 2001); and a median lethal dose (or LD_{50}) was provided by a 2-day exposure to a lead (Pb) ion concentration of 1 mg/L (Mardarevich et al., 2001).

The acute toxicity of 17 differently substituted benzaldehydes to *D. magna* was studied by Jin et al. (1998). The bioreactive toxicity of these compounds was quantitatively dependent on the nature of the substituted groups.

The larvicides spinosad and diflubenzuron significantly affect *D. magna* and *D. pulex* life spans; *D. magna* is more severely affected than *D. pulex* by diflubenzuron (Duchet et al., 2011).

CHAPTER

11

Reproduction

11.1 ANATOMICAL BACKGROUND

Female Cladocera have paired ovaries and males have paired testes, along both sides of the abdomen. In the ovary, there are several four-cell groups: one cell develops into an egg, while the others are *nurse cells* that are resorbed during egg formation (Weismann, 1877a, 1877b). The process of oogenesis was investigated more recently by Makrushin (1966–1980).

Spermatogenesis was described by Wingstrand (1978). The sperm of different species were found to be significantly different in form, sometimes even in representatives of the same genus.

Parthenogenetic females either lay many eggs into the brood pouch (i.e. they are polyembryonic) or normally lay only two eggs (most chydorids). In polyembryonic species, the number of eggs is not fixed and may decrease if the food supply is insufficient.

Cladocera are one of the rare groups of animals, together with aphids, that reproduce principally through parthenogenesis. With the exception of *Leptodora*, they have no larval stages in the free-living period of their life cycle.

11.2 CYCLICITY

Although they may be quite long, periods of parthenogenetic reproduction are interrupted by gamogenetic (bisexual) reproduction, which results in the formation of dormant eggs that serve for the latent period of the life cycle, termed *diapause*. The period of parthenogenetic reproduction along with the following period of gamogenetic reproduction was termed a *cycle* by Weismann (1880). If there is only one cycle per year, the population is called *monocyclic*; if there is more than one cycle, then it is called *dicyclic*, *tricyclic*, or *polycyclic*. When there is no period of bisexual reproduction, it is called *acyclic*. Males emerge from eggs that show no external differences from the usual parthenogenetic eggs (Banta and Brown, 1924).

According to Vereshchagin (1912), this cyclicity depends on the latitude (i.e. on the length of the ice-free period), but varies in different species. Frey (1982) considered the generation change further and noted that in the tropics gamogenesis is rather irregular and is unrelated to any particular season. He also noted that in the northern United States, as in Europe, gamogenesis occurs prior to winter for

Physiology of the Cladocera.
DOI: http://dx.doi.org/10.1016/B978-0-12-396953-8.00011-2

most species. Further south, gamogenesis in "northern species" becomes less intensive and tends to occur in spring, whereas in "tropical species" gamogenesis occurs irregularly.

Under laboratory conditions, the period of parthenogenetic reproduction, i.e. when only diploid females are produced, may be almost indefinitely long. Continuous parthenogenetic reproduction of some Cladocera spp. has been observed in the laboratory for up to 13 years (Banta, 1925). In culture, over 700 uninterrupted parthenogenetic generations have been recorded for *Daphnia pulex* and *Moina macrocopa* (Banta and Brown, 1929).

Part of *Daphnia* population may pass into gamogenesis while the rest of it continues parthenogenetic reproduction, as shown, for example, with reference to *Daphnia longispina* by Manuilova (1959). In addition, parthenogenetic reproduction may be resumed after the formation of gamogenetic eggs and ephippia, as has been shown in *Daphnia* (Sklayrova, 1938; Green, 1956a; Makrushin, 1968) and *Moina* (Makrushin, 1968). For *Daphnia*, *Simocephalus*, and *Moina*, the same female may, in the same brood, produce (1) parthenogenetic females, (2) parthenogenetic females and males, or (3) parthenogenetic females, gamogenetic females, and males (Papanicolau, 1910; Agar, 1920).

11.3 PARTHENOGENETIC REPRODUCTION

Parthenogenetic reproduction yields progeny that immediately start an active life of their own upon liberation from the parental brood pouch. There are no free-living larval stages in Cladocera, except in *Leptodora*. After an indefinite number of parthenogenetic generations, a cladoceran species may pass into gamogenetic (bisexual) reproduction, during which dormant eggs are produced, either surrounded by an ephippium (a reinforced part of the shell; in anomopods) or without a special case (in ctenopods).

The oocytes growing in the ovaries derive their fat from the fat body, and thus the content of fat in the fat body is inversely proportional to the development of the oocytes (Sterba, 1956a). Fat reserves (especially triglycerol) are transferred to the ovaries late in each intermolt period (Tessier et al., 1983).

Although Cladocera seem to be able to reproduce through an almost indefinitely long number of parthenogenetic generations, it has been shown that inbreeding *D. longispina* after the 279th, 278th, 442nd, and 570th generation results in an increasingly lower hatching rate of sexual eggs and the production of dwarfs, sterile individuals, specimens that produce nonviable parthenogenetic eggs, fewer young, and of mainly or all-male young (Banta and Wood, 1937).

Addition to the culture medium of gonadotropin, oxytocin, prefison, or prednisolone modified the life cycle parameters of *Daphnia magna* (Vinokurova, 1977). Following the addition of gonadotropin, oxytocin, or prednisolone, *Daphnia* reach maturity earlier. Embryogenesis is accelerated by gonadotropin, whereas prefison retards maturation and embryogenesis.

11.3.1 Fecundity

Cladocera may produce dozens of litters during their lifetime; thus, one specimen may potentially produce large numbers of progeny. According to Wojnarovich (1958, 1959), this can be up to 15 litters in *D. magna*, totaling up to 1055 juveniles. For one specimen of *D. magna*, Shpet (1968) recorded the potential progeny to be 10^6 individuals. *Moina* species are even more prolific.

In polyembryonic cladocerans, the number of parthenogenetic eggs produced is directly proportional to body size (Green, 1954), varies with season, and depends on environmental conditions. Dormant eggs are less numerous: one is present in ephippium of chydorids, one or two in daphnids, one or two in *Moina*, and up to 11 in *Eurycercus lamellatus* (Smirnov, 1971).

Dependence of Fecundity on Food Abundance: Lag Effect

Cladocera species have either a few large parthenogenetic eggs in a single brood (Aloninae and most Chydorinae) or numerous small eggs. The diversity of life histories within Cladocera, with respect to predation and food resources, was analyzed by Lynch (1989).

Polyembryonic families include the Daphniidae, Moinidae, Sididae, Eurycercidae, and some of the Macrothricidae. In polyembryonic daphnids, the number of eggs per individual is highly variable and may reach c. 40, depending on environmental conditions. For *Simocephalus*, the number of eggs in a single female may reach 40, but is 20 on average (Green, 1966a).

Except for *Archepleuroxus*, chydorids bear two parthenogenetic eggs or a single latent egg. When they have inadequate food, chydorids produce one parthenogenetic egg or none at all. A very rare exception was observed in Kosino Lakes (Moscow): two *Chydorus sphaericus* females contained three parthenogenetic eggs (instead of the usual two), a phenomenon never previously seen during decades of work with thousands of samples from every continent. A still rarer case was reported by Michael and Frey (1984.p. 94) for the usually two-egged *Disparalona rostrata*: "quite a number of *D. rostrata* females were carrying from 4 to 10 small parthenogenetic eggs. This is the first time in our examination of many thousands of parthenogenetic females from many different species."

The size of juveniles within the same brood may vary, as noted with reference to *Simocephalus* (Shkorbatov, 1953) and *D. magna* (Green, 1956a). Nonsynchronous development of embryos was also noted for *Daphnia*, *Evadne*, and *Podonevadne* by Rivier (1974). Moreover, Ramult (1926) found variations in the vitality of eggs within the same brood following exposure to solutions of mineral salts.

Reproduction and molting in Cladocera are profoundly influenced by the photoperiod (Parker, 1966). It is thought that the photoperiod influences production of a corresponding neurohumor. It was reported that marine representatives of Polyphemoidea (*Pseudevadne* and *Evadne*) release neonates during total darkness, i.e. between 10 PM and 4 AM (Onbé, 2002).

Egg production in *D. magna* is reduced at low concentrations of calcium (Ca) (Hessen and Rukke, 2000a, 2000b).

Effect of Resource Quantity

Planktonic Cladocera exist under conditions of fluctuating food stocks due to factors including the periodic development of algae. When food is in short supply, they produce smaller (as shown e.g. for *D. pulex* by Lynch, 1989) and fewer eggs. Tessier and Consolatti (1991) also observed that the quantity and weight of *Daphnia* neonates depend on the level of available food and the C:N ratio of the food: they called this "adaptive plasticity in reproduction."

For reactions to starvation, see also Chapter 4, section 4.6.

Reproductive Constraint at Low Food Concentrations

Under unfavorable low-food conditions, *D. pulex* produces fewer eggs e.g. two, or even one (Pyatakov, 1956). When there is an extremely large reduction in the number of eggs and the formation of only one egg during the intermolt period, the ovaries function alternately. The fewer eggs produced are also larger. The clutch size is smaller when food supply is limited, as shown for *Daphnia* (e.g. by Gliwicz and Noavida, 1996). As shown under experimental conditions by Gliwicz and Guisande (1992), when food is scarce, *Daphnia* produce small clutches of comparatively large eggs; in contrast, eggs produced with abundant food are larger, and offspring hatched from larger eggs can survive for longer during periods of starvation.

Effect of Resource Quality

Vitamins have been shown to affect reproduction. In *D. pulex*, the number of progeny was highest at a vitamin B_{12} concentration of 0.75 g/L; when deprived of vitamin B_{12}, they did not produce viable progeny (Keating, 1985). Following the addition of pantothenic acid to food consisting solely of *Chlamydomonas*, the life span of *Daphnia* increased by a factor of three (Fritsch, 1953).

11.3.2 Impact of Environmental Factors

The most intensive *M. macrocopa* reproduction was recorded at 28°C by Terao and Tanaka (1928b). Within the temperature range to which they were exposed, the highest fecundity was measured at >30°C in *Latonopsis* cf. *australis* and at 20–30°C in *Diaphanosoma brachyurum* (Chaparro-Herrera et al., 2011).

Effect of Xenobiotics on Fecundity

Since the first publications on toxicological experiments, numerous data have been reported on fecundity decreasing upon exposure to xenobiotics. For example, following exposure to 25–50 nM Cd^{2+}, *D. magna* fecundity decreased from c. 6.8 to 2 neonates/female, and at 100 nM Cd^{2+}, it declined to zero (Baillieul and Blust, 1999).

11.3.3 The Physiology of Parthenogenetic Eggs and Embryos

The physiological traits of eggs and their early developmental stages will be described next.

The density of *D. magna* eggs is about 0.37 mg dry weight (DW)/mm^3, irrespective of food availability to the females (Trubetskova and Lampert, 1995). The eggs are very strong and the egg envelopes of *Simocephalus* have been shown to withstand centrifugation at $1058 \times g$ (Hoshi, 1950a).

The size (volume) of parthenogenetic eggs varies within the same species and depends on the season and the length of the females (Green, 1956a). In *Bosmina longirostris*, Kerfoot (1974) found that the females carry small parthenogenetic eggs in spring-summer; in late fall, larger eggs are produced that contain double the yolk volume and produce larger young. This cycle is, however, not strongly associated with nutritional conditions. Egg size in *D. magna* was largest under conditions of low food availability and small at starvation level in the study of Trubetskova and Lampert (1995).

The chemical composition of *Daphnia hyalina* eggs 9.3% DW of nitrogen (N), 53.6% DW of carbon (C), and 1.2 % DW of phosphorus (P) (Baudouin and Ravera, 1972). The P content of *D. magna* eggs did not depend on its level in food and varied only slightly, within c. 14–14.6 mg P/g DW, with large variations in the C:P ratio (Becker and Boersma, 2005).

The color of parthenogenetic eggs is brownish, but may vary within the same species. Eggs may be green ("ova plerumque viridis") (Müller, 1785, p. 81; Sars, 1862), greenish-blue, orange, or brown in *Daphnia* (Konovalov and Konovalova, 1955) and blue or violet in *Moina* (Papanicolau, 1910). Clearly depending on the type of food available, parthenogenetic eggs of *Acroperus harpae* may be gray, green, orange, or violet, all of which develop into viable progeny.

Hemoglobin (Hb) is passed into developing oocytes in the *Daphnia* ovary, and the transfer is especially intensive in the last few hours prior to egg laying (Dresel, 1948). Two possible explanations for this were suggested by Dresel: either the Hb molecules can pass through the cell membranes or they must be broken and rapidly resynthesized in the oocytes. Before parthenogenetic eggs are laid, they receive Hb from the blood at the end of each instar, as shown for *Daphnia* by Green (1965b); the parthenogenetic eggs of *Simocephalus* also contain Hb (Hoshi, 1957). The Hb content in eggs is 0.17 mg Hb/g DW from pale *D. magna* and 0.53 mg Hb/g DW from red *D. magna* (Kobayashi and Nezu, 1986). During their development, the Hb content of embryos decreases (Fox, 1948), but embryos with their Hb blocked by carbon monoxide (CO) remain normally active (Fox, 1948). The Hb content is higher in early embryos of *D. magna* than in the hemolymph of adults, and in an anoxic environment a longer time is required for Hb deoxygenation in early embryos than in adults (Kobayashi and Takahashi, 1993). The Hb content in a single *D. magna* embryo was measured as 0.12 μg (Kobayashi et al., 1987). Moreover, the oxygen affinity of Hb in *D. magna* embryos is higher than in adults (Hoshi et al., 1977).

The carotenoid content of cladocera eggs depends on the quality of the food consumed by the adult (Green, 1957). Developing eggs are colored green or blue by carotenoid protein; in contrast, in fully developed embryos that are about to be released, the fat globules become orange colored due to the release of carotenoid from the bound protein, as was observed by Green (1957) in many littoral and planktonic species. About half of the mother's carotenoids are transferred to the eggs of each brood (Herring, 1968). This pigment is not utilized by the developing eggs or embryos and its presence does not enhance either the viability or fertility of the eggs. Carotenoids have never been observed in *Leptodora* eggs (Green, 1957).

The hatchability of *D. magna* eggs is best (100%) at an oxygen tension c. 15 torr (approximately 2000 Pa), and it dropped at c. 12 torr (approximately 1600 Pa). Hatchability was also highest at temperatures of 10–27°C and abruptly decreased at 30°C; 50% hatchability was recorded at 29°C for embryos from pale animals and at 30°C for embryos from red animals (Kobayashi et al., 1987).

After being laid into the brood pouch, the parthenogenetic eggs of most cladocera develop using the resources contained within the egg, independent of the maternal metabolism. The eggs may, therefore, be cultivated in a watch glass (Krogh, 1939; Kobayashi et al., 1987; Sobral et al., 2001). They develop well if removed no earlier than the first hour after transfer to the brood chamber (Krogh, 1939).

However, monoids and polyphemids are an exception (Weissman, 1877a, 1887b) in that *Polyphemus* females produce small eggs and possess placenta (the original German term is *Nährboden*) to supply nutrition for the eggs developing in the brood chamber. Patt (1947) presented cytological evidence of this process (Fig. 11.1). A kind of placenta is also present in *Moina*.

In other Cladocera, parthenogenetic eggs may be successfully cultivated in water, outside the brood chamber; this has been demonstrated for *Simocephalus*, *Scapholeberis*, *Eurycercus* (Rammner, 1933; Hoshi, 1951b), *Daphnia* (Obreshkove and Fraser, 1940a, 1940b), *Ceriodaphnia reticulata* (Shuba and Costa, 1972), and *Chydorus* (Green, 1961b).

As shown for *Simocephalus vetulus*, lipid oxidation predominates in early embryos, whereas free-living individuals rely on carbohydrate oxidation (Hoshi, 1951).

The sources of material for the development of cladoceran eggs are the yolk and oil. In a developing egg, the initial large fat globule diminishes in size and splits (Green, 1957). Later in the developed embryo, the fat is distributed as small globules in the fat cells. Relatively larger eggs convey to the embryos a greater relative amount of triacylglycerol that is left over from embryonic development (Goulden et al., 1987).

The egg membrane is slightly permeable to water (Krogh, 1939). The osmotic pressure in parthenogenetic and fertilized winter eggs is 57–72 mM and reaches 216 mM within

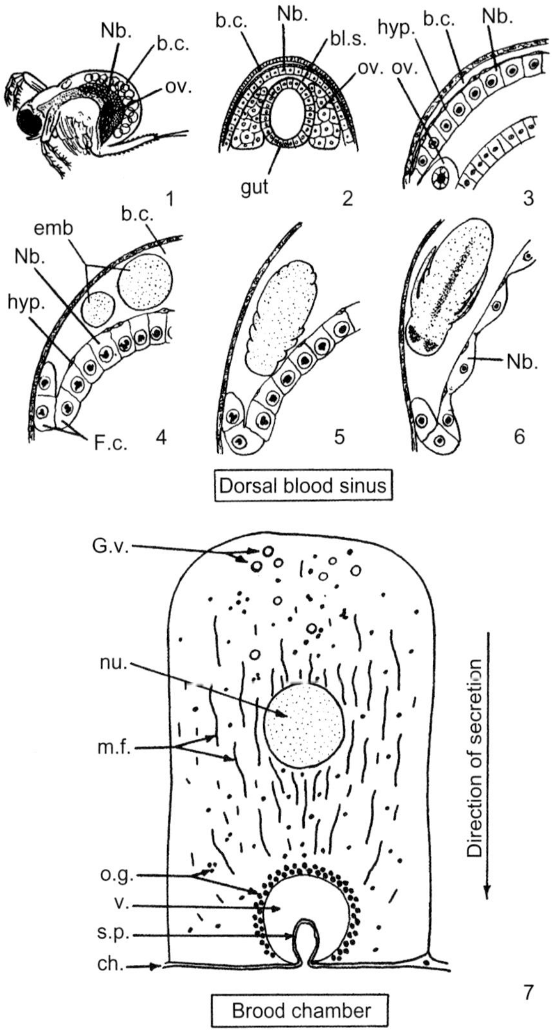

FIGURE 11.1 Placenta of *Polyphemus* during development. 1, a female; 2–6, development stages; 7, a typical Nährboden cell; b.c., brood chamber; bl.s., dorsal blood sinus; ch., hypodermis and chitin lining brood chamber; emb., embryo; f.c., marginal fold cells of Nährboden; G. v., Golgi bodies; hyp., hypodermis cell; m.f., mitochondria; Nb., Nährboden; nu., nucleus; o.g., osmiophilic bodies; Ov., ovary; s.p., secretion plate; v., secretion vacuole. *Source: Patt (1947).*

50–80 h (Krogh, 1939). In Cladocera with an open brood chamber, the total osmotic concentration of fluid in the embryo is supported at a level of about 5‰; during embryogenesis, this does not decrease below 4‰ (Aladin and Valdivia Vilar, 1987). When they start feeding, hypertonia abruptly increases because the total osmotic concentration of the hemolymph initially depends on the amount of ingested salts.

The respiration rates of *D. magna* eggs and early embryos are about a third of those of neonates and adults (Glazier, 1991). Thus, in brooding females they take a small proportion of the total oxygen consumption. It was reported in *Daphnia* embryos that oxygen consumption per unit weight increases at later developmental stages, in contrast to the post-embryonic period (Eremova, 1991).

The peak of RNA concentration occurs at the last stages of *D. pulex* embryo development and is followed by a gradual decrease during juvenile development (Gorokhova and Kyle, 2002), whereas the DNA concentration is highest in the early stages of postnatal development.

In their early stages, *Simocephalus* embryo development is mainly due to fat metabolism; at later stages, close to the emergence of juveniles, the respiratory quotient (RQ) shifts toward predominantly carbohydrate (i.e. glycogen) metabolism (Hoshi, 1950b, 1951): at the gastrula stage, the RQ is 0.74–0.8; in an embryo liberated from the egg's envelope, the RQ is 0.88; and in newborn juveniles, it is 0.99. In the embryo, glycogen is firmly bound to protein and fat; the glycogen content increases in the organs of the developing embryo and then drops in thereleased young (Hoshi, 1951a). Hoshi believed that free-living *S. vetulus* young predominantly use carbohydrate as a metabolic substrate.

In *Simocephalus*, the longest survival under anaerobic conditions occurs during gastrula, suggesting that anaerobic metabolism predominates in this period (Hoshi, 1957).

It has been suggested that *Ceriodaphnia* eggs can delay hatching in the presence of chemical cues from predators (Blaustein, 1997).

Finally, it may be added that *D. pulex* parthenogenetic eggs are reported to have remained viable (at least some of them) and to have hatched after passing through the digestive tract of a fish (*Carassius auratus*) (Samchyshyna, 2002).

Effects of Extreme Limits of Environmental Factors and Xenobiotics on Eggs and Embryos

At the extreme limits of temperature, Cladocera do not reproduce. During several weeks at 5°C, no reproduction occurred in *C. sphaericus* and *Pleuroxus denticulatus* (Keen, 1971). Food limitation results in decreased fecundity and interrupted reproduction.

A pH above 10, which often occurs in lakes, is detrimental for *Daphnia galeata* egg development (Vijverberg et al., 1996).

The effect of xenobiotics has mainly been tested in adult cladocerans. However, it has been shown, with reference to *D. magna*, that eggs and embryos are sensitive to cadmium (Cd), copper (Cu), dodecyl benzyl sulfonate, and 3,4-dichloroaniline (Sobral et al., 2001).

Mercury (Hg) at concentrations of 0.32 μg/L and higher is highly toxic to embryos, as shown with reference to *D. magna* (Khangarot and Das, 2009). The number of *D. magna* embryos was also affected by 15 aniline derivatives in a dose-dependent manner, although no morphological abnormalities were induced (Abe et al., 2001). The total number of offspring was slightly decreased in *D. magna* exposed to 17α-ethynylestradiol and norethindrone (Goto and Hiromi, 2003). In *D. magna*, 4-nonylphenol can be directly toxic and embryotoxic, depending on its concentration (LeBlanc et al., 2000): this substance interferes with the metabolic elimination of testosterone. Exposure of maternal *Daphnia* to testosterone caused abnormalities in the development of their embryos. However, 4-nonylphenol-induced abnormalities are different from those caused by testosterone. The concentration threshold for the induction of developmental abnormalities is rather low (c. 44 μg/L).

Maternal exposure of *D. magna* to the fungicide propiconazole at 0.25 mg/L led to developmental arrest at later stages of embryonic maturation, and thus to embryonic abnormalities and death (Kast-Hutcheson et al., 2001). Methyl farnesoate, pyriproxyfen, and fenoxycarb [ethyl 2-(4-phenoxyphenoxy) ethylcarbamate; a juvenile hormone analog] all disrupt ecdysteroid-regulated aspects of embryo development in *D. magna*; exposure of ecdysteroid-responsive cells from *D. magna* embryos to 20-hydroxyecdysterone reduces cell proliferation (Mu and LeBlanc, 2004).

The insecticide tanrek blocks the oocyte development in *D. magna* and is directly toxic to newborns and adults (Papchenkova, 2012).

Hanazato and Dodson (1992) considered the effect of the *Chaoborus* kairomone and the insecticide carbamyl on *D. pulex* neckteeth formation to be synergic.

11.3.4 Abortion of Eggs

A considerable proportion of the parthenogenetic eggs within a brood may be nonviable. In *D. cucullata* and *D. galeata*, 20–35% of females carried a number of nondeveloping eggs and the maximum proportion in such females was 70% (Boersma and Vijverberg, 1995). These authors noted that "it might be a more common phenomenon than is supposed" and mention nutritional deficiency as a possible cause. Threlkeld (1979, p. 611) noted that "[i]t is unclear how widespread such egg abortion is." According to more recent evidence, this is a widespread and quantitatively significant event under normal, natural conditions. In addition, some of the embryos may be resorbed during development.

In *Daphnia*, egg mortality and the abortion of embryos seem to be caused by nutritional deficiency (Gajewskaja, 1949; Boersma and Vijverberg, 1995). In *D. rosea*, under natural conditions eggs degeneration can reach 30% of their total number when there is an inadequate concentration of algal food (Redfield, 1981).

Resorption of embryos under extreme conditions has been observed in *D. pulex, D. pamirensis, Evadne*, and *Podonevadne* (Rivier, 1974), and Burgis (1967) recorded the presence of "sterile" (nondeveloping) eggs in *Ceriodaphnia*. For *Bythotrephes*, this has been demonstrated to be a natural process, not necessarily related to water pollution (Rivier, 2005).

In daphnids, parthenogenetic eggs have been shown to develop successfully outside the brood chamber (Obreshkove and Fraser, 1940a, 1940b); therefore, when liberated into the brood chamber, they do not receive any nutrition from the female. Egg mortality must therefore be caused by inadequate food during the period of egg formation.

Egg abortion is aggravated by extreme or toxic conditions. Lesnikov (1967) noted *Daphnia* pressing eggs out from the brood pouch following a prolonged exposure to phenols or oil. A noticeable percentage of eggs were also aborted by *D. magna* in the presence of a decomposing mass of the blue-green alga *Microcystis* (Barros et al., 2000) or the surfactant, laurox-9 (Shcherban, 1980), In addition, 3,4-dichloroaniline caused abortion or resorption of eggs by *Daphnia* (Ivanova et al., 1989). It has also been shown that *Daphnia* can lose eggs as a result of postabdomenal movements that clean blue-green algae from their limbs (Bednarska and Ślusarczyk, 2011).

11.3.5 Sex Hormones

Sex hormones are derivatives of steroids (i.e. they are steroid hormones): these include the male hormones, testosterone and androsterone, and the female hormones, estron (progesterone) and estradiol. Both growth and reproduction are steroid hormone-dependent processes. Although the presence of sex hormones may be assumed from general biological considerations, the first papers to directly demonstrate the presence and role of sex hormones in daphnids were only published at the beginning of 1994, although some experimental work was done earlier by Dehn (1950, 1955). Dehn (1955) was probably the first to draw attention to fat in food as a factor controlling cladoceran population dynamics. While experimenting with *Moina*, she found that higher quantities of fat, which occur in declining algal populations, stimulate the appearance of males. Dehn (1950, 1955) also found that feeding *Moina* with whole or defatted yeast with an addition of an extract (containing ergosterin) resulted in the production of 30% males, as well as females containing resting eggs.

Female Hormones

Estrogenic effects on *D. magna* were determined using the model estrogen diethylstilbestrol (DES; a nonsteroid vertebrate estrogen) (Baldwin et al., 1995). Chronic exposure to 0.50 mg/L DES reduced molting frequency and the fecundity of second-generation daphnids. Over a period of 21 days, 4-nonylphenol treatment decreased the number of offspring in the first and second generations, whereas DES had a similar effect on the second generation only. Exposure of female *D. magna* to DES and to methoprene (an insect juvenile hormone analog) stimulated development of the abdominal process (a characteristic of females) (Olmstead and LeBlanc, 2000).

The number of young produced by *Daphnia* doubled in a solution containing gonadotropin (i.e. the urine of nongravid cows, at 0.3–0.7 International Units (IU)/mL in the culture medium) in comparison with medium containing gonadotropin inactivated by heating at 68°C for 1 h (Čehović, 1954a, 1954b).

The normal content of glucocorticoids in *D. magna* is 8.4–12.7 pmol/g hydrocortisone and 6.9–10.1 pmol/g corticosterone (Polunina, 1999); when estradiol was added to the culture medium, these values changed to 8.3–22.1 pmol/g and 14–20 pmol/g, respectively.

Hydrocortisone at 0.25 mg/L increases the *D. magna* life span, as well as its heart rate and the number of eggs in the brood pouch, compared with controls (Kudikin, 2001). Fecundity is also increased by progesterone.

Male Hormones

The biotransformation of testosterone in *D. magna* was investigated by Baldwin and LeBlanc (1994a) (Fig. 11.2). Numerous metabolites are produced at phase I of the biotransformation: polar metabolites are preferentially excreted, whereas nonpolar ones are preferentially retained. Testosterone and phase I metabolites are excreted in glucose-conjugated forms and 2α-hydroxytestosterone as a sulfate conjugate. Nonpolar phase I metabolites of testosterone are not sulfate conjugated. These results demonstrate that *Daphnia* can convert polycyclic compounds into multiple polar and nonpolar metabolites.

Under conditions of chronic exposure to 0.50 mg/L DES the rate of elimination of hydroxylated metabolites of testosterone in *D. magna* is significantly reduced and that of testosterone glucose conjugates is elevated, thus indicating an alteration to their steroid metabolism (Baldwin et al., 1995). In *D. magna*, the elimination of several of the glucose-conjugated testosterone metabolites also decreased in proportion to the pentachlorophenol concentration; sulfate-conjugated metabolites also decreased (Parks and Le Blanc, 1996).

Testosterone metabolism in *D. magna* is also modified by xenobiotics. Nonylphenylpolyethylene glycol at 5.0 mg/L inhibits metabolic elimination of testosterone in *D. magna* (Baldwin et al., 1997). Interference of testosterone metabolism in *D. magna* by the xenobiotic 4-nonylphenol was also studied by Baldwin et al. (1997). The presence of 25 μg/L or 100 μg/L 4-nonylphenol in solution disrupted the testosterone metabolic pathway, resulting in the elimination of testosterone and the accumulation of androgenic derivatives. Gibble and Baer (2003) also reported that 100 μg/L 4-nonylphenol in solution modifies the formation of secondary morphological characteristics in *D. magna*.

Testosterone metabolism is enhanced by tributyltin at concentrations approaching lethal

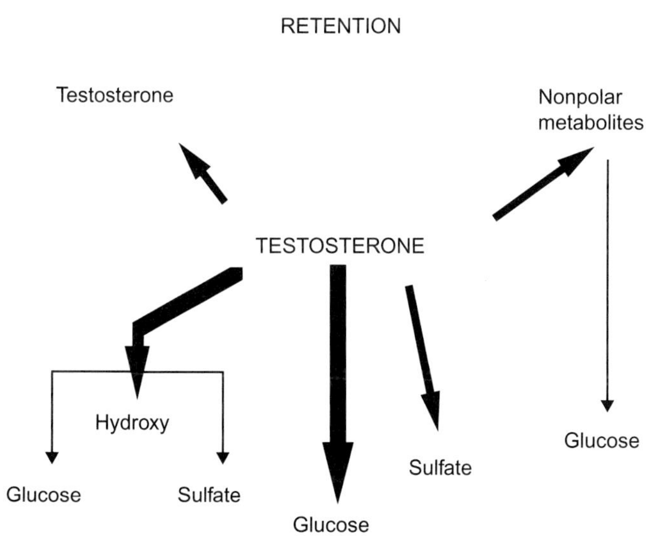

FIGURE 11.2 **Testosterone metabolism by *Daphnia magna*.** *Source: Baldwin and LeBlanc (1994b).*

levels (c. 2.5 μg/L); this was estimated by the elevated production of hydroxylated, reduced/dehydrogenated, and glucose-conjugated testosterone metabolites (Oberdörster et al., 1998). Exposure of *D. magna* to 1.3 μg/L tributyltin increases the rate of elimination of oxido-reduced, hydroxylated, and glucose-conjugated testosterone derivatives (LeBlanc and McLachlan, 2000). These authors showed that tributyltin causes modifications to the metabolism (Fig. 11.3) that result in an increased production of oxido-reduced derivatives that are preferentially retained *D. magna* tissues.

It has been shown experimentally that oxytocin increases the fecundity of *D. magna* (Polunina, 1999; Nikitina and Polunina, 2000). Quantitative differences in the concentrations of hydrocortisone and corticosterone in the body of *D. magna* were determined to be both seasonal and caused by exogenous oxytocin, prednisolone, testosterone, and estradiol (Nikitina and Polunina, 2000).

Olmstead and LeBlanc (2002) discovered a hormone controlling the formation of males in *Daphnia*. Exposure of *D. magna* cultures to solutions of methyl farnesoate (MF) at concentrations greater than 30 nM stimulated the production of males. MF is produced by the mandibular organ in crustaceans. The sensitive period in ovarian egg development occurs after 12–24 h. The development of embryos in *D. magna* takes two molt cycles: during the first cycle, an ovicell develops in the ovary; following molting, the eggs are transferred to the brood chamber, where they develop into embryos; and these are released just prior to the next molt. The beginning of the intermolt period in adult females is indicated by the presence of a molted exoskeleton.

The production of males may be influenced by various substances: 20-hydoxyecdysone (a crustacean hormone) at concentrations of 1 or 10 μg/L increased production of all-male broods in *D. pulex*, but a concentration of 100 μg/L reversed the effect (Peterson et al., 2001).

According to Banta and Brown (1924), in *M. macrocopa* the critical period in which eggs within the ovary may be induced to develop into either females or males is confined to 2–4 h before the eggs leave the ovary. Agents that reduce the maternal metabolism induce the production of males. The critical time mentioned above is the period when the spindle is formed during embryogenesis.

The exposure of sensitive ovarian eggs to 400 nM MF results in the production of 100% males, whereas exposure to 52 nM MF produces mixed broods. Similar to MF, the exposure of *D. magna* to pyriproxyfen or methoprene causes *D. magna* oocytes to develop into males (Olmstead and LeBlanc,

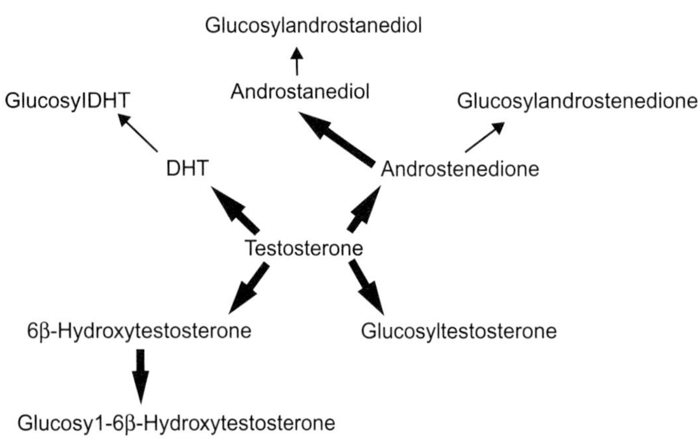

FIGURE 11.3 Metabolic pathways of testosterone in *D. magna* that are stimulated by tributyltin exposure (shown in bold). DHT, 17β-hydroxy-5α-androstan-3-one. *Source: LeBlanc and McLachlan (2000).*

2003). Tatarazako et al. (2003) exposed *D. magna* females less than 24 h old to MF, methoprene, two phenoxyphenoxy derivatives (pyriproxyfen and fenoxycarb), and juvenile hormone III; all resulted in the progeny being dominated by males (Fig. 11.4). Exposure to pyripoxyfen and fenoxycarb at a concentration of 330 ng/L resulted in the production of almost all males. Other substances cause the production of males only at higher concentrations (over 3.7 μg/L).

In *Moina* and *Ceriodaphnia* spp., male neonates appeared following treatment with fenoxycarb (Oda et al., 2005a). In *D. magna*, they appear when the mothers are exposed to juvenile hormones and the insecticides kinoprene, hydroprene, epofenonane, and fenoxycarb (Oda et al., 2005b). Exposure of *Daphnia* to MF also results in the formation of males and to an increased Hb content in the body (Rider et al., 2005). Clones that produced males also showed increased Hb production, whereas non-male-producing clones produced no Hb in response to the hormone. The authors therefore concluded that both physiological processes are regulated by the same signaling pathway.

Kim et al. (2006) obtained males in *Daphnia* and *Bosmina* species following MF treatment, despite the fact that, in three of their species, males had not been previously observed. Following this method, Sinev and La-Orsri Sanoamuang (2011) exposed females of tropical chydorids to 100 nM MF and obtained males, which are usually rare in nature.

The induction of male formation by externally added substances indicates the possibility that stimulation by substances that accumulate seasonally in natural water affects the appearance of males.

Xenobiotics may also induce the formation of males. The production of males increased in *D. pulicaria* in the presence of atrazine (at 2–2.3 ppb; Fig. 11.5) (Dodson et al., 1999), and in *Moina* and *Ceriodaphnia* in the presence of fenoxycarb (Shigeto Oda et al., 2005).

Natural Factors that Influence the Formation of Males

Various factors that cause the onset of gamogenesis also lead to the formation of males (listed in Section 11.4). The formation of males is induced by increased temperature (Fries, 1964), a decrease in day length to 12 h (Stross and Hill, 1965), and crowding. Males are produced by *Moina* following food shortages (Stuart and Cooper, 1932), and males and gamogenetic females are produced by *Moina* in a medium supplemented with metabolites that accumulate in their culture (Orlov and Cherepanov, 1986).

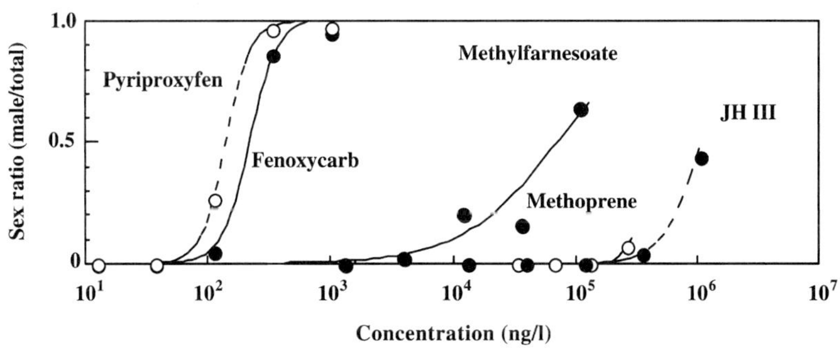

FIGURE 11.4 **Increased proportion of males is caused by exposure to various hormonal substances.** *Source: Tatarazako et al. (2003).*

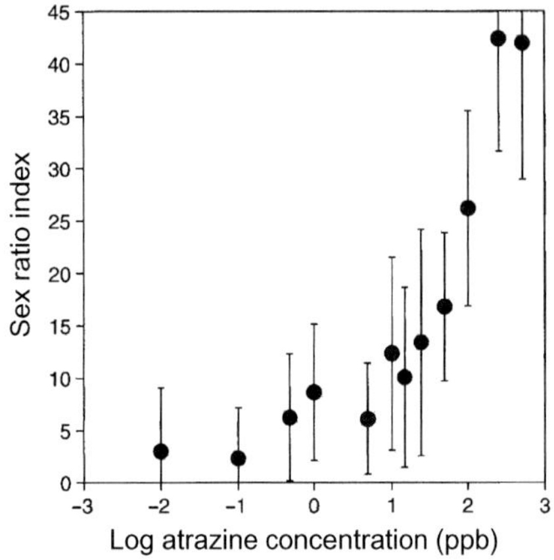

FIGURE 11.5 **Increase in male formation in the presence of atrazine.** *Source: Dodson et al. (1999)*

Fat is known to be accumulated by algae under conditions of depleted nutrients and by senescent algal cultures. Thus, two processes seem to occur in nature: (1) the accumulation of fat by cladocerans due to feeding on fat-rich algae and (2) stimulation of male production by the algal fat.

Formation of Males Following Exposure to Chemical Compounds

Very early studies by Banta and Brown (1930) found that chloreton, phenylurethane, and potassium cyanide (KCN) retard the development of *M. macrocopa* and increase the percentage of males in their progeny. Substances that disrupt the endocrine system have been shown to also perturb sex determination. Due to these effects, an increased ratio of males was produced in *D. magna* that were exposed to solutions of dichlorodiphenyltrichloroethane (or DDT), methoxychlor, and 4-nonylphenol (Baer and Owens, 1999). The premature appearance of males may also lead to reduced formation of ephippia. The development of morphological traits that are characteristic of a particular sex was shown to be dependent on biologically active substances; for example, the exposure of males to androsterone (a steroid vertebrate androgen) stimulated development of the first antennae (Olmstead and LeBlanc, 2000).

A combination of food deprivation and crowding also resulted in the production of males and ephippia by *D. magna* (Olmstead and LeBlanc, 2001b), while treatment with methoprene resulted in the production of males and inhibited the production of dormant (resting) eggs.

11.4 GAMOGENETIC REPRODUCTION: DIAPAUSE

At the onset of unfavorable conditions, Cladocera form males and gamogenetic females, and switch to bisexual reproduction. However, male-producing and non-male-producing females may coexist, e.g. as noted for *D. longispina* by Manuilova (1959) and for *D. pulex* by Innes (1997). This suggests the presence of different clones within a single population of a species, as shown by Carvalho and Hughes (1983) by differential photoperiodic induction of ephippia formation in *D. magna* (Fig. 11.6).

The onset of bisexual reproduction may be caused by different factors (see, e.g. Spaak, 1995): food shortage (Tauson, 1931), changes in the food quality (fat accumulation by declining algal populations) (Dehn, 1948, 1950, 1955), photoperiod (Buikema, 1968), crowding (Banta and Brown, 1929), and temperature are factors that are frequently mentioned, with explanations provided for each.

Gamogenetic reproduction was observed to be caused by changes in the photoperiod (in *Daphnia*) (Stross, 1965, 1969c) and by the accumulation of metabolic products in the medium (in *Moina brachiata* and *M. macrocopa*, but only by those from the same species) (Lopatina and Zadereev, 2007). In *Scapholeberis*

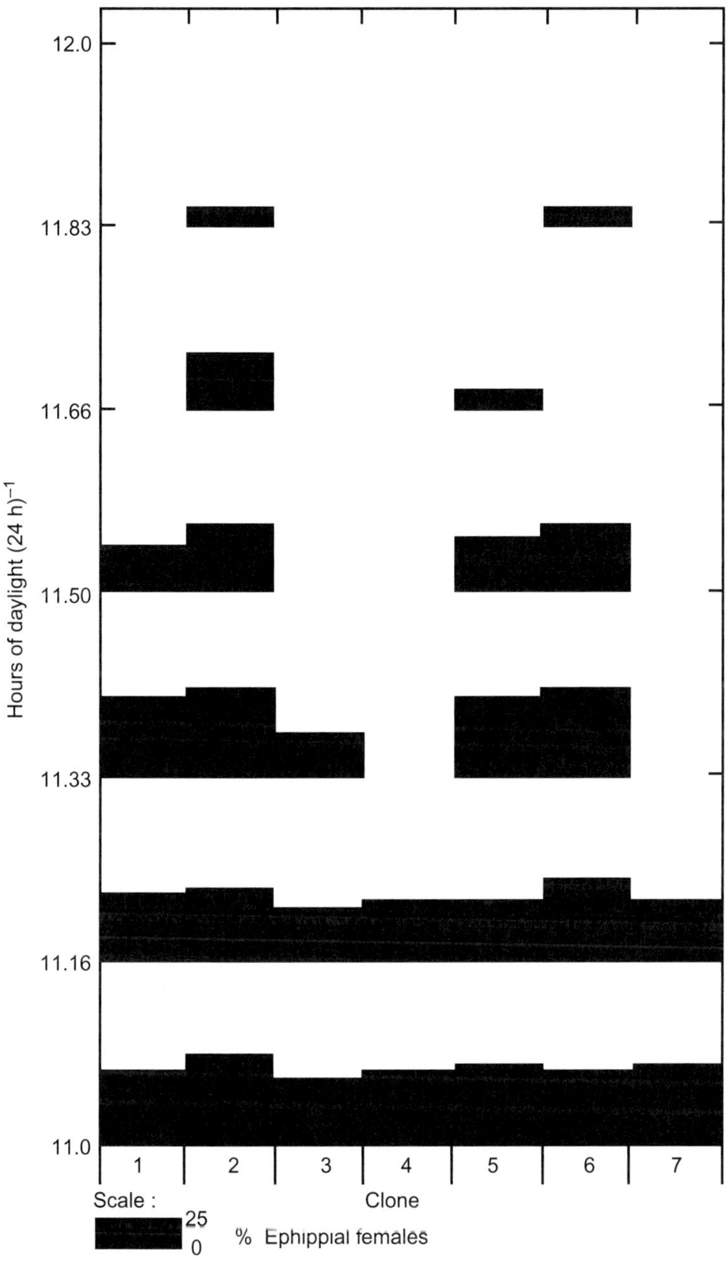

FIGURE 11.6 Differential induction of photoperiodic ephippia formation by different *Daphnia magna* clones. *Source: Carvalho and Hughes (1983).*

armata, ephippial females appeared under conditions of a short photoperiod (11.4 h light; vs. 15.5 h light) (Berner et al., 1991), usually after one or more parthenogenetic broods. This is likely to be a result of the combined action of several factors. It has been shown experimentally for *D. magna* that formation of gamogenetic eggs is induced by a combination of three factors: food limitation, short-day photoperiod, and chemically induced crowding (Kleiven et al., 1992).

Latent eggs are contained in the ephippium in anomopods or are liberated in ctenopods. The ephippium takes various forms in different species. In *Lathonura* and *Alonopsis*, it consists of an upper-posterior part of a shell, which is reinforced by special glands (Fig. 11.4) (discovered by Makrushin, 1970). In *Daphnia*, the same species and a clone may produce both buoyant and non-buoyant ephippia, as shown in *D. pulicaria* (Cáceres et al., 2007).

Protective melanistic coloration of ephippia varies from 0.5% to 96.5% of its surface, as shown for *D. pulicaria* (Gerrish and Cáceres, 2003), and is caused by genetic and environmental components. In most daphnids, ephippia with resting eggs mainly float on the water surface (Makrushin et al., 1990), whereas in other anomopods they sink or are glued to underwater surfaces (Fryer, 1972; Makrushin et al., 1990). In macrothricids, at least, the ephippia are attached to substrata, thus ensuring "continuity of an already established colony" (Fryer, 1972, p. 79).

Within the same species, the sequence of parthenogenetic and bisexual reproduction may differ in different water bodies. Manuilova (1951) notes that *D. longispina* in ponds (of the Yazhelbitsy Fish Farm, Russia) showed a gradual increase in bisexual reproduction from its initiation until the middle of July, when parthenogenetic reproduction was completely discontinued. In other ponds, the same species reproduced both parthenogenetically and bisexually throughout the summer. The latter was also the case for *C. reticulata* in ponds of the Yazhelbitsy Fish Farm, in contrast to those in Lake Kolomenskoe (Tverskaya Oblast, Russia).

It has also been observed that a female of *Simocephalus*, *Daphnia*, or *Moina* may produce parthenogenetic females, parthenogenetic females and males, or parthenogenetic females, ephippial females, and males within a single brood (Papanicolau, 1910; Agar, 1920; Orlov and Cherepanov, 1986; Zadereev et al., 1998). A single female *D. pulex* may produce only females, only males, or a mixture of both sexes (Innes, 1997). Cladocerans may therefore pass not only from the production of parthenogenetic eggs to the production of latent eggs in ephippia but may also backtrack from ephippia formation to formation of parthenogenetic eggs, as reported for *Moina* and *Daphnia* (Makrushin, 1969).

Ephippia formation has been experimentally induced in *D. magna* at low food levels, high culture densities, and short-day photoperiods (12 h:12 h light:dark periods) (Carvalho and Hughes, 1983). *D. schoedleri* switches to gamogenesis when there is an increased content of neurosecretory hormones in its body (Alekseev, 2010). In energetic terms, the formation of an ephippial clutch is equivalent to laying eight to nine parthenogenetic eggs (Zadereev et al., 1998).

Gynandromorphism should also be discussed. The presence of individuals with mixed female and male features (gynandromorphic) has been reported for daphnids, chydorids, and *Leptodora* (Frey, 1965). Experimentally, the formation of gynandromorphs was caused in *D. magna* by MF treatment (Olmstead and LeBlanc, 2007): especially high numbers of gynandromorphs were produced at concentrations of c. 12 μg/L, which is intermediate between the concentrations that produce 100% females and those that produce 100% of males.

The presence of hybrids between species is a rather common occurrence (Agar, 1920; Spaak, 1995, 2000; Flössner, 2000; Korovchinsky and Boikova, 2009). Naturally occurring hybridization was recorded in *Simocephalus* by Hann (1987) and in *Daphnia* by Hobæk et al. (2004). Shan (1971) experimentally demonstrated the possibility of crossing *Pleuroxus denticulatus* and *P. procurvus*.

11.4.1 The Physiology of Males

Almost all of facts summarized so far have concerned females. However, the size and behavior of males is different, which implies physiological differences. Unfortunately, data on the physiology of males are very scarce.

The life span of males is considerably shorter than that of females because, as shown with reference to *D. magna* (MacArthur and Baillie, 1929), they are more sensitive to temperature changes. *Daphnia* males have a higher beat frequency of their thoracic appendages than do females (Peñalva-Arana et al., 2007), a higher oxygen consumption (Vollenweider, 1958), higher heart rates (Fig. 6.8) (MacArthur and Baillie, 1929; Maynard, 1960), and a higher Hb content (Kobayashi, 1970).

Data on the growth of males is comparatively limited; it was briefly summarized by Frey and Hann (1985). In chydorid males, there are two immature instars and one mature instar (Frey and Hann, 1985). After their third (mature) instar, *Eurycercus longirostris* males live for up to 3 weeks (Hann, 1980). In *D. magna*, males live almost as long as females—up to 45 days at 28°C, and longer at lower temperatures (MacArthur and Baillie, 1929). When exposed to 10–50 mM NaOH or 1–10 mM KCN, the survival time of *Daphnia* males is 75–85% that of females (MacArthur and Baillie, 1929).

Males have a strong drive for attachment to females, as described e.g. by Van Damme and Dumont (2006). In the presence of females, *C. sphaericus* males attach not only to females but also to other males. Sometimes, males (in the presence of females) attach to several other males, making chains of two to three individuals (Smirnov, 1971).

There are several ways by which males can recognize conspecific females.

By Touch

In *Moina*, Goulden (1966, 1968) reported that males recognize the gamogenetic female of the same species by contacting their antennules with the surface of the female ephippial shell, which has various different sculptural characteristics in different species. The grasping antennules of *Moina* males are also somewhat different in different species. The same explanation was suggested for *Bosmina* by Kerfoot and Peterson (1980).

By Chemical Sensing

The attraction of males by glycoproteins was demonstrated in *Daphnia obtusa* and *Ceriodaphnia dubia* by Carmona and Snell (1995). The attractant (pheromone) was present in the area around the ovaries, where it exceeded the background concentration by two or more times. According to Carmona and Snell, after the male attaches to the female's valve, it then introduces its first antenna and touches the ovary's surface with its sensory papillae. More than one type of glycoprotein was found to be present at this site, thus indicating a potential mechanism for thee recognition of conspecific females by males.

11.4.2 The Physiology of Dormant Eggs

The following notes are clearly incomplete; however, they characterize the current state of investigations in this field. Resting (dormant) eggs are produced by all cladocera. In the absence of males, gamogenetic eggs are usually resorbed, either in the ovary or in the ephippium (Zhukova, 1955). In the absence of males, empty ephippia may be produced.

The same female may form several ephippia, one after another, as has been observed in *Moina* (Murakami, 1961) and *Eurycercus*. After gamogenesis, parthenogenetic reproduction may be resumed (Sklayrova, 1938; Green, 1956a).

The term *pseudosexual reproduction* has been applied to *Daphnia*, some races of which produce ephippial eggs and ephippia in the absence of males that develop without fertilization (Banta, 1925; Schrader, 1925). Such races have been called *thelytokous*, with the ephippial eggs being known as *pseudosexual eggs*.

During their formation in *Acroperus, Alona affinis*, and *Pleuroxus truncatus*, resting eggs are enveloped in a special membrane formed by the paired protoephippial gland (described by Makrushin, 1972). In *Lathonura* (Macrothricidae) and chydorids, Makrushin (1970) discovered glands that contribute to the formation of ephippia (Fig. 11.7). These glands accumulate a secretion during the formation of the resting eggs, which is maximal at molting. In *Lathonura*, the gland surrounds the inner surface of the hindgut, whereas in *Acroperus* it is situated in the dorsal surface of the trunk, inside the brood pouch.

Resting eggs are well protected from mechanical damage. In *D. pulex*, the envelope of resting eggs is 2.2 μm thick, whereas the outer wall of nonresistant eggs is only 0.35 μm (Seidman and Larsen, 1979). The shell composition of resting eggs of *Daphnia* includes crystalline calcium phosphate and magnetic material; their distinctive chemical composition and honeycomb structure are thought to ensure survival under harsh conditions (Kawasaki et al., 2004b). The distribution of elements over the integument of *D. magna* ephippia was found not to be homogeneous (Kawasaki et al., 2004b): P, Ca, and K were present mostly over the embryos, whereas sulfur (S) is distributed over the eggshell.

Ephippia formation in *Daphnia* has been shown to depend on illumination, not, however, on its intensity or polarization (Buikema, 1968), but rather on a short-day photoperiod, in combination with the population density (Stross, 1969a, 1969b, 1987; Stross and Hill, 1968). It has also been observed that *D. magna* does not form dormant eggs in complete darkness (Ślusarczyk et al., 2005).

Resting eggs may survive in ephippia for many years. Living Cladocera were obtained from sediments that had been dry for 13 years (Crispim and Watanabe, 2001). Much longer periods of viability have also been reported: resting eggs of *Daphnia* have been known to remain viable for more than a century (Cáceres, 1998; Kawasaki et al., 2004b) and even in some populations for longer than 200 years (Cáceres, 1998). Resting eggs remain viable after periods of drying, freezing, and exposure to the digestive system of predators (Kawasaki et al., 2004b). For example, latent *Alona guttata* eggs remained viable after passing through the digestive tract of a duck (Proctor, 1964).

The respiration rate of the resting eggs (i.e. of the developing embryos within resting eggs) was determined to be c. 2.5 μg O_2/mg DW/h in *Leptodora kindtii* at 4°C and increased to c. 6 μg O_2/mg DW/h at 6°C; in *Bythotrephes longimanus*, the respiration was c. 0.5 μg O_2/mg DW/h and did not change within a temperature range of 2.4–6°C (Andrew and Herzig, 1984).

Drying

Resting eggs may survive desiccation; it may be recalled that this fact was used by Sent-Iler (1860) and by Sars (1886, 1888) for rearing living specimens from dry lake sediment. This efficient method remains in current use (Van Damme and Dumont, 2010).

The number of resting eggs per unit area of the bottom of a water body may be very high, but is rarely actually counted. In *Daphnia*, the number (per m^2) of diapause eggs may reach 25,000 (*D. pulicaria* in Oneida Lake, New York) or 80,000 (*D. galeata mendotae*) (Cáceres, 1998); numbers of *Moina* ephippia may reach 600,000 (data from Japan) (Murakami, 1961). In the author's experience, floating ephippia of *Moina* that were driven by wind to one side of a pond (in the park of the Agricultural Academy, Moscow) could be collected in unlimited quantities (using 8-L pails). However, after long periods of desiccation, the percentage of hatching eggs was

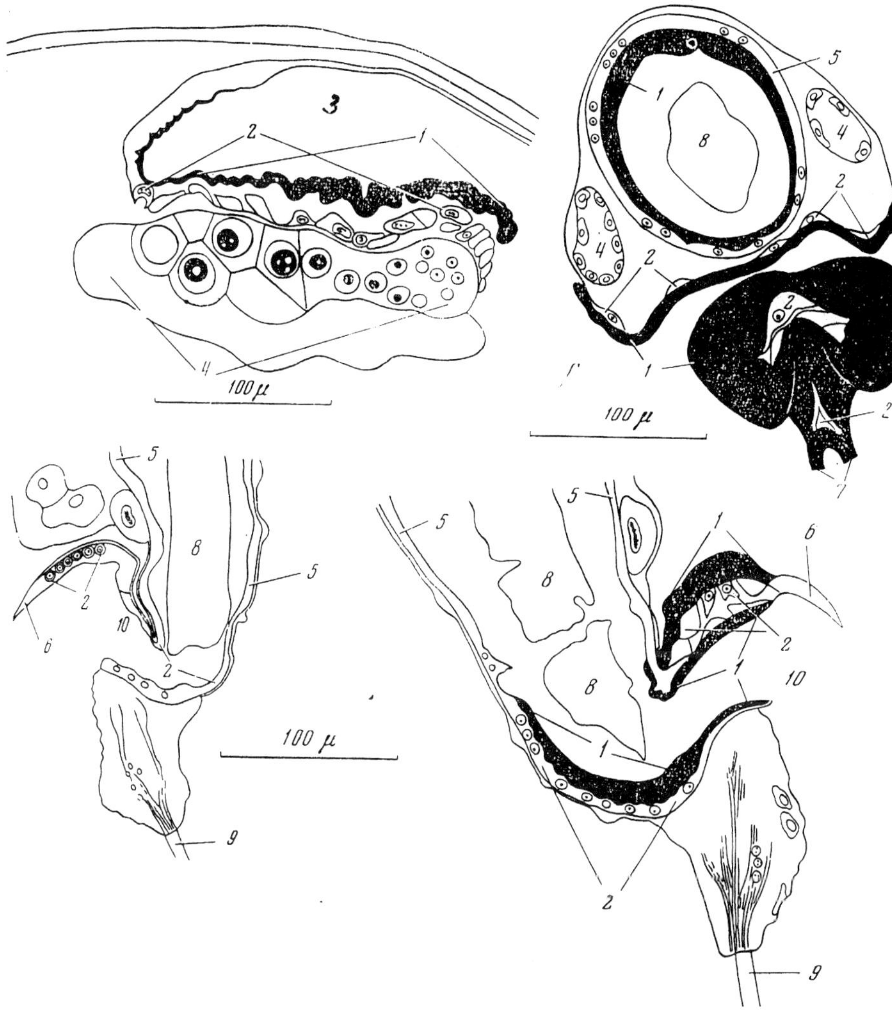

FIGURE 11.7 **Glands that participate in ephippia formation in *Lathonura* and *Alonopsis*.** Upper left, *Alonopsis elongatus* female prior to molting. Latent egg is outside the optical section. Upper right and bottom right, *Lathonura* gamogenetic female. Bottom left, *Lathonura* parthenogenetic female. 1, secretory gland; 2, gland cells; 3, brood chamber; 4, ovary; 5, gut; 6, claws; 7, bases of setae natatoriae; 8, gut contents; 9, seta natatoria; 10, anus. *Source: Makrushin (1970).*

observed to decrease, as was shown in *D. pulex* by Midzuno et al. (1960) and in *Moina* by Murakami (1961).

Dried and rehydrated ephippial eggs of *D. pulex* start developing when illuminated with light of wavelength 350–475 μm and energy of 2×10^6 ergs/cm^2 (2 kJ/m^2), with a wavelength of 410 μm being the most efficient (Davison, 1969). For 100% activation, resting eggs had to remain in the dark for at least

5 days. Drying turned out to be unnecessary for the development of resting eggs of *Streblocerus serricaudatus* (Fryer, 1972).

Makrushin (1978, 1980) attributes the ability of resting eggs to withstand desiccation to several properties of the yolk: its homogeneity, fine structure, and absence of vacuoles. Since living progeny may emerge from latent eggs of various ages, generations at different periods within a biological community may be mixed.

Freezing

Under European temperate conditions, latent eggs hatch after ice thawing within a rather brief period (Cáceres, 1998). Diapause eggs may withstand long-term freezing. Latent daphnid eggs can withstand freezing for 2 years and those of *Bosmina obtusirostris* can withstand repeated freezing (Makrushin et al., 1990).

Resting eggs, at least those of mid-latitude Palearctic cladocera, can withstand freezing down to about −50°C, judging by observations made in an area with this average minimum winter temperature (Yakutia in Siberia, Russia); the diversity of such fauna in this region is no lower than that of warmer European areas. In addition, mud samples collected in West Greenland and kept frozen at −18°C for 18 years produced *Chydorus*, *Alona*, *Macrothrix*, and *Daphnia* (Meijering, 2003).

Freshwater Cladocera ephippia can also withstand seawater, but under such conditions do not hatch and their resting period is extended (Meijering, 1970).

A certain period of freezing is thought to be necessary before the start of ephippia development. Illumination is also necessary for this process. Both factors seem to be different for different species and clones (Stross, 1966). The hatching rate of ephippial eggs of *Moina mongolica* increased with incubation for up to 62 days under conditions of alternative wet-dry cycles, and illumination was also necessary for hatching (Lu et al., 2000).

Cryoprotectants

Dormant eggs contain more glycerol and more Hsp60 (a heat shock protein) compared with subitaneous eggs (Pauwels et al., 2007). Glycerol is involved in energy storage and is also a cryoprotectant. The heat shock protein is thought to assist in maintaining structural integrity and inhibiting cellular metabolism. Hoshi (1957) noted that sexual eggs contain no Hb.

Dormant *D. magna* eggs contain approximately double the amount of trehalose than their asexual eggs, which increases their tolerance of freezing (Putman et al., 2011).

Hatching of a resting egg occurs due to rupture of the outer, inelastic membrane (chorion) by the increased size of the developing embryo and subsequent uptake of water through the embryonic membranes (as observed in *Daphnia* by Selvaraj, 1979), or due to the osmotic uptake of water, which stretches the elastic embryonic membranes and breaks the outer inelastic membrane (Davison, 1969; Fryer, 1996). The swelling is osmotic and can be retarded by sucrose solutions (Davison, 1969).

Resting eggs always produce females, as has been shown e.g. in *Pleuroxus denticulatus* by Shan (1969). The hatching of resting eggs depends on environmental conditions; for example, illumination favors the hatching of dormant eggs. *P. denticulatus* ephippial eggs stored in water in the dark for 84 days were induced by fluorescent illumination to hatch within 6–15 days; earlier times occurred at longer photoperiods (Shan, 1970). The hatching yield decreases at higher light intensities: in the light, up to 50% of *Daphnia* ephippial eggs hatched within 100 days, and only 0–2% in the dark. Moreover, several weeks of storage at room temperature was necessary before latent eggs could respond to light (Pancella and Stross, 1963) and previous desiccation reduced the hatching rate. Light perception by resting eggs is thought to involve carotenoids (Davison, 1969).

Exposure to decreased temperature and 12 h:12 h light:dark cycles resulted in the best hatching rates in *Daphnia carinata* (Tsitsilas and Barry, 2002). Moreover, it was shown that more resting eggs of *Acroperus* and *Alona rectangula* hatch at 10–20°C than at higher temperatures; and a photoperiod of 16 h light/day is generally more favorable (Vandekerkhove, Declerck, Brendonk, et al., 2005). However, these authors found no such preference, in other species, e.g. *C. reticulata* and *Daphnia parvula*.

Hatching of *Pleuroxus denticulatus* resting eggs occurs earlier at a higher illumination levels and longer photoperiods; however, their hatching rate decreased under continuous illumination at >600 foot-candles (fc; approximately 6458 lx) (Kuo-cheng Shan, 1970). Resting eggs removed from dark ephippia hatched earlier. However, a period of dormancy and drying is not always necessary for the further development of resting eggs. In *M. macrocopa*, the majority of resting eggs started development without drying within 2–24 days of their formation (Wood, 1932). It was then shown experimentally that dormant (sexual) eggs of *D. longispina*, as well as those of *M. macrocopa*, do not need drying for further development (Wood and Banta, 1937). These authors believed that altered osmotic concentration of the medium is responsible for the initiation of development of dormant eggs in *D. longispina*.

Failure of dormant eggs of *Daphnia* to hatch might be caused by degradation of biochemical components that are critical for responding to hatching stimuli and renewal of their development (Vanvlasselaer and De Meester, 2008).

Xenobiotics (as has been shown with reference to the fire retardant FireTrol 934) decrease the hatchability of resting eggs of *D. curvirostris* (Angeler et al., 2005).

11.5 IMPACT OF XENOBIOTICS ON REPRODUCTION

11.5.1 Inhibitory Effects

Generally, xenobiotics inhibit reproduction and reduce the number of progeny in Cladocera. Therefore, measurements of *D. magna* numbers have been used for the assessment of water quality and bottom sediments (Romanenko et al., 2011).

Cadmium (as cadmium chloride; $CdCl_2$) levels as low as 0.01 mg/L negatively influence the growth of *M. macrocopa* and *Macrothrix triserialis* populations (Garcia et al., 2004.). In *M. mongolica*, algae (*Chlorella*) preexposed to Cd caused a reduction in the number of neonates, which increased in subsequent broods; this effect was also caused by the lowest Cd concentration tested, 8.5×10^{-21} g Cd per *Chlorella* cell (Wang et al., 2010).

In contrast, Cu incorporation into food *Chlorella* did not cause a decreased brood size for the first brood of *M. mongolica*; however, a decrease was observed in all subsequent broods (Wang et al., 2007).

Ni reduced the reproduction rate of *D. magna* (a decrease of 33% in the total number of offspring per female) and reduced the number of juveniles in the first brood by 21–33% (Evens et al., 2009). Moreover, *D. carinata* maturation was delayed by the presence of lanthanum (La) at levels below the lethal concentration (Barry and Meehan, 2000).

The number *D. pulex* progeny decreased in the presence of cobalt (at 5 μg Co/L; provided in algal food) but was restored by the addition of vitamin B_{12} (Keating, 1985). The highest number of progeny was reached at 0.75–1.00 μg/L vitamin B_{12}.

At concentrations of 0.04–2.66 g/L, Banlen (a herbicide, consisting of 2-methyl-4-chlorophenoxyacetic acid and 2-methoxy-3,6-dichlorobenzoic acid) generally increased the number of eggs per brood in *D. magna*, but led to a high percentage of aborted eggs in subsequent generations (Trofimova, 1979). At a concentration of 0.66 g/L, this process was not accompanied by the production of deformed juveniles

M. macrocopa fecundity was reduced by 97% in the presence of 2 μg/L endosulfan (an insecticide) (Chuah et al., 2007).

D. magna withstood exposure to 0.1–1000 pg/L 2,3,7,8-tetrachlorodibenzo-p-dioxin (TCDD), but its reproduction rate decreased after 8 days (Wu et al., 2001).

Low concentrations of anionic surfactants (alkyl benzosulfonate, alkyl monosulfate, and sulfanole chloride) caused abortion of eggs and embryos in *D magna*, and sometimes resorption of their gonads (Shcherban, 1979). These surfactants are more toxic than are surfactants based on polymers. At superphosphate concentrations of >0.5 g/L, *Daphnia* embryos develop but cannot break their egg integuments and therefore perish (Sklayrova, 1938). Exposure to acetylcholinesterase-inhibiting pesticides (dimethoate and pirimicarb) caused reproductive damage to *D. magna*, thus decreasing the number of offspring (Andersen et al., 2006).

11.5.2 Stimulatory Effects

In contrast to their effects at high concentrations, low concentrations of various substances may stimulate vital functions or increase life span. This phenomenon is known as *hormesis*. With reference to *Ceriodaphnia affinis*, it was shown that its life span and fecundity increases at low concentrations of potassium dichromate ($K_2Cr_2O_7$), ethanol, or the antioxidant SkQ1 [(10-(2,5-dihydroxy-3,4-dimethylphenyl)decyl)triphenylphosphonium] (Gershkovich et al., 2012), tested at 1 μg/L Cr, 0.02–0.002 mg/L, and 0.03–3 μg/L, respectively. At 1 μg/L Cr, the average life span decreased compared with control. In addition, ethanol concentrations of 0.125% increased the reproduction rate of *Chydorus ovalis* compared with controls without alcohol (Yatsenko, 1928).

It was also shown that 25 μg/L $ZnCl_2$ increased the number of *Moina irrasa* progeny over six broods (Zou, 1997) and chronic exposure to 36 g/L fluoxetine increased fecundity in *D. magna*.

Saponin is lethal for *D. magna* at 5 mg/L [100% lethal dose (LD_{100}) in 2 h] (Naberezhnyi and Gorbatenkiy, 1973). At lower concentrations, it leads to a decrease in life span. At 0.2 mg/L and 2 mg/L, it initially has a stimulatory effect, i.e. there are more numerous progeny than in the control.

Cadmium chloride increased the total acid phosphatase activity in *D. magna*, although the activity of individual molecular forms of this enzyme either increased or decreased (Tsvetkov et al., 1997).

Although brief, this section is important because it describes how pollutants that are initially toxic and inhibitory may be diluted and can then stimulate various vital processes. (See also Chapter 10, section 10.5, concerning *Ceriodaphnia*).

11.5.3 Transgenerational Transfer of Xenobiotics

The transgenerational transfer methylmercury was studied in *D. magna* by Tsui and Wang (2004a). It takes 2.5–3 days for methylmercury assimilated in the gut to be transferred to eggs in the brood chamber. *D. magna* quickly acclimates to Hg, but animals that recover from Hg stress are more vulnerable to Hg toxicity (Tsui and Wang, 2005).

For selenium (Se), it was found that about 19–24% of Se in the F_0 generation is transferred maternally via reproduction to the F_1 generation (Lam and Wen-Xiong Wang, 2006); for Cu, c. 6.5% of the mother's Cu is transferred to the offspring within 7 days (Zhao et al., 2009).

11.6 PARASITES AND PARASITIC CASTRATION

Cladocera are attacked by both predators and parasites. Cladocerans parasites were noted by numerous early authors, such as Leydig (1860). Frequent and diverse *Daphnia* diseases caused by bacteria, sporozoa, fungi, and *Saprolegnia* were noted by Metchnikoff (1892). Parasites and epibionts of Cladocera have been investigated, and extensive information is summarized by Green (1974). Endoparasites comprise bacteria, fungi, Sarcodina, Plasmosporidia (Subclass Microsporidia), Cestoda, and Nematoda (Green, 1974). Heavy infestations of both epibionts and endoparasites decrease cladocera egg production. Epibionts are shed with each molting.

Metchnikoff (1888) described *Daphnia* that perished due to infestation by the bacterium *Pasteuria ramosa*. Sometimes, phagocytes ingest spores of this bacterium; this was described in detail, and it was confirmed that *Daphnia* egg production discontinued after infestation (Ebert et al., 1996; Auld et al., 2010).

As observed by Metchnikoff [1892 (1947)], spores of *Metschnikowiella bicuspidata* (syn. *Monospora bicuspidata*, Ascomycetes) that are consumed by *D. magna* are liberated from their coat and penetrate into the body cavity through the wall of the intestine, where they are surrounded, ingested, and destroyed by leukocytes (phagocytosis).

Microsporidia that are developing inside the body, especially in the ovary, decrease fecundity in Cladocera (Green, 1974; Voronin, 1986; Voronin and Makrushin, 2006), resulting in a general decline in the population. They have been found in *Alonella*, *Bosmina*, *Chydorus*, *Daphnia*, *Ilyocryptus*, *Moina*, *Pleuroxus truncatus*, *Scapholeberis*, *Sida*, and *Simocephalus*. In the summer of 1985, up to 97% of *B. longirostris* in the Volga Delta were infested with *Coelosporidium* (Gorbunov et al., 1995). A similar infestation of *Bosmina* was observed in the Rybinsk Reservoir (Russia) in the 1960s.

D. magna infestation with the microsporidium *Octosporea bayeri* induced a higher proportion of males in the progeny, and less sperm was produced in parasitized adult males (Roth et al., 2008).

Infestation of eggs by *Aphanomyces daphniae* (Saprolegniaceae) in the brood chamber of *D. magna* has also been reported (Stazi et al., 1994).

CHAPTER

12

Locomotion

12.1 ANATOMICAL BACKGROUND

Cladocera muscles do not form large compact masses and are mainly discernible individually. The general structure of the muscular system was studied in planktonic *Daphnia* by Binder (1929, 1931) (Fig. 1.3). Since these early reports, no other investigators have attempted detailed, general investigations in this critically important field. Abundant, but incomplete, data has been published by Fryer (1963, 1968) for some chydorids, macrothricids (Fryer, 1974), and daphnids (Fryer, 1991) (see Figs. 1.1 and 1.2).

Muscle fibers comprise the cross-striated skeletal muscles, the dilatators of the esophagus and anus, and the constrictors of the rectum; the longitudinal muscles of the gut consist of smooth fibers (Binder, 1929).

Littoral cladocerans have diverse forms: globular, lenticular, and elongated. In one case (*Graptoleberis*), lateral compression is combined with dorsoventral flattening and elongation. Most of these species crawl and only occasionally swim, which probably influences their hydrodynamic properties. In contrast to littoral species, the pelagic species are laterally compressed and are mostly not in contact with the substrata.

Organs involved in locomotion, as well as in the movements of the inner organs, are set in motion by muscles; exoskeletal structures, in combination with the muscles, form mechanical tools for the various vital actions of the animal and its component parts. The attachment of muscles to the integuments in *Daphnia pulex* was investigated histologically by Schultz and Kennedy (1977), who also studied *Daphnia* tendons and showed that they are located within muscle cells and not within the epidermis.

Thus, the system of locomotion clearly differs between crawling and swimming species. The former represent the ancestral muscular system, but the available data for crawling species are mostly fragmentary (Fryer, 1963, 1968, 1974).

The largest cladoceran muscles are the dorsal muscles and the three longitudinal ventral muscles that extend along each side of the body from the head to the postabdomen. The lowest ventral muscle rises at the posterior and is attached in the upper zone of the abdominal region in *Daphnia* (according to Binder, 1931; Green, 1956c) and *Eurycercus* (Green, 1956c), and to the postabdomen in chydorids; the uppermost ventral muscle

Physiology of the Cladocera.
DOI: http://dx.doi.org/10.1016/B978-0-12-396953-8.00012-4

descends and is attached in the postabdomen. The crossing over of these muscles allows the possibility of postabdomenal movements in a medial plane.

From the lateral viewpoint, three large antennal (paired) muscles can be seen in the daphnian head: the two anterior muscles are the antennal abductors, and the third muscle is the antennal levator (Binder, 1931). There is also a small flexor antennula that is well represented e.g. in chydorids but reduced in *Daphnia* females.

The labrum is set in motion by levator labri; there are also other small muscles within the labrum. In addition, the adductor muscle of the carapace extends from valve to valve under the gut.

In crawling species, movement is principally achieved by the action of muscles of the thoracic limbs, postabdomen, and antennae, but in pelagic, free-swimming species this is done by means of the antennal muscles.

The muscle system has been studied in a few species, but variations among different families have not been reported; in general, this is rarely commented upon. No detailed comparative investigation has been made in differently specialized species. Investigations of transformations in the muscle system and relevant skeletal structures in different genera are therefore highly desirable. Such studies would contribute to our understanding of fundamental issues regarding the phylogeny and mode of life of Cladocera.

12.2 ENVIRONMENTAL BACKGROUND

Littoral Cladocera live among various substrata that offer all types of surfaces. Frequently, there are filamentous algae of various diameters commensurate with the crawling appendages of chydorids. There are also small and large clumps of organic debris and there may be both fine-grained and coarse-grained sediments of organic and mineral material on the bottom of water bodies.

Cladocera may stick to the surface film of water, from which they are not generally able to free themselves. The only reported exception was observed by Green (1956c): *Camptocercus lilljeborgi*, which possesses a unique system comprising a long, reinforced abdomen and a postabdomen containing specialized muscles, can rotate its postabdomen outside the shell and thus free itself from the surface film.

Pelagic Cladocera usually live their entire life cycle without making any connection to a substrate. The aquatic environment is viscid (Zaret, 1980) and may be characterized by the following parameters (Eq. 12.1).

$$\text{Reynolds number } (Re) = \rho \mathrm{u}\mathrm{a}/\eta = \mathrm{u}\mathrm{a}/v \quad (12.1)$$

where

ρ is the water density (-1 g/cm^3);
u is the flow rate in relation to the body;
a is the length of a swiming animal;
η is the water viscosity (=0.010 ps at 20°C); and
v is kinematic viscosity (=η/ρ; for water, v = η cm^2/sec.
For cladocera, length "a" may be assumed to be 0.1 cm.
For small chydorids, "u" may be taken to be 5 mm/sec.

Under these conditions, for most chydorids,

$Re = 0.5$ cm/sec $\times$ 0.1 cm/0.01 cm^2/sec = 5 (which is much lower than for rapidly swimming and larger fish).

It may be assumed that for all Cladocera the *Re* ranges from 0.1 to 100. At slow swimming velocities, Cladocera movement depends on the viscosity of the aquatic environment, which in its turn directly depends on temperature. Little is known about the hydrodynamics of small moving objects, such as cladocera; this is therefore a promising field of investigation (as noted by Fryer, 1968, p. 347).

12.3 MOVEMENT TRAJECTORIES

Cladocera are slightly heavier than water and sink if they make no effort to swim. It has been shown experimentally that living *Daphnia* sink more slowly than predicted by Stokes' law, and more slowly than anesthetized specimens (Dodson and Ramcharan, 1991; Gorski and Dodson, 1996). The difference may be due to their form and to the feeding current produced by the *Daphnia*. A potential hydrostatic role for oil drops that are situated at certain places within the body is evident but has not been especially studied in Cladocera.

Pelagic Cladocera and most of the littoral species are always moving. Benthic species of the genus *Ilyocryptus*, on the other hand, either move slowly or do not crawl for long periods, and some species may be immobile for long periods e.g. *Lathonura* (before suddenly jumping) and *Drepanothrix* (both belonging to the family Macrothricidae). *Scapholeberis* stays, at least for some of the time, attached to the surface film.

The movement of Cladocera species is rarely unidirectional, and their tracks are mostly complicated and tortuous; however, until recently, methods for recording three-dimensional movement had not been developed (Uttieri et al., 2005).

12.3.1 Littoral Cladocera

Littoral cladocerans exploit various kinds of movement. Particular species may crawl and occasionally start swimming. Species living on the bottom may run rapidly or swim along tortuous trajectories, or they may be slow or even immobile for long periods. Fryer (1968) observed crawling (running on and between substrata) and scrambling (among clumps of debris and algal filaments) movements. The reasons for such tortuous movement have not been studied, although Scourfield (1905, p. 2) suggested that such movements may depend on the absence of landmarks from which the animal might take a bearing: "It is the difficulty of keeping a straight course owing to the want of fixed points or datum lines, for both horizontal and vertical directions, from which bearings may be taken."

Crawling is performed by means of thoracic limb I and is aided by the postabdomen. In most Aloninae (Chydoridae), their antennae also participate in crawling and are used when such species occasionally switch to swimming. In Chydorinae (Chydoridae), the antennae are pressed to the body while crawling and are only used while swimming. Crawling Cladocera only move forward (never backward).

The gaits of crawling Cladocera have not been investigated.

12.3.2 Pelagic Cladocera

The trajectories of planktonic cladocerans are mainly linear and consist of sinking and refloating ("hop-and-sink" behavior), as well as a progressing vector (Lochhead, 1961). Changes in *Daphnia* swimming speed in response to light intensity was discussed by Ringelberg (1964). However, swimming behavior may be variable and individualistic. Fryer (1991, p. 14) notes: "Even large, heavily built species, such as *Daphnia magna*, do not merely 'hop' essentially in the vertical plane, but can swim horizontally, dive steeply and rapidly head first, pursue a meandering course that can be changed with great rapidity, swim with the body inclined to one side, and orientate and navigate with great precision."

The observed swimming behavior of *Daphnia* spp. in three-dimensional (3D) space is strongly affected by the light level, food availability, and the volume of an observation

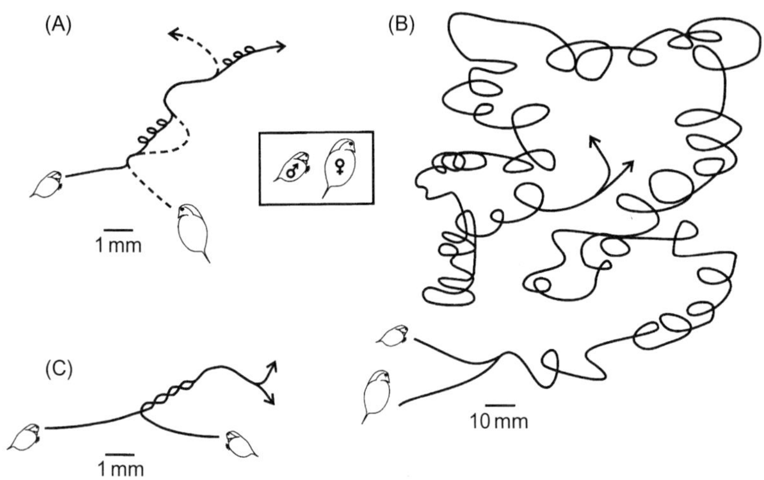

FIGURE 12.1 Trajectories of mating *Daphnia*. A, male unsuccessfully pursuing female. B, male successfully grasping female C, two males pursuing one female. *Source: Brewer (1998).*

chamber (Dodson et al., 1997). In the dark, *Daphnia* spp. swam slowly and exposure to infochemicals from fish changed neither their distribution pattern nor swimming speed (Larsson et al., 2000). Using video recording and computer analysis of their motion, O'Keefe et al. (1998) obtained digitized 3D video records of *Daphnia* movement. Large differences were observed in the swimming speeds of individuals belonging to different clones of the same species. Uttieri et al. (2004) noted that a random component is predominant in *D. pulex* movements.

Daphnia trajectories were recorded using a digital image analysis system (Baillieul and Blust, 1999) or a Critter-Spy, i.e. a high resolution 3D recording system (Seurat et al., 2004), while a 2D representation of the movements of mating *Daphnia* was obtained by Brewer (1998) (see Fig. 12.1).

Hebert (1978b) seems to have been the first to relate the swimming ability of *Daphnia* to the size of the antennal muscles, and made special measurements. He pointed out the relationship between their swimming capacity and the size of their antennal muscles. He also calculated the ratio of the length of the first antennal abductor (which he terms the *adductor*) to the carapace length in seven *Daphnia* species and found it to vary from 0.15 (in *D. nivalis*) to 0.33 (in *D. projecta*). He also noted that the force of the stroke depends on the cross section of the muscle. The next step in evaluating the swimming capacity involved measuring the stroke rate of the antennae. Hebert (1978b, p. 318) also correctly highlighted the usefulness of this measurement for understanding swimming in Cladocera: "a thorough investigation of locomotion in *Daphnia*, with particular regard to the relationship between muscle structure and propulsion in media with varying viscosities, would be well worthwhile."

12.3.3 Swimming Parameters

Movement may be measured as absolute units or in body lengths.

Littoral Species

The swimming rate of small chydorids is 2.4–7 mm/sec and is 17 mm/sec for 2-mm long *Eurycercus lamellatus* (Smirnov, 1971, 1975). As expressed in the number of body length per second, it is slowest in *Paralona pigra* (syn. *Chydorus piger*) (=4.5) and fastest in

Alonella exigua (=47), but is much slower in *Alonella excisa* (=19) (Fryer, 1968). In most littoral chydorids (with a body length of 0.3–1 mm), the swimming speed is from 10 to 20 body lengths/sec (Fryer, 1968); the rate of antennal strokes in chydorids is 252–406/min (Smirnov, 1971).

Pelagic Species

The rate of antennal beats in *Daphnia* ranges from 75 to 250/min (McMahon and Rigler, 1963; Fryer, 1991) but is variable and is usually interrupted by periods of rest (Fryer, 1991, p. 12). In *Bosmina longirostris*, the stroke rate varies within 5–40 Hz (Zaret and Kerfoot, 1980). It is thought that the antennal rate increases in gravid *D. magna* due to both its increased weight and the necessity of counterbalancing the displaced center of gravity (Fox and Mitchell, 1953; Lochhead, 1961). In swimming *Ceriodaphnia reticulata*, the beat rate of their antennae is c. 1 every 2 sec (Seely and Lutnesky, 1998).

According to Kerfoot et al. (1980), the speed of Cladocera varies from 0.01 to 1.0 cm/sec. The swimming speed of planktonic *Daphnia longispina* is 1.4 cm/sec at a current speed of 6 cm/sec (Luferova, 1962). At a moderate current speed, most *Daphnia* swam against the current, with a swimming speed of 0.6 cm/sec. Szlauer (1965) observed that the swimming speeds of adult and young *Daphnia* are different, which leads to their different distributions.

In *Ceriodaphnia quadrangula*, the swimming speed in pale and red specimens was about 43 cm/min at a low oxygen concentration and 25°C; increasing oxygen concentration was followed by an increase in swimming distance. In red specimens, when the hemoglobin (Hb) was blocked by carbon monoxide, the swimming activity increased at increasing oxygen concentrations in almost the same way as occurred in pale specimens (Kobayashi and Yoshida, 1986).

Under conditions of scarce food, hungry *Daphnia* show a tendency to wander, compared to satiated ones; this reaction starts after a delay, when the intestine becomes empty (Szlauer, 1962). It has also been observed that the movement of *D. obtusa* grown on severely phosphorus (P)-limited green algae (*Scenedesmus*) is comparatively sluggish (Sterner et al., 1993).

Trajectories

The behavior of littoral Cladocera is extremely diverse; there are sluggish ones with prolonged immobility and to rapid ones that run along various trajectories. The various trajectories of Cladocera, especially those of the littoral species, have been little studied and make a promising field of investigation.

Pelagic cladocerans e.g. *Daphnia*, exhibit more strictly defined movements, which normally consists of sinking, refloating, and moving forward. The swimming tracks of *Bosmina* and *Polyphemus*, and their turning rates, were compared in a study by Young and Taylor (1990). The tracks of *Polyphemus* were mostly oriented orthogonally to those of *Bosmina*, thus providing them with a greater probability of encountering prey. Kerfoot (1978) filmed the trajectories and akinesis (dead-man response) of *Bosmina* when attacked by *Cyclops* (Fig. 14.1).

12.4 MUSCLE PHYSIOLOGY

Information about the physiology of cladoceran muscles is scarce. Sollman and Webb (1941) exposed striated (i.e. voluntary) muscles of *D. magna* to mecholyl, pilocarpine, and physostigmine and observed little or no effect. Epinephrine and nicotine stimulated muscular activity in *Daphnia*, but curare first increased and eventually arrested all movements. These authors also applied electrodes directly to the shell of *Daphnia* and gave them a single electric shock. This resulted in a single contraction of their postabdomen and antennae, and a few beats of their thoracic limbs.

Flückiger (1952) observed the action of the swimming antennae muscles of *Daphnia* to be retarded by L-adrenaline bitartrate, D-adrenaline bitartrate, L-noradrenaline bitartrate, L-ephedrine hydrochloride, dihydroergotamine methanesulfonate, and oxyphenylethanomethylamine tartrate. However, no clear activity was observed in the muscles of their thoracic limbs.

A swarm of *Daphnia* emits a weak electric field, originating from their muscular activities, which can be perceived by fish (Freund et al., 2002).

The biomechanics of cladoceran movements comprise levers, joints, and methods of muscle attachment to the skeleton.

12.4.1 Joints

Movement of the Cladocera, as well as some aspects of their food handling, is related to the structure of their joints. Generally, the following kinds of joints are present in the body:

1. cylindrical joints, in which freedom of movement is controlled by their mechanical potential (e.g. joints of the antennal branches);
2. concertina-like joints (such as those at the base of the antennae);
3. telescopic joints (as in segments of the abdominal part of the body); and
4. a lever joint (between the body and the postabdomen), which is more developed in bottom-living chydorids.

12.4.2 Levers

It is obvious that the general body form of cladocerans is related to their locomotion. It is notable that the posterior part of the body is transformed into a special region, the postabdomen, which is bent at 90–180° to the body axis.

The postabdomen is the principal lever that pushes the animal (thus liberating it from obstacles) and cleans its food-collecting apparatus; it is especially well developed in bottom-living cladocerans, particularly in chydorids. Concerning the major joint of *Eurycercus*, Fryer (1963, p. 341) noted that "[e]nergy losses at joints are reduced to the absolute minimum as only a single 'joint' is involved" (i.e. between the postabdomen and the rest of the body). To understand the action of these levers, both their exoskeletal structure and the muscles that control their action should be considered. The resulting effort depends on where the muscles are attached and the length of the postabdomen. *Camptocercus lilljeborgi*, with its very long postabdomen, makes a unique, extreme case. Green (1956c) observed that its postabdomen may move outside the shell and liberate the animal from obstacles, including from adhesion to the surface film of the water. Green also discusses this problem with respect to his observations on the structure of the postabdomen and of postabdominal muscles in *Eurycercus* and *D. magna*. These studies provide a starting point for further useful investigations. Little, if any, work has yet been done on this subject.

Chydorids crawl on substrata using thoracic limb I, use their postabdomen for pushing, and occasionally swim by means of their antennae. In some chydorids (e.g. *Alona affinis*), the antennae are used in crawling on substrata and so may therefore be considered to form a pair of levers.

By immobilizing *Daphnia* with D-tubocurarine chloride and measuring their oxygen consumption by means of a Cartesian diver, O'Conner (1950) found that muscle activity consumes 32% of their total metabolism and during intensive activity may take up most of their oxygen consumption.

12.5 IMMOBILIZATION

Now, we will discuss more about immobilization. Both littoral and pelagic Cladocera are very mobile. Their investigation often requires either retarding their movements or complete immobilization, as practiced by various authors. There are two ways to immobilize cladocera: *physical* (e.g. attaching them to something, as mentioned in Chapter 2, "Methods") and chemical immobilization.

12.5.1 Chemical Immobilization (Anesthesia or Narcotization)

Immobilization may be caused by low temperatures or by exposure to carbon dioxide. In a solution saturated with carbon dioxide, *Alona*, *Bosmina*, and *Chydorus sphaericus Scapholeberis*, and *Simocephalus* lost mobility within c. 0.1–1 min, and *Daphnia* in 14 sec (Nikitinsky and Mudrezowa-Wyss, 1930; Mudrezowa-Wyss, 1933). Sollman and Webb (1941) observed that carbon dioxide produced general depression of *D. magna*, including decreased movement, and the *Daphnia* sank to the bottom.

Of practical importance is narcotization with carbon dioxide (using a saturated liquid) prior to preservation (e.g. with formalin) (Infante, 1978). Under these conditions, the animal, e.g. *Diaphanosoma*, is preserved in a dilated state.

In experimental practice, *Daphnia* have been immobilized with a chloretone solution (Anderson, 1933), propylene phenoxetol (1% aqueous solution) (Young and Downing, 1976), and urethane (Postnov and Philippova, 1988); and *Moina* has been immobilized with urethane (Hubareva and Svetlichnyi, 1998; Svetlichny and Hubareva, 2002a, 2002b, 2004). *D. magna* was efficiently anesthetized by bubbling halothane, isoflurane, and enflurane through the culture medium (McKenzie, Calow, and Nimmo, 1992); this was confirmed by a lack of movement in response to stimulation with strong light.

A mixture of chloroform and 96% ethanol at a 1:10 (v/v) ratio efficiently immobilizes cladocera (*Daphnia* and *Bosmina*), whereas rotifers are unaffected (Straskraba, 1964). Thus, planktonic crustaceans can be separated from rotifers, allowing quantitative measurements to be recorded separately. Gliwicz (1968) immobilized the pelagic cladocera *Bosmina*, *Chydorus*, *Daphnia*, and *Diaphanosoma* using physostigmine salicylate (eserinum). This chemical caused complete paralysis of their appendages at a concentration of 50 mg/L within 3–5 min; it acts by depolarizing and thus inhibiting neuromuscular synapse activity by inhibition of cholinesterases (responsible for acetylcholine breakdown). The author noted that this substance is rapidly degraded in water.

Anaesthetization of *Simocephalus* and *Daphnia* with 1% urethane was used by Philippova and Postnov (1988) to measure the energy required for movement; *D. obtusa* (10% solution) and *Ceriodaphnia dubia* have been anesthetized with tricaine methanesulfonate solution (Carmona and Snell, 1995).

Cyclophosphane also causes paralysis in *D. magna*, which develops as a result of acute energy deficiency (Ivnitskii et al., 1998). These authors found a high protective activity by transporting forms of succinic acid and a sensitizing activity by amino acids.

12.6 FATIGUE AND STRESS

Fatigue was observed by Clarke (1930) in experiments with reactions of *Daphnia* to dimming of light. Having clearly reacted to the first three stimuli the experimental animals then demonstrated a response of about 1/3 magnitude and needed a 3-hour period of rest (complete darkness in these experiments), before a normal response was produced.

In response to environmental stress, heat shock proteins (Hsps) are formed. Hsps were investigated in *D. magna* by Bond et al. (1993). In *D. magna* exposed to 34°C, both Hsps glutathione *S*-transferase were detected (Bond and Bradley, 1995).

Stress induced in *D. magna* by exposure to predators was demonstrated by increased Hsp levels (Pauwels et al., 2005). After exposure to predators, the level of Hsp60 increased; this returned to normal levels within 24 h.

12.7 IMPACT OF XENOBIOTICS ON LOCOMOTION

With increasing salinity, there is an immediate decrease in the swimming velocity of *D. magna*, followed by acclimation and a return to the normal swimming velocity (Baillieul et al., 1998). The swimming activity of *D. magna* is also decreased under conditions of acute, sublethal cadmium (Cd) stress (3.5–5.0 g/L) (Wolf et al., 1998), and exposure to 10–100 nM Cd^{2+} (Baillieul and Blust, 1999) and tributyltin chloride or polychlorinated biphenyls (Schmidt et al., 2006). Baillieul and Blust (1999) also demonstrated in *D. magna* that the frequency of antennal beats decreases from c. 4.6 to 3.7 beats/sec at a Cd^{2+} concentration of 50 nM. A decreased average swimming velocity has also been demonstrated in *D. magna* exposed to above 5 ppb copper (Cu) using image analysis (Untersteiner et al., 2003).

Compared with the effect of single compounds, there was an approximately additive effect of exposure to a mixture of polychlorinated biphenyls and tributyltinchloride on *D. magna* swimming behavior, whereas there was a synergistic effect on their reproduction (Schmidt et al., 2005).

Immobilization by Cu was measured as the number of immobilized organisms collected in the field after 48 h of exposure (Bossuyt and Janssen, 2005). It was lowest in *Scapholeberis mucronata* (5.3 μg/L Cu), *D. longispina*, and *B. longirostris*, and highest in *Disparalona rostrata* (c. 71 μg/L Cu), *Pleuroxus truncatus*, and *D. magna*. Intraspecies differences were observed only in *Ctenodaphnia*. In the presence of organic exudates of *Anabaena*, the Cu toxicity (estimated by immobilization) significantly decreased in *Ceriodaphnia cornuta* (Chouderi et al., 2009).

The time to immobilization of daphnid species in response to various pesticides (Munn et al., 2006) or the levels of immobilization in 48 h by metals (Deleebeeck et al., 2008) have been used as a measure of toxicity.

CHAPTER

13

Nervous System and Sense Organs

13.1 ANATOMICAL BACKGROUND: SENSE ORGANS

On the subject of Cladocera, Leydig (1860, p. 3, translated from German) remarked that "[t]he nervous system, as the center of the animal organism, deserves, similar to in other animals, a complete and precise investigation."

Early data mainly dealt with the most discernible cephalic region. The first detailed general description of the nervous systems of *Daphnia* and *Moina* was made by Spangenberg (1877), who preferred to use young, recently molted, specimens in his studies. He also dissected specimens that, according to his description, were placed in a weak wood vinegar solution for several days. The material was then macerated for 2 days in a mixture of alcohol, water, and glycerol, and then stained with osmic acid. This study was followed by an investigation of the nervous systems of *Bythotrephes, Leptodora, Sida*, and *Simocephalus* by Samassa (1891) (Fig. 13.1). This author preserved cladocerans in osmic acetic acid and used histological methods for their analysis. The principal traits of the nervous system in *Daphnia* were revealed by Fischel (1908) by means of intravital staining with alizarin (Fig. 13.1). The general structure of the nervous system of Cladocera was also discussed by Leder (1913).

More recently, Weiss et al. (2012) used modern methods for a comparative investigation of the nervous systems of three species of *Daphnia*. These authors concluded that the function of the frontal filament (ventral frontal organ) is unclear and that the function of the dorsal frontal organ is probably secretory. They note that "however, it remains to be investigated which substances are secreted." By Nissl staining the head of *Daphnia* spp., they also revealed, "bulged cells" in the dorsal area and near the antennule, whose function is thought to be secretory. These authors do not cite the study by Angel (1967), who indicated the presence of similar cells in *Daphnia magna*, but at least the cell near to the antennule corresponds to that described by Weiss et al. (2012). Angel called them *storage cells* in the figure, but no comments were made in the text.

The nervous system of *Leptodora* was studied in excellent detail by Kirsch and Richter (2007), who applied confocal scanning microscopy. The ventral nerve cord consists, as shown e.g. with reference to *Sida* and *Leptodora*, of a double chain of ganglia joined by longitudinal connectives and transverse junctions, thus producing a ladder-like appearance. Each ganglion sends out

Physiology of the Cladocera.
DOI: http://dx.doi.org/10.1016/B978-0-12-396953-8.00013-6

nerves to the appendages. The largest anterior ganglion is the brain (the suprapharyngeal ganglion), which supplies the eye and the ocellus with nerves.

Central parts of the nervous system send nerves to various specific organs: the gut, heart, and sense organs. According to Fischel (1908), the brain supplies the ganglion opticum and the occipital organ with nerves. The frontal organs have been examined by several authors but their functional role is unclear. Fischel also demonstrated ganglion at the base of setae natatoriae (dorsal setae at the boundary between the abdomen and postabdomen) and nerves supplying the heart and thoracic limbs. Some attempts at staining following Fischel's methods have been unsuccessful.

The cephalic region of the nervous system is shown in Fig. 1.6. The brain and the adjacent optic ganglion were described in detail by Lidvanov and Lvova (2003). These authors arrived at the conclusion that the nervous

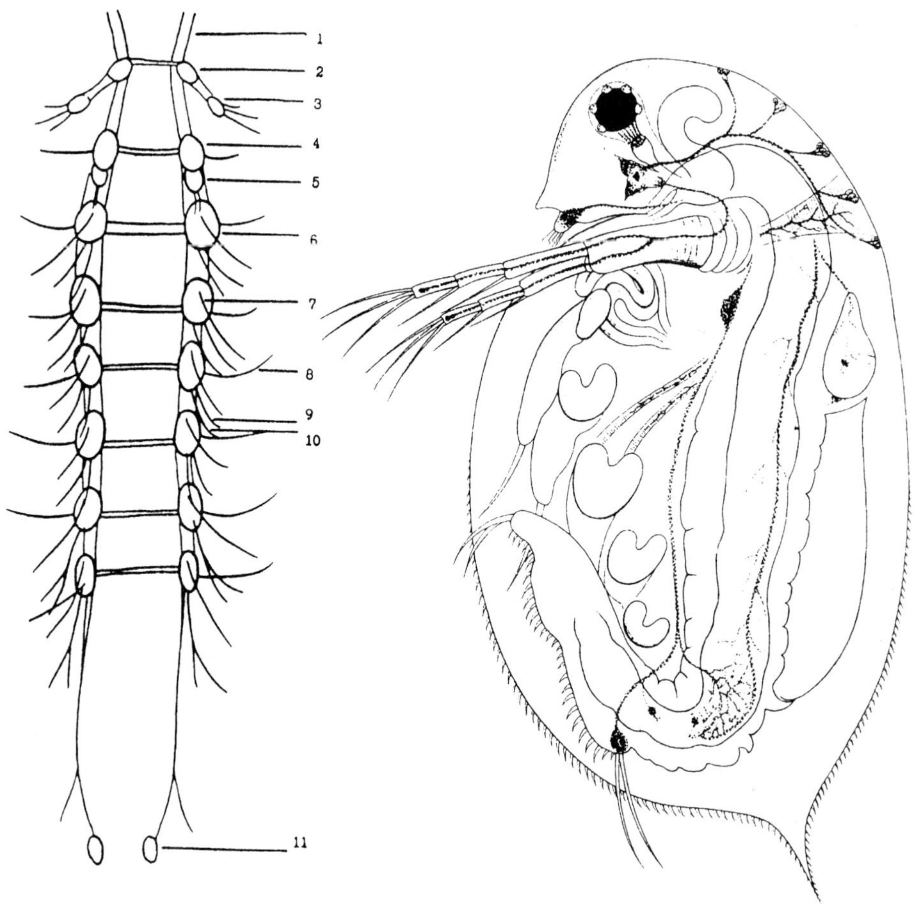

FIGURE 13.1 **Nervous system.** Left, *Sida crystallina*. Right, *Daphnia*. *Source: left, redrawn from Samassa (1891); right, Fischel (1908).*

system of *Daphnia* is homologous to that of Anostraca. In *Daphnia*, most neurons of the central nervous system are multifunctional, thus indicating that the system is primitive. They also discovered that there are no structures for the liberation of neurohormones into the hemolymph and that neurosecretions (the synthesized neuropeptides) are liberated via axons.

Angel (1967) observed that many cells in the central nervous system exhibit a glycogen cycle.

Following on from the ideas of Hutchinson (1953), it may be said that all environmental factors act together on all sense organs. Hutchinson (1953, p. 156) states that "there are an enormous number of interconnecting, internal things." This immensely complicates any estimation of the resultant effect by scientists; however, cladocerans themselves can do it, thus performing every function and movement in their own way.

13.2 NEUROSECRETION

The major functions of the nervous system comprise general neurosecretory regulation and general control of the activities of sense organs. Further, as shown in daphnids, the cladoceran nervous system is cholinergic. Nervous impulses in Crustacea are transferred by acetylcholine (Ach) from nerve endings to particular organs. It is known (Prosser and Brown, 1967 [1962]) that Ach is synthesized in the presence of ATP and cholinesterase. Daphnid cholinesterase shows the characteristics of a pseudocholinesterase, since it prefers propionylthiocholine to Ach (Vesel et al., 2006). The characteristics of pseudocholinesterase (cholinesterase II) include inhibition by diisopropylfluorophosphate, prostigmine, physostigmine (eserine), tetraethyl pyrophosphate, and hexcaethyl tetraphosphate (Prosser and Brown, 1967 [1962]).

13.2.1 Neurosecretory Cells of the Nervous System

Sterba (1957) distinguished special cell groups in the cephalic part of the *Daphnia* nervous system that differ in their structures. He could clearly distinguish them from neighboring cells by their Gomori-positive staining and also distinguished different cell groups by their varying secretions: he identified cells in the protocerebrum, the deuterocerebrum, a segment of antenna II, and a segment of the mandible (Fig. 13.2). Sterba found that secretions are liberated from these cells without accumulation in a special reservoir, and that the amount of secretion is greater at the beginning of egg maturation and very low immediately before the eggs are transferred to the brood chamber. During these periods molting takes place.

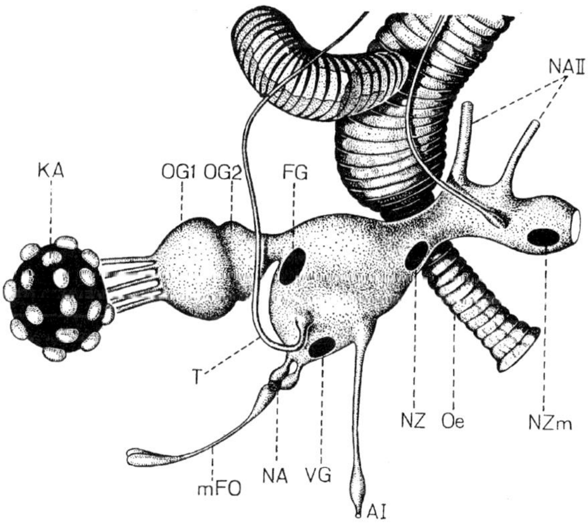

FIGURE 13.2 Neurosecretory areas in the anterior region of the *Daphnia* nervous system. AI, antenna I; FG, neurosecretory cells of "the frontal group;" KA, compound eye; mFO, median frontal organ; NA, eyespot; NAII, nerves of antenna II; NZ, neurosecretory cells on the circumenteric connective; NZm, neurosecretory cells on the mandibular ganglion; Oe, esophagus; OG1, optical ganglion 1; OG2, optical ganglion 2; T, tegmentarius, the nerve to lateral frontal organ; VG, neurosecretory cells of "the ventral group." *Source: Sterba (1957).*

Four distinct cell groups were identified by Angel (1967) in the supraesophageal ganglion of *Daphnia* (Fig. 13.3), and Halcrow (1969) revealed areas of presumed neurosecretory activity in the nervous system of *D. magna* (Fig. 13.3) after staining with paraldehyde fuchsin. Fuchsinophilic granules are found in the following regions of the nervous system: scattered throughout the optic ganglion; on the anterior and ventral surfaces of the brain; on circumenteric connectives; and at the junctions of the second ventral commissure with ventral nerve cords. The presence of these groups was confirmed and reinvestigated by Bosch de

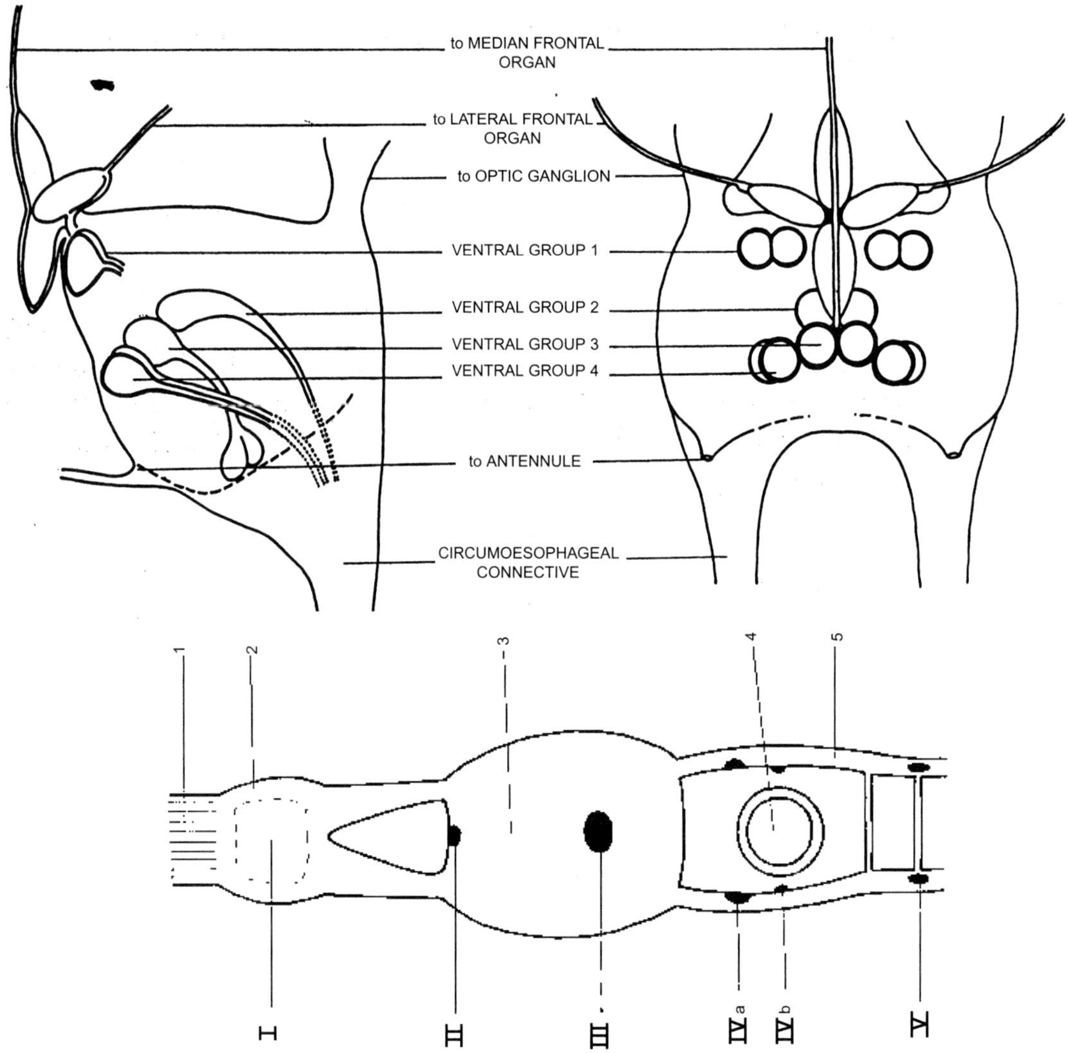

FIGURE 13.3 **Neurosecretory areas in the anterior region of the *Daphnia* nervous system.** 1, nerves to compound eye; 2, optic ganglion; 3, brain; 4, lumen of gut; 5, circumenteric connective. *Sources: upper, Angel (1967); lower, Halcrow (1969).*

Aguilar (1969, 1972), who found the following groups of neurosecretory cells in the cephalic ganglion of *Daphnia*: frontal; ventral; a group of esophageal connectives; groups of postesophageal commissures; cells of the mandibular ganglion; and neurosecretory cells at the site of connection of the ventral chain and of nerves of the second pair of thoracic limbs (Bosch de Aguilar, 1969); and the parapharyngeal group (Bosch de Aguilar, 1972). In *Podon intermedius*, neurosecretory cells were found in the protocerebrum, at the boundary between the protocerebrum and the deuterocerebrum, at the base of the antennal nerve, and in the tritocerebrum (Bosch de Aguilar, 1971).

In the nervous system of *D. magna*, a hyperglycemic hormone and immunoreactive neurons are found (Zhang et al., 1997). Their wide distribution is shown in Fig. 13.4.

A well-developed histaminergic system was identified in *D. magna* and *D. pulex*, including in the visual system, by McCoole et al. (2011) (Fig. 13.5); histamine takes part in the negative response of *Daphnia* to ultraviolet radiation (UVR), but exposure to cimetidine (a blocking agent) inhibits this negative response. Further study led to the discovery of signaling dopamine, octopamine, and serotonin, as well as receptors and transporters for each amine (McCoole et al., 2012). These studies demonstrate that *Daphnia* possess a nervous system that controls the functioning of various, distant structures.

13.2.2 Neurosecretory Substances

By observing the state of neurosecretory cells and the corresponding stages of molting and reproduction, Bosch de Aguilar (1969, 1972) concluded that the esophageal group produces a factor that inhibits molting, the mandibular group controls development of eggs, and the ventral group produces a factor that inhibits formation of the ephippium. Neurosecretory cells have also been found in the brain of *Simocephalus* (Zahid et al., 1980).

Supposed neurotransmitters or neuromodulators (and their metabolites or precursors) were discovered in *D. magna* by Ehrenström and Berglind (1988). *D. magna* synthesizes and transforms these substances which include 3,4-dihydroxyphenylalanine (l-DOPA), dopamine, noradrenaline, adrenaline, tyramine, epinine, 3-methoxyturamine, 3,4-dihydroxyphenylacetic acid (DOPAC), l-tryptophan, 5-hydroxytryptophan, 5-hydroxytryptamine, and 5-hydoxyindolacetic acid. The range of test concentrations for these substances was 0.01–32.8 ng/mg protein. The quantity of catechols l-DOPA, dopamine, and DOPAC in *Daphnia* showed diurnal variations, reaching a diurnal peak at 8 AM, although other biogenic amines did not exhibit diurnal variations.

Tonkopiy et al. (1994a) demonstrated that M-cholinolytics decrease the toxicity of armine, aminostigmine, and arecoline for *D. magna*. On the basis of experiments with cholinomimetics, H-cholinolytics, and M-cholinolytics, it was concluded that *Daphnia* possess a well-developed cholinergic system, containing cholinoreceptors that is analogous to that of mammals. In *D. magna*, Podosinovikova et al. (2001) inhibited the dopamine system by blocking the D_2-receptors with haloperidol. After this treatment, *Daphnia* were exposed to solutions of cholinoblockers (amedine, amizil, atropine, cyclodol, norakin, pentifin) and their antihaloperidol activity was estimated. The swimming behavior of *Daphnia* has been shown to become erratic at a dopamine concentration c. 100 mM (Peñalva-Arana et al., 2007).

13.3 SENSE ORGANS

13.3.1 Anatomical Background

For a long time, only two kinds of sense organs were recognized in cladocera: the eyes

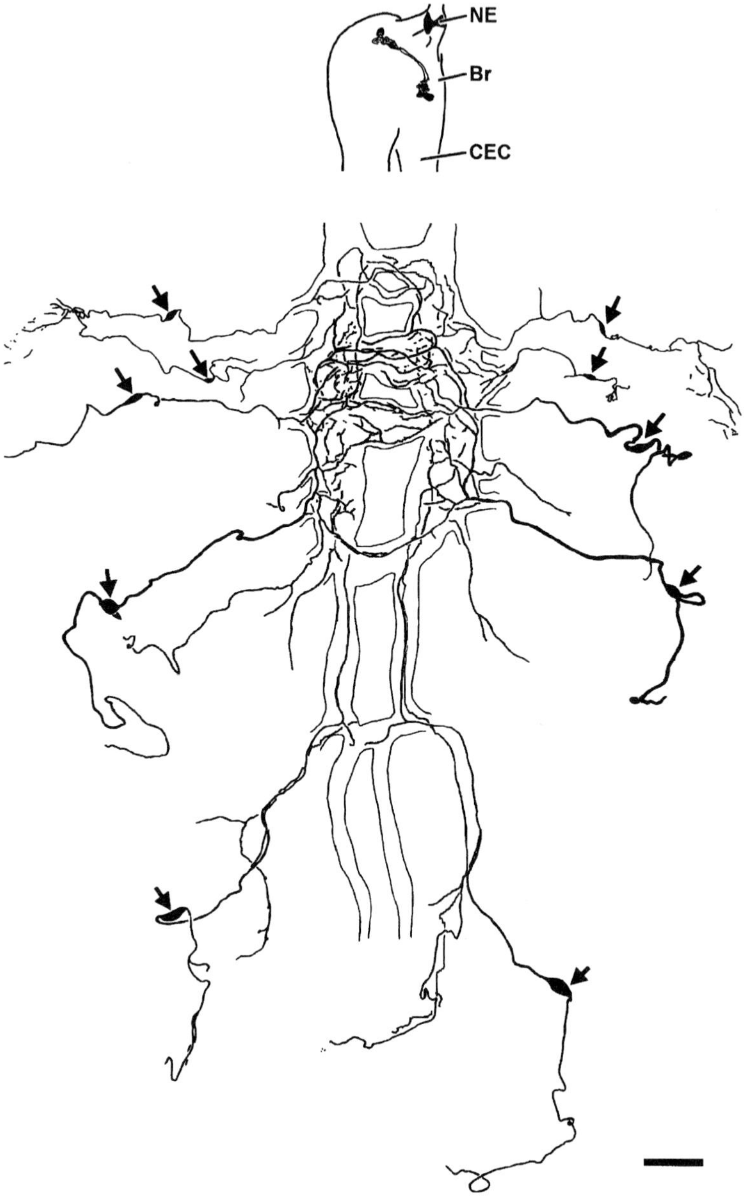

FIGURE 13.4 Distribution of hyperglycemic hormone-reactive neurons originating in the thoracic ganglia of *Daphnia magna*. Br, brain; CEC, circumesophageal connective; NE, nauplius eye. Scale bar 50 μm. *Source: Zhang et al. (1997).*

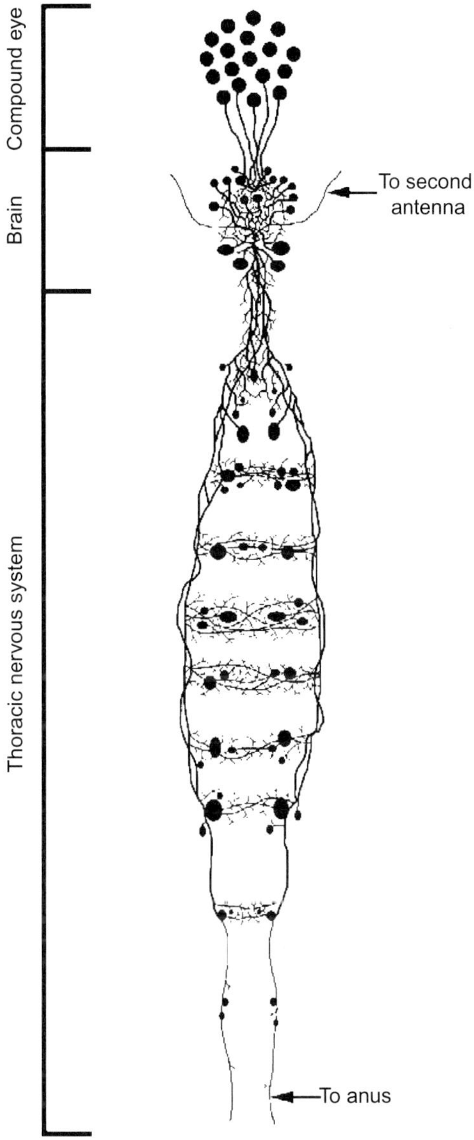

FIGURE 13.5 Schematic of *Daphnia* nervous system showing histamine-like immunoreactivity sites (filled circles). *Source: McCoole et al. (2011).*

(Figs. 1.1, 1.2, 13.1, and 14.1) and the esthetascs on the antennules. Now various other sense organs are also known to be present, or are suspected of being so, in cladocerans. While some sense organs, such as eyes, are anthropomorphically understandable, most of the sense organs of Cladocera have nothing in common with those of mammals, although they may provisionally be designated as organs of chemical sense (chemoreceptors) and tactile organs (mechanoreceptors). The latter may be involved in the perception of gravity and of equilibrium. The presence of sense organs that perceive vibration, e.g. stridulation of aquatic organisms, is also suspected and Dumont and Van de Velde (1976) suggested that one of the functions of head pores may be vibration detection.

The olfactory setae (esthetascs) were examined and compared by Scourfield (1896), who also noted that there are nine in female anomopods, Sididae, and *Leptodora*, six in female *Holopedium*, and five in female Polyphemidae. Much later, Rieder (1987) described the esthetascs in detail (see Fig. 13.6).

Sensilla were identified on the thoracic limbs of chydorids and macrothricids by Fryer (1963, 1974), Smirnov (1967) (Fig. 13.7), and Dumont and Silva-Briano (1997). No such sensilla are present on the limbs of daphnids. Generally, on the body (limbs included) of a cladoceran there are numerous papilliform or setiform structures that are suspected to have a sensory function.

In the head, there is a lateral frontal (parietal) sense organ (Gicklhorn, 1931b) (Fig. 13.1) and each of its cell clusters is supplied by a nerve arising from the cerebral ganglion (Gicklhorn, 1931b; Fryer, 2004). Its function may be light perception or it may be a pressure gauge (Fryer, 2004). The median (ventral) frontal organ is similarly innervated (Fryer, 2004).

Cladocera can perceive a wide range of external stimuli such as various kinds of irradiation, including daylight, colored light, and polarized light, plus various chemical, olfactory, and mechanical stimuli.

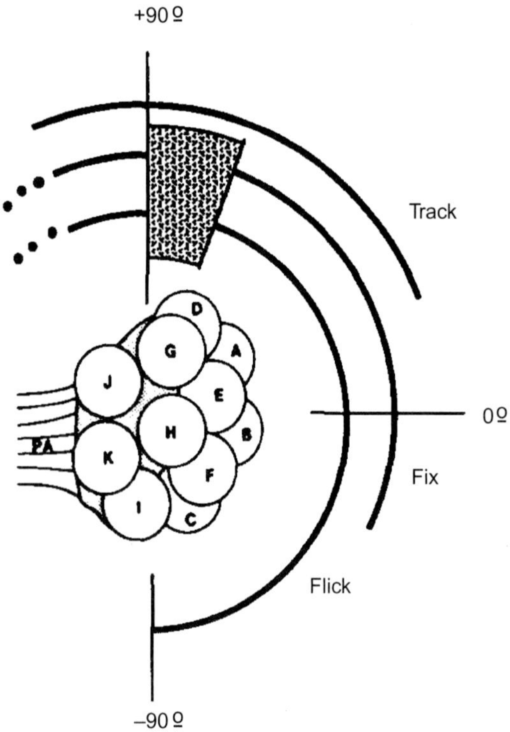

FIGURE 13.6 Behavioral regions on the compound eye of *Daphnia*. A–K, lenses; fix, fixation; PA, photoreceptor axons. *Source: Consi et al. (1990).*

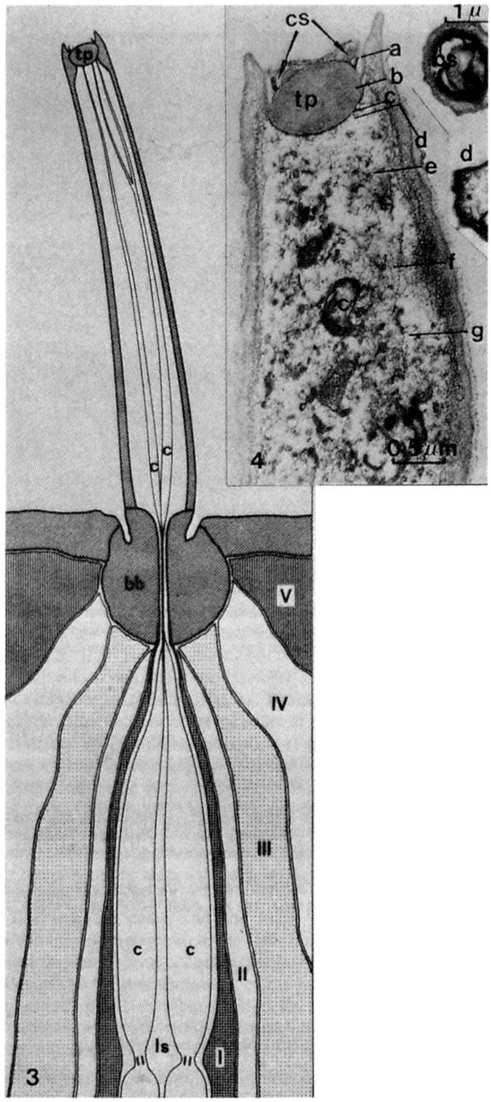

FIGURE 13.7 Structure of an esthetasc of *Daphnia magna* and of its terminal part. The inner space (Is) and the basal bead (bb) of each of nine olfactory setae is surrounded at its base by five sheath cells (I–V). The receptor cilia (c) extend toward the terminal pellet, tp. a–g, cross sections at different levels. *Source: Rieder (1987).*

13.4 VISION

13.4.1 Anatomical Background of Vision

Most cladocera possess an eye and an ocellus (a pigment spot), both of which are black. The eye consists of integumental cells, an intercellular substance, lenses, and receptor cells containing pigment (Figs. 1.1, 1.2, and 1.6) (Gueldner and Wolff, 1970). The structure of the eye is compound, or ommatidium; a *Daphnia* ommatidium is shown in Fig. 13.8. The ocellus is a pigment spot. Both the eye and the ocellus may be developed to various extents or may be absent. The general trend is toward an increased ocellus in bottom-living forms, in contrast to pelagic species.

The relative size of the eye is different in different species. It is very large in *Dadaya* and in polyphemids it constitutes a major part of the total body volume. In some bottom-living

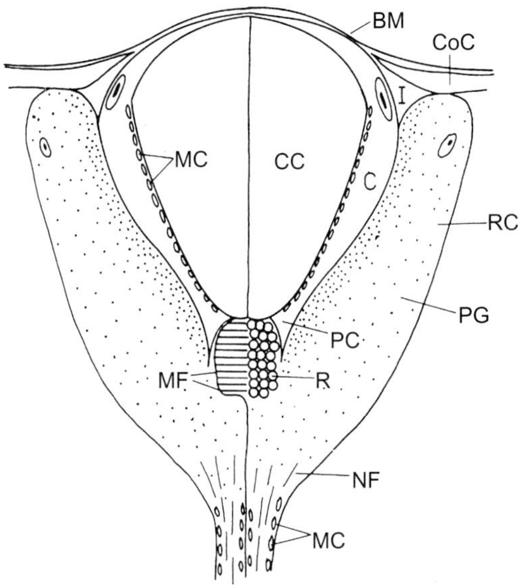

FIGURE 13.8 **Structure of a *Daphnia* ommatidium.** BM, basilar membrane; C, cone cell; CC, crystal cone; CoC, covering cell; MC, mitochondria; MF, microvilli; NF, neurofilaments; PC, process of cone cell; R, rhabdom; RC, retinula cell. *Source: Ringelberg (1987).*

species, which obviously live in the dark (in interstitial spaces), the eye may be small or absent, whereas the ocellus may be increased in size, or both may be absent. In contrast, in planktonic species the ocellus is usually small or absent. The ocellus may be normally very small (as in *Daphnia* species) or absent (as in most *Moina* species). Abnormal, eyeless specimens of *Moina* and *Simocephalus* have been reported (Banta, 1921; Banta and Brown, 1922). Fryer (1953) identified a specimen of *D. pulex* that had no complex eye, although eyes were present in its embryos. Mutant *Daphnia* have been reported that lack black pigment in the eye (Young and Downing, 1976). Both the eye and ocellus and their function have been studied using various experimental methods, including surgical removal (extirpation).

In various chydorids, the eye is situated just on the cephalic ganglion. As direct observation shows, it is either immovable (as in females of *Alona affinis*, *Alonella nana*, *Camptocercus fennicus*, *Disparalona rostrata*, *Graptoleberis testudinaria*, *Picripleuroxus laevis*, and *Pleuroxus trigonellus*) or flutters within a very limited amplitude (as in *Acroperus*, *Chydorus sphaericus*, and *Pleuroxus truncatus*). The muscles of the eye seem to be short.

In *Eurycercus*, the eye is situated at some distance from the brain and is moved by thin muscles, as shown clearly by Leydig (1860) (Fig. 13.1). It flutters, and makes limited rotations, and is sometimes turned up by c. 180°.

As reported for daphnids, the eye is situated at some distance from the brain and is constantly rotated by means of thin muscles. Three pairs of muscles move the eye: the levator, the lateralis, and the depressor (Binder, 1931). These muscles are inserted ventrally, dorsally, and laterally onto the eye (Consi et al., 1987). It should also be mentioned, for comparison, that the eyes of vertebrates, including fish, are also supplied by six muscles: four rectus oculomotorius muscles and two oblique oculomotorius muscles.

Downing (1974) discovered that the *Daphnia* eye is suspended in position by a membrane that forms a water-tight seal between the eye and the hemocoel.

The eye in preserved specimens may be discolored by alkaline solution and the number of lenses (ommatidia) may then be easily counted. This number may be up to 10 in *Penilia* and chydorids, 44–81 in other ctenopods (Korovchinsky, 2004), 28 in *Eurycercus lamellatus*, 22 in *Moina micrura* (Smirnov, 1975), and 22 in *Daphnia* (Heberdey and Kupka, 1942; Downing, 1974; Frost, 1975; Fryer, 2004).

Comparative investigations into the topography of the eye and its muscles, and of eye movements, although they are possible and would be very informative, have never been done in bottom-living Cladocera (Chydoridae, Macrothricidae, and Ilyocryptidae).

Bottom-Living Cladocera

Few, if any, studies have been made into the vision of bottom-living and littoral Cladocera. Occasional reports have been made on color vision (see below) in the littoral species *Pleuroxus truncatus* (syn. *Peracantha truncata*) and *Scapholeberis mucronata*, which are attracted by blue illumination (Peters, 1927), and on *Ceriodaphnia, Kurzia, Moina, Pseudochydorus, Sida* and *Simocephalus* (Smith and Baylor, 1953).

Pelagic Cladocera: Experimental Approaches and Direct Observation

Most experimental evidence concerns planktonic species. The *Daphnia* eye rolls almost ceaselessly (i.e. is in a state of rotatory tremor) as it is pulled by special muscles, and is thought to scan visual information in this way. It was assumed that such eye movements are related to the mechanism of vision, that illumination is assessed by *Daphnia* in this way, and that the necessary angle in relation to the source of light is retained (Waterman, 1961). Frost (1975) determined four kinds of eye movement in *D. pulex*:

1. tremors at 16 Hz;
2. a scanning movement at 4 Hz;
3. large fast eye movements; and
4. optokinetic nystagmus, produced by moving striped patterns around a daphnid.

The first two kinds of movement were observed in diffuse light and the third kind in the presence of spots of light crossing the visual field. Three other types of activity by the *D. magna* compound eye were discovered by Consi et al. (1990): a "flick" caused by a flash of light, "fixation" in response to a stationary light stimulus, and "tracking," i.e. following a moving stimulus (Fig. 13.6). When a light stimulus is situated at about 80° dorsal to the eye's axis, there is no response (this is the null area).

When *D. schoedleri* were cultivated in blue light, no pigment sensitive to red light was formed (over a period of 44 days); however, if returned to white light, the normal level of this pigment was reached after 90 days (McNaught, 1971).

13.4.2 Environmental Background

Solar light rapidly becomes extinct in water, especially long-wavelength (red) light (see, e.g. Hutchinson, 1957, Ch. 6). Red light (720 nm) extinction is complete at c. 4 m depth in pure water (Dodson, 2005). In water, polarized light has great ecological importance; it is reflected from various particles, including food algal cells.

13.4.3 Surgical Methods of Investigation

Eyeless daphnids were used for investigations into the effects of light on cladocera, despite the fact that such specimens (*Daphnia* and *Simocephalus*) very rarely occur in nature.

Extirpation of the Eye and Ocellus

For investigating the reactions of eyeless *Daphnia*, Schulz (1928) removed the eye through a puncture hole made by a fine needle. For this procedure, the daphnid was placed with its frontal side upon a slide and excess water was removed so that the daphnid's head was outside the water. When the integument above the eye was punctured, the eye emerged above the integument and could be separated out directly or was shed with the carapace at the next molting; this was confirmed by Downing (1974), who repeated this method. The percentage of successful operations was satisfactory and, after two molts, all damage caused by the operation was healed and the animals could be used for experimentation.

Extirpation of the complex eye for experimental purposes was also described by Harris and Mason (1956), who copied the methods of

Schulz (1928). In their experiments, a *Daphnia* was placed into a moistened groove made in paraffin, on its back and slightly tilted to the side. The carapace was then punctured above the eye with a fine pin. If the hole was successfully made, the eye emerged above the puncture. It was then removed and the animal was returned to the culture medium. After some time, it became evident whether the animal had survived the operation; if so, such animals were used for experimentation. Generally, these animals survived for several weeks and reproduced even more intensively than normal animals.

In a similar fashion, Schulz (1928) separated and removed the ocellus. Several animals survived the operation, molted, and were used in experiments into the effect of light on various processes. Later, Consi and Macagno (1985) also removed the ocellus from *D. magna* immobilized with carbonated water using a microbeam from a dye laser (Candela SL1-500; 450 nm) coupled to a Zeiss microscope. Several laser pulses at 26–27 kV were used to delete the ocellus. The animals recovered after 2–4 days and were used in experiments. Such animals had the same spectral sensitivity as controls.

In the operated animals (Schulz, 1928), the reaction to light was identical to that of intact *Daphnia*, and Schulz (1928, p. 543) concluded that "the ocellus of daphnias as a rudimentary organ does not play a noticeable part any more in the life of these animals." In Consi's (1985) experiments, animals that had the ocellus removed had the same spectral sensitivity as normal animals.

After extirpation of the eye and the ocellus, the reactions of daphnias to light remain essentially the same due to their general light sensitivity (Schulz, 1928; Harris and Mason, 1956; Fryer, 2004). However, completely eyeless *Daphnia* do not exhibit a dorsal light reflex (Schulz, 1928). These blinded *Daphnia* are also more sensitive to vibration (Harris and Mason, 1956). In summary, *Daphnia* generally do well without eyes.

In addition, *C. sphaericus* successfully propagated in the dark over a period of about 2 months, and they had only slightly different characteristics (e.g. number of progeny and longevity) from controls kept in the light (Smirnov, 1971).

13.4.4 Taxes

Reaction to Light: Phototaxis

Daphnia have been found to be primarily negatively phototropic and positively geotropic, but a reduction in light intensity reversed the signs of tropism (Clarke, 1930). This reaction persisted for a few minutes, showing that *Daphnia* exhibit labile behavior in relation to illumination. If illuminated from the side, daphnid turn toward the more strongly illuminated side.

Wavelength peaks of light sensitivity are different in different species: in *D. magna*, they occur at 440, 470, 520, and 640 nm (Wechsler and St. John, 1960); and in *D. retrocurva*, they occur at 370, 435, and 570 nm, and less so at 685 nm (McNaught, 1966; McNaught and Hasler, 1966).

The threshold of visual sensitivity in *Daphnia* is 10^{-4}–10^{-5} lx (McNaught and Hasler, 1964).

The response of *Daphnia* to illumination is dynamic. These reactions are not rigidly fixed and vary depending on changes in the environment and on the condition of the animal (Loeb, 1924). A reduction in light intensity reverses these reactions for a short time (Clarke, 1930), and periodic changes of phototactic signs have also been observed under constant low illumination (Clarke, 1932). This may be understood as exploratory behavior. Experimental reversal of the reaction of *D. magna* to illumination was also observed by Chernykh and Panasyuk (1964). The basic observations by Ringelberg (1964, 1987) were that under conditions of decreasing intensity of illumination, *D. magna* move toward the light source; at increasing intensity, they move away from it. A continuous decrease in light intensity led to upward swimming.

De Meester and Dumont (1988) identified three phenotypes of *D. magna* with respect to light: positive phototaxis, negative phototaxis, and random wandering between areas of low and high light intensity. They consider these phenotypes to be genetically determined. Thus, reactions to light may be individual: some specimens may be strongly negatively phototactic and others strongly positively phototactic, even within a single natural *D. magna* population (Dumont et al., 1985). *Daphnia* specimens may deviate by up to 30° from the direction perpendicular to the polarization plane (Waterman, 1960, 1961). However, shoaling *Daphnia* congregate at the water surface, i.e. contrary to the aforementioned tendency (Fryer, 2004). *Scapholeberis* also attach to the surface film of water under bright illumination and *Holopedium* ascends to the water surface by day and moves to deeper water by night (Fryer, 2004).

Starved *Daphnia* tend to move, whereas satiated ones stay at the same level. Starved *D. pulex* swim upwards, but starved *D. magna* sink (Szlauer, 1962b).

The orientation of a daphnid depends on the direction of light. When illuminated from above, daphnias maintain a position with their dorsal surface turned toward the light (Schulz, 1928; Harris and Mason, 1956), an orientation known as the *dorsal light reflex*. When illuminated from below, the daphnias swim upside down, i.e. with their dorsal side turned toward the source of light (Schulz, 1928; Harris and Mason, 1956). These results deserve further investigation.

As a result of being attracted by light, some Cladocera can also be collected in light traps, especially *Bosmina coregoni*, *Bythotrephes*, *Eurycercus*, and *Leptodora* (Szlauer, 1971).

It is thought that the muscles that move the eye also transfer the stimulus to muscles of the antennae: if daphnid eye muscles are strained differently at each side of the eye, then the antennal muscles act to change the position of the body (Harris, 1953; Jander, 1959). Movement of the *D. magna* complex eye may also be stimulated by its intermittent illumination (but not by the illumination of any other part of the body) (Young, 1974). If the complex eye is illuminated from above, then maximum movement is caused by the short-wavelength range of the spectrum; if illuminated from the side, then maximum movement is caused by the yellow-green range.

Genostrychnine intensifies the photokinetic activity of *D. pulex*, and chlorpromazine renders its occasional movements less frequent (Rimet, 1965).

Extraocular light sensitivity also exists (Ringelberg, 1987). Light is sensed through integuments, as well as the eyes, as shown for *D. pulex* by Scheffer et al. (1958). However, light perception and phototactic movements occur in *Daphnia* when both the compound eye and the eye of the nauplius are extirpated (Ringelberg, 1987). Fryer (2004, p. 46) also noted that "eyeless individuals manage their affairs reasonably well!"

Reversal of Phototaxis

The reaction of *Daphnia* to light may be reversed if some of the large constellation of factors that affect their behavior change (Loeb, 1924). For example, reversal of phototaxis in *C. sphaericus* and *C. ovalis* by acids was shown by Bryukhatova (1928). Clarke (1932) also demonstrated that phototaxis in *Daphnia* may be reversed by changes in external or internal factors.

Skadovskiy (1939, 1955) noted that the reaction of *D. pulex* to light may be different depending on the environmental circumstances. In daphnias settling in summer to lower water levels in a eutrophic water body the positive reaction to light becomes more intensive and they move to warmer and better aerated layers, whereas, with insufficient alimentation their negative reaction to light increases and they settle into colder water layers with more abundant food. In the presence of a sufficient quantity of undissociated carbonic acid molecules phototaxis becomes positive. Positive phototaxis also occurs in a situation where there is insufficient oxygen. In negatively phototactic daphnias, the oxygen consumption rate is considerably higher than in positively phototactic ones.

Various factors may be responsible for the reversal of phototactic reactions in *Daphnia*. The positive phototactic response in *D. magna* decreased as food became limited and this response was also dependent on the particular clone of this species being investigated (De Meester and Dumont, 1989). The *D. magna* photoreaction also differs depending on their previous adaptation to temperature (Lobashev and Ivanova, 1947).

The phototactic reaction also changed in *D. magna*, *Evadne*, and *Podon* following the addition of $CdCl_2$ ($1{:}10^2$–$1{:}10^5$ w/v) (Smirnova, 1960) to their environment; because it binds sulfhydryl groups, phototaxis could be restored by the addition of cysteine ($1{:}10^2$–$1{:}10^4$ w/v), a sulfhydryl group donor. The positive phototactic reaction was changed to negative in Evadne and Podon if $CdCl_2$ was added to the water (Smirnova, 1960), and in *D. pulex* if illumination was increased from 1300 to 5274 lx (Rimet, 1961).

13.4.5 Perception and Effect of Polarized Light

Cladocera can detect polarized light, which influences their orientation and direction of movement, i.e. they exhibit polarotaxis (Waterman, 1961; Young and Taylor, 1990). In the aquatic environment, a source of polarized light is the light reflected by algal cells and other particles, which are potential food items. Thus, sensitivity to polarized light plays an important role in obtaining food. *D. magna* and *D. pulex* collected much more food in a zone of polarized light than in a zone of nonpolarized light of the same intensity (Verkhovskaya, 1940).

It has been shown that several species (both littoral *Ceriodaphnia*, *Kurzia*, *Moina*, *Pseudochydorus*, and *Simocephalus* spp. and pelagic, *Bosmina*, *Daphnia*, and *Leptodora* spp.) that Cladocera illuminated by a vertical beam of polarized light swim perpendicular to the polarization plane (Baylor and Smith, 1953; Waterman, 1960, 1961; Hazen and Baylor, 1962). Gromov (1992) also observed that *Daphnia* can distinguish between vertical and horizontal beams of polarized light; in the vertical beam, *Daphnia* movements are reduced by half. Following the addition of sodium bromide (NaBr), the response of *D. pulicaria* to linearly polarized white light (90° e-vector orientation) became more random (Goksen and McNaught, 1995). These authors suggest that this is due to a disturbance in the transmission of nerve impulses, as Br^- ion blocks chloride (Cl^-) channels.

13.4.6 Perception of Colored Light

Ewald (1914) found *D. pulex* to be sensitive to colored light, with two maxima: in the green-yellow and blue-violet regions. Lumer (1932) exposed *D. pulex*, *D. magna*, *Moina*, and *Leptodora* to a graded series of light levels of different wavelengths but equal intensity. All of the tested animals moved toward the orange light (620–640 nm). *Moina* was also attracted by green light (540 nm) with approximately the same efficiency; its secondary maximum was in the blue range (440 nm). Each species tested showed its own characteristic reaction to light. The littoral species *Pleuroxus truncatus* (syn. *Peracantha truncata*) and *Scapholeberis mucronata* aggregated in the zone of blue light (Peters, 1927).

Sensitivity of the *Daphnia* complex eye was shown to be highest to green light; it also perceives light through its integuments, but the maximum sensitivity in this case is to blue-violet wavelengths (Scheffer et al., 1958).

In most cladocerans, four visual pigments have been found, with maximum sensitivity to light of wavelengths 370, 430, 560, and 670 nm (red) (McNaught, 1971). In oligotrophic lakes, the environment is predominantly blue but during eutrophication a redder environment is formed. In the ommatidia of *D. magna*, Smith and Macagno (1990) identified four spectral classes of photoreceptors with peak sensitivities at 348, 434, 525 and 608 nm for the dorsal ommatidia; in contrast, there was only a small difference in peak sensitivities for the ventral ommatidia.

Both the normal and the eyeless *Daphnia* were particularly attracted by yellow and green light in Schulz's experiment (1928). Color preference was also studied in *Daphnia carinata*. For this, animals were placed in a vessel surrounded by black paper, with four openings illuminated by lights of different color (Maity and Saxena, 1979). After illumination for 15 min, the distribution of daphnias was recorded: *D. carinata* preferred light of the following colors, in decreasing order: yellow, orange, red, violet, blue, and green.

D. magna showed peaks of light sensitivity at wavelengths 440, 470, 520 (blue), and 640 (red) nM (Wechsler and St. John, 1960). Gromov (1992) also observed the influence of light wavelength (or color) on the locomotory activity of *Daphnia*, as seen by their attraction to light. The locomotory activity of mature *Daphnia* decreased within the range 400–525 nm (violet-green).

When Cladocera were illuminated with a vertical beam of light, for which the wavelength was controlled using colored filters, they responded by swimming upward when the light changed from blue to white and downward when the light changed from yellow to white. This response to color change was observed in both littoral *Ceriodaphnia*, *Kurzia*, *Moina*, *Pseudochydorus*, *Sida*, and *Simocephalus* and *pelagic Daphnia*, *Bosmina*, and *Leptodora* spp. (Smith and Baylor, 1953). *Moina* was almost paralyzed by a sudden exposure to blue illumination. Responses in *Ceriodaphnia* were preceded by a considerable time lag.

When illuminated from above with red light (about 600 nm) at a uniform intensity over the aquarium surface, *Bosmina*, *Ceriodaphnia*, *Daphnia*, and *Moina* and were observed to generally dance upright, with a small horizontal vector in their movements, and thus stayed in the same area (Smith and Baylor, 1953). Under blue light (about 500 nm), these cladocerans were distinctly agitated, leaning forward with a large horizontal vector, and thus swimming from one place to another. According to Stearns (1975), *Daphnia* tend to move vertically between 440 nm (blue) and 735 nm (red), and horizontally at 440 nm (violet) and under white light.

If light falls on the *D. magna* eye through the top of the head, the action spectrum peaks at low wavelengths, but if it falls on the eye through the side of the head, the action spectrum peaks in the yellow-green region (Young, 1974).

A negative effect of blue light has also been observed. *Sida* show a negative reaction to blue light (Peters, 1927) and *Moina* is extremely sensitive to blue light, being almost immobilized by it (Smith and Baylor, 1953). Prolonged illumination with blue light (of about 500 nm) killed cladocerans (Smith and Baylor, 1953).

13.4.7 Photoperiod

The photoperiod, i.e. length of the day (i.e. light) in relation to the duration of darkness, plays an important role in controlling the life cycle and onset of gamogenesis (bisexual reproduction) in Cladocera (Fig. 11.4). In culture, *Pleuroxus denticulatus* switched to gamogenesis under the influence of photoperiod: it was more prominent with a long day length in populations originating from the southern USA and at a short day length in northern populations (Shan, 1974). *P. procurvus* and *P. truncatus* started gamogenesis only under the short day conditions. It has also been shown in *Daphnia* that the formation of males may be induced by a decrease in day length to 12 h (Stross and Hill, 1965).

13.4.8 Effect and Perception of Ultraviolet Radiation

The effect of UVR has been studied by many researchers, and it has been shown that UVR seriously endangers Cladocera. At naturally occurring intensities, it has lethal effects e.g. on *Daphnia middendorffiana* (Luecke and O'Brien, 1983) and *D. pulex* (Wübben and Vareschi, 2001). Hurtubise et al. (1998) determined the median lethal dose (LD_{50}) of UVR for *Ceriodaphnia reticulata*, *D. magna*, and *Scapholeberis kingii* using a solar simulator: the LD_{50} at 96 h, ranged from 4.2 to 84 $\mu W/cm^2$. *Scapholebris kingii* turned out to be highly sensitive, and *D. magna* and *C. reticulata* moderately so. *Bosmina meridionalis* showed no change in mortality over the whole range of UVR tested in comparison with the controls and in contrast to both *Ceriodaphiunia dubia* and *D. carinata* (Wübben et al., 2001).

The damage and repair processes induced by UVR in Cladocera may be interconnected. Following exposure to damaging UVR (280–320 nm), *Daphnia*, in the presence of both long wavelengths and visible radiation, exhibited a large increase in survival due to stimulation of photoenzymatic repair (the induction of which is favored by UV-A radiation, 320–400 nm) (Williamson et al., 2001). The lethal effect was increased when *Daphnia longispina* were exposed to UV-B than to UV-A; exposure to UV-A caused significant oxidative stress, as detected by increased malonaldehyde, whereas exposure to UV-B was followed by increases in both malonaldehyde and catalase (Vega and Pizarro, 2000). The addition of ascorbic acid reduced mortality caused by UV-A, but not by UV-B. Therefore, a protective mechanism against photooxidative stress is presumed to exist. UVR is also deleterious to newborn *D. galeata* at the water surface, and this effect decreases with increasing depth (Winder and Spaak, 2002).

Under UVR illumination, *D. pulex* is negatively phototactic (Peters, 1927). *D. magna* illuminated by UV light also exhibit negative phototaxis (at a maximum spectral sensitivity of 349 nm), whereas phototaxis to visible light (120–600 nm) is positive (Storz and Paul, 1998). Many species living at the water surface and at high altitudes develop melanistic coloration, probably as a protection against UVR.

It has been shown experimentally that downward migration provides effective protection for *D. pulex* (Vareschi and Wübben, 2001) and *D. galeata* (Winder and Spaak, 2002). Siebeck (1978) compared the effect of UVR on *D. pulex* (light brown) and *D. galeata* (uncolored) and found that the former was 1.5 times better protected from UV. Melanic clones of *D. pulex* and *D. middendorffiana* also survived exposure to 20 W/m^2 near UV light for twice as long as unpigmented ones (Hebert and Emery, 1990). The melanic morph of *D. longispina* survived better than the hyaline morph under solar UVR (De Lange et al., 2000). Melanin in the carapace reduces UV penetration.

In the presence of UVR, a large proportion of *D. pulicaria* migrate downward, compared with that are specimens shielded from UVR (Leech and Williamson, 2001; Leech et al., 2005). UV is therefore a factor that induces downward migration of *Daphnia* spp. whereas species with a more highly pigmented carapace stay closer to the water surface when exposed to UVR (Rhode et al., 2001).

It has also been shown in *D. pulex* and *D. tenebrosa* that melanin plays a major role in protection against UVR (Hessen, 1999). In the absence of blue light and UV, *Daphnia* do not reconstitute their carapace melanization after molting. The concentration of protective melanin in *Daphnia* increased after the ice break-up in North Finland, i.e. in the period of maximum underwater UV intensity (Rautio and Korhola, 2002).

The detrimental effects of UVR on *D. magna* and *D. tenebrosa* increased in the presence of a reduced calcium content in the solution (Hessen and Rukke, 2000b).

In *D. catawba* exposed to UVR, the respiration rates increased by 31.8% (at sublethal irradiation of 2.08 kJ/m^2 UVB + visible photorepair radiation) and decreased by 70.3% (at 4.16 kJ/m^2) (Fischer et al., 2006). The enhanced respiration rates were attributed to energetic costs associated with the repair of damaged cell components.

With increasing temperature, UV-induced DNA damage is increased to *D. pulicaria* at the water surface, as well as DNA repair rates; thus net DNA damage is greater at lower temperatures, where survival may also be lower (Macfadyen et al., 2004). Photoprotection may therefore be more effective than photoenzymatic repair, and more effective under low temperatures.

D. magna exposure to UVR did not induce a change in catalase activity and caused a slight increase in glutathione transferase activity; however, no changes in enzyme activity were recorded at different oxygen levels, although survival increased at lower temperatures (Borgeraas and Hessen, 2000).

The activity of antioxidants (catalase, glutathione transferase, and superoxide dismutase) was studied in *Daphnia* spp. in order to assess their role in UV photoprotection (Borgeraas and Hessen, 2002a). In alpine populations of *D. longispina*, there was a positive correlation between the water absorbance and catalase activity, which could be related to photoinduced hydrogen peroxide production. The activity of superoxide dismutase was high in *D. longispina* from a lowland humic pond. The activity of glutathione dismutase was low in a melanic *D. pulex* group. No difference in antioxidant activity was found between the alpine melanic and nonpigmented *D. longispina* populations.

Following UV exposure, catalase was found to be more active in *Daphnia* in the presence of organic matter, whereas the opposite was shown for glutathione transferase (Hessen et al., 2002).

A positive reaction to UVR (i.e. radiation with wavelengths of <400 nm), was reported for littoral *Pleuroxus truncatus* (syn. *Peracantha truncata*) and *Scapholeberis*. However, *Daphnia* had the highest eicosapentaenoic acid content in ponds with the highest UVR exposure, and sublethal damage of the gut preceded UVR-induced mortality (Zellmer, Arts, Abele et al., 2004).

Photoinduced toxicity in *D. magna* (estimated by immobilization) to various chemicals liberated by the pulp and paper industry was ranked as pyrene > anthracene > retene (Huovinen et al., 2001). The toxicity of these chemicals is thought to result from internal photosensitization reactions caused by UVR. UV-damaged cells are thought to release histamine. Indeed, small doses of antihistamines (α-benz-hydrylether-β-piperidoniethane, 2-phenylbenzylaminoethyl-imidazolin, or α-phenylamino-β-aminoethane) do suppress negative phototaxis (Poupa, 1948). The minimum effective dose for the first substance is 2.5 $\mu g/mL$. At higher doses, phototaxis became positive. Histamine takes part in the negative response of *Daphnia* to UVR (McCoole et al., 2011), and exposure to cimetidine (a blocking agent) inhibited the negative response of *Daphnia* to UV exposure.

Interestingly, completely eyeless *Daphnia* are repelled by UV illumination (Schulz, 1928).

UVR also affects food algae, thus influencing their consumers indirectly. It has been shown (Leu et al., 2006) that UVR exposure modifies the lipid content of green algae: the content of 18:1ω9 decreased, C18 polyunsaturated fatty acids (PUFAs) increased, and the ratios C:P and N:P decreased. In *D. magna* fed with these algae, the content of 18:1ω9 decreased but the essential PUFA content did not.

See also Chapter 3, section 3.2.1 on melanin.

Exposure of *D. magna* containing embryos to low-frequency laser radiation of 633 nm for 60 sec resulted in the production of larger juveniles (Osipova et al., 2010).

13.4.9 Perception and Effect of X-Rays

Irradiation with X-rays has a negative effect on Cladocera. In *Simocephalus*, it decreases the respiration rate and inhibits growth, possibly through changing cell permeability (Obreshkove and King, 1932a). In young *Simocephalus*, it also induces invagination of the brood pouch, which becomes worse at each molting (Obreshkove and King, 1932b).

Baylor and Smith (1958) showed that *D. magna* swim irregularly in a horizontal beam of red light and swim to the bottom in a vertical X-ray beam; in contrast, outside this beam they swim back up. These authors assumed that this action is a result of X-ray-induced free radicals having different actions on visual pigments in the compound eye and the nauplius eye. In *Moina macrocopa*, X-ray irradiation is reported to be destructive to the facets of the compound eye (Fuchikawa et al., 1995).

13.5 EFFECTS OF ELECTROMAGNETIC FIELDS

Cladocera are sensitive to electric fields. Electric fields are a normal environmental constituent and Skadovskii (1955) showed that in the bottom layers of water bodies the electric potential may change by 0.6-0.8 V over 1 cm.

Although they are normally negatively phototactic, *Daphnia* swam to the anode in an electric field with an electric current density of 0.5–1 ma/cm^2 (Clarke, 1932; Clarke and Wolf, 1933). Exposure to electromagnetic radiation of 50 Hz for 8 h over a period of 30 days delayed *D. magna* maturation and decreased fecundity, but was not lethal (Somov and Malinina, 2003). However, a low-frequency electromagnetic field (with a frequency of 50 Hz) accelerated *D. Magna* maturation, whereas an electric field of 500 Hz negatively affected both survival and maturation. It also increased the proportion of nonviable progeny (Krylov, 2008) and reduced both the number and size of the offspring (Krylov, 2010).

In a weak strychnine solution (i.e. a saturated strychnine solution diluted 1:2000), *Daphnia* became positively phototactic and also swam to the cathode (Clarke and Wolf, 1933).

13.6 CHEMORECEPTION

13.6.1 Anatomical Background

Chemoreception in cladocera is ascribed to the sensory papillae (olfactory setae and esthetascs) on the antennules. The chemoreception function of these sensory papillae has been discussed by Gicklhorn and Keller (1926). A detailed description of the structure of olfactory setae was provided by Rieder (1987) (Fig. 13.7). The inner space (is) and the basal bead (bb) of each of the nine olfactory setae of *D. magna* are surrounded at the base by five sheath cells (I-V), and the receptor cilia (c) extend to the very end of the seta (to the terminal pellet; tp).

Sensilla were originally discovered on the thoracic limbs of *E. lamellatus* (Fryer, 1963) and later identified in many chydorid and macrothicid species (Smirnov, 1971). Sensilla on the

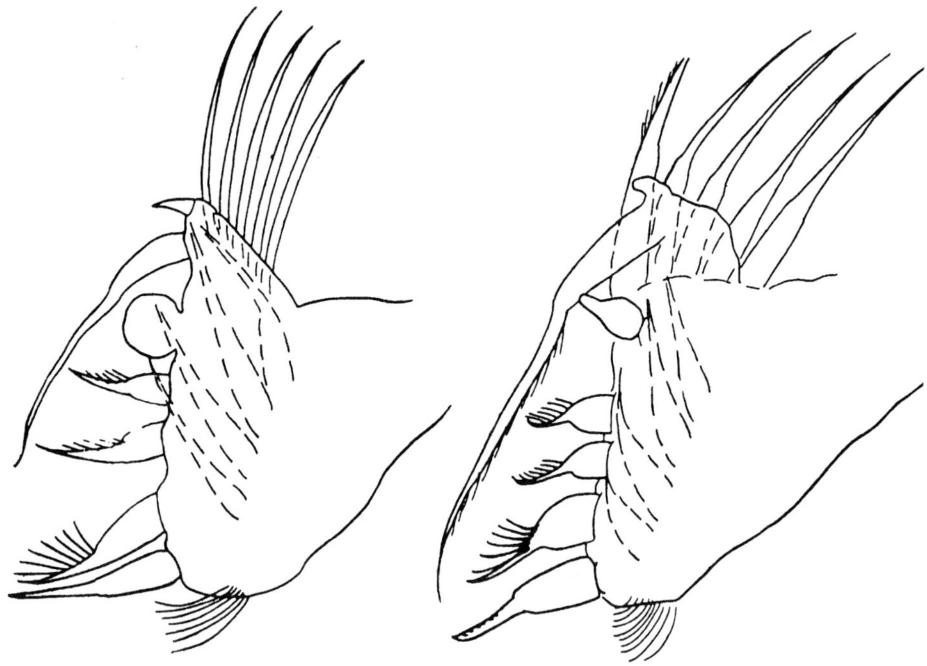

FIGURE 13.9 **Esthetascs on thoracic limb IV.** Thoracic limb IV of *Alona affinis* (bottle-like, right), of *Alona quadrangularis* (globular, left). *Source: Smirnov (1971).*

thoracic limbs may be finger-like, globular, or bottle-like (Fig. 13.9). Dumont and Silva-Briano (1997) described the distribution of sensory structures on the trunk limbs of chydorids in detail.

Chemoreception obviously combines taste and olfaction. The notion of taste has been applied to Cladocera only once, by Kerfoot and Kirk (1991), who reported that small *Bosmina* demonstrate some ability to discriminate between food particles by means of taste, in contrast to *Chydorus* and larger cladocerans. According to Hardy and McDougal (1894, p. 3), deglutition (swallowing) "starts at the mouth as a result of the stimulation of the sensory surfaces there"; however, these sensory surfaces are not commented on further or named.

In crawling chydorids, esthetascs on the antennules face outward and contact the substratum (shown e.g. in Fig. 1.3); thus, they perceive the substrate quality.

13.7 MECHANORECEPTION

The setae of various Cladocera seem to include a sense of touch in their function. In chydorids crawling on substrata, antennules with their esthetascs are in contact with the various clumps and filaments they encounter. The outer distal lobes of their thoracic limbs I "are capable of independent movements" (Fryer, 1963, p. 362) and one may observe how a moving specimen touches its surroundings with the setae of these lobes.

The presence of mechanical wave receptors in *D. pulex* was reported by Szlauer (1964). Later, Meyers (1985) studied in detail the setae at the base of the swimming antennae of *D. magna* and concluded that they are rheoreceptive and "mediate gravity perception" during the sinking phase of their movement.

Goulden (1966, 1968) suggested that copulating males of *Moina* species recognize gamogenetic females of their own species using the sensory papillae and setae on their grasping antennules and by feeling the surface sculpture of the female carapace (future ephippium), which is different in different species. One of the functions of head pores, and of the attached tissues, is also thought to involve mechanoreception (Dumont and Van de Velde, 1976).

In Cladocera, *hearing* was not completely excluded by Leydig (1860, pp. 41 and 42, translated from German): "Whether the organ of hearing belongs to sense organs of cladocerans is doubtful ... it is quite possible that one day it will be shown that the antennules of daphnids form the organ of hearing."

13.7.1 Orientation in Space

Littoral and bottom-living chydorids are oriented in relation to their substrata in various ways. They may crawl on the substrate, mainly at a right angle to the substrate's surface, although a cladoceran may also be located on the top or the side of a clump of debris. They may also swim over and between various littoral substrates. Of the macrothricids, *Lathonura rectirostris* remains immobile for rather long periods and then makes sudden jumps, whereas *Drepanothrix dentata* lies on the surface of bottom debris with its ventral side up and only its thoracic limbs moving. The diverse methods of movement of littoral cladocera, and their different orientations to the substrata, have been insufficiently studied and offer a wide scope for further observations. There are also specialists that mostly live attached to the substrate by their dorsal side, i.e. *Sida* and *Simocephalus*, the former by their dorsal organ and the latter by means of hooked terminal setae on their antennae.

Planktonic cladocerans are oriented with their longitudinal axis at some angle to the vertical, i.e. with the head upward and forward. In *Bosmina*, the movement phase lasts for c. 25 msec (Zaret and Kerfoot, 1980). Strokes of its antennae pull the body forward.

Swimming is controlled by a complex set of environmental factors, with illumination being a prominent influence. *D. magna* and *D. pulex* swimming is oriented by the predominant direction of light. If they are illuminated with light of the same intensity from all sides, they rotate (Ringelberg, 1963, 1964); when the light intensity is then increased from one direction, the animal resumes its normal position. Under more normal conditions, if the illumination intensity decreases then the *D. magna* swimming rate increases (Ringelberg, 1964); however, below a certain light intensity, *Daphnia* swim normally, independent of the angular distribution of light.

13.7.2 Control of Body Posture

The body of planktonic cladocerans (*Bosmina* and *Daphnia*) is generally oriented with its dorsal side upward and backward. According to Szlauer (1962a), *gravitation* is the main factor controlling the orientation of *D. magna* movements. Due to gravity, the animals sink and their position is frequently corrected by strokes of their antennae. There are alternating short periods of movement and rest. According to Jacobs (1964), the elongated head of some *Daphnia* counterbalances the part of the body that is behind the antennae.

The position of the *center of gravity*, as noted by Scourfield (1900b), is one factor influencing the position of a cladoceran body in space. He observed that the position of the body in space differs between *Daphnia* (head and the dorsal side obliquely upwards) and *Simocephalus* (head and the ventral side obliquely upwards). Scourfield also changed the center of gravity by occasionally attaching a drop of glue to the animal or introducing an air bubble under the shell. In such cases, he observed the animal to

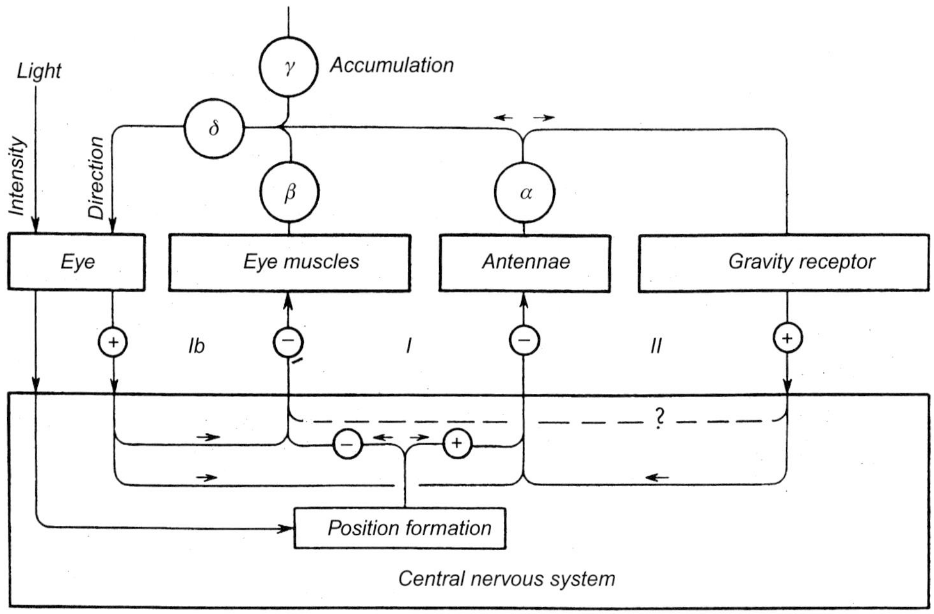

FIGURE 13.10 **Scheme of the *Daphnia* orientation system.** I, phototaxis; II, geotaxis; δ, direction of light falling on the eye, depending on α, β, and γ. *Source: Jander (1966).*

swim abnormally. Changes to the orientation of the body due to displacement of the center of gravity may be observed e.g. in gravid *Eurycercus* compared with juvenile specimens of the same species. Female Eurycercus that contain numerous embryos in the brood pouch swim with their head upwards and move with their dorsal side forward (Smirnov, 1971, p. 80), whereas younger specimens are oriented with their back upwards. No special observations of this kind have been made in representatives of other cladoceran genera.

Buoyancy

Buoyancy also depends on the specific weight of cladocerans; it is modified by oil inclusions (hydrostatic effect) and the development of slime envelopes. Planktonic cladocerans also possess thin exoskeletons. The obvious hydrostatic role of oil drops in the body of cladocerans has not been investigated.

Immobilized small Cladocera sink at a rate of 2–3 mm/sec, or as a percentage of body length, 350–550% for Chydoridae and 250–72% for *Sida* (Daphniidae) (Smirnov, 1971). The likely perception of gravity by the antennal setae of *Daphnia* was suggested by Bidder (1929) and Grosser et al. (1953). The position of the body in space may also be controlled by the tension receptors of the oculomotor muscles transferring stimuli to the antennal muscles (Harris and Mason, 1956). Harris (1953) also noted that a different type of muscle located on different sides of the eye causes strokes of the antennae that contribute to daphnid rotation. Jander (1966, 1975) suggested a more complex system of support for orientation in *Daphnia*, comprising commands from the eyes, the antennae, and a gravity receptor (which is still unidentified) (Figs. 13.10 and 13.11).

In the dark, *C. ovalis* and *C. sphaericus* become homogeneously distributed in water; illuminations is followed first by upward

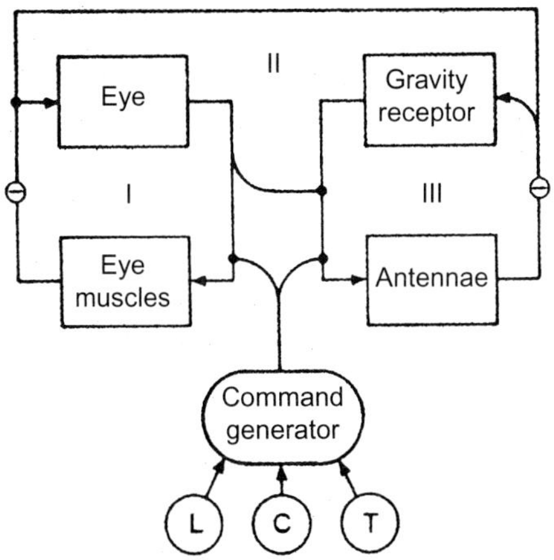

FIGURE 13.11 Later schematic of the *Daphnia* orientation, showing three negative feedback loops. C, chemical information; L, light intensity information; T, temperature information. *Source: Jander (1975).*

swimming and then they become concentrated at the bottom (Bryukhatova, 1928). In complete darkness (as instantaneous photographs have demonstrated), daphnias maintain a position corresponding to that adopted in the presence of illumination from above (Jander, 1966). In addition, after extirpation of both the eye and ocellus, the reactions of daphnias remain essentially the same owing to a general sensitivity to light (Schulz, 1928; Harris and Mason, 1956; Fryer, 2004).

Thus, orientation of the daphnid body obviously depends on a combination of several varying stimuli.

It may also be noted that locomotory responses may differ in individuals of different sexes. Thus, *D. pulex* females run away from a thin rod, whereas males approach it and may even become attached to it (Szlauer, 1964). Movement related to predators principally involves the strategy of akinesis (mainly in littoral species; Chapter 14, section 14.3.1) or the strategy of escape (mainly in pelagic species; Chapter 14, section 14.3.2).

13.8 ENDOGENOUS RHYTHMS OF ACTIVITY

In continuous darkness, *D. magna* is least active at midnight and maximally active at noon (Harris, 1963), whereas under constant illumination their activity period was revealed to be about 28 h. A circadian time of 27 h 50 min in *D. magna* was also reported by Ringelberg and Servaas (1971). Under natural conditions, a similar circadian activity rhythm was found in *D. pulex*: its activity was low at night, maximum in the morning, and then declined thereafter (Stearns, 1975).

13.9 EFFECT OF XENOBIOTICS

There are only a few examples of experiments related to the effect of xenobiotics on movement. In the presence of 0.02 mg/L Cu^{2+}, *D. magna* became less positively phototactic (Kien et al., 2001). Cr (as K_2Cr2O_7) inhibits phototaxis in *D. carinata*, as shown by a negative linear correlation between the phototactic index and Cr^{6+} concentration (Wu et al., 2005).

CHAPTER

14

Behavior

14.1 DIFFERENCES IN BEHAVIOR AMONG SPECIES

Types of behavior characterize the lives of different Cladocera species. It is likely that each of the numerous littoral and pelagic exhibits species-specific reactions. Representatives of the same genus may be strikingly different, e.g. *Chydorus gibbus* is strictly bottom-living, whereas other species of this genus tend to swim in open water. The well-known urge of *Sida crystallina* to attach itself to any surface was studied quantitatively by Szlauer (1973). Males show a strong drive for attaching to females, as reported in *Chydorus sphaericus* (Smirnov, 1965b, 1971; Van Damme and Dumont, 2006).

14.1.1 Littoral and Bottom-Living Species

The modes of locomotion related to substrata include crawling, attachment to various substrates, and attachment to the surface film of water (Scourfield, 1894, 1900a, 1923, 1929). Crawling is characteristic of the littoral forms, is combined with occasional swimming over substrata, and is diverse, ranging from burying into the bottom mud to running over the surface of various bottom substrata. The action of pushing through narrow passages was termed *scrambling* by Fryer (1968).

Crawling over a flat substrate is characteristic of *Graptoleberis testudinaria*, which also attach themselves to the substrata by sucking (Fryer, 1968) and are thus able to glide along the surface. The ventral side of this chydorid is flat, the outline is densely supplied with plumose setae (Fig. 1.4), and the suction force is achieved by pumping water from inside the valves.

Movements are notably slow in the bottom-living *Ilyocryptus*. After producing progeny they remain in the same place surrounded by their young (Chirkova, 1972a, b). There are also some species that live on the surface of mud with their ventral side up (*Ilyocryptus* and *Drepanothrix*). Another species that lives on the mud surface, *Lathonura*, finds a convenient place, arranges itself ventral side up, and can then be observed to be immovable for hours, with only its thoracic limbs moving.

Cladocerans that crawl on the bottom substrata can stay in any position while feeding on the substrate. In bottom-living cladocerans, the form of the shell is usually streamlined, frequently containing teeth at a ventroposterior

Physiology of the Cladocera.
DOI: http://dx.doi.org/10.1016/B978-0-12-396953-8.00014-8

angle. This angle suggests that they serve to direct away the water flow formed at the posterior end of the animal, but the actual hydraulics have never been specifically studied.

14.1.2 Pelagic Cladocera

Pelagic cladocera, on the other hand, may live all of their life cycle without making contact with the substrata. They support their body in the water by the strokes of their antennae (Fryer, 1991). There are two kinds of strokes: slow, sufficient for keeping the sinking body in place; and rapid, used to travel through space.

Cladocera movement depends on various environmental factors; it is highly variable and dynamic but the physiological mechanisms that transform the stimuli into the responses are not always obvious. Hutchinson (1953) highlighted the random stimulation of all *Daphnia* sense organs by turbulence.

In rising temperatures, daphnias swim upwards, whereas in falling temperatures they tend to sink (Gerritsen, 1982). The redox potential also influences the direction of movement. In the presence of catechol (E_h, otherwise termed E'_o, = +0.33 V) daphnids move upwards, and in the presence of cysteine (E'_o = −0.14 V) they move downwards (Baylor and Smith, 1957). The transitional level is about E'_o = +0.045 V.

The behavior of females and males of the same species is clearly different. For *Daphnia pulex*, Szlauer (1964) observed that females fled from a thin rod while males swam toward it and sometimes attached to it. The mating behavior of *Daphnia pulicaria* was investigated by Brewer (1998) (Fig. 12.1), who showed that male swimming was faster than and orthogonal to that of females. Video recording of the swimming trajectories of females and males revealed that males pursued and contacted both females and males, the latter being contacted more often, although for a shorter time. *D. pulicaria* males can detect a female at a maximum distance of 6.4 mm (i.e. four body lengths), principally via mechanoreception of fluid disturbances created by the swimming females. The trails of swimming *Daphnia* were visualized and photographed by Brewer (1998) by using a Schlieren optical path and a smooth density gradient.

Surface Film

Some cladocerans may attach themselves to the surface film of water with their ventral side up (e.g. *Scapholeberis* and *Dadaya*) and thus feed for long periods in an "upside-down" position on particulate matter and probably also on the unimolecular film on organic matter. Both behavior and the structure of the ventral side of the valves in *Scapholeberis* were examined by Scourfield (1894, 1900b) and later reported in more detail by Dumont and Pensaert (1983) (Fig. 3.11). *Scapholeberis* sometimes swims within the water column, in a similar way to other Cladocera, but once below the surface film it is reluctant to leave it (Fryer, 1991). Fryer (2006) also noted that despite this characteristic behavior there is no sign of reshaping of any of its organs in response to such an inverted position.

Terrestrial Environment

Bryospilus crawls in moist, moss-like growths on the trunks of tropical trees (Frey, 1980). Despite having a morphology characteristic of other chydorids, it refuses to swim if placed in water.

14.2 MIGRATION AND SWARMING

14.2.1 Vertical Migration

Vertical migration is known to occur in both littoral species (as indicated for *Acroperus* and *Eurycercus* by Szlauer, 1962a, 1963a) and many pelagic daphnids, including *Daphnia* species (Bainbridge, 1961; McNaught and Hasler, 1964;

Wright et al., 1980). Cladocera can withstand the changes of pressure experienced during vertical migration.

Under conditions of decreased illumination, littoral *Daphnia magna* and *Ceriodaphnia reticulata* demonstrate a rapid upward movement (Meyers, 1980). Vertical, but delayed, ascent was also shown in *C. sphaericus* and *Pseudochydorus globosus* (if they were not attached to a substratum), and their ascent was stimulated by decreased oxygen levels.

Changes in the eye of *Daphnia longispina*, such as an increased pigmented surface under laboratory conditions of constant illumination, were found to precede its ascent during vertical migration; this therefore forms the expression of an endogenous rhythm (circadian rhythm) (Cellier et al., 1998; Cellier-Michel and Berthon, 2003). Vertical migration may, however, occur without participation of the eye; this has been shown both in the dark and in specimens with an extirpated eye. Szlauer (1963a) demonstrated that daily vertical migrations of *D. magna* occur in the absence of changes in illumination or in the dark. Daphniae that have had their eyes removed can still migrate vertically in response to differences in light intensity (Harris and Mason, 1956). Daily vertical migrations in *D. magna* may also be caused by fish kairomines (Loose et al., 1993).

Gliwicz (1986) highlighted the dependence of Cladocera population density on lunar cycles, resulting from a combined effect on the vertical migration of Cladocera and the feeding intensity of the fish.

14.2.2 Swarming

Swarming has been observed in both littoral (Lastochkin, 1930) and pelagic (Birge, 1898; Künne, 1926; Dumont, 1967; Johnson and Chua, 1973) cladocerans, including *Bosmina*, *Ceriodaphnia*, *Daphnia*, *Moina*, *Polyphemus*, and *Scapholeberis*. Swarms may range from a few centimeters to a few meters across (Young and Taylor, 1990). Sometimes such shoals consist of mixtures of various species. Kotov (2000) observed a shoal in Lake Glubokoe (Moscow region, Russia) that predominantly consisted of *Scapholeberis*, but in which about 10% was made up of *Bosmina*, *Ceriodaphnia*, *Pleuroxus truncatus*, and other chydorids. In experiments, *Bosmina* swarming behavior was observed only in the light and with abundant food (Jacobsen and Johnsen, 1988).

A central question here is: what stimuli induce individual cladocerans to gather into swarms? The mechanism may involve visual landmarks or chemical stimuli (e.g. attractive odors), or it may simply be behavioral. It was also observed that *Bosmina* and *Polyphemus* are reluctant to leave a swarm and change their swimming behavior in the marginal zone of a swarm (Young and Taylor, 1990). *D. magna* has a tendency to form and maintain aggregates in response to chemical cues from fish and invertebrate predators (Pijanowska and Kowalczewski, 1997). In addition, *Polyphemus pediculus* was found to excrete 1-heptadecene into the surrounding water; this substance may be involved in the formation, maintenance, and recognition of swarms (Wendel and Jüttner, 1997). Thus, swarms may create a special physiological environment that, in turn, influences the members of the swarm. A swarm may also create its own general water movements, which would not otherwise take place. As they are uniformly orientated, the members of a swarm can create general water movements within the swarm. Individuals within a swarm are integrated into it and the swarm may move as a single entity within a water body (Young and Taylor, 1990).

14.2.3 Hydrostatic Pressure

Cladocera have been found at depths of up to 150 m (as summarized by Smirnov, 1971).

For example, *Eurycercus* and *Camptocercus* have been recorded at 150 m (Zschokke, 1911). In Lake Baikal, *Alona labrosa* was abundant at c. 70 m and *A. setosocaudata* was present at depths of up to 133 m (Vasilyeva and Smirnov, 1975). Thus, cladocerans can live at considerable hydrostatic pressures. A water depth of 10 m approximately corresponds to 1 bar (1 atm = 1.013 bar = 14.696 psi). Another characteristic is their ability to withstand changes in hydrostatic pressure during vertical migration. It has been demonstrated experimentally that during vertical migration (in response to light) *D. magna* is unaffected by high pressure (Lincoln, 1970).

14.3 EMOTIONAL REACTIONS

Cladocera have immediate reactions to various stimuli, shown as a temporary increase in heart rate by about 20% (Smirnov, 1965b) following e.g. a slight mechanical irritation. In addition, upon seeing specimens of its own species, oxygen consumption in *Polyphemus pediculus* was shown to change: it decreases in parthenogenetic females containing eggs, and increases in females containing developed embryos, newborn, gamogenetic females, and males (Butorina, 1979, 1980). In newborns, the increase can reach 69%.

The mating behavior of *Daphnia pulicaria* was investigated by Brewer (1998) (Fig. 12.1).

Cladocera also exhibit an immediate reaction to disturbance: akinesis in most littoral species (i.e. they faint) or an escape reaction in planktonic species (they "flee in fright").

14.3.1 Akinesis

When disturbed, chydorids and macrothricids stop moving, fall to the bottom, and remain immovable, sometimes for a long time. If you shake a glass vessel containing littoral cladocerans, you will see that some of them stop their usual movements and sink to the bottom. This behavior is called *akinesis* (sham death or "dead-man response"), and is an important reaction in arthropods, related to the fact that predators will only attack moving prey (Kerfoot et al., 1980). It is manifested principally in littoral cladocera (Smirnov, 1977), whereas planktonic species use a different method to avoid predators: they flee from danger (they manifest "the escape reaction").

When disturbed, Chydorids, *Drepanothrix*, *Lathonura* (Smirnov, 1977), and *Ilyocryptus* (Mordukhai-Boltovskoi and Chirkova, 1973; Fryer, 1974, p. 142; Smirnov, 1977) play dead for a few seconds. *Eurycercus lamellatus* and *Lathonura* remain immovable for more than 3 min, while this may continue in *Chydorus* for several minutes (Kerfoot, 1978). In these motionless specimens, the thoracic limbs may either stop or continue beating, e.g. in *Ilyocryptus* (Mordukhai-Boltovskoi and Chirkova, 1973). In *Lathonura*, either the thoracic limbs and heart both stop beating or only the thoracic limbs stop during akinesis. Akinesis may continue for rather a long time until the animal resumes its usual movement for no obvious reason.

In planktonic *Bosmina*, it was impossible to discern its brief akinesis by direct observation. However, using cinematographic photography, Kerfoot (1978) (Fig. 14.1) found that *Bosmina* use brief akinesis (sham death) when attacked by a predator, and passively sink by 2–4 times body length at 0.6–0.8 mm/sec, just sufficient for the attacking predatory cyclops to miss its prey (Kerfoot, 1978; Kerfoot et al., 1980).

I did not manage to find any references to investigations on the physiological background of akinesis in arthropods. It seems that attempts to provide a deeper explanation of this widely known fact have not been made. In addition, quantitative and qualitative ratings of the stimuli that cause akinesis have never been made.

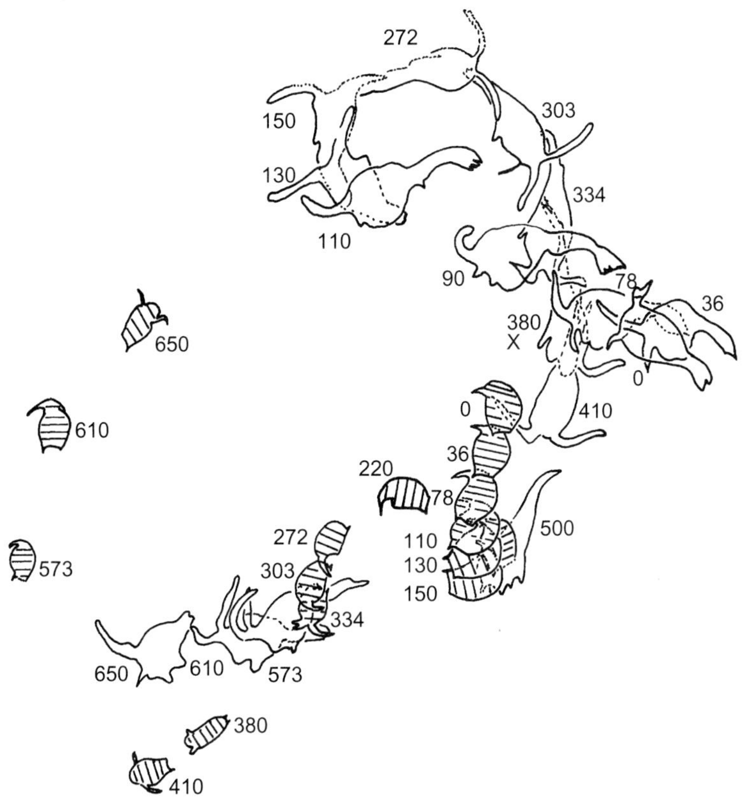

FIGURE 14.1 **Trajectories and akinesis (dead-man response) in *Bosmina* attacked by *Cyclops*.** An attacked *Bosmina* in the state of akinesis sinks by several its diameters. *Source: Kerfoot (1978).*

14.3.2 Escape Behavior

Most pelagic cladocerans (*Ceriodaphnia, Daphnia, Diaphanosoma, Leptodora, Polyphemus, Sida,* and *Simocephalus*) escape predators by fleeing and do not manifest akinesis, as shown e.g. by Szlauer (1964). The escape capacity of sidids was discussed by Korovchinsky (2004).

Szlauer (1964) observed that female *Daphnia* avoid a tube or glass wand of a certain diameter when it is used to model a predator by moving it through a *Daphnia* culture at a certain speed. Their escape ability was recorded as the inverse of the number of specimens sucked into a tube. Interestingly, males behaved differently: they approached the object and some even attached themselves to it. The highest escape ability was manifested by *Diaphanosoma*, and the lowest by *Bosmina* and *Chydorus* (Szlauer, 1965). By applying a similar method, Drenner et al. (1978) studied capture frequencies (by a siphon tube) for different cladocerans in relation to their distance from the tube. Captures frequency of cladocerans was greater than that of copepods.

Somersaulting has been observed in *Daphnia* both in response to a predator and to chemical cues from crushed conspecifics (Pijanowska and Kowalczewski, 1997). *D. magna* reacts to these stimuli by escaping, sinking to the bottom, or making aggregates; specimens that experience these cues were more successful in avoiding predators (Pijanowska and Kowalczewski, 1997).

14.4 IMPACT OF XENOBIOTICS ON BEHAVIOR

At sublethal concentrations, xenobiotics disturb the normal behavior of cladocera in various ways. Upon exposure to deltamethrin, *D. spinulata* jerks as if prodded with a needle; its escape reaction became less intense and is followed by erratic swimming (Alberdi et al., 1990). Upon exposure to lindane, *D. magna* failed to migrate in a directed manner (Goodrich and Lech, 1990), whereas upon exposure to carbaryl (1 ppb), *D. pulex* escape-like behavior increased and it exhibited extreme escape behavior at an acutely toxic level (40 ppb) (Dodson et al., 1995).

Phototaxis was depressed in *D. magna* by bacitracin and lincomycin, whereas aminosidine increased the reaction to light and erythromycin did not alter phototaxis (Dojmi di Delupis et al., 1992). *D. magna* lost mobility either completely or partly on exposure to dimethoate or pirimicarb (Andersen et al., 2006). Their recovery following single-pulse exposures was also studied.

Hyperactivity was observed in *D. magna* exposed to pesticides, followed by reduced activity and death (Matthias and Puzicha, 1990).

CHAPTER

15

Ecophysiology

15.1 PHYSIOLOGICAL BACKGROUND OF THE LIMITS OF ENVIRONMENTAL FACTORS

Their morphological and physiological potential enables Cladocera to live in a variety of situations. Most Cladocera are freshwater animals but some are marine species or live in saline lakes. Their ability to live at different salinities depends on their capacity for osmotic regulation (discussed in Chapter 8)

Various different Cladocera species inhabit a wide variety of environments: various substrates, open water, microtelmata (accumulations of a few cubic centimeters of water in epiphytic plants) (Smirnov, 1988), interstitial capillary water (Dumont, 1987, 1995; Sabater, 1987), surface films, moist, and moss-like growths on tree trunks in tropical forests (*Bryospilus*) (Frey, 1980).

All aquatic invertebrates are adapted to the natural dynamics of their environment. Investigation into the distribution of Cladocera species in relation to the ranges of environmental factors has revealed that at least some species are present within certain limits of those factors (Mäemets, 1961; Røen, 1962; Fryer, 1993); their limited distribution may depend on physiological differences between the species. Therefore, physiological mechanisms must be examined to provide explanations for these phenomena.

There exist acidophilic species—the macrothricid genera *Acantholeberis, Streblocerus,* and some Australian endemic chydorids—that are confined to acid coastal dune lakes (Smirnov and Timms, 1983). On the other hand, alkaline lakes are inhabited by a few other species of Cladocera.

It should be remembered that water in nature comprises a complex solution of diverse inorganic and organic substances, including biologically active and toxic ones (the latter may be partly of natural origin, e.g. phenols). Dodson (2005, p. 176) summarize the situation thus: "Aquatic organisms live in an olfactory field of dozens, if not hundreds of biologically significant chemical signals. ... For example, *Daphnia* can probably sense at least a dozen different chemical signals. These signals provide information about whether to accept or reject food and about the presence of macrophytes, specific predators, competitors, congeneric individuals, and males." The presence of complex and variable chemical factors should also be mentioned, as well as the fact that

Physiology of the Cladocera.
DOI: http://dx.doi.org/10.1016/B978-0-12-396953-8.00015-X

factors react and interact in solution, either alleviating or enhancing the action of particular components.

Primary and secondary producers are sensitive to the presence of dissolved organic acids, vitamins, and so on, which are variously distributed in space and time. These are "complex chemical landscapes," according to Keller et al. (2001), and it should be added that such landscapes, both experimental (Bowles et al., 2002) and natural, are highly labile.

Physical factors are also in a state of continuous change, and reactions may be different to constant and fluctuating factors. Thus, Rider et al. (2005, p. 15) formulated the dependence of physiological processes on the environmental conditions as: "Environmental signals regulate a variety of key processes in animal physiology. In many genera, environmental signals such as changes in day length or temperature signal the onset or termination of reproduction. These environmental signals typically stimulate the release of neuroendocrine signaling molecules that initiate endocrine cascades culminating in physiological responses to the environmental cue."

Cladocera species integrate various environmental factors, which results in a certain quantity of progeny. Depending on the combination of numerous and not precisely discernible factors, the result may be a peak or a decline in abundance. In a particular water body, certain species produce peaks of abundance, although in different years such peaks are different in height and their timing may be significantly different.

15.1.1 Temperature

Cladocera have been observed to develop in temperate latitudes in warm seasons, i.e. from May till September in the northern hemisphere. It was therefore inferred that they prefer warm waters. Most (but not all) macrothricids prefer warm water subtropical and tropical situations. In *Daphnia longispina*, the growth rate and maturation time remain constant within 14–20°C (Sarviro, 1985). In *Daphnia magna*, the temperature range is 11–30°C, and the thermal sensitivity for avoidance responses is about 1.5°C (Lagerspetz, 2000). In a temperature gradient, most *D. magna* have been shown to collect together at 18–20°C (Skadovskiy, 1955); at 20°C for adults and 25°C for juveniles (Chernykh and Panasyuk, 1964); or at about 24°C, depending on their previous acclimation (Verbitsky and Verbitskaya, 2011).

The upper limit of temperature for *Chydorus sphaericus* was determined to be 34°C (Mortimer, 1936) or about 32°C (Bogatova, 1962). For daphnids, the upper limit of temperature [100% lethal dose (or LD_{100})] has been determined as 32°C for *D. cucullata*, 35° for *D. magna*, 36°C for *D. pulex*, and 31°C for *Scapholeberis mucronata* (Mortimer, 1936); 30°C for *Daphnia* and *Ceriodaphnia* (Mallin and Partin, 1989), and 36°C for *D. longispina*. *Moina macrocopa* is inhibited by a temperature of 36°C (Maksimova, 1969) and perishes at 39.5°C (Brown and Crozier, 1927) or 41°C (Maksimova, 1969). *D. magna* is eliminated from natural waters at this temperature, and *D. pulex* at slightly over 30°C (Brown and Crozier, 1927; Goss and Bunting, 1976); so, there are clearly differences between species (LaBerge and Hann, 1990). For *Bosmina longirostris*, the tolerant zone for life activities is up to 36°C, but the zone of normal life activities is 11–23°C, according to Verbitsky and Verbitskaya (2002).

It is likely that tropical macrothricids can withstand somewhat higher temperatures, although this has not been investigated. *Latonopsis* and *Alona cambouei*, for example, have been found in hot springs in India at 34–36°C (Padhye and Kotov, 2010). *Macrothrix* spp. are abundant in Iraq in pools that are normally hot under the normal weather conditions of that country.

A further increase in temperature eventually causes protein denaturation (Brown and Crozier, 1927). For *Daphnia middendorffiana*, the upper level of temperature was found to be 26°C, above which inactivation of respiratory enzymes started (Yurista, 1999); however, it was shown, with reference to *D. magna* and *D. pulex*, that Cladocera may survive sudden temperature changes over the range of 5–30°C (Goss and Bunting, 1976).

The lower limit of temperature tolerance seems to be about 0°C, since *C. sphaericus* remained active and survived at this temperature, and did not avoid melting snow (Smirnov, 1971). The life span of *D. longispina* at 5°C is much longer than at 20°C (Munro and White, 1975). Some cladocerans stay active and reproduce at a temperature of about 0°C (Rivier, 1986, 1992). They are known to exist on arctic islands and in tundra, where they are exposed, at least for some of the time, to very low temperatures. Rivier (1986) described winter communities from many ice-covered lakes and reservoirs in which *Daphnia* and *Bosmina* contained embryos. She also reported the presence in winter of *Daphnia* with numerous embryos at the bottom of ice-covered Lake Siverskoe, at a low level of oxygen (ca. 2 mg/L O2).

Generally, rising temperature is followed by a rising intensity of activity. For example, the heart rate of *D. magna* increased from 60 to 400 beats/min when the temperature rose from 5°C to 28°C (Mac Arthur and Baillie, 1929) (Fig. 6.6). It has also been noted that daphnias swim upward in rising temperatures, whereas they tend to sink at falling temperatures (Gerritsen, 1982).

Heating water from 15°C to 25°C, with a subsequent reduction to 20°C, resulted in larger body sizes of *Ceriodaphnia quadrangula* young, shorter longevity and earlier maturation of exposed juveniles (Romanovskiy and Loganova, 2010). In *Simocephalus*, Verbitsky and Verbitskaya (2009) also identified the after-effects of previous temperature changes, e.g. increased of the population growth after previous cooling.

15.1.2 Some Physiological Events Occur at Low Temperatures

Farkas et al. (1984) assumed that *D. magna* cannot overwinter in an active state due to a failure to adapt their phospholipid composition and the physical state of their membranes to low temperatures. Docosapolyenoic acid is not present in *Daphnia* phospholipids and the level of polyenoic acid does not increase when the temperature decrease from 20°C to 10°C.

Schlechtriem et al. (2006) cultivated *D. pulex* at 11°C and 22°C for 1 month on a highly unsaturated fatty acid (HUFA)-free diet. Long-term exposure to cold temperature caused a significant increase in eicosapentaenoic acid (EPA; 20:5ω3) and conversion of C18 fatty acid precursors to EPA and arachidonic acid (20:4ω6) was observed. Schwerin et al. (2009) observed compensatory control of physiological processes in the cold: in *D. pulex*, most cold-repressed proteins comprise enzymes involved in protein digestion (trypsins, chymotrypsins, astacin, and carboxypeptidases); and cold-induced proteins include several vitellogenin and actin isoforms and AAA + ATPase. The increased actin concentration in cold-acclimated animals may contribute to the preservation of their muscular performance.

Following ultraviolet (UV)-B irradiation, *Daphnia* spp. showed increased survival and removal rates of light-dependent DNA damage at 10°C than at 20°C (Connelly et al., 2009).

According to Gainutdinov et al. (2000), the thermostability of *D. magna* is modified by the addition of glutamate, glycine, or adenosine 3′,5′-cyclic monophosphate (cAMP), with the changes being slow, strong, or fluctuating, depending on the concentration of substance added. The authors interpreted this process as "coding of the chemosensory information by the nervous system," although it is unknown which organs of Cladocera are involved in temperature perception.

15.1.3 Oxygen Concentration

Some Cladocera, e.g. *Ceriodaphnia laticaudata*, can withstand low levels of oxygen (Fox, 1945; Burgis, 1967). Accordingly, this species developed in great abundance during the first year of existence of the Cherepovets Reservoir (in the Upper Volga basin) in response to a general oxygen deficit caused by decomposition of flooded vegetation and forest litter (Smirnov, 1966). In subsequent years, not a single specimen of *C. laticaudata* was found, although *C. pulchella* and *C. quadrangula* became dominant instead.

15.1.4 Acidity-Alkalinity Ranges

Most cladocera prefer a neutral pH. The upper limit of pH was determined for chydorids as 10.6 (Bogatova, 1962), while a lower limit of pH 3.7 has been reported for *Acantholeberis curvirostris, Alonella excisa, C. sphaericus, Ceriodaphnia setosa* and *Ophryoxus gracilis* (Kitaev, 2007). Havas and Likens (1985) showed that *Daphnia catawba* die below pH 5.0, but that *Holopedium* can survive even at pH 4.0.

D. pulex usually lives in circum-neutral conditions, and Weber and Pirow (2009) investigated its response to acid stress. At pH 7.8 and normocapnia (i.e. a normal partial pressure of CO2) its extracellular pH is 8.33 and PCO2 is 0.56 kPa, and bicarbonate concentration in its hemolymph is 20.9 mM. Acidic conditions caused a slight acidosis (ΔpH was 0.16–0.23), a 30–65% bicarbonate loss, tachycardia, hyperventilation, and hypermetabolism. At pH 5.5, defense mechanisms were activated, extracellular PCO2 decreased to 0.33 kPa and the animals tolerated only short-term exposure to this pH. These authors also suggested a model of whole-animal CO2 transport.

Preference for an acid medium is especially clear in the macrothricids *Streblocerus serricaudatus* and *A. curvirostris*. The latter retains sodium (Na) better than does *D. magna*, which is improved in acidic medium: the rate of sodium loss at pH 3 is lower than at pH 7 (Potts and Fryer, 1979).

15.1.5 Illumination

Photoperiod (see Section 13.4.7) and seasonal changes of illumination obviously influence the population dynamics of Cladocera. However, some Cladocera can live in the dark.

Species of the same genus may differ greatly in their reactions to environmental conditions and in their behavior. For example, different species of *Ceriodaphnia* are known to tolerate different oxygen limits (Burgis, 1967), different *Moina* species either inhabit freshwater or live in saline waters, *Daphnia* species are ecologically different, and *Chydorus* species either swim or are confined to the bottom (e.g. *C. gibbus*). It is remarkable that the reactions of Cladocera to external factors are so labile and that the values that characterize cladocerans reactions to various factors are so variable. The numerical values of the physiological parameters discussed throughout this text are therefore not absolute. Such values depend on:

1. age and sex;
2. previous adaptation;
3. interaction with other factors (synergies and antagonistic interactions); and
4. the presence within a morphologically defined species of genotypes (clones) that have different tolerances to environmental factors.

This is why the characteristics of environmental factors are relevant to a discussion of the geographic distribution of a particular species.

15.1.6 Adaptation

It should be noted that the values given for ranges of stability to certain factors, and any such figures, are not constant but instead depend on other variables and on previous acclimation. This has been shown e.g. for copper (Cu) by Bossuyt et al. (2005), and for acclimation to different previous temperatures by Shkorbatov (1982): temperature shock in *Daphnia* acclimated to 28°C occurs at a higher temperature than that of *Daphnia* acclimated to 8°C. However, beyond a certain limit physiological temperature adaptation may be replaced by a regular diapause (Mitchell, Halves, and Lampert, 2004).

Survival may also depend on the previous environment of the specimens being tested. Shkorbatov (1953) highlighted the effects of prolonged physiological adaptation of cladocera, e.g. *Simocephalus vetulus* taken from a pond survived in deoxygenated water much longer than did specimens taken from a river. In addition, Stroganov (1983) emphasized the fact that individuals also differ in their stability, as observed in *D. magna* in solutions of sodium pentochlorophenolate.

15.1.7 Genotypes

Green (1956) found "racial and clonal differences" within the same *Daphnia* species in the ability of individuals to synthesize hemoglobin (Hb) and thus of their adaptation to low oxygen levels in the habitat. Quantitative estimations of the response of *D. magna* to the impact of environmental factors may also vary depending on genetic factors (Barata et al., 2000). In *D. pulex*, 92 genotypes have been found (among 227 specimens); in *S. vetulus*, 18 genotypes were identified (among 112 specimens), which showed physiological differences (LaBerge and Hann, 1990). In 25 tested clones of *D. magna*, thermoresistance of the heart, when tested at 38°C, was variable: some clones exhibited a high mean thermoresistance, whereas in others it was low (Pashkova et al., 1998). *D. longispina* lineages have been shown to differ in their resistance to Cu (Martins et al., 2005). Thus, it is likely that the elimination of vulnerable lines may lead to a loss of genetic diversity.

15.2 SYNERGISM AND ANTAGONISM AMONG ENVIRONMENTAL FACTORS

Numerous dynamic factors can act "in concert," as Steinberg et al. (2010) noted, and the independent action of one factor may be modified by other factors. Antagonism (or synergism) between ions in solution modifies their individual actions in certain species. Tauson (1924a, 1924b) was probably the first to demonstrate this in cladocerans. She observed a decrease in the death of various cladocerans in the presence of certain combinations of cations in comparison to treatment with single ions. Winberg (1933) demonstrated experimentally that the destructive action of the potassium ion (K^+) on *C. sphaericus* is decreased in the presence of the sodium ion (Na^+), but is unaffected by the presence of calcium ions ($Ca2^+$). In contrast, in the case of *Chydorus ovalis* $Ca2^+$ also moderates the negative action of K^+.

Pairs of metal ions may be either antagonistic (as estimated by toxicity in *D. magna*), e.g. Al + Mo, Cr + Co, and Cr + Mg, or synergistic, e.g. Cr + Se, Cr + Zn, and Fe + Se, as demonstrated by Tomasik et al. (1995). The relationships between ions also depend on their concentration. Yakovlev et al. (2000) found that the combination of Zn + Cd is antagonistic toward *D. magna* survival, whereas Zn + Cu are synergistic. According to these authors, there are also still more complex interactions.

Combinations of metal ions are more dangerous for *D. magna* than are isolated ions at the same concentration. It has also been shown by Dediu et al. (1995) that combinations of metals are more dangerous than separate metals at the same concentration, as shown for Cu + Zn, Cd + Zn, Hg + Zn, Cu + Cd, Cd + Hg, and Cu + Hg in *D. magna*. Clifford and McGeer (2009) have shown that the zinc (Zn) toxicity toward *D. pulex* is alleviated by Ca^{2+} whereas, in *D. magna*, in the presence of Ca, the Cd, Ni, Pb and Zn uptake is suppressed and elevated Na concentrations increased the uptake rates of Cd and Pb (Komjarova and Blust, 2009). Selenium uptake by *D. magna* increased significantly as sulfate concentrations decreased (Ogle and Knight, 1996). In addition, Ca and Hg each inhibit the accumulation of the other by *D. magna* (Teles et al., 2005).

Glyphosate reduces the acute toxicity of Ag, Cd, Cr, Cu, Ni, Pb, and Zn (but not of Hg and Se) toward *Ceriodaphnia dubia* (Tsui et al., 2005). The effect on *C. dubia* of spinosad and R-11 (an adjuvant used with pesticides) was also found to be synergistic (Deardorff and Stark, 2009).

There are two possible explanations for these observations: effects on the direct toxicity of particular ions or the more complex effect of deviation from the natural equilibrium of ions in fresh or brackish water, to which certain species are adapted.

15.3 LIPID PATHWAY FROM ALGAE VIA CLADOCERA TO FISH

Müller-Navarra et al. (2004) advocate, with good reasons and reference to *Daphnia* data, that the HUFAs EPA (C20:5ω3) and docosahexaenoic acid (C22:6ω3) levels provide a strong predictor of zooplankton growth. These fatty acids are conservatively transferred up the food web to fish and to man.

In temperate latitudes, diatoms (with their stock of lipids) dominate in the spring, followed by blue-green and green algae, the latter of which mostly contain a stock of starch. This change has profound alimentary significance, as algal lipids are assimilated with little change. Following extensive analytical measurements, Sushchik (2006) found that the growth of planktonic Cladocera *(Daphnia)* is controlled by EPA and nitrogen (N) availability; a general conclusion was also made that EPA and N levels indicate the quality of their food within a water body. The seasonal distribution of EPA in different links of the trophic chain was determined from May till October by Sushchik et al. (2008) in phytoplankton (in May the diatom *Stephanodiscus* was dominant), zooplankton, and fish *(Carassius)* in a reservoir near Krasnoyarsk (Russia). The quantitative distribution of EPA in zooplankton generally resembled that in phytoplankton (with a peak of c. 120 g/L in phytoplankton in May). In general, the quantity of EPA in a cyprinid fish *(Carassius)* followed that in zooplankton. The predominant fatty acids in the fish were 16:0, 18:1, and 22:6ω3. Sushchik (2001) had previously observed that *Daphnia* development is controlled by polyunsaturated fatty acids (PUFAs) derived from diatoms and blue-green algae. Sometimes, the diatoms were lacking in 20:5ω3 or other PUFAs, and the blue-green algae contained most of the essential 18:3ω3.

Kainz et al. (2004) observed that the concentration of an essential fatty acid (docosahexaenoic acid) in the food chain decreased in the cladoceran link in all of the lakes studied. It should also be noted that in lakes in which diatoms are the dominant primary producers, as indicated by a sediment consisting of diatomite (e.g. in Kamchatka, in Iceland), the trophic chain is based principally on lipids.

15.3.1 Lipids in Freshwater Fish

Fats are known to be incompletely transformed by their consumers; the composition of their fat is similar to that of their food items. For their normal development, Fish mainly require three long-chain PUFAs (Sargent et al., 1999): docosahexaenoic acid (22:6ω3), EPA (20:5ω3), and arachidonic acid (20:4ω6). Following consideration of the experimental data, Herodek and Farkas (1967) suggested that most of the fatty acids in fish originate in the crustaceans they consume.

As seen from the fat composition of fish, cladocerans, and algae, few modifications are made to fatty acids in the pathway from algae via crustaceans to fish. The extent of fatty acid modification at the algae-crustacea and crustacea-fish links is the focus of further investigations.

With good reason, Arts et al. (2001, p. 122) indicate that fatty acids "are important "drivers" of ecosystem's health/stability," and may determine a humanity's past and future health. In fish, man consumes the algal (slightly modified) fat. The fish smell not actually from fish; rather, it is from algal fat.

15.4 ENVIRONMENTAL CONDITIONING BY CLADOCERA

15.4.1 Biofiltration (Clearance)

As a mass group, the cladocera within a water body filter great volumes of water (see also Chapter 4, section 4.1, "Feeding"). Their filtering rate has been determined to be 0.9–5.15 mL/individual (ind.)/24 h in 0.7–1.8 mm long *D. pulex*, increasing with size in a curvilinear fashion (Richman, 1958); in 2-mm long *D. pulex*, it is about 5 mL/ind./24 h in the daytime, increasing at night up to about 20 mL/ind./24 h (Haney, 1985). As summarized in ctenopods by Korovchinsky (2004), it is 1–248 mL/ind./24 h in *Diaphanosoma brachyurum* s.l., 10–58 mL/ind./24 h in *Holopedium*, 2–657 mL/ind./24 h in *Sida*, and 32–252 mL/ind./24 h in *Penilia*. In *D. magna*, it is 0.24–3.56 mL/ind./h depending on the food concentration (Porter and Orcutt, 1980). In one example, the clearance rate (i.e. filtering rate) of a *Daphnia hyalina* population was about 50% of the lake volume per day during most of the growing season (Balayla and Moss, 2004). Cladocera thus clarify water and their feces fall to the bottom.

15.4.2 Impact on the Gaseous Conditions of the Environment

Cladocera consume oxygen and release carbon dioxide. The extent of their influence may be assessed from their rate of oxygen consumption and the quantity of Cladocera per unit volume of water.

15.4.3 Enrichment of the Environment with Organic Matter

Cladocera release considerable quantities of N, phosphorus (P), and carbon (as dissolved organic C). While N is released as a final product of protein metabolism, the source of liberated P seems to be digested compounds from lipids and phosphoric acid. As a result of their abundance in the aquatic environment, Cladocera make a large contribution to the complex solution that is natural water. This contribution can be quantitatively estimated by consideration of the available data. For such quantitative estimations, see also Chapter 7, section 7.2.

Cladocera also release a variety of organic and inorganic compounds into the water (as noted e.g. by Fish and Morel, 1983). The mean release of urea by well-fed *Daphnia* was estimated to be 0.36 μg/mg/h and that of ammonia to be 0.76 μg/mg/h (Wiltshire and Lampert, 1999); in contrast, starved *Daphnia* liberated 0.06–0.1 μg/mg/h of urea and 0.45 μg/mg/h of ammonia. The released urea is presumed to influence the development and morphology of green algae.

Fish and Morel (1983) determined that *D. magna* release about 40 pmol/organism/h of organic Cu-binding compounds, thought to consist of molecules larger than amino acids. *D. magna* also release a chemical (an organic substance of small molecular mass) into the environment that induces colony formation in the green alga *Scenedesmus acutus* (Lampert et al., 1994). Yasumoto et al. (2005) also demonstrated that *D. pulex* releases aliphatic sulfates that increase the size of *Scenedesmus* cenobia.

Immediately after molting of a single *D. pulicaria*, the activity of β-N-acetylglucosaminidase in the culture medium increased by 40–100 times (Vrba and Machaček, 1994); this molting enzyme sustained its activity for >2 days.

Within swarms of *M. macrocopa*, the concentrations of various forms of N and P increase, protein and carbohydrate fractions appear, and various amino acids are present, as well as palmitic, tridecanoic, and stearic acids (Kalacheva et al., 2000).

Polyphemus pediculus releases 1-heptadecene into the surrounding water; 7 ng/L was found within a swarm vs. 1.5 ng/L some distance away (Wendel and Jüttner, 1997).

B. longirostris tends to increase the N:P ratio in the surrounding water, thus increasing phosphorus limitation (Balseiro et al., 1997). Alkaline phosphatase and a small amount of acid phosphatase are released by *D. magna* into the environment (Boavida and Heath, 1984; Zhao et al., 2006). The liberation of enzymes by animals to their aquatic medium may be a common occurrence, as it has also been shown for fish (Kuzmina et al., 2010).

15.4.4 Infochemicals

The substances released by invertebrates, as far as they are involved in chemical communication in the aquatic environment, are known as *infochemicals*. Thus, formation of cenobia in the green alga *Scenedesmus* is influenced by a chemical released by *D. magna* during its nondigestive metabolism (von Elert and Franck, 1999). This infochemical has been characterized as an olefinic low-molecular-weight carboxylic acid.

The quantitative scale of these processes may be estimated by considering the aforementioned data.

15.5 THE IMPACT OF EXTREME LIMITS OF ENVIRONMENTAL FACTORS AND OF XENOBIOTICS

15.5.1 Impact of Extreme Limits of Environmental Factors

Temperature

All of the evidence indicates that the winter minimum temperature is not disastrous for the cladoceran fauna, judging e.g. by the rich fauna in Yakutia, which is within the zone of global minimum temperatures. Egg latency protects the local species of Cladocera.

Ultraviolet Radiation

UV radiation (UVR) is thought to affect cladocera that inhabit water bodies at high altitudes and high latitudes. These Cladocera develop black pigmentation, which is thought to be protective. Indeed, exposure to sublethal UV irradiation resulted in damage to the intestine in a large proportion of subarctic *Daphnia* (Zellmer and Arts, 2005). It was shown by Zellmer et al. (2006) that following exposure to sublethal solar UVR, *D. pulex* suffers damage to the intestinal system similar to that induced by fasting, although the function of its digestive enzymes (amylase and cellulase) is not much altered.

15.5.2 Impact of Xenobiotics

Inorganic Xenobiotics

It was shown in *D. magna* exposed to tritiated water with an activity of 5×10^{2}–5×10^{8} Bq/L that the number of broods, the size of each brood, life span, and growth rate decreased within five generations; embryos developed disproportionately; and their eggs decomposed (Gudkov and Kipnis, 1995).

SILVER

Silver (Ag) is one of the most toxic elements. *D. magna* and *C. dubia* have been shown to be highly sensitive to $AgNO_3$ but less so to AgCl (Rodgers et al., 1997).

ALUMINUM

Daphnia catawba and *Holopedium* were not found to be particularly sensitive to aluminum (Al) (Havas and Likens, 1985).

ARSENIC

It has been shown that trivalent arsenic (As) is more toxic than pentavalent arsenic to *Daphnia carinata* (He W. et al., 2009).

CADMIUM

It was determined experimentally that *C. dubia* is more sensitive than *D. magna* to cadmium (Cd; as $CdCl_2$) (Suedel et al., 1997). The uptake rate and killing rate of Cd for *D. magna* increased with an increase in temperature from 10°C to 35°C (Heugens et al., 2003).

COPPER

The presence of Cu in water is directly toxic to cladocerans. It also decreases the oxygen consumption in *Daphnia* (Knyazeva, 1994). The effect of Cu depends on the clone, as shown for *D. longispina* (Lopes et al., 2005). Cu toxicity decreases in the presence of Na (De Schamphelaere et al., 2007) or organic matter in the medium, especially humic acids, as has been shown for *D. magna* (De Schamphelaere et al., 2004) and *C. dubia* (Sang Don Kim et al., 1999).

MERCURY

The scale of mercury (Hg) contamination may be illustrated by the case of Lake Managua (Nicaragua), where about 40 tons of elemental Hg was introduced from a chloralkali industry during the first 12 years of its activity (Lacayo et al., 1991). One result was the complete disappearance of *Bosmina* from the plankton. In the aquatic situation, Hg is transformed into methylmercury, which is even more toxic. According to Tsui and Wang (2004b), the uptake of Hg and methylmercury by *D. magna* is proportional to their ambient concentration, and efflux rates are comparable; release takes place via excretion, egestion, molting, and neonate production. In bioaccumulation, methylmercury is predominant. In lakes in Wisconsin (USA), the methylmercury concentrations were 1–211 ng/g in *D. pulex*, *D. galeata mendotae*, and *D. ambigua*, and 40–419 ng/g in *Holopedium* (Back and Watras, 1995). A low concentration of mercuric chloride ($HgCl_2$; a daily rate 0.01 μg Hg^{2+}) given with food to *C. affinis* stimulated reproduction, but survival and longevity decreased (Gremiachikh and Tomilina, 2010).

POTASSIUM

Juvenile *D. magna* exhibit a very varied sensitivity toward potassium dichromate ($K_2Cr_2O_7$) (Klein, 2000), as noted by their immobilization.

URANIUM

It was determined after 21-days exposure of *D. magna* to free uranyl that the chemical toxicity of uranium (U) predominated over its radiotoxicity (as estimated by decreased somatic growth and reproduction) (Zeman et al., 2008).

Comparative tests of Cd and Zn on *D. magna, D. pulex, D. ambigua,* and *C. dubia* demonstrated that *D. magna* is significantly more tolerant to these metals, or their combinations, than other daphnids (Shaw et al., 2006). Acclimation of *D. magna* to Zn was demonstrated, which influenced determinations of Zn toxicity (Muyssen et al., 2002).

Organic Xenobiotics

Acetylsalicylic acid is more toxic to *D. longispina* than to *D. magna* in acute assays, but is more toxic to *D. magna* under chronic exposure (Marques et al., 2004a). Its metabolites were also tested; these turned out to be most toxic to *D. longispina* (Marques et al., 2004b). Nonviable neonates and aborted eggs were produced in addition to the direct toxicity observed.

Fipronil (a phenylpyrazole insecticide) is more toxic for *C. dubia* than its optical isomers (enantiomers), racemate, or photodegradation product (fipronil desulfinyl) (Konwick et al., 2005).

PHENOL

Of the cladocera tested, *S. vetulus* was the least sensitive to the presence of phenol and *Sida crystallina* was the most sensitive (Luferova and Flerov, 1971b). These authors mainly attribute the latter effect to the higher (almost double) beating rate of the thoracic limbs in *Sida*. The motor activity of *D. longispina* increased at increasing phenol concentrations, but abruptly decreased at a concentration of 70 mg/L (Luferova and Flerov, 1971a). Na transport in *D. magna* is inhibited by 2–4 dinitrophenol (Stobbart et al., 1977).

Enantiomers of the same compound (present in pesticides) were found to have drastically different toxicities when tested on *D. magna* and *C. dubia* (Liu et al., 2005).

The survival of cladocerans exposed to toxicants also depends on the period between repeat exposures (Naddy et al., 2000). *D. magna* can withstand the acutely lethal organophosphate insecticide chlorpyrifos (at c. 1 μg/L), provided there is adequate time for recovery between exposures (Naddy and Klaine, 2001).

The effects of different xenobiotics are also discussed in sections describing their particular functions.

15.6 CLADOCERA IN WATER QUALITY TESTING

There are three main roles of cladocera to consider: their use in measuring pollution levels, drinking water quality, and in testing human body fluids.

First, Cladocera spp. have found a role in the system of indicators of the saprobic level of water (i.e. of pollution with organic substances) (Kolkwitz and Marsson, 1909) and in subsequent modifications of this system. According to these authors, they appear to decrease pollution due to self-purification in both β-mesosaprobic and oligosaprobic zones. Obviously, their distribution depends on hydrochemistry.

Hrbácek and Hrbácková (1980) discussed the use of planktonic Cladocera for the estimation of water eutrophication. Andronnikova (1996) described mixtures of zooplankton characteristic of oligotrophic, eutrophic, and polyhumic lakes.

Second, *Daphnia* was the first organism used to determine the quality of drinking water (Naumann, 1929). Cladocera have frequently been used for this purpose (see e.g. Anderson, 1944; Lesnikov, 1967). *C. dubia* has also been used for the biotesting of water (Flerov et al., 1988).

The characteristics and conditions for using daphnids in toxicity testing have been described by Lesnikov (1967), Adema (1978), Leeuwangh (1978), Ten Berge (1978), and Pieters (2007). Lesnikov (1967) suggested estimation of the condition of daphnias by scoring their oogenesis, embryogenesis, coloration, and oil drop status, as well as fullness of the gut.

In the wide field of aquatic toxicology, the cladocera (mostly daphnids) are frequently used as test objects or *sentinel species* (Lesnikov, 1967; Isakova and Kolosova, 1988) and *Daphnia* has been described as an excellent model organism for studying multiple environmental stressors (Altshuler et al., 2011). *Daphnia* was especially recommended for phenol assessment (Tumanov et al., 1988). *D. magna* can also be used to evaluate the quality (toxicity) of bottom sediments (Terra et al., 2010; Romanenko et al., 2011).

The suitability of Cladocera as indicators of various environmental conditions has been discussed by Walseng and Schartau (2011). The sensitivity of *D. magna*, *D. pulex*, and *D. cucullata* to various chemicals was found to be similar (Canton and Adema, 1978), although Lesnikov (1967) found that *D. pulex* is somewhat less sensitive and different lines (or clones) of this species noticeably differ in their sensitivity. Of course, pollutants act as a mixture of interacting ingredients, the effect of which is unpredictable, as particular substances can react with each other, leading to partial neutralization or entering into synergistic relationships.

The reaction of *D. magna* to a wide range of pesticides has also been determined (Frear and Boyd, 1967). The pesticides tested ranged from extremely toxic [median lethal dose (LD_{50}) of carbophenothion − 9 ppt (parts per trillion)] to noticeably less toxic (LD_{50} of prometone = 35 ppm). Tumanov et al. (1988) suggested a quick method for measuring pollution using previously prepared calibration curves of *Daphnia* immobilization by particular organophosphates at different concentrations.

The use of *D. magna* as an ecotoxicological indicator is allowable if its genotype and previous adaptation status are taken into consideration, as discussed by Baird et al. (1989). A fast toxicity test using parthenogenetic *D. magna* eggs in vitro was also suggested by Sobral et al. (2001) and Krylov (2011). The rate of formation of males in *D. magna* was used to measure the presence of pollutants with hormone-disrupting effects (Tatarazako and Oda, 2007).

Of the littoral species, the use of *C. sphaericus* was suggested for assessing water quality (Pieters et al., 2008) or sediment toxicity (Dekker et al., 2006). For *C. sphaericus*, LD_{50} at 96 h was determined by Dekker et al. (2006) to be 46 mg ammonia (NH_3) /L.

Third, an attempt has also been made to use *Daphnia* for testing the toxicity of human physiological liquids: urine, blood serum, hydroptic liquid, and pleural fluid, as well as various biologically active substances (Billiard, 1925). This author determined the average survival time of *Daphnia* in urine to be 7 min (Billiard, 1926). If the daphnias lived longer, the urine was considered hypotoxic; if shorter, it was considered to be hypertoxic. In patients with lung inflammation and a high temperature, the urine contained albumin and the daphnias lived for an abnormally long time (Billiard and Perrot, 1926).

CHAPTER

16

A Cytological Perspective

Margaret J. Beaton and Carli M. Peters

Biology Department, Mount Allison University, Sackville, NB, CANADA

Despite their miniature size, as a group, cladocerans have been the focus of intense ecological and life history study for nearly two centuries. Literature on their feeding/growth rates far outnumbers that focused on the physiological constraints of the group, or their genetics. Cytological work has been even less prominent, despite the earliest publications occurring only a few decades after Latreille named the order. These first studies generally examined embryological development in *Daphnia*, *Leptodora*, and *Moina* (for example, Weismann, 1874; Grobben, 1879; Lebedinsky, 1891) and were published in German. As a result, complete translations are not readily available, severely impeding accessibility for many. However, these papers include detailed and valuable drawings of tissue and cellular structures; for example, Grobben's (1879) survey of development in *Moina* was thought to be the impetus for later studies "on cell lineage and on formative substances" (Witschi, 1934). Later investigations concentrated on cell size and basic morphology, as well as on specific structures such as the labrum and nervous system (e.g. see Sterba, 1957). The first cytogenetic examinations were completed long before the links between chromatin, genes, DNA, and heredity were established and consisted of gross organization of chromosomes.

Today, researchers are applying a burgeoning assortment of microscopy and molecular tools to address questions of genome structure/organization and the pivotal roles that genes play in cellular growth and development. These studies will contribute to our understanding of the genetic and developmental bases for an array of ecological and physiological traits in cladocerans. The ability to pinpoint the location of factors in cells as well as measuring upregulation of proteins, for example, will strengthen our understanding of intra- and intercellular processes, and by extension the physiology and biology of the organisms. The highly advertised release of the nuclear genome sequence for a specific clone of one of the recently diverged species of the *Daphnia pulex* lineage, *D. arenata* (Colbourne et al., 2011, plus supporting online material) has revealed some remarkable findings. Like its body size, the genome of this model organism is diminutive, but it still manages to contain many more

Physiology of the Cladocera.
DOI: http://dx.doi.org/10.1016/B978-0-12-396953-8.00016-1

genes than are found in the human genome (> 30,000, which is 5000–10,000 more). The inflated gene count appears to be a result of duplications as well as the presence of novel genes; about one-third of the identified genes have no known homology with the proteome of any other species. Furthermore, using cDNA libraries and microarrays, Colbourne et al. (2011) found that expression of much of the transcriptome changes with altered environmental (biotic and abiotic) stresses. This represents an enormous step forward in our quest to understand the links between genetics, ecology, and life history strategies of *Daphnia*. In addition, the project led to chromosome analyses that have provided us with further insights into their structure and organization (see below).

Many cladoceran species make ideal study organisms due to their ease of maintenance in culture and their parthenogenetic life cycle. However, in most species, the body carapace, which restricts growth except at ecdysis, combined with a diminutive body and similarly small genome size, has generally hindered cytological research on the group. Members of the genus *Daphnia* (especially *D. magna* and *D. pulex*) have been at the center of much of this research, so comprehensive comparisons are often lacking and coverage of some topics is quite limited. To that end, the available literature on cladoceran genome size variation, cytogenetics, endopolyploidy, and cytological studies of various tissues will be examined. One goal of this chapter is to identify gaps in our cytological knowledge of the order.

16.1 GENOME SIZE AND POLYPLOIDY

16.1.1 Background

Studies estimating the amount of chromatin present in a nucleus predate the discovery of the structure of chromosomes (for a brief review see Gregory 2002, 2005). For example, Swift (1950) found a constant and characteristic amount of DNA per haploid chromosome set for each of two plant species. He also demonstrated that cells in somatic tissues regularly contain a discrete multiple of the haploid amount. From Swift's work, the term *C-value* (meaning genome size) was defined as the quantity of DNA that is contained in a species' haploid chromosome complement. Mirsky and Ris (1951) provided the first two crustacean estimates (for a decapod and a barnacle), but the first cladoceran estimate was not published until much later (Rasch, 1985).

Genome size is thought to vary by more than 200,000-fold across eukaryotic species! This range shrinks at lower taxonomic levels (excluding incidents of polyploidy), so comparisons across members of the same order tend to show contracted differences. Gregory's (2013) animal genome size database includes records from more than 1500 invertebrates, 318 of which are crustaceans (representing 278 species). Such meager coverage for a group comprising nearly 70,000 extant species demonstrates our limited knowledge in this area. As estimates accumulate, patterns in the data that seem obvious today may become murky tomorrow and patterns may appear next week that were not obvious yesterday.

Among crustaceans, C-values range from 0.14 to 64.62 pg, with a mean of 4.45 ± 0.43 pg (Gregory, 2013). This represents a greater than 400-fold range, the widest range of published estimates among invertebrate groups; members of the Platyhelminthes exhibit the second largest invertebrate range (slightly less than 350-fold) (Gregory, 2013). To place the crustacean range in perspective, 64.62 pg and 0.02 pg represent the largest and smallest published invertebrate estimate (Gregory, 2013). So, where do cladoceran genomes fit within the crustacean range of DNA contents?

16.1.2 Overview of the Techniques

The genome size, or amount of DNA in the haploid cell, has been determined via several methods over the years, including biochemical separation, photometric estimates (typically using the DNA-specific Feulgen reaction or flow cytometry), and even genome sequencing. The methods differ in their ease of use and cost, as well as the expertise, time, and equipment required.

Early eukaryotic measurements were produced via biochemical analysis. Genome size was predicted from the molecular weight for a large tissue sample divided by a cell number estimate; this provided imprecise but strong evidence of the constancy of DNA (Hardie et al., 2002). More recent methods have utilized the properties of the Feulgen reaction, which selectively stains nucleic acids, in concert with densitometry. During the staining protocol, mild acid hydrolysis (typically using fixed tissue) cleaves purine bases, exposing free aldehyde groups (Chieco and Derenzini, 1999). Subsequent staining with Schiff reagent allows aldehyde groups to bind the pararosaniline present in the reagent, turning it from colorless to magenta. The intensity of the stain is proportional to the number of aldehyde sites available and, therefore, to the amount of DNA present. A key step in the reaction is acid hydrolysis; extending the duration of this step leads to increased breakage and loss of sections of the DNA strands, thus decreasing the available staining sites and likewise the intensity of the stain.

Developments in densitometry techniques to measure staining intensity of Feulgen-treated nuclei have also progressed from flying spot to scanning stage, and most recently to image densitometry. The difference between the level of light able to pass through a nucleus compared to the background is a measure of its transmittance (T), which can be converted into optical density (OD) by the simple transformation, $OD = \log_{10}(1/T)$ (Hardie et al., 2002). Arbitrary OD levels can then be converted into to a DNA amount (typically in picograms, pg) with the inclusion of "standards" (cells for which the DNA content is known) in each staining procedure.

Flying spot and scanning stage densitometry both systematically measure OD across a nucleus, with the amount of light passing through a small aperture recorded at each location. Multiple measurements collected across a single nucleus can be integrated to produce an OD measure for that nucleus. The flying spot and scanning stage methods differ in their mechanics. The former method uses a beam of light that moves across the nucleus, whereas a stationary beam and a motorized stage gliding the slide (and thus the nucleus) across the beam's path is employed in the latter (Hardie et al., 2002). Image densitometry calculates the OD in the same manner but employs captured images. The benefit of this method is that each image may include dozens of nuclei, all of which can be processed simultaneously, thus tremendously accelerating the rate of acquisition of nuclear measurements.

Finally, flow cytometry has occasionally been employed to estimate the DNA content of cells. This technique involves creating a suspension of cells that have their nuclei marked with one of a number of DNA-specific fluorescent dyes [e.g. 4′,6-diamidino-2-phenylindole (DAPI) and propidium iodide). The fluorescently tagged cells are passed through a laser beam of the appropriate wavelength for the chosen dye, and a detector records the brightness as the cells pass through the beam (Hardie et al., 2002). This method can process thousands of cells at a time, but a drawback of flow cytometry is that cells from a tissue must be isolated from each other. As with Feulgen densitometry, the inclusion of a suspension of cells of known DNA content allows the intensity of fluorescence to be converted to a measure of the DNA amount.

16.1.3 Evaluation of Techniques

A close inspection of Gregory's (2013) database reveals both perplexing and intriguing results for cladoceran measurements. In total, 62 C-value estimates have been published, representing 49 species and including multiple estimates for five species of *Daphnia* (*D. arenata*, *D. magna*, *D. pulex*, *D.pulicaria*, and *D. tenebrosa*). First, substantial intraspecific variation is evident, which is probably due to the method of estimating genome size. Second, cladocerans rank among the smallest of the crustacean (and indeed, the invertebrate) genomes. Of the 318 crustacean estimates listed in Gregory's database, 101 are less than 1.00 pg and include all of the cladoceran as well as a number of amphipod, barnacle, copepod, and ostracod estimates.

While interspecific differences are routine, the genome size is generally predicted to be constant and stable for members of a species. Some differences reported within a species are probably due to the acquisition methods and experimental error. We can consider the magnitude of this error in an examination of cladoceran estimates. From 1985 to 2009, many species of *Daphnia*, and an array of other cladocerans, were the subject of DNA investigations (Rasch, 1985; Beaton, 1988; Beaton and Hebert, 1989; Beaton, 1995; Dufresne and Hebert, 1995; Korpelainen et al., 1997; Vergilino et al., 2009). Of the surveyed species, *D. pulex* was by far the most intensively studied, with determinations based on multiple methods (Table 16.1). Differences across estimates appear to be attributable to the method used for determining the C-value. Scanning Feulgen densitometry (by far the most commonly used method in this group) seems to have consistently resulted in an overestimation, if the other estimates are correct. Flying spot densitometry and flow cytometry both produced estimates slightly larger (0.23–0.33 pg) than predictions based on the genome sequencing initiative (0.204 pg). However, Colbourne et al. (2011) remarked that the 200-Mb draft genome assembly probably represented 80% of the entire genome, with the remainder comprising some duplicated genes and heterochromatic regions. Furthermore, a clone of *D. arenata* (not of *D. pulex*) was used. This would shift the estimate originating from genomic sequencing from 0.204 to 0.256 pg. Based on this predicted

TABLE 16.1 Selected *Daphnia* Genome Estimates for Comparison of Methods

Species	Method of Estimation	C-Value (pg)	Reference(s)
D. pulex	Flow cytometry	0.28–0.33	Korpelainen et al. (1997)
D. pulex	Flow cytometry	0.23	Vergilino et al. (2009)
D. pulex	Flying spot Feulgen densitometry	0.23	Rasch (1985)
D. pulex	Scanning Feulgen densitometry	0.37–0.38	Beaton (1989, 1995)
D. arenata	Flow cytometry	0.24	Vergilino et al. (2009)
D. arenata	Scanning Feulgen densitometry	0.42	Beaton (1995)
D. pulex/*D. arenata*	Genome sequencing	0.204	Colbourne et al. (2011)
D. tenebrosa	Flow cytometry	0.29	Vergilino et al. (2009)
D. tenebrosa	Scanning Feulgen densitometry	0.53	Dufresne and Hebert (1995)
D. tenebrosa	Scanning Feulgen densitometry	0.58	Beaton (1995)

genome size, measurements produced using flow cytometry and flying spot densitometry are both slight underestimates.

Flow cytometry for *D. arenata*, *D. tenebrosa*, and *D. pulex*, consistently produce values that are 54-62% of those obtained using scanning densitometry. Since the genome size predictions for other cladoceran members were determined by scanning microdensitometry, their DNA contents probably also represent overestimates and are even smaller than those reported by Beaton (1988, 1995). Consistent overestimation means that the reported C-values, while suspect for determining precise DNA contents, should still allow comparisons within a technique (but certainly not between techniques). It is possible that the tissues and species used as the standards (often blood smears from a variety of fish species (Beaton, 1988)) were too dissimilar and biased the conversion from pixels to DNA content. Regardless, the general finding that cladoceran genomes are minute, based on the scanning densitometry estimates, is probably not affected by the bias. As such, current estimates for *Bosmina longirostris* and *Scapholebris kingii* (0.19 and 0.16 pg, respectively) are among the smallest for a crustacean, with only two cyclopoid copepods possessing C-values of about 0.14 pg (Gregory, 2013).

16.1.4 Patterns in Interspecific Variation and Ecological Correlates with Genome Size

Gregory (2013) compiled 62 cladoceran C-values from 13 genera and 49 species. The nearly fourfold range in estimates forms a bimodal frequency distribution (Fig. 16.1), with most C-values being about 0.25 or 0.4–0.45 pg. More than 90% of the available estimates are from representatives of the suborder Anomopoda (80% of which are from one of the three *Daphnia* subgenera) and only one or two estimates are drawn from three of the other suborders. With such poor representation across the group, few conclusions can be drawn and there are no obvious patterns that might indicate a genome size increase or decrease from basal to derived groups (Table 16.2). In addition, C-values show similar ranges across the three *Daphnia* subgenera.

In an examination of DNA content variation among species of *Daphnia*, Beaton (1995) could find no association between genome size and habitat preference; the ranges for species preferring pond environments were

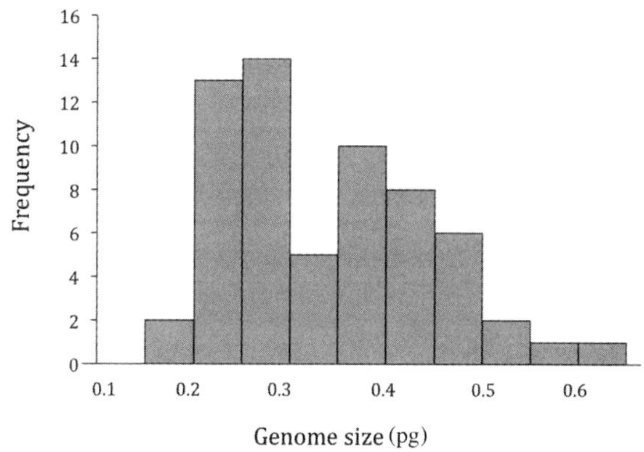

FIGURE 16.1 Distribution of genome size estimates of cladoceran members.

TABLE 16.2 Ranges of C-Value Estimates

Order	Genus/ Subgenus	No. Species	C-Value Range (pg)
Anomopoda		45	0.16–0.53
	Ctenodaphnia	13	0.21–0.53
	Daphnia	16	0.24–0.58
	Hyalodaphnia	7	0.22–0.36
Ctenopoda		2	0.22–0.47
Haplopoda		1	0.28
Onychopoda		1	0.22

approximately the same as those that prefer lake habitats. Since pond-dwelling cladocerans tend to be larger than those found in lakes, an association between body size and DNA content might not be expected. However, eukaryotic genome size is positively correlated with cell size, nuclear volume, cell doubling rate, and so on (Cavalier-Smith, 1985). When a limited number of representatives are chosen, a relationship between genome size and body size appears strong among cladocerans; the large-bodied *Eurycercus glacialis* (0.63 pg) and *D. magna* (0.4–0.53 pg) have among the highest C-values for the group, whereas the tiny *B. longirostris*, *Scapholebris kingii*, and *Ceriodaphnia reticulata* (0.19, 0.16, and 0.23 pg, respectively) possess the group's smallest estimates (Beaton, 1988). However, using a larger dataset for members of *Daphnia*, a much weaker relationship was found. Some of the largest members of the genus (for example, *D. cephalata*, *D. longicephala*, and *D. magniceps*, with respective estimates of 0.27, 0.27, and 0.30 pg) each possess a genome size comparable to the small-bodied members (for example, *D. ambigua* and *D. parvula*, with estimates of 0.24 and 0.28 pg) (Beaton, 1995). There can be two ways to achieve a large body: increasing cell size (and by extension genome size) or increasing cell numbers. The weak observed relationship points to the employment of the latter option by at least some species.

Beaton (1995) identified two ecological correlates with genome size: the ability to alter head shape (inducible defenses) and egg volume. Members of *Daphnia* that are capable of forming an inducible head defense possess small genomes, whereas those whose head shape is canalized tend to have larger genomes. This correlation is not perfect, but Beaton (1995) found it to be statistically significant. Head defenses (neckteeth, spines, helmets, and crests) are all formed from the epidermal sheet situated just under the carapace via an increase in cell numbers (Beaton and Hebert, 1997). In *D. pulex*, for example, neckteeth are present only during an animal's first two or three instars, but exposure to the pertinent predator kairomone is required from early embryogenesis onwards to attain the most pronounced structure (Miyakawa et al., 2010). Therefore, there must be a critical period in development during which induction is initiated, but since the embryonic period is quite brief, neckteeth formation must be rapid. The other induced head shapes are similarly created in just a few instars, and the fast response time is probably necessary to minimize the time that animals are vulnerable to predation. Since cell cycle length is positively correlated to DNA content across many taxa (e.g. Bennett, 1972; Shuter et al., 1983; Cavalier-Smith, 1985), the logical extension seems to be that for species under variable levels of predation pressure, a smaller genome would be advantageous, by allowing higher cell proliferation rates during key developmental periods, and thus allowing prompt formation of the defensive structures.

The second strong ecological correlate observed with genome size was a positive one with egg size (Beaton, 1995). This relationship has also been observed for selected taxa, for example, among actinopterygian fish (Hardie and Hebert, 2004), as well as members of the rotifer, *Brachionus* (Stelzer et al., 2011). This may be a surprising result, as the size of the egg can also be a reflection of the amount of yolk laid down. However, Hardie and Hebert (2004) noted that in fish, egg size limits fecundity and such a limitation would also occur in cladocerans. A trade-off between brood size and neonate size may occur. Among small-bodied organisms, the size of the brood pouch may severely limit the number of eggs, but in larger species, many small eggs or fewer larger ones are viable options. Some of the largest species of *Daphnia* (such as *D. cephalata* and *D. longicephala*) have opted for large clutches of small eggs, leading to small neonates.

Finally, it must be noted that polyploid clones have been found for some members of *Daphnia*. Ploidy assignments have been made using a combination of isozyme patterns and DNA content measurements. The use of in situ hybridization, as performed by Colbourne et al. (2011), may offer an alternative means of confirming ploidy assignments when inferences made by allozyme and DNA content analysis are inconclusive. Adamowicz et al. (2002) summarized the geographical range of polyploid representatives of the *D. pulex* complex as being the dominant form at latitudes above 70°N and 46–54°S (in Argentina). Triploids are believed to be the dominant polyploid level in the Canadian arctic (with tetraploids observed less frequently), whereas only tetraploids were recovered in Argentina (Adamowicz et al., 2002; Vergilino et al., 2009).

16.2 CYTOGENETICS

16.2.1 Background

Historically, karyotypes have been generated from cells that are arrested in metaphase, when the chromosomes are in a condensed state. To ensure that this brief period in the cell cycle is observed, a population of cells may be treated with a dilute colchicine solution to inhibit spindle formation. Once a population of cells is fixed onto a slide, the material can be treated with one of several DNA stains (aceto-orcein is a simple and commonly used dye) prior to visualization.

There are likely to be at least two reasons why cytogenetic work on cladocerans has been hampered. First, as a whole, this group possesses miniscule chromosomes owing to their condensed genomes packaged into reasonable chromosome numbers. Second, the choice of tissue and methods employed to obtain karyotypes have been linked to variable counts (Zaffagnini and Trentini, 1975). The production of a suspension of cells for examination is not an easy proposition with cladocerans. Creating a cell suspension from adults is fraught with difficulties, including the separation of cells, as well as the confounding issue of rampant somatic polyploidy in members of this group (Beaton and Hebert, 1989). Using newly deposited subitaneous eggs (sometimes referred to as *summer eggs*), a technique that is frequently employed (Zaffagnini and Sabelli, 1972; Zaffagnini and Trentini, 1975; Trentini, 1980) will generate cells that can be squashed, but pinpointing the ideal interval for egg collection is critical and the number of viable smears will be reduced at this developmental stage. Alternatively, whole animals can be fixed, blocked, and sectioned with a microtome, but the resulting counts can be variable when chromosomes in a cell do not lie within a single plane (Zaffagnini and Trentini, 1975). Finally, preparations of chromosome spreads can be obtained from early embryos or from the epidermis underlying the carapace. In *Daphnia* (and presumably in other cladocerans), the epidermis is a single sheet of diploid cells, with polyploid cells restricted to the tissue margins (Beaton and Hebert, 1989, 1994a). A synchronized cycle is often observed in the epidermis, with a brief window of opportunity to examine metaphase chromosomes occurring a few hours postecdysis (the first 12 h typically). Although this method allows examination of large numbers of mitotic figures, tissue collection is time sensitive.

16.2.2 Cladoceran Karyotypes

A minute genome, combined with diminutive body size and a constrained growth pattern, has probably contributed to the paucity of published cytogenetic work for this group. In addition, information regarding centromere location and chromosome arm length has been difficult to obtain (because the chromosomes are so small!). With the exception of a handful of studies published since 1995, the bulk of information has been limited to chromosome numbers. Overall, diploid counts in the group have been obtained for members of only two of the suborders: Anomopoda and Onychopoda (Table 16.3). Chromosome complements of many other members of the group are certainly required.

At least some of the published images have been obtained from colchicine-treated tissues. Trentini (1980) and Zaffagnini and Trentini (1975) obtained chromosome spreads of parthenogenetic eggs using a squash method. A similar technique, but employing colchicine-treated embryos to create a suspension of metaphase-arrested cells, was used for several *Moina* spreads (Wenqing et al., 1999; Zhang et al., 2009). Beaton and Hebert (1994b) obtained spreads using thoracic epithelium of mature females treated with a weak colchicine solution.

16.2.3 Suborder Anomopoda

Daphnia has been far and away the most intensely surveyed genus of the group, with counts for 21 species. Conservation of chromosome numbers in each of the *Daphnia* subgenera has been found, with diploid counts of 20, 20, and 24 for members of *Ctenodaphnia*, *Hyalodaphnia*, and *Daphnia*, respectively (Table 16.3). It should be noted that *Daphnia curvirostris* was historically classified within the subgenus *Daphnia* (previously referred to as the *pulex* group) based on morphological characters, so its diploid complement of 20 was considered a puzzle. This led Beaton and Hebert (1994b) to question its placement in the subgenus. Subsequently, molecular analyses have confirmed that this species is most closely aligned with members of the subgenus *Hyalodaphnia*, concordant with chromosome numbers (Lehman et al., 1995; Colbourne and Hebert, 1996; Kotov et al., 2006).

There have been both lower and higher reported chromosome complements for *D. magna* and *D. pulex*, but these are generally considered to be unreliable. For example, those obtained from tissue sections have generally been refuted and dismissed, sometimes by the authors themselves (Chambers, 1913 and Kühn, 1908, based on my translation of Zaffagnini and Trentini, 1975). A diploid count for *D. magna* by Rossetti (1952, in Trentini, 1980) represents the only aberrant set of counts for which we have no evidence favoring their rejection, aside from the overwhelming number of studies contradicting his counts (Mortimer 1936, in Trentini 1980; Zaffagnini and Trentini 1975; Trentini 1980; Beaton and Hebert 1994b). All aberrant records have been placed at the end of Table 16.3 simply for reference.

The sequencing initiative by the Daphnia Genomics Consortium succeeded in characterizing the nuclear genome of the water flea, *D. pulex* (*arenata*), including a detailed description of chromosome structure in this species. Colbourne et al. (2011) categorized the chromosomes (ranging in length from about 1 to 6 μm) into three broad classes based on size: the first group included only one chromosome pair, with the second group comprising the next three largest pairs. Only about 25% of the total chromosome area was found to be heterochromatic and G-bands were only found on the largest four chromosome pairs. While no attempt was made to identify the position of the centromere, the three largest pairs, at least, appear to be meta- or subtelocentric based on their image (Colbourne et al., 2011). The chromosome ends were characterized as telomeres made of long stretches of repeated TTAGG, the same sequence used by *Bombyx mori* and similar to that found at the ends of vertebrate chromosomes. Finally, fluorescence in situ hybridization (FISH) was used to locate the ribosomal RNA (rRNA) arrays (Colbourne et al., 2011). Using two probes, Pokey, a transposon that

TABLE 16.3 Diploid Chromosome Counts for Cladocerans

Suborder/Family	Genus/Species[a]	2*n*
ANOMOPODA		
Daphniidae	*Ceriodaphnia*	
	C. laticaudata[18,19]	14
	C. reticulata[19]	14
	C. pulchella[19]	20
	Ctenodaphnia	
	C. angulata[20]	20
	C. cephalata[20]	20
	C. exilis[20]	20
	C. longicephala[20]	20
	C. lumholtzi[20]	20
	C. magna[11,18,19,20]	20
	C. salina[20]	20
	C. similis[20]	20
	Daphnia	
	D. catawba[20]	24
	melanized sp. nov.[20]	24
	D. middendorffiana[17,20]	24
	D. minnehaha[20]	24
	D. obtusa[18–20]	24
	D. pulex[7,8,11,14,17,20]	24
	D. pulicaria[20]	24
	D. schodleri[20]	24
	Hyalodaphnia	
	H. curvirostris[19,20]	20
	H. galeata mendotae[20]	20
	H. hyalina[18]	20
	H. longispina[19]	20
	H. rosea[20]	20
	Daphniopsis tibetana[23]	24
	Simocephalus	
	S. vetulus[18,19]	20
	S. exspinosus[19]	20
Moinidae	*Moina*	
	M. affinis[18]	20
	M. affinis[24]	26
	M. macrocopa[9,21,24]	22
	M. micrura[24]	24
	M. mongolica[22,24]	24
	M. rectirostris[24]	24
	M. rectirostris[13]	30
Onychopoda	*Bythotrephes longimanus*[2]	4?
Polyphemidae	*Polyphemus pediculus*[3]	8
ABERRANT COUNTS		
Anomopoda	*Ctenodaphnia magna*[14]	28–32
	C. magna[12]	8 F
	D. pulex[4]	8 (7–10) F
	D. pulex[10]	8 F, M; 9 S
	D. pulex[15]	20
	D. pulex[16]	16
	D. pulex[6]	8 10 M, S
	S. vetulus[5]	8–10 M
	M. paradoxa[1]	8
	M. rectirostris[1]	8

(*Continued*)

[a]*Numbers represent count source: 1, Weismann and Ishikawa (1889) in Makino (1951); 2, Weismann and Ishikawa (1891) in Makino (1951); 3, Kuhn (1908) in Mackino (1951); 4, Kuhn (1908) in Trentini (1980); 5, Chambers (1913); 6, Taylor (1914) in Trentini (1980); 7, Fanghaut (1921) in Trentini (1980); 8, Schrader (1925) in Trentini (1980); 9, Allen and Banta (1928) and (1929) in Mackino (1951); 10, Rey (1934); 11, Mortimer (1936) in Trentini (1980) ; 12, Lumer (1937) in Trentini (1980); 13, von Dehn (1948); 14, Rossetti (1952) in Trentini (1980); 15, Ojima (1954); 16, Bacci et al. (1961); 17, Zaffagnini and Sabelli (1972); 18, Zaffagnini and Trentini (1975); 19, Trentini (1980); 20, Beaton and Hebert (1994b); 21, Wenqing et al. (1995); 22, Wenqing et al. (1999); 23, Zhao et al. (2004); 24, Zhang et al. (2009).*

?, questionable count; F, during oogenesis; M, during spermatogenesis; S, during mitosis of somatic cells.

inserts into a specific site of the large subunit rDNA (Eagle and Crease, 2012) and a region of the intergenic spacer (IGS), Colbourne et al. (2011) confirmed that the rRNA genes were present as a single tandem array on one chromosome pair, with the transposon inserted along the entire length of the array.

Chromosome numbers for *Ceriodaphnia* have been published for only three members, with two diploid complements observed: 14 and 20 (Table 16.3). It is interesting to note that *C. reticulata* ($2n = 14$) and *C. pulchella* ($2n = 20$), which can co-occur, exhibit differences in both body and egg size (Burgis, 1967). What might not be expected is that *C. pulchella* has the larger chromosome complement, but a smaller egg and body size. With no phylogenetic assessment of the genus found in the literature, we cannot speculate on the basis for the chromosome variation.

With only two species of *Simocephalus* studied, diploid chromosome numbers appear to be conserved at 20 (Table 16.3). Chambers (1913) reported an aberrant diploid count of 8–10, but as it was based on tissue slices, the count is considered unreliable. *Simocephalus* and *Ceriodaphnia* have been linked as sister taxa based on a phylogenetic reconstruction using three genes (deWaard et al., 2006). Chromosome counts for members of these two genera ($2n = 20$ and 14, respectively) show the largest separation observed across the Anomopoda to date (Trentini 1980). A single diploid chromosome count of 24 has been recorded for *Daphniopsis*, a sister genus to *Daphnia* (Table 16.3). Zhao et al. (2004) further established that the diploid karyotype for *Daphniopsis tibetana* has three pair of metacentric (M) and nine pair of telocentric (T) chromosomes ($2n = 24 = 6M + 18T$). This arrangement is reminiscent of the findings of Colbourne et al. (2011) for *D. pulex* (*arenata*). Overall, among the members of the family Daphniidae, only three reliable counts have been reported, $2n = 14$, 20, and 24.

With just six species of *Moina* surveyed, diploid chromosome numbers in the genus appear to be restricted to 22–26, although counts for *M. rectirostris* (syn. *Moina brachiata*) have been reported as 8 and 30, in addition to 24, and a count of 8 was reported for *M. paradoxa* (Table 16.3). The highest and lowest counts must be regarded as suspect; von Dehn's (1948) estimate ($2n = 30$) was based on tissue sections, and a description of how Weismann and Ishikawa (1889, in Makino, 1951) collected their data ($2n = 8$) could not be obtained. Similarly, the diploid count of $2n = 8$ for *M. paradoxa* should be considered suspect unless confirmation for the species is possible. Zhang et al. (2009) obtained a diploid count of 24 for *M. micrura*, *M. mongolica*, and *M. rectirostris* with diploid karyotypes of $10M + 14T$, $6M + 18T$, and $10M + 14T$, respectively. Their spreads of *M. affinis* revealed $2n = 26$ chromosomes (in contrast to Zaffagnini and Trentini, 1975, who obtained a count of 20), with a karyotype of $12M + 14T$. Finally, for *M. macrocopa*, a diploid karyotype of $2n = 22 = 10M + 12T$ has been described (Zhang et al., 2009).

The basis of the discrepancies observed among species of *Moina* (especially for *M. affinis*) must be investigated further. Errors in the counts, misidentification of individuals, or the presence of cryptic species complexes are possible explanations. Obtaining karyotypes from additional individuals, populations, and species should be undertaken to definitively establish the correct counts.

16.2.4 Suborder Onychopoda

Only two early counts have been reported for this suborder and these should be viewed with some skepticism. The diploid counts of four and eight need to be confirmed and several more representatives of this suborder surveyed before any generalizations are possible.

16.3 ENDOPOLYPLOIDY

Endoreplication, or somatic polyploidy, is observed when DNA replication occurs in a cell without cytokinesis. Separating the process of DNA replication from one or more steps of cell division can lead to a single reconstituted cell (rather than two daughter cells) that possesses twice as much chromatin than expected. Repetition of this abbreviated cycle produces an enlarged cell and nucleus containing 2^n copies of the nuclear genome (where n is the number of completed endocycles or endomitotic cycles). Endocycles bypass mitosis entirely, so chromosomes do not condense, creating multistranded chromatin as seen in larval *Drosophila* salivary gland cells. Endomitotic cycles, on the other hand, typically include prophase, metaphase, and anaphase, but during telophase a single nuclear membrane is formed around all of the chromatin (see Lee et al., 2009 for a brief comparison). As a consequence, the latter process will result in multiple sets of the haploid chromosome complement (endopolyploidy).

This sweeping description of endoreplication hints at the occurrence of some accident or abnormality, as is the case with some cancerous tumors. However, endopolyploidy is a ubiquitous phenomenon, and is routinely a necessary and normal act in the developmental process of most multicellular organisms. Trophoblast cells of rodents, many plant embryo suspensor cells, and the giant neurons of the sea hare, *Aplysia californica*, are but a few examples of this widespread phenomenon (Nagl, 1978). In general, polyploid cells range from 4 to 10^6 C (where C represent the DNA content of a single chromosome set) and those cells with the highest levels are often associated with secretory roles, although the neurons of *Aplysia* (approximately 10^6 C) contradict this pattern. The widespread nature of somatic polyploidy could be an evolutionary mechanism for allowing abrupt shifts in traits or physiological responses (Beaton and Hebert, 1997).

Some of the first reports of endopolyploidy in cladocerans came from examinations of labrum and fat cells (Cannon, 1922; Jaeger, 1935; Sterba, 1956a, 1957b). More recently, Beaton and Hebert (1989, 1997, 1999) established the extent of polyploidy for multiple *D. magna* and *D. pulex* tissues, with several substantive patterns revealed. First, most examined tissues in mature females were found to contain multiple polyploid cells, each with DNA content ranging from 4 C to 2048 C (Tables 16.4 and 16.5). The Feulgen staining pattern of many of the moderately polyploid

TABLE 16.4 Typical Number of Fat and Epipodite Cells per Abdominal Appendage Pair of *Daphnia pulex* and *D. magna*

Appendage Number	Fat Cells		Epipodites	
	D. magna	*D. pulex*	*D. magna*	*D. pulex*
1	41–42	16–17	107–108	45–46
2	46–47	26–27	122–123	58–59
3	125–126	49–50	123–124	82–83
4	109–110	40–41	137–138	85
5	45–46	26–27	130–131	86

Summarized from Beaton and Hebert, 1989.

TABLE 16.5 Summary of Occurrence of Endopolyploidy in Adult Female *Daphnia*

Tissue	Number of Polyploid Cells		Levels Observed (C)	Specific Comments
	D. magna	*D. pulex*		
Epidermis	ND	35	8–64	16–32 C is most common
Epipodites	621	357	8–32	16 C is most common
Fat cells, limbs	371	161	64–256	128–256 C are more common
Gut	All	All	4–16	4 C for almost all nuclei
Labrum	8	8	64–2048	distinct patterns for species
Rostrum	156	83	8–128	32–64 C is most common
Shell gland	300	102	8–256	16 C is most common

Number of polyploid cells represents the counts for the entire tissue (for example, for all five pairs of limbs). ND, cell counts (and corresponding ploidy levels) were not determined.
Summarized from Beaton and Hebert, 1989.

cells (32–64 C) appears to be far more heterogeneous than those at 2–16 C (Beaton, unpubl.; Sterba, 1956a, 1957b). Two of the tissues, the gut and the epipodites (the leaf-like lobes located at the base of each of the five pairs of thoracic appendages), contain only polyploid cells and the labrum contains cells reaching the highest levels (1024–2048 C). Overall, Beaton and Hebert (1997) estimated that nearly half of each individual's DNA was packaged in a polyploid state; with no mitotic figures recorded from polyploid cells other than those in the gut, it is assumed that polyteny does not occur in *Daphnia*.

Second, both tissue- and species-specificity was found with regards to levels and number of cells involved (Tables 16.4 and 16.5). For example, in newly mature *D. magna* and *D. pulex* females, the labrum possessed only eight polyploid cells, but levels differ between species (Beaton and Hebert, 1997). In addition, for a given species, epipodites of mature females contained cells that were between 8 C and 32 C, but the number of cells in each epipodite differed both among thoracic limbs and between species (Table 16.4).

Third, the number of polyploid nuclei in a tissue, once established (typically before instar three or four), remained constant with age (Beaton and Hebert, 1999). When individual polyploid cells in the epidermis are destroyed via laser ablation they are not replaced (Beaton, pers. comm.), suggesting that a genetic control to establish these cells is in place. The inference can also be made that polyploid cells in most tissues cannot exit the endocycle and return the typical cell cycle to restore cell numbers. The exception to this rule, however, was observed in gut cells, which appear to reach the tetraploid level by instar 2, and then return to a normal cell cycle, which allows tissue growth without further ploidy increase (Beaton and Hebert, 1999). In general, polyploid cells do not return to the cell cycle once they have opted for the endomitotic cycle, so this finding is particularly intriguing. The significance of digestive cells maintaining a low ploidy level has not been investigated.

Finally, Beaton and Hebert (1999) found tissue specificity in *D. pulex* with regards to increasing ploidy levels (Table 16.5). During the first instar, 4 C was the common level found in most tissues, but specific labral cells had already reached 64 C. Over subsequent instars, the levels in labral cells continued to increase (although a doubling of the DNA content with each instar was not observed), and two of the eight polyploid nuclei reached 2048 C by instar 14 when the study ended. The tissues most involved in feeding and growth, i.e. the epidermis (which lays down the new carapace each molt), the labrum, and the fat cells of the limb central core, all showed substantial ploidy level increases across the instars. By comparison, the cells located in the epipodites typically remained at 16 C or 32 C from the sixth to the thirteenth instar.

The functional significance of possessing polyploid cells in some tissues can be inferred. For example, a band of polyploid cells along the underside of the rostrum are probably important for chemoreception. The presence of 20–25 polyploid cells distributed along the ventral margins and dorsal fold in the epidermal sheet that lies adjacent to the carapace may act as the primary developmental centers to control the growth rate of the animal. Similar types of cells in the cephalic epidermis have also been hypothesized to act in this manner; mitotic rates in diploid cells near polyploid cells were higher than in cells in more distant positions (Beaton and Hebert, 1997). Further evidence supporting this comes from studies in which ablation of selected polyploid epidermal cells in the head of helmeted *D. lumholtzi* led to a 10–40% decrease in the size of the helmet in the next instar (Beaton, pers. comm.). It would be interesting to know how ploidy levels (and polyploid cell number) compare for members of the other three suborders, as well as for those with distinct physical characteristics, such as *Sida*, *Holopedium*, and *Leptodora*. Each of these species presumably uses the epidermal tissue for different functions. In *Holopedium*, this tissue presumably lays down its distinctive jelly coat, whereas at least some of the epidermis must form the "sucker" found on the dorsal surface of *Sida*; in addition, the carapace seems to have "fused" with the thorax in *Leptodora* (Olesen et al., 2003). The function of polyploidy in some tissues will be considered further in the next section.

16.4 CYTOLOGICAL OBSERVATIONS FOR SPECIFIC TISSUES

16.4.1 Nuchal Organ and Epipodites

As a rule, cladocerans can actively regulate the osmotic pressure of their fluids to maintain homeostasis of their water content (Aladin and Potts, 1995). Within the order, 95% of the species are confined to freshwater, 3% are found in brackish water, and only 2% are restricted to marine habitats (Bowman and Abele, 1982, in Weider and Hebert, 1987). Osmoregulatory strategies of the group are as diverse as the environments that they inhabit. Hyperosmotic regulation of hemolymph is found in freshwater animals and those living in slightly brackish, whereas hypoosmotic regulation occurs in marine individuals (Aladin and Potts, 1995). A combination of hyper- and hypoosmotic regulation is found in cladocerans from both brackish and highly mineralized waters (Aladin and Potts, 1995).

The two primary structures utilized by the group for osmoregulation are the epipodites and the nuchal, or neck organ (Aladin and Potts, 1995). Although they had commonly been thought of as a gill, Koch (1934) suggested that the selective permeability to silver stain suggested that the epipodites have an ion-regulatory role. The nuchal or neck organ has been described in a variety of larval and adult crustaceans, including branchiopods, copepods, and malacostracans (Aladin and Potts, 1995). Among malacostracans, the nuchal or neck organ has been suggested to serve in chemoreception, mechanoreception, and baroreception. Branchiopod neck organs may have evolved from these sensory functions; as with these other functions, salt uptake is also only effective where a thin cuticle is present (Aladin and Potts, 1995). It seems probable that the neck organ first evolved in a branchiopod nauplius to allow salt uptake in freshwater environments. This organ is now retained in adult cladocerans that have raptorial limbs and no longer possess epipodites (Aladin and Potts, 1995). Members of the five orders of Cladocera, Ctenopoda, Anomopoda, Haplopoda, and Onychopoda may use one or both of these ion-regulating systems (Table 16.6). Among the Ctenopoda, epipodites are well developed in some members (*Sida*), but are small for others (*Penila*). Furthermore, in *Sida* it has been modified, allowing it to also function as an attachment device (Olesen, 1996; Peñalva-Arana and Manca, 2007). In Anomopoda, such as *Daphnia*, the nuchal gland persists from embryo to the first instar, although it may become inactive just a few hours after release from the brood pouch (Benzie, 2005). This group possesses epipodites upon release from the brood pouch. In contrast, haplopods possess only the nuchal gland, which develops into a broad shield containing numerous mitochondria-rich cells. Similarly, members of Onychopoda possess only a well-developed nuchal gland (Aladin and Potts, 1995).

TABLE 16.6 Variation in Osmorgulatory Structures Present in Suborders

Suborder	Epipodites	Nuchal Organ
Anomopoda	Present	Limited to embryo and first instar
Ctenopoda	Present in varying sizes across species	Modified to function for attachment
Haplopoda	Absent	Present in all stages, surrounding the head within the shield
Onychopoda	Absent	Well developed and incorporated in head shield

Sandwiched between the outside medium and the hemolymph, both structures are covered by an extremely thin cuticle (Aladin and Potts, 1995). The ultrastructure of the neck organ and epipodite are very similar: cells of both structures contain dense cytoplasm with numerous mitochondria (Aladin, 1991). Furthermore, their cell structure is unaffected by the type (hyper- or hypoosmotic) of ion regulation employed. Structurally, these tissues share similarities with ion-transporting cells of the branchiopods, cells of similar function in the gills of other crustaceans, those in insects and fishes, and the salt-secreting glands of reptiles and birds (Aladin and Potts, 1995).

The structure of the nuchal organ in *Daphnia* has been well studied. First instar individuals have a nuchal organ that, upon external observation, appears as an expanded portion of a dorsal ridge of the head. In *D. himalaya*, a groove appeared to mark the margin of the structure, which has a mediodorsal hole (Peñalva-Arana and Manca, 2007). Schwartz and Hebert (1984), using a simple silver staining technique, reported that the position of the organ could be used as a diagnostic trait between the subgenera, *Ctenodaphnia* and *Daphnia*. Benzie (2005), on the other hand, asserted that its location is not specific to a subgenus, but it is situated at the border between cephalic and dorsal shields. Internally, lying just below the cuticle, the nuchal organ consists of two cell types (referred to as *dark* and *light*) that fill a large portion of the hemocoelic space (Halcrow, 1982). The higher proportion of microvilli on the apical surface of both cell types makes them noticeably different than the surrounding squamous cells. The darker of the two cell types has microvilli that sometimes branch and may be smaller in diameter compared to those of the light cells. Deep within both cells and at the bases of microvilli, are numerous mitochondria with prominent cristae. It is possible to define the border of the nuchal organ with permanganate staining: the cuticle overlying it is densely stained, whereas the thicker surrounding cuticle is not (Halcrow, 1982). In *D. magna*, increased membrane surface area (microvilli) and numerous mitochondria are common features of nuchal organ cells that have osmoregulatory or ion-transporting roles in other crustaceans. This structure remains functional as an ion-transporter for a brief period of time; only approximately 12 h in the free-swimming first instar animals (Halcrow, 1982). Due to its rapid loss of function, it has been suggested that the nuchal organ is most beneficial in these animals during their development in the brood pouch, when limb movement is severely limited or nonexistent. These cells remain large in older animals, but have reduced numbers of microvilli and mitochondria (Halcrow, 1982). Beyond the first instar, epipodite epithelial cells presumably serve the primary ion-regulatory role in *D. magna* (Halcrow, 1982). Early research suggested a respiratory function for epithelial epipodite cells; however, since the cuticle of this tissue in *D. magna* is approximately 4-5 times thicker than is found in the surrounding limb epithelia, they probably represent poor gas exchange surfaces (Goldmann et al., 1999). Two cell types, dark and light, are found arranged in a jigsaw-type pattern in the epipodite epithelia (Kikuchi, 1983). Dark cells have many infoldings, large mitochondria, and a complex tubular system, and show an accumulation of chloride ions, suggesting an ion-transporting function. The light cells have strong infoldings of both basal and lateral cell membranes and tubule-like protrusions toward the blood space, but an osmoregulatory function has not been suggested (Kikuchi, 1983; Goldmann et al, 1999).

The presence of two cell types in the epipodites raises some interesting questions. Beaton and Hebert (1999) found that *D. pulex* epipodites routinely maintained nuclei with two ploidy levels, 16 C and 32 C (with occasional 8 C and 64 C cells) from the sixth through to the thirteenth instar. Could the cells with differing ploidy levels correspond to the light and dark cell types? In addition, how is the polyploid nature of these cells important for their function? Furthermore, we should ask if osmoregulatory capability is sustained beyond the sixth or seventh instar, since ploidy levels (and number of cells, for that matter) do not appear to increase, despite the continued increase in body size. It seems logical that, without ploidy increases, as the animal grows each epipodite cell must experience a greater osmoregulatory load, and at some point the control of ion concentration could be compromised.

The nuchal organ of *Leptodora* is a saddle-shaped area of integument almost completely encircling the head (Halcrow, 1985). The structure is similar to that of other branchiopods in mitochondrial abundance, varied cell types, surface area, and distinct integumental boundary. However, the microvilli that cover almost the entire apical surface of the epidermal cells in the nuchal organs of *Artemia* and *Daphnia* are absent in *Leptodora*. This structural difference may be due to the organ's larger surface area in *Leptodora* or to a difference in metabolic activity, but is not believed to result from functional differences. Unlike the nuchal organ of *Artemia* and *Daphnia*, it persists in *Leptodora* adults, probably because epipodites are lacking on their stenopodial limbs (Halcrow, 1985).

Marine members of the Onchopoda retain an undiminished neck organ throughout their lives (Meurice and Goffinet, 1983). Light, scanning, and electron microscopy of four mature and immature representatives (*Podon intermedius*, *Evadne nordmanni*, *E. spinifera*, and *E. tergestina*) revealed that the neck organ hangs as a mass from the dorsal carapace in the middle of the interantennary cephalic shield (Meurice and Goffinet, 1983). The cells composing the organ have an apical zone with a dense cytoplasm that surrounds the nucleus and a basal zone containing many mitochondria distributed throughout a lacunar system (Aladin, 1991). Dense bundles of microtubules are found in the apical cytoplasm and many free polysomes fill the cytoplasmic interstices. A small number of cisternae near the nucleus make up the rough endoplasmic reticulum (Aladin, 1991). Juxtaposed plasma membranes are held together by septate desmosomes, while a thin basement membrane defines the ventral cell mass of the organ (Aladin, 1991).

The most interesting features of the *Podonidae* neck organ are the small number and the structure of the cells that comprise the tissue. The organ is composed of at most 12 cells, compared to 50 to 60 cells in *Artemia*, and the ultrastructure of all cells are similar, in contrast to the two cell types found in *Daphnia* (Meurice and Goffinet, 1983). Comparisons between environmental salinities and hemolymph concentrations indicate that a hypoosmotic regulating mechanism is utilized by marine gymnomeran Cladocera.

A third osmoregulatory structure utilized by many cladocerans is the closed brood pouch. When the brood pouch contains developing eggs and embryos, the osmolarity of the marsupial fluid equals that of the hemolymph, but once the embryos develop a neck organ, the fluid ion concentration rises to that of the surrounding sea (Aladin, 1991). The ability to regulate the embryonic environment and protect developing embryos allows Cladocera to inhabit high-osmolarity water (Aladin, 1991).

16.4.2 Labrum and Fat Cells

Three structures that are essential in digestion and energy storage, the labrum, gut, and fat cells, have been the subject of some study. In *Daphnia*, the digestive tract is basically an extended tube, consisting of a foregut, midgut, hindgut, and a pair of intestinal ceca. The walls of both the foregut and hindgut are composed of cuboidal epithelial cells (about 5 μm tall) lined with chitinous intima (Schultz and Kennedy, 1976a). The midgut can be distinguished from foregut and hindgut based on cell structure: midgut cells are taller (about 30.5 μm), the nucleus is basally located, mitochondria are found anteriorly, and the apical membrane possesses long microvilli. Cell density along the gut remains relatively constant across age classes despite increases in the length of the tract as animals grow. Since the ploidy level of all gut cells stabilizes at 4 C, it is likely that these cells exit their endomitotic cycle and return to a mitotic cell cycle (Beaton and Hebert, 1999). This is quite unusual, but has not been studied and further investigation is required.

Early descriptions of the labrum depicted the structure as glandular and organized into two groups of cells in *Simocephalus sima* and *Acanthocercus* (Cunnington, 1903 and Schödler, 1846; both in Cannon, 1922). In *S. sima*, the four large distal cells (whose shape were likened to a hollow bowl) were thought to be replaced when they "lose their secretory power" by the proximal group of cells, which were smaller and thought to be epidermal in origin (Cunnington, 1903; in Cannon, 1922). The distal cells were linked to the outside by a duct leading to the inner surface of the labrum. Cannon (1922) similarly found two groups of cells in the labrum of *S. vetulus*: the proximal group was separated into two lateral parts of 20 epidermal cells each, both situated near the nerve of the first antennae. It is likely that these cells correspond to the polyploid cells of *Daphnia* that are located along the underside of the rostrum (Table 16.5). The distal group of cells consisted of eight large cells (four per side) arranged in pairs, with anterior cells connected to the posterior ones. The two posterior cells "embraced" two "duct cells" that were slightly larger than muscle cells (Cannon, 1922), so presumably were 2–4 C.

Sterba (1957b) recognized three gland cell types in an examination of *Daphnia* (presumably *D. magna*), which he referred to as head bottom, replacement, and main cells (the names of which are based on a partial translation of his work). He found that the multiple head bottom cells, which are proximally located, reached 128 C and the medial paired replacement and distal main cells reached 2048 C. Sterba (1957b) also noted that the cells within each cell type were linked via plasma bridges. Beaton and Hebert (1989, 1999) similarly observed polyploid nuclei in *D. magna* and *D. pulex* arranged in three groups (Table 16.5). From proximal to distal, the first group comprised four cells (maximum of 512 C), the second group had two extremely large cells (maximum of 2048 C), and the most distal group of two cells reached 256 C in *D. pulex* and 1024 C in *D. magna*. In *D. pulex*, the largest of these cells appeared to be Y-shaped, suggesting a similar structure to that found in the labrum of the conchostracan, *Caenesteriella* (Larink, 1992). None of the four most distal cells in *D. magna* exhibited the Y-shape or folded conformation found in *D. pulex*, despite possessing commensurate ploidy levels. Overall, there is congruence in the pattern of polyploidy observed in the labrum (Sterba, 1957b; Beaton and Hebert, 1989, 1999). This tissue appears to possess cells with the highest ploidy level in the animal.

Lipid and glycogen can constitute a massive portion of the volume of a daphnid's fat cells, which are situated in the body core along the length of the gut and in the endopodites of thoracic appendages to the base of the epipodites. The level of lipid reserves in these cells appears to be intimately related to an organism's reproductive status. Prior to peak ovary growth, the fat cells hold maximal lipid stores; as developing parthenogenetic oocytes fill with yolk, lipid, and glycogen levels in fat cells diminish (Jaeger, 1935). These cells are the most likely sites of vitellogenin synthesis, a principle component of yolk (Zaffagnini and Zeni, 1986). In addition to lipid droplets, the cytoplasm of the fat cells contain numerous free ribosomes, a well-developed rough endoplasmic reticulum (indicating active protein synthesis), mitochondria, and small Golgi complexes with enlarged cisternae (Zaffagnini and Zeni, 1986). Sterba (1956a) noted that the mitochondria are uniformly distributed throughout the cytoplasm until lipid stores fill the cells, at which point they arrange in chains (my translation) and are moved to the cell margins. The structural organization of fat cells in parthenogenetic *D. obtusa* females appeared to be similar to those of adult female insects in the vitellogenic phase and to subepidermal fat cells of reproducing females of the amphipod, *Orchestia gammarellus*. These similarities support the hypothesis that the fat cells of *Daphnia* synthesize vitellogenin (Zaffagnini and Zeni, 1986).

The nucleus in each fat cell contains an irregularly shaped, large nucleolus, which occasionally appears as two or more pieces. In *Daphnia*, these cells are known to be highly polyploid (Jaeger, 1935; Sterba 1956a). Beaton and Hebert (1989) found fat cells in the limb central core of *D. magna* and *D. pulex* females (typically corresponding to instar 4 or 5 for well-fed animals) that contained up to 256 times more DNA than is found in haploid cells. This level agreed with a study of fat cells completed by Sterba (1956a). An examination of ploidy shifts in *D. pulex* associated with growth and development revealed that the number of core limb cells was unchanged with instar, but that ploidy levels changed with food availability (Beaton and Hebert, 1999). Body size and nutritional status apparently influence the ploidy level present in this tissue; this is an expected finding, since fat cells store lipid reserves for potential oocytes. When food supplies were low, inadequate glycogen and lipid supplies were amassed, resulting in decreased egg production or the requirement of additional lipid synthesis by the ovaries during the period of sexual reproduction.

Fat cells have also been implicated with epipodite cells in the synthesis of hemoglobin (Hb), the primary oxygen-carrying molecule in cladocerans. In contrast to malacostracan crustaceans that use hemocyanin as their oxygen carrier, many branchiopods utilize Hb as their respiratory protein. The site of Hb synthesis had been pinpointed for members of several invertebrate phyla, such as Annelida and Nematoda and one insect (Bergtrom et al., 1976), but not for a crustacean until the fat cells of *D. magna* were identified as a primary manufacturing site. Goldmann et al. (1999) hybridized a probe based on a previously determined Hb cDNA sequence (Tokishita et al., 1997) to sections of adult specimens. Both fat and epipodite cells produced mRNA signals for Hb; stronger staining in epipodites compared to fat cells (Goldmann et al., 1999) may reflect relative levels of production.

16.5 OOGENESIS

Cladocerans have evolved two forms of reproduction that may alternate depending on environmental conditions. For much of the growing season, a population will be composed of nearly all females that produce parthenogenetic eggs, which develop immediately. Early accounts asserted that eggs were created ameiotically, so simple mitotic divisions (apomixis) were invoked to retain their diploid condition. Under adverse conditions, a switch to the second mode of reproduction occurs, and in most cladocerans, germ cells undergo meiosis to produce haploid gametes that must be fertilized to reconstitute the diploid state. The eggs resulting from this process do not develop immediately, but undergo a period of diapause prior to completing embryogenesis. We must remember that sex determination in this order is under environmental control. In cyclically sexual cladocerans, males are produced from clutches of parthenogenetic eggs before the sexual eggs can be formed.

Contrary to early research, a recent examination of the "parthenogenetic" process by Hiruta et al. (2010) revealed that, in *D. pulex*, these eggs are not produced mitotically but instead undergo an abortive meiosis. Using histological and immunochemical analyses, they observed two divisions, with paired homologous chromosomes aligning at the equatorial plate during the first division (as happens during meiosis). Bivalents were split to form two half bivalents that began separation, but reassembled as a diploid equatorial plate after anaphase I. During the second division, one of the separated sister chromatid sets migrated and was elevated above the egg surface to form a tiny polar body-like daughter cell, reminiscent of meiosis II (Hiruta et al., 2010). Hiruta and Tochinai (2012), on examining spindle assembly in *D. pulex*, revealed a distorted spindle shape during abortive meiosis. Contrary to the normal mitotic silhouette, spindles appeared barrel-shaped and lacked centromeres. Furthermore, γ-tubulin appeared to be distributed along the spindle microtubules; in contrast, during mitosis this protein is concentrated around the poles (Hiruta and Tochinai, 2012).

16.5.1 Formation of Parthenogenetic Eggs

Paired ovaries lie along the length of the intestine in the thorax of cladocerans. The location of the germarium in the tubular ovary appears to vary across the suborders; it is located in the posterior region in the

Anomopoda and Haplopoda (Rossi, 1980; Kato et al., 2012) and anteriorly in the Ctenopoda (Rossi, 1980). Among the Onychopoda, Jorgensen (1933) found the germarium of *Evadne nordmanni* to be near the anterior of the ovary, but Rossi (1980) positioned it at the posterior end in *Bythotrephes longimanus*. Within the ovaries are oocyte clusters, comprising one oocyte and three nurse cells. Zaffagnini and Lucchi (1965, in Zaffagnini, 1987) reported that, in *D. magna*, intercellular bridges arising from incomplete cytoplasmic divisions connect the cells within each cluster, indicating one oogonium to be the origin of the four cells.

The oocyte is the second or third cell to exit the germarium, and the four-cell cluster forms a single row, although in *Daphnia* the nurse cells are shifted dorsolaterally or laterally (Rossi, 1980; Zaffagnini, 1987). At this time, the oocyte's cytoplasm appears homogeneous and the nucleus contains a large nucleolus. Somatic cells in the area form an incomplete follicle surrounding just the oocyte (Zaffagnini, 1987). The presence of follicular cells may be unique to *Daphnia*; their absence has been noted in members of all suborders (for example, *Bythotrephes*, *Leptodora*, *Moina*, and *Sida*) (Rossi, 1980).

In *Leptodora*, *Moina*, and the onychopods, embryos develop in a closed brood pouch and are nourished by a maternal glandular structure, the Nährboden, which was first described by Weismann (1877, in Patt, 1947). Using *Polyphemus pediculus* as a model, Patt (1947) confirmed Weismann's findings. The structure can be seen for the first time during embryogenesis as a line of cells situated between the gut and future brood pouch. Patt (1947) reported that as the animal grows the structure also grows via cell expansion. In these species, the Nährboden appears not to be functional when females produce amphigonic eggs. When parthenogenetic eggs are deposited into the brood pouch, Nährboden cells increase in size to a maximum just before embryos possess limb buds. These observations almost certainly indicate that the structure is composed of polyploid cells, and DNA contents should be examined.

Initially, the oocyte of *Daphnia* is indistinguishable from nurse cells. The cytoplasm appears homogeneous and the nucleus contains a large, central nucleolus (Zaffagnini, 1987). As the oocyte initiates growth, cytoplasmic RNA concentration increases; as the oocyte begins to differentiate, the RNA concentration decreases slightly. At the onset of vitellogenesis, the oocyte appearance is clearly different from the nurse cells; the cell is much larger, the cytoplasm appears more granular, its RNA concentration drops further, and lipid droplets become visible (Zaffagnini, 1987). The presence of oil droplets is one of the first outward sign that distinguishes the oocyte from nurse cells.

During vitellogenesis, the ooplasm accumulates yolk globules via pinocytosis, and as individual globules enlarge, they concentrate in the cell periphery. Lipid droplets also increase in size and number, and mitochondria are sometimes observed to cluster around them (Zaffagnini and Lucchi, 1965, in Zaffagnini, 1987). Microvilli develop on the surface of the oocyte near the follicular cells and pinocytotic vesicles and endoplasmic reticulum form in the immediate vicinity (Zaffagnini, 1987). Kessel (1968) observed periodic accumulations of yolk bodies in the cisternae of the rough endoplasmic reticulum. The nucleus also experiences change; as the nucleolus grows, its interior becomes less dense. Furthermore, the nucleolar RNA concentration increases throughout differentiation until vitellogenesis, indicating active RNA synthesis.

The function of the nurse cells, therefore, seems less clear. As vitellogenesis proceeds, the nurse cells progressively shrink and degenerate. Simultaneously, RNA-rich bodies appear around each nucleus before being transferred to the oocyte. However, since the oocyte actively manufactures RNA and the nurse cells do not synthesize lipid or yolk, the nutritive function of the nurse cells doesn't appear to be essential (Zaffagnini, 1987).

16.5.2 Formation of Amphigonic Eggs

Generally one or two sexually produced eggs are formed at a time. The ovaries of examined members of Onychopoda and Anomopoda (*Evadne*, *Podon*, and *Daphnia*) appear similar in parthenogenetic and sexual females (Zaffagnini, 1987; Egloff et al., 1997). The amphigonic oocyte is formed with three nurse cells in the same manner as a parthenogenetic oocyte. The nucleus and nucleolus of both oocyte types also appear to be similar. However, in marine onychopods, the brood chambers of gamogenic (sexual) and parthenogenetic females differ. Cells of the former are cuboidal or hexagonal and may contribute to the egg's chitinous layer, whereas cells of the latter are transparent and amorphous (Kuttner, 1911 and Rivier, 1968; both in Egloff et al., 1997).

Several characteristics have been reported for *Daphnia* that distinguish sexual from parthenogenetic eggs: the amphigonic oocyte has a more granular cytoplasm and contains no lipid droplets; and the yolk globules do not accumulate in the periphery, but fill the ooplasm uniformly with many tiny yolk balls that have phospholipid composition and concentration differences compared to a parthenogenetic oocyte (Zaffagnini and Minelli, 1968, in Zaffagnini, 1987). Furthermore, although the nurse cells of both types of oocytes seem to be similar in both appearance and RNA content, those of the sexually produced oocytes cease growth by time that vitellogenesis is initiated (Zaffagnini, 1987). Finally, upon completion of vitellogenesis, the amphigonic egg occupies nearly the entire ovary volume (one egg per ovary).

16.6 CONCLUDING REMARKS

While a great deal of cytological information has been amassed for selected cladocerans, there is still a great deal of work left to be done. For much of this chapter, we have attempted to highlight some cytological patterns that have emerged from the last hundred or so years. However, some prominent gaps in our knowledge are evident, even from a cursory examination. Much of our understanding of the subject is based on studies of one or two species of *Daphnia* and occasionally of one additional cladoceran member. Only four genome size estimates are known for representatives from outside the suborder Anomopoda. Our understanding of chromosome numbers is also limited to Anomopoda, except for two ancient and unsubstantiated estimates. Many more surveys of DNA content and chromosome number, which represents some basic science, is required. The ultrastructure of some tissues has been carefully completed for several tissues (but again with heavy emphasis on *Daphnia*). However, questions of the function and development of some structures will benefit substantially from the burgeoning molecular tools that are available to researchers today. A much clearer understanding of the mechanism used for creating parthenogenetic eggs has only recently been described, due in no small part to the developments in immunohistochemistry. The function of nurse cells needs to be reevaluated via gene expression and in situ hybridization studies. Similarly, molecular studies focused on structures such as the Nährboden of *Polyphemus* will clarify the types of material that cells of this tissue provide to parthenogenetic embryos and how they function when amphigonic eggs form. There is clearly much left to be investigated.

CHAPTER

17

Immunology and Immunity

Stuart K. J. R. Auld

School of Natural Sciences, University of Stirling, Stirling, Scotland, UK

17.1 INTRODUCTION

Infectious diseases are a challenge for cladocerans, as they are for almost all living organisms. Cladocera are host to numerous parasites and pathogens from a wide variety of taxa, including amoeba, bacteria, fungi, nematodes and microsporidians (Green, 1974; Stirnadel and Ebert, 1997; Goren and Ben-Ami, 2013) (see Table 17.1), and the fitness consequences of parasitic infection also vary greatly according to the combination of host and parasite species. Cladoceran-parasite systems have been pivotal in improving our understanding of the causes and consequences of host immunity to infection, both in the laboratory and in the wild. This is especially true of species from the genus *Daphnia* (Ebert, 2005, 2008) and, as a consequence, much of the research discussed in this chapter concerns various *Daphnia* species.

The father of modern immunology, Elias Metchnikoff, noted that "[t]he common water flea or *Daphnia* seems quite suitable for studies on pathological processes and may be able to throw some light on many general questions in medicine" (Metchnikoff, 1884). In this early work, Metchnikoff carefully observed *Daphnia magna* that were infected with a yeast parasite (which he referred to as *Monospora bicuspidata*; now known as *Metschnikowia bicuspidata*). He made detailed drawings of parasite spores and host immune cells throughout the duration of the infection (until host death), and observed *Daphnia* cells enveloping yeast spores and yeast spores destroying *Daphnia* cells, i.e. the battle between host and parasite. More recent research on cladocerans has extended Metchnikoff's work by exploring variation in host immune activity and linking measures of host immune activity with infection outcomes and both host and parasite fitness (Mucklow and Ebert, 2003; Mucklow et al., 2004; Auld et al., 2010; Decaestecker et al., 2011; Duneau et al., 2011; Graham et al., 2011; Pauwels et al., 2011; Auld et al., 2012a, Auld et al., 2012b).

Below, I take a step-wise look at cladoceran defenses against disease, not all of which involve the host systemic immune system. I start by looking at how shifts in host behavior and phenology allow the avoidance of parasites and reduce the likelihood of infection, and then go on to describe how host barrier and systemic defenses contribute to preventing

Physiology of the Cladocera.
DOI: http://dx.doi.org/10.1016/B978-0-12-396953-8.00017-3

TABLE 17.1 Parasites of Cladocera, Infected Tissues, Host Species and Known Distribution

Parasite	Group	Infected Tissue	Host	Locations	Reference(s)
Amoebidium parasiticum	Amoeba	Epidermis	*Ceriodaphnia laticaudata*	Italy	Green (1974)
			Daphnia obtusa	Italy	Green (1974)
			Moina macrocopa	Czech Republic	Green (1974)
			Moina micura	Cameroon	Green (1974)
			Simocephalus vetulus	Greenland	Green (1974)
Pansporella perplexa	Amoeba	Hemolymph	*Daphnia longispina*	UK	Stirnadel and Ebert (1997)
			Daphnia magna	UK	Stirnadel and Ebert (1997)
			Daphnia pulex	UK	Stirnadel and Ebert (1997)
Pasteuria ramosa	Bacteria	Hemolymph	*Ceriodaphnia dubia*	USA	Auld et al. (unpubl. ms)
			Ceriodaphnia dubia	USA	Auld et al. (unpubl. ms)
			Daphnia dentifera	USA	Duffy et al. (2010); Auld et al. (2012)
			Daphnia longispina	UK	Stirnadel and Ebert (1997)
			Daphnia magna	Finland, Germany, UK	Green (1974); Stirnadel and Ebert (1997); Carius et al. (2001); Mitchell et al. (2004)
			Daphnia pulex	UK	Stirnadel and Ebert (1997)
			Daphnia pulicaria	USA	Duffy et al. (2010)
			Simocephalus vetulus	Czech Republic, Israel, UK	Goren and Ben-Ami (2012); Green (1974)
Spirobacillus cienkowskii	Bacteria	Hemolymph	*Daphnia atkinsoni*	Israel	Goren and Ben-Ami (2012)
			Acroperus harpae	Macedonia	Green (1974)
			Bosmina longirostris	UK	Green (1974)
			Chydorus sphaericus	Finland, UK	Green (1974)
			Daphnia ambigua	UK	Green (1974)
			Daphnia curvirostris	Uganda	Green (1974)
			Daphnia dentifera	USA	Green (1974); Duffy et al. (2010)
			Daphnia hyalina	Czech Republic, UK	–
			Daphnia laevis	Uganda	–

(Continued)

TABLE 17.1 (Continued)

Parasite	Group	Infected Tissue	Host	Locations	Reference(s)
			Daphnia longispina	Uganda	Green (1974)
			Daphnia magna	Finland, Israel, UK, Ukraine	Goren and Ben-Ami (2012); Green (1974)
			Daphnia obtusa	UK	Green (1974)
			Daphnia pulex	Finland	Green (1974)
			Pleuroxus aduncus	UK	–
			Pleuroxus trigonellus	UK	–
			Pleuroxus uncinatus	UK	–
			Sida crystallina	UK	–
			Simocephalus vetulus	Israel, UK	Goren and Ben-Ami (2012); Green (1974)
			Simocephalus expinosus	Czech Republic, UK	–
Aphanomyces daphniae	Fungi	Body cavity	*Daphnia hyalina*	–	–
			Daphnia pulex	–	–
Blastulidium paedophthorum	Fungi	Hemolymph, embryos	*Bosmina obtusirostris*	UK	Green (1974)
			Camptocercus lilljeborgi	Denmark	Green (1974)
			Ceriodaphnia megalops	UK	Green (1974)
			Chydorus sphaericus	France	Green (1974)
			Daphnia ambigua	UK	Green (1974)
			Daphnia longispina	UK	Green (1974)
			Daphnia magna	UK	Green (1974)
			Daphnia obtusa	France	Green (1974)
			Eurycercus lamellatus	UK	Green (1974)
			Pleuroxus laevis	Denmark	Green (1974)
			Scapholeberis mucronata	UK	Green (1974)
			Sida crystallina	UK	Green (1974)
			Simocephalus vetulus	France, UK, USA	Green (1974)
Metschnikowia bicuspidata	Fungi	Hemolymph	*Alona affinis*	UK	Green (1974)

(*Continued*)

TABLE 17.1 (Continued)

Parasite	Group	Infected Tissue	Host	Locations	Reference(s)
			Bosmina longirostris	UK	Green (1974)
			Ceriodaphnia reticulata	UK	Green (1974)
			Ceriodaphnia rotunda	Czech Republic	Green (1974)
			Daphnia dentifera	USA	Duffy et al. (2010)
			Daphnia longispina	UK	Green (1974) Stirnadel and Ebert (1997)
			Daphnia magna	Israel, Russia, UK	Goren and Ben-Ami (2012); Metchnikoff (1884); Stirnadel and Ebert (1997)
			Daphnia pulex	UK	Stirnadel and Ebert (1997)
			Ilyocryptus sordidus	UK	Green (1974)
			Scapholeberis mucronata	UK	Green (1974)
Glugoides intestinalis	Microsporidia	Gut wall	*Bosmina longirostris*	UK	Green (1974)
			Daphnia dentifera	USA	Green (1974)
			Daphnia magna	UK	Green (1974)
			Daphnia obtusa	UK	Green (1974)
			Daphnia pulex	Albania, Czech Republic, Macedonia	Green (1974)
			Daphnia longispina	UK	Stirnadel and Ebert (1997)
			Eurycercus lamellatus	Albania, Macedonia	Green (1974)
			Simocephalus vetulus	UK	–
Gurleya vavrai	Microsporidia	Epidermis	*Daphnia pulex*	UK	Green (1974); Stirnadel and Ebert (1997)
			Daphnia longispina	UK	Green (1974) Stirnadel and Ebert (1997)
Hamiltosporidium sp.	Microsporidia	Fat body, gonads	*Daphnia magna*	Finland, Israel, Sweden	Goren and Ben-Ami (2012); Haag et al. (2011)
Octosporea bayeri	Microsporidia	Fat body, ovaries	*Daphnia magna*	Czech Republic, Finland, UK	Ebert et al. (2001); Green (1974)

(Continued)

TABLE 17.1 (Continued)

Parasite	Group	Infected Tissue	Host	Locations	Reference(s)
Echinuria uncinata	Nematode	Hemolymph	*Daphnia magna*	Canada, UK	Green (1974)
			Daphnia pulex	Canada, UK	Green (1974)
			Simocephalus vetulus	Canada, UK	Green (1974)
			Daphnia obtusa	UK	Green (1974)
Caullerya mesnili	Unknown	Gut wall	*Daphnia galeata*	Austria, Germany, Switzerland	Bittner et al. (2002)
			Daphnia hyalina	Austria, Germany, Switzerland	Wolinska et al. (2006)
			Daphnia longispina	UK	Stirnadel and Ebert (1997)
			Daphnia magna	UK	Green (1974) Stirnadel and Ebert (1997)
			Daphnia obtusa	UK	Green (1974)
			Daphnia pulex	Denmark, UK	Green (1974); Stirnadel and Ebert (1997)
Polycaryum laeve	Fungus	Connective tissue	*Daphnia pulicaria*	USA	Johnson et al. (2006)

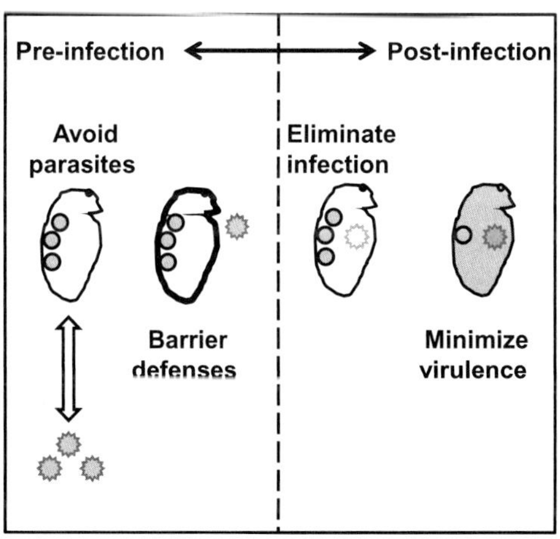

FIGURE 17.1 **Stages of cladoceran defense against parasitism.**

infection (Fig. 17.1). In cases in which infection occurs, I discuss how hosts can limit the negative fitness impacts of infection by shifting allocation into different life history traits. Finally, I examine some evolutionary and coevolutionary consequences of cladoceran-parasite interactions.

17.2 PREVENTING INFECTION

17.2.1 Avoiding Parasites

The likelihood of parasitic infection depends on (1) the probability of the host encountering infectious parasite transmission stages and (2) the efficacy of host barrier and immunological defenses following parasite exposure (henceforth, *intrinsic resistance*) (Combes, 2001) (see Fig. 17.1). Parasite avoidance is thus an effective mechanism for reducing the likelihood of parasitic infection and its associated disease, especially since parasite transmission stages are often unevenly distributed over both space and time. Nevertheless, avoidance is frequently ignored as a mechanism of host immunity.

The transmission stages for many cladoceran parasites, including *Metschnikowia bicuspidata*, *Pasteuria ramosa*, and *Spirobacillus cienkowskii*, lie in the sediment at the bottom of ponds and lakes (Decaestecker et al., 2004), where they await consumption by susceptible hosts. Potential hosts may minimize the risk of infection by keeping close to the water's surface. Indeed, *D. magna* were found to be more likely to suffer parasitic infection if they were negatively phototactic (i.e. repelled by light) and spent most of the time at the bottom of the water column near the mud; infection was reduced when *Daphnia* were positively phototactic (i.e. attracted to light) and spent most of their time at the top of the water column near the surface (Decaestecker et al., 2002). However, being positively phototactic comes at a cost: *Daphnia* that are attracted toward the water's surface are more probably to fall victim to visual predators, such as fish. This trade-off between susceptibility to parasites and vulnerability to predation can thus maintain diversity in *Daphnia* and prevent the evolution of universal host resistance to parasitism (Decaestecker et al., 2002).

The risk of suffering parasitic infection varies over time as well as space. Epidemics of *P. ramosa* occur in the fall in North America (in *Daphnia dentifera*, *D. pulicaria*, and *Ceriodaphnia dubia* (Grippi, Auld, and Duffy unpubl. data; Duffy et al., 2010) and in the summer in Europe (in *D. magna*, *D. pulex*, and *D. longispina*) (Stirnadel and Ebert, 1997; Little and Ebert, 1999; Mitchell et al., 2004; Duncan et al., 2006; Auld et al., 2012b); *Metschnikowia bicuspidata* and *S. cienkowskii* epidemics occur in the fall in North America (in *D. dentifera* populations) (Cáceres et al., 2006; Duffy et al., 2010; Duffy et al., 2012); and epidemics of *Caullerya mesnili*, peak in the fall and winter (in various European *Daphnia* populations) (Wolinska et al., 2006). As host population density increases, the likelihood of an individual host consuming an infectious parasite transmission stage also increases, as does the probability of parasite transmission from diseased to healthy hosts. Sickness thus begets more sickness.

Some genotypes of *D. magna* switch from asexual to sexual reproduction during *P. ramosa* epidemics in the wild (Duncan et al., 2006) or when surrounded by high densities of infected hosts in the laboratory (Duncan et al., 2009). Moreover, parasite-susceptible *D. magna* genotypes are most likely to switch from an asexual to a sexual reproductive mode (Mitchell et al., 2004). These sexually produced eggs are encased in a tough casing, and they hatch in the following year (i.e. after the current epidemic has passed). Therefore, by reproducing sexually, *P. ramosa*-susceptible *D. magna* mitigate the fitness consequences of infection by ensuring that their offspring avoid the current epidemic and instead face a different parasite population in the future (to which they may be more broadly resistant).

17.2.2 Barrier Defenses

Upon encountering parasite transmission stages, a cladoceran's first line of defense consists of both the carapace and the epithelium. To infect, a parasite must attach to and (usually) penetrate host tissues (Fig. 17.2, stages 2 and 3); the parasite can only establish an infection and proliferate after it has overcome at least one of these barrier defenses (Figs. 17.1 and 17.2, stage 4). Since invertebrates have a single open body chamber (the hemocoel), it is prudent for them

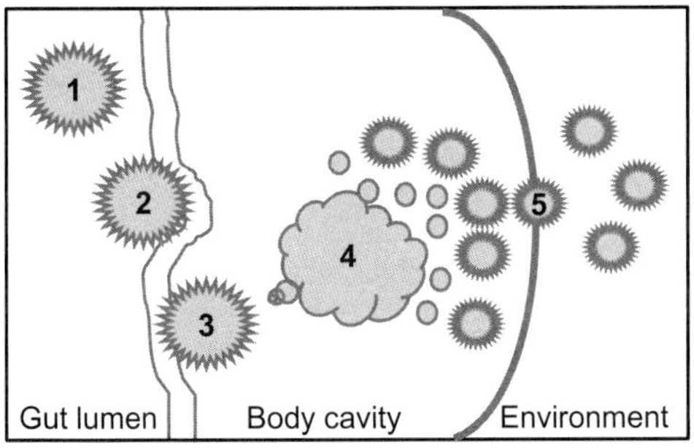

FIGURE 17.2 **A typical infection sequence for a cladoceran parasite.** 1, Host exposure to the parasite transmission stage. 2, Attachment to the gut epithelium. 3, Penetration into the host's body cavity. 4, Proliferation within the host's body cavity. 5, Parasite transmission into the environment.

to invest heavily in keeping parasites out. Conversely, vertebrates can more easily isolate parasitic infections in certain tissues and are thus better at fighting disease-causing organisms within the body (Theopold et al., 2004).

Daphnia research has been key in improving our understanding of host barrier defenses, in particular, by revealing that barriers are not necessarily unsophisticated walls against parasites and that they instead vary in efficacy depending on the particular immune challenges. Studies using the parasite *P. ramosa* have shown that spore attachment to the esophageal gut epithelium is crucial to the infection process: *P. ramosa* spores were found to successfully bind to the esophagus and then infect certain genotypes of *D. magna* and a genotype of *D. dolichocephala*; and infections only occurred where attachment was successful (Duneau et al., 2011). *P. ramosa* failed to attach to and infect individual genotypes of *Daphnia arenata*, *Daphnia barbata*, *Daphnia lumholtzi*, and *Daphnia similis* (Duneau et al., 2011); however, since just a single genotype was studied for each of these species, we cannot draw broad conclusions regarding their ability to prevent *P. ramosa* attachment.

By taking a classical genetics approach (using parental, F_1, and F_2 generations, backcrossed and selfed genetic lines), Luijckx et al. (2011b) found that *D. magna* display either complete resistance or complete susceptibility to a purely clonal *P. ramosa* strain and that resistance is inextricably linked to the parasite attachment mechanism. This shows that *D. magna* resistance to this particular *P. ramosa* strain (which stems from the ability of the host to prevent spore attachment to the gut) depends on a single host gene. Further, it suggests that the resistance gene has two alleles and that resistance is dominant over susceptibility (Luijckx et al., 2011b). However, both *D. magna* and *D. dentifera* can exhibit strong genetic specificity with *P. ramosa* i.e. infection success depends on the specific combination of host genotypes and parasite isolates (Carius et al., 2001; Luijckx et al., 2011a; Auld et al., 2012c), so resistance to one *P. ramosa* strain may come at a cost of susceptibility to another. Since *Daphnia* frequently encounter multiple *P. ramosa* genotypes in the wild, resistance in general is likely to be a quantitative trait.

In any case, the ability of *P. ramosa* to attach to and penetrate host barrier defenses is a major determinant of infection outcome. Further supporting this, *D. magna* genotypes mount a strong cellular response (a systemic immune response that occurs within the body cavity) only to *P. ramosa* isolates that can cause infection, and the magnitude of this cellular response increases with the dose of infectious parasite spores (Auld et al., 2012a). These findings suggest that the cellular immune response is an indicator of the capacity for parasite spores to successfully overcome host barrier defenses in the gut (Auld et al., 2012a).

17.3 THE INNATE IMMUNE SYSTEM

Once the barrier defenses have been breached, parasites face the next round of host defenses: the systemic immune system.

Vertebrates possess an immune system with both innate and adaptive components, whereas invertebrates have only an innate immune system (Schmid-Hempel, 2011). Adaptive immune defenses consist of highly specific responses to an ongoing immune challenge (Janeway et al., 2001; Schmid-Hempel, 2011) and are therefore comparatively slow to respond to an immune challenge. Conversely, innate immune responses are much less specific—they are thought to distinguish between broad classes of parasite, e.g. Gram-positive bacteria, Gram-negative bacteria, and fungi (Janeway and Medzhitov, 2002; Ferrandon et al., 2003) (see later Sections 17.5.1 and 17.6.2). Innate immune responses can, however, be mounted relatively soon after immune challenge: the *D. magna* cellular immune response can occur within hours of an immune challenge (Metchnikoff, 1884; Auld et al., 2010).

Like other invertebrates, *Daphnia* innate immunity genes fall into three functional groups: (1) pathogen recognition, (2) immune regulation, and (3) attack. Although much of what we know about invertebrate immunology stems from studies of Dipteran insects (in particular *Drosophila melanogaster* and *Anopheles gambiae*), recent investigations of the *D. arenata*, *D. magna*, *D. pulex*, and *D. parvula* immune system have expanded our knowledge greatly; indeed, *D. pulex* was the first crustacean genome to be sequenced (Colbourne et al., 2011). This work has revealed many similarities between the innate immune systems of these cladocerans and other invertebrates, as well as some interesting differences.

McTaggart et al. (2009) found 82 putative immune genes from 21 families in *D. pulex*. This is fewer than in *D. melanogaster*, more than in *Apis mellifera*, and similar to numbers in *Tribolium castanaeum*. Furthermore, *D. pulex* immune genes undergo more adaptive evolution than genes that are not thought to be associated with the immune system, pointing toward a coevolutionary history between *D. pulex* and its parasites (McTaggart et al., 2009). It is important to realize that many of these genes are putative immune genes; they are similar to genes that are known to be involved with an immune response in other organisms, but their link with *Daphnia* immunity has yet to be demonstrated.

17.3.1 Pathogen Recognition

Pathogen recognition receptors (PRRs) detect molecular patterns on the surface of both prokaryote and eukaryote pathogens and complete the first step of initiating an innate immune response. *Daphnia* species possess multiple types of PRRs, which are discussed individually below.

Gram-negative binding proteins (GNBPs) are a major class of PRR. GNBPs recognize and bind to β-1,3-glucans and activate immune defenses (including the prophenoloxidase cascade, which is known to be present in *D. magna*). Although they bind principally to Gram-negative bacteria and fungi, it is important to note that GNBPs are also known to bind to Gram-positive bacteria in *Drosophila* (Gobert et al., 2003; Filipe et al., 2005). *D. pulex* has 11 GNBP copies—more than *A. gambiae* (7), *A. aegypti* (7), *D. melanogaster* (3), *T. castanaeum* (3), or *A. mellifera* (2) (McTaggart et al., 2009), and five of these GNBPs have also been sequenced in *D. parvula* (McTaggart et al., 2012a). Ten of the 11 *D. pulex* GNBPs are grouped together, along with a GNBP gene from *Eisenia foetida* (an annelid), suggesting they are relatively ancient; the remaining gene is more derived and groups with *GNBP* genes from *A. gambiae* (McTaggart et al., 2009).

Another invertebrate PRR group includes peptidoglycan recognition proteins (PGRPs). Peptidoglycan is a key constituent of the cell wall of Gram-positive bacteria and PGRPs are known to bind to this peptidoglycan and initiate immune cascades involved in phagocytosis, generation of immune cytotoxins, and generation of antimicrobial peptides in *Drosophila* (Hultmark, 2003). Remarkably, *D. pulex* does not possess any PGRPs (unlike *A. gambiae*, *A. aegyptii*, *D. melanogaster*, *T. castaneum*, and *A. mellifera*) (McTaggart et al., 2009). We can only speculate that *D. pulex*'s numerous GNBPs compensate for this and provide defense against Gram-positive bacteria (McTaggart et al., 2009).

Thioester proteins (TEPs) are also important PRRs in many species. Seven *TEP* genes were found on the *D. pulex* genome, including the well-studied alpha-2-macroglobulin (α2m) gene, which inhibits protease production in Gram-positive bacterial cells, while promoting phagocytosis in vertebrates (deBoer et al., 1993). *Anopheles gambiae* TEPs bind to pathogenic cells and await phagocytosis (Levashina et al., 2001), although TEPs may have other nonimmunological functions in other organisms. A study of *D. magna*, *D. pulex*, and *D. rosea* found strong evidence of positive selection in the bait region of the α2m gene, which is consistent with the hypothesis that coevolutionary interactions between *Daphnia* and Gram-positive bacterial pathogens have fostered an arms-race evolution in α2m (Little et al., 2004). However, α2m transcription in *D. magna* does not increase after exposure to *P. ramosa* (Decaestecker et al., 2011) (see Section 17.6.1).

Scavenger receptor-A proteins (SR-As) are a conserved and structurally similar to a group of PRRs found on both cell surfaces and in soluble form; they have two roles. SR-As provide a housekeeping role by "mopping-up" cellular debris (Munier et al., 2004). They can also bind to and promote the encapsulation of bacteria: the *D. melanogaster* protein Tequila/graal and the orthologous *A. gambiae* protein Scrasp1 are upregulated following an immune challenge (Danielli et al., 2000; Munier et al., 2004). The *D. melanogaster* protein Corin is also upregulated following a bacterial or fungal immune challenge (Irving et al., 2001). *D. pulex* has six SR-As: two of these proteins are orthologous to Tequila/graal/Scrasp1 (Dappu-SCV3 and Dappu-SCV4), and one is orthologous to Corin (Dappu-SV2). Therefore, Dappu-SCV2, Dappu-SV3, and Dappu-SV2 may all have an immune role (McTaggart et al., 2009), although this remains to be tested.

Down syndrome cell adhesion molecules (Dscams) are important for arthropod neural development, as well as being PRRs (where they are principally involved in phagocytosis (Watson et al., 2005). Arthropod *Dscam* genes have an immunoglobulin (IgSF) domain, like their vertebrate counterparts, and are especially interesting because they contain clusters of variable exons that can be alternatively spliced during transcription, thus facilitating the generation of tens of thousands of unique transcripts (Du Pasquier, 2005; Watson et al., 2005; Brites et al., 2008; Brites et al., 2011). This ability to generate a multitude of transcripts is consistent with *Dscam* genes being PRRs; it has been hypothesized that these alternative transcripts may be functionally equivalent to vertebrate antibodies (Watson et al., 2005).

17.3.2 Immune Regulation

Signal transducers play a pivotal role in developing immune responses. For example, the Toll pathway and associated Toll receptors interact with GNBPs and PGRPs elicit the appropriate attack responses to particular pathogens (including phagocytosis). Seven Toll receptors have been found in *D. pulex* (McTaggart et al., 2009) and three have been sequenced in *D. parvula* (McTaggart et al., 2012a). Many of the *D. pulex* Toll receptors show similarities to those of *D. melanogaster*. In particular, *D. pulex* Dappu-TOLL1 and Dappu-TOLL3 are within the same clade as *D. melanogaster* *Dm*-Toll-1, which is upregulated following exposure to fungi and bacteria (Lemaitre et al., 1996; Michel et al., 2001), and Dappu-TOLL2 and Dappu-TOLL4 cluster with *Dm*-Toll-9 and are similar to mammalian Toll like receptors.

17.3.3 Pathogen Attack

Pathogen attack responses are the best-known components of cladoceran systemic immune responses. They consist of both cellular and humoral defenses, which are discussed below.

Circulating hemocytes are a key component of antiparasite defenses in numerous invertebrates (Ataev and Coustau, 1999; Elrod-Erickson et al., 2000; Kraaijeveld et al., 2001; Canesi et al., 2002). Invertebrate hemocytes (Fig. 17.3) can either phagocytose (or encapsulate) parasitic organisms or generate immune cytotoxins such as reactive oxygen species and reactive nitrogen species (Strand, 2008). At the very beginning of this chapter, I discussed Metchnikoff's (1884) study of *Daphnia* cellular responses to the yeast *M. bicuspidata*. He found that once the needle-like transmission spores of the parasite had penetrated the host's body cavity, numerous ameboid hemocytes bound firmly to the spores to isolate the parasite and that the parasite often swells and becomes yellow-brown. Metchnikoff concluded that these hemocytes caused this change in the parasite spores and neutralized the infection; he supported this claim with observations of *M. bicuspidata* spores that had partially penetrated the gut epithelium. Of spores in the process of penetration, only the sections that were within the host's body cavity became covered with hemocytes and underwent digestion; sections of the spore that remained in the gut lumen were free of hemocytes and appeared undigested (Metchnikoff, 1884). However, this host cellular immune response was not always effective: sometimes *M. bicuspidata* spores did not fall victim to the host's hemocytes, leading to the establishment of infection (see Section 17.6.1).

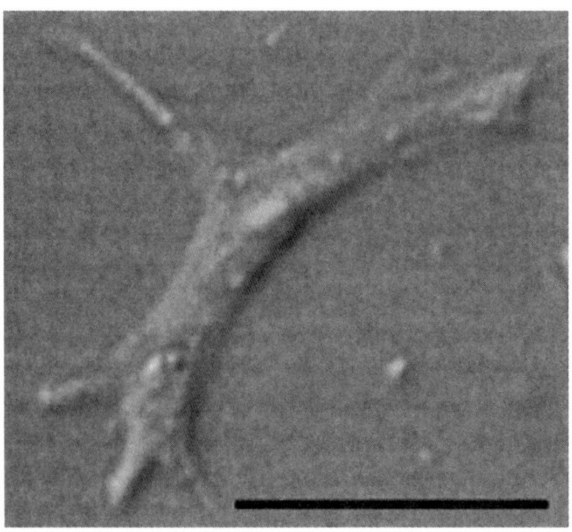

FIGURE 17.3 Differential interference contrast image of a *Daphnia magna* ameboid hemocyte. Scale bar, 5 μm. *From Auld et al. (2010).*

Nitric oxide (NO) is a highly oxidative free-radical gas that can severely damage enzymes and DNA, making it generally toxic to a variety of parasitic organisms (Nappi and Ottaviani, 2000; Rivero, 2006). NO is generated in vertebrates, invertebrates, and plants by nitric oxide synthase (*NOS*) genes. *NOS* genes are either constitutively expressed [constitutive NOS (cNOS)] or their expression is induced upon encountering an immune challenge [inducible NOS (iNOS)]. However, its highly oxidizing nature means that NO can be toxic to the host that produces it, i.e. it can cause immunopathology (in which host disease is caused by host immune activity). NO is thus a highly potent weapon, but the host risks potentially severe collateral damage (Rivero, 2006). Unlike all other investigated invertebrates, *D. pulex* has two *NOS* genes (other invertebrates possess only one), and these two genes are very different in structure: the resulting proteins are only 44% identical (Labbé et al., 2009). McTaggart et al. (2009) hypothesized that this could be due either to a relaxation from a selective constraint or to positive selection that has promoted evolutionary divergence between the two *NOS* genes. In either case, duplication of the *NOS* gene did not occur recently.

The prophenoloxidase (PO) cascade, which involves the generation of PO, is another cytotoxic immune defense known to be important in numerous invertebrates (Soderhall and Cerenius, 1998). *D. magna* produce more PO when they experience wounding (Mucklow and Ebert, 2003) and PO has been shown to limit parasite success in multiple parasite species (Mucklow et al., 2004; Pauwels et al., 2011; see Section 17.6.1).

Chitinases are another potentially important immune effector. Chitinases hydrolyze chitin, a polysaccharide found in arthropod exoskeletons, as well as the cell walls of fungal spores and amoeba cysts. Proteins similar to chitinase have been found to be upregulated in *A. gambiae* when the host is exposed to Gram-positive bacteria (Shi and Paskewitz, 2004), and there is evidence of rapid adaptive evolution in a plant chitinase, probably caused by coevolution with a fungal parasite (Bishop et al., 2000). Chitinases are thus a useful weapon for both animals and plants against many pathogenic organisms. Seventeen chitinase genes were found in the *D. pulex* genome, which is a similar number to that found in *A. gambiae* (13), *A. aegyptii* (19), *D. melanogaster* (16), and *T. castanaeum* (16), but considerably more than in *A. mellifera* (5) (McTaggart et al., 2009). However, we have yet to determine which, if any, of these chitinase genes have an immunological function in *D. pulex*.

17.4 MECHANISMS TO LIMIT THE SEVERITY OF INFECTION

Parasites, by their very nature, reduce the fitness of the hosts they infect. This parasite-induced decline in host fitness is termed virulence. As discussed above, cladocerans have evolved elaborate defense systems to prevent infection (e.g. immune systems). However, preventing infection is just one component of these defenses; the other component involves reducing virulence, which often involves hosts shifting resource investment into different life history traits.

A prime example of a virulence-limiting life history shift occurs in *D. magna*. When *D. magna* are at the early stages of infection with the sterilizing parasite *P. ramosa*, they accelerate to produce a clutch of offspring before the parasite fully sterilizes the host (termed fecundity compensation) (Ebert et al., 2004; Vale and Little, 2012). This is also true of *D. magna* exposed to the horizontally transmitted microsporidian, *Glugoides intestinalis*: although it does not fully sterilize its hosts, *G. intestinalis*-infected *D. magna* alter their life history to produce more of their offspring early in life (Chadwick and Little, 2005). By inducing fecundity compensation, *D. magna* can reduce the virulence suffered from parasitic infection without necessarily limiting parasite proliferation.

17.5 EVOLUTION AND COEVOLUTION IN CLADOCERAN-PARASITE SYSTEMS

Cladoceran hosts exhibit abundant genetic variation in both their susceptibility to infection and the virulence they experience as a result of parasitism. Indeed, when genetic variation for parasite resistance has been tested in wild *Daphnia* and *C. dubia* populations, it has been found in every case (Auld et al., in prep.; Ebert, 2005), suggesting that parasites are an ever-present selective force in these host populations. Furthermore, there is abundant evidence that parasite epidemics select for host resistance, thus driving rapid evolution in natural host populations (Duncan and Little, 2007; Duffy et al., 2012; Auld et al., in press). Yet, despite the ubiquity of parasites, susceptible host genotypes persist and are sometimes favored by selection. There are two main hypotheses for the maintenance of host variation for susceptibility: genetic specificity and the costs of resistance.

17.5.1 Genetic Specificity

Genetic specificity occurs when the likelihood of successful infection depends on the precise combination of host and parasite genotypes (Schmid-Hempel and Ebert, 2003; Lambrechts, 2010); no one host genotype is resistant to all local parasite genotypes and no parasite genotype is infectious to all local host genotypes (Fig. 17.4). *Daphnia*-parasite systems provide some excellent examples of genetic specificity: for both *D. magna* and D. dentifera, the likelihood of suffering infection with *P. ramosa* depends on which host genotypes are exposed to which parasite isolates (Carius et al., 2001; Auld et al., 2012c), with particular *P. ramosa* strains either achieving 100% or 0% infection success depending on the hosts they are exposed to (Luijckx et al., 2011a). Moreover, genetic specificity can occur at the species level: in both *D. magna* and *D. dentifera*, infection depends on the combination of the *Daphnia* genotype and the parasite species (Decaestecker et al., 2003; Auld et al., 2012c).

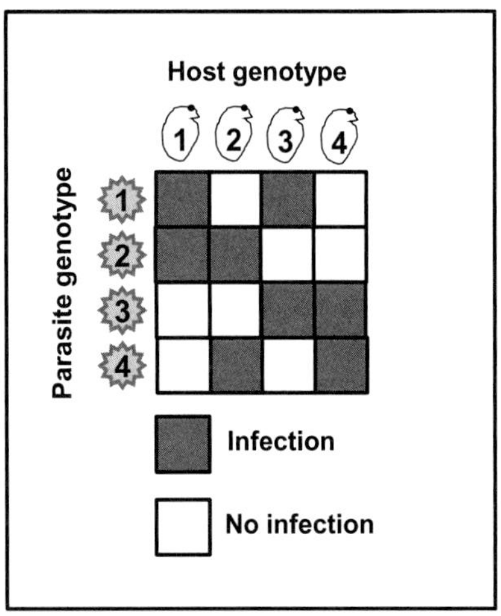

FIGURE 17.4 **Genetic specificity.** Host genotypes have that same mean resistance to infection (50%) and parasite genotypes have the same mean infectivity (50%), but the likelihood of infection depends on the combination of host and parasite genotypes.

Genetic specificity is important because it can maintain both host and parasite genetic diversity and drive host-parasite coevolution (providing the parasites cause a high degree of virulence). This is because each parasite genotype (or species) can only select against a subset of host genotypes (and vice-versa). It also means that individual parasite genotypes select against the most common host genotype (and vice-versa), leading to negative frequency-dependent selection and a cycling of both host and parasite genotypes (Hamilton, 1980). This form of genotype cycling is referred to as Red Queen dynamics after the Red Queen in Lewis Carroll's *Through the Looking Glass*, who had to constantly run in order to stay still (Carroll et al., 1971). A recent study looked for evidence of Red Queen dynamics in hatching dormant stages (resting eggs) of *D. magna* and *P. ramosa* from sediment cores that represented decades of time. Consistent with the Red Queen hypothesis, they found *P. ramosa* isolates were best adapted to contemporary *D. magna* genotypes (i.e. had the highest infection successes) (Decaestecker et al., 2007).

However, genetic specificity is not universally present in cladoceran systems. For example, there is no evidence for genetic specificity in the *D. dentifera-M. bicuspidata* system (Duffy and Sivars-Becker, 2007). Remarkably, there is no evidence for genetic variation in *M. bicuspidata* for infection success: *M. bicuspidata* samples taken from different populations at different times do not vary in their ability to infect *D. dentifera* (Duffy and Sivars-Becker, 2007). Despite this, there is considerable genetic variation for *D. dentifera* resistance to *M. bicuspidata* (Cáceres et al., 2006; Duffy and Sivars-Becker, 2007; Auld et al., 2012c, Duffy et al., 2012). What could be responsible for maintaining this variation? The answer stems from the fact that resistance to parasitism often comes at a cost.

17.5.2 Costs of Resistance

Initially, one would expect organisms with the strongest resistance to parasites to have the highest fitness and be favored by natural selection, i.e. for parasite-resistant genotypes to dominate host populations. However, immunity is not free; it can come at the expense of one or more other fitness-related traits. Indeed, these are termed the *evolutionary costs of resistance*, or *resistance trade-offs* (Sheldon and Verhulst, 1996; Kraaijeveld and Godfray, 1997).

A recent study found that while *D. dentifera* populations that suffered large *M. bicuspidata* epidemics evolved increased resistance to *M. bicuspidata*, populations that experienced relatively small epidemics evolved increased susceptibility (Duffy et al., 2012). Further, parasite-mediated disruptive selection has also been observed in this system: an *M. bicuspidata* epidemic was found to simultaneously favor both the most resistant and the most susceptible host genotypes in the population, and select against those of intermediate resistance (Duffy et al., 2008). Both of these studies point toward an evolutionary cost of resisting *M. bicuspidata* infection; indeed how else could susceptible hosts be favored?

Hall et al. (2010) determined a simple underlying mechanism for this cost, which hinges on the filtration rate of the host: *D. dentifera* genotypes with a high filtration rate obtain more algal food and can therefore produce more offspring, but they are also more likely to consume *M. bicuspidata* spores and suffer infection (Fig. 17.5). This trade-off influences overall host resistance by mediating exposure to parasite transmission stages; it does not concern the costs associated with the evolution of particular host immune defenses, i.e. intrinsic resistance (Hall et al., 2010). That is not to say that there are not evolutionary costs associated with intrinsic resistance in the *D. dentifera-M. bicuspidata* system—there may well be—but such costs have not yet been found. Since overall resistance is a complex trait that depends on both host exposure and intrinsic resistance, one might expect the magnitude of any evolutionary costs of resistance to vary over time or space or both.

It is important to note that the evolutionary costs of resistance are not a general phenomenon in cladoceran-parasite systems. *D. magna* that are susceptible to the microsporidian *Octosporea bayeri* are not more fecund in the absence of infection (Altermatt and Ebert, 2007), nor are *D. magna* that are susceptible to *P. ramosa* (Little et al., 2002).

Evolutionary costs of resistance are not the only type of cost: there may also be costs associated with activating defense mechanisms (e.g. immune responses). Kraaijeveld and Wertheim (2009) liken evolutionary costs to the costs associated with building an army and activation costs to the costs associated with sending the army to war. Studies have tested for an activation cost of resistance in both the *D. magna-P. ramosa* (Little and Killick, 2007; Labbé et al., 2010) and the *D. dentifera-M. bicuspidata* (Auld

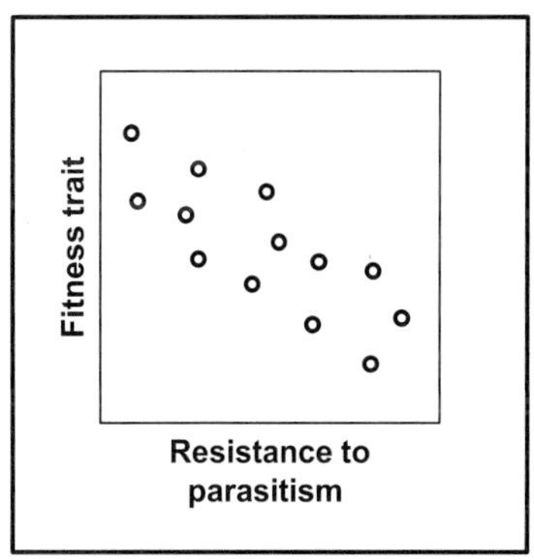

FIGURE 17.5 **Evolutionary cost of resistance.** There is a trade-off between resistance and another fitness-related trait (i.e. fecundity).

et al., 2012c) systems by comparing the fitness of hosts that were exposed to, but resisted, infection from a particular parasite with hosts that were not exposed to the parasite. While Little and Killick (2007) found evidence for a cost of resistance in terms of *D. magna* survival when exposed to *P. ramosa,* more extensive study in this system failed to uncover activation costs (Labbé et al., 2010). No activation costs of resistance in terms of host fecundity or survival were found in the *D. dentifera-M. bicuspidata* system (Auld et al., 2012c).

Why are activation costs of resistance so rare in these particular cladoceran-parasite systems? The explanation probably lies in the infection process. If attachment of spores to the gut wall explains most of the variation in infection likelihood (as it does in the *D. magna-P. ramosa* system; see above), then host resistance does not stem from the mobilization of costly systemic immune responses.

17.6 LINKING IMMUNOLOGY TO IMMUNITY: STUDIES INTO *DAPHNIA MAGNA* AND *PASTEURIA RAMOSA*

17.6.1 Ecological Immunology

Much immunological work is conducted on a limited number of host genotypes and parasite strains in the laboratory, but in the wild, variation abounds. Students of evolutionary ecology have long been interested in how and why some hosts resist parasitic organisms and others do not. This has given rise to the field of ecological immunology—the study of the variation in host immune strategies and their underlying costs and benefits (Rolff and Siva-Jothy, 2003; Schmid-Hempel, 2003; Sadd and Schmid-Hempel, 2009; Graham et al., 2011). Cladocerans (*Daphnia* in particular) and their parasites are well poised for eco-immunological study for numerous reasons. First, the fact that many *Daphnia* species are cyclically parthenogenetic (see Chapter 11) means that genotypes can be propagated clonally and thus that one can examine the relative contributions of genetic and environmental factors to immune strategies and infection outcomes. Second, there is a rich body of research examining variation in susceptibility to various parasites both within and across host species (Green, 1974; Stirnadel and Ebert, 1997; Little and Ebert, 1999; Wolinska et al., 2006; Duffy and Sivars-Becker, 2007; Duffy et al., 2010; Goren and Ben-Ami, 2013). Third, there is an ever-growing body of genetic resources, in particular from the recently sequenced *D. pulex* genome (McTaggart et al., 2009; Jansen et al., 2011). Finally, one can easily pair controlled laboratory experiments with observational study of immune phenotypes in the wild—indeed, *Daphnia* research is making key contributions to the emerging field of wild immunology (Auld et al., 2012b).

Evolutionary ecologists have often assumed that a larger immune response leads to improved immunity and greater host fitness in the face of parasites. This is largely true in many host-parasite systems. However, research with *D. magna* and *P. ramosa* has emphasized that immune responses can also be a consequence of parasitic infection: upon exposure to *P. ramosa, Daphnia* that mount strong cellular immune responses go on to suffer infection with sterilizing bacterium and thus experience genetic death (i.e. an end to their fitness potential) (Auld et al., 2010; Auld et al., 2012a). Further, both wild-caught and laboratory reared *D. magna* have a greater baseline number of circulating hemocytes than their healthy counterparts when suffering a *P. ramosa* infection (Auld et al., 2012b).

Immune gene expression studies have also highlighted the danger of assuming that greater immune activity causes host resistance to infection. When *P. ramosa* (a Gram-positive bacterial parasite) was exposed to both resistant and susceptible *D. magna* genotypes, there was no significant increase in expression of α2m relative to controls (Decaestecker et al., 2011). As discussed earlier, α2m is a PRR known to target Gram-positive bacteria in *Drosophila*. While this brings into question the role of α2m in defending *D. magna* against *P. ramosa*, it is important to note that TEP proteins can be highly specific to particular immune challenges: *D. melanogaster TEP6* (referred to as the macroglobulin complement-related gene) is known to be highly specific to the yeast *Candida albicans* (Stroschein-Stevenson et al., 2006), so *D. magna's α2m* gene may play a role in defending against a parasite or parasites other than *P. ramosa*.

Another study demonstrated that *P. ramosa*-infected hosts had higher constitutive PO levels, but in this case, PO activity was negatively associated with parasite burden in infected hosts (Pauwels et al., 2011). This study, along with those examining the *D. magna* cellular response, provides a compelling example of how immune responses can indicate catastrophic failures in other more important components of the host's immune system. Immune failure is particularly disastrous for *D. magna* exposed to *P. ramosa* because the parasite sterilizes its host and thus renders it a genetic dead-end. Immune activity and immunity are not always one and the same: a strong immune response can herald the onset of parasitic infection and a rapid decline in host fitness.

17.6.2 Priming

It has long been thought that immune specificity (i.e. the ability to distinguish between very similar infectious agents) and immune memory (i.e. the ability of an immune system to react more rapidly and effectively to challenges it has previously encountered) are properties of the adaptive immune system only, and thus only possible in vertebrates (Janeway et al., 2001; Schmid-Hempel, 2011). However, it has now been shown that *D. magna* can "remember" immune challenges from specific parasite genotypes and modulate their defenses accordingly. *D. magna* that were exposed to and successfully resisted *P. ramosa* were more likely to resist infection when reexposed to infectious spores 48 h later (McTaggart et al., 2012b). Even more remarkably, upon exposure to *P. ramosa*, *D. magna* are known to be able to pass on strain-specific resistance to their offspring (Little et al., 2003): when a mother is exposed to a particular *P. ramosa* isolate, her offspring have increased resistance to that particular isolate (but not to other *P. ramosa* isolates). Cladoceran immunity therefore has the potential to reach across generations.

17.7 SUMMARY

Cladocerans, and *Daphnia* in particular, have taught us much about the nature of immunity. Early work by Metchnikoff gave us insight to the importance of hemocytes in invertebrate immunity, and more recent analyses of the newly sequenced *D. pulex* genotype have uncovered many putative immune genes. Much of what we understand about the ecology and evolution of infectious disease also stems from studies conducted in cladoceran-parasite systems. In particular, they have helped us to elucidate the mechanisms that maintain host (and sometimes parasite) genetic diversity and drive host evolution and host-parasite coevolution.

The challenge now is to bridge the gap between mechanistic immunology and whole-organism measures of infection traits—to examine variation in immune activity within and across cladoceran species and to link it to measures of parasitism. Which immune responses are effective against which parasites? What are the costs associated with maintaining or mobilizing different immune responses? What is the mechanism underlying transgenerational immunity? And finally, how do particular immune functions shape coevolution between cladocerans and their various parasites.

CHAPTER

18

The Genomics of Cladoceran Physiology

Dörthe Becker[1], *Kay Van Damme*[1], *Elizabeth Turner*[2], *Joseph R. Shaw*[2], *John K. Colbourne*[1] *and Michael E. Pfrender*[3]

[1]Environmental Genomics Group, School of Biosciences, University of Birmingham, Birmingham, UK, [2]School of Public and Environmental Affairs, Indiana University, Bloomington, IN, USA, [3]Department of Biological Sciences, Environmental Change Initiative, University of Notre Dame, Notre Dame, IN, USA

18.1 INTRODUCTION

Understanding the tolerance of organisms to environmental change, and the underlying mechanisms that define the thresholds of physiological capacity (e.g. Hofmann and Todgham, 2010) are major objectives for modern biological investigation. Against the background of a rapidly changing world, there is an increasing research emphasis on linking organism-environment interactions to physiological responses and phenotypic plasticity. To obtain a thorough understanding of the origin and maintenance of phenotypic variation of populations in the natural world (i.e. how environmental stimuli affect individual fitness and the evolutionary potential of populations), we need to decipher the molecular machinery that regulates phenotypic responses to environmental conditions. These responses span the short-term physiological changes of acclimation to long-term genetic alterations of adaptation. Linking organismal change across these disparate time frames requires delving deeply into the genetic mechanisms underlying phenotypic plasticity. A plastic (flexible) phenotype is the outcome of modification to the functional genome, for example by altering transcription or translation (e.g. Whitehead, 2012). As Whitehead (2012) noted, "environmental change that emerges across generations can also be accommodated by plasticity, or alternatively may drive structural change of genomes by adaptive and demographic evolutionary processes that sort allele frequencies within populations across generations."

Making connections between the functional genome and the interaction with the environment that give rise to the phenotype is a challenging task. One central difficulty is that environmentally sensitive physiological traits are rarely controlled by single, or a few, genes

Physiology of the Cladocera.
DOI: http://dx.doi.org/10.1016/B978-0-12-396953-8.00018-5

of major effect. They are often highly polygenic traits (West-Eberhard, 2005; Aubin-Horth and Renn, 2009; Ayroles et al., 2009; Flint and Mackay, 2009), which dictates that entire regulatory networks need to be considered. Studying the genetic basis of cladoceran physiology provides an ideal context for understanding the generality of genomic mechanisms underlying physiological responses. Cladocerans, with their pronounced phenotypic plasticity and particular mode of reproduction, are useful organisms to study genome evolution (Colbourne et al., 2011). Only through a better understanding of the phenotype (e.g. morphology and physiology) and the genomic underpinnings of phenotypic variation can linkages be forged between genes and physiological functions.

Functional genomics allow a new approach to explore the mechanisms that drive physiological change and help to unravel key processes that enable organisms to cope in their natural environment (Aubin-Horth and Renn, 2009; Bell and Aubin-Horth, 2010). This area of investigation is embodied by the emerging field of ecological and evolutionary genomics (EEG), which seeks to link gene functions to phenotypes and ecological factors (e.g. Renn and Siemens, 2010; Pavey et al., 2012). Gene-environment interactions can be best studied in animals with well-known ecologies, on a physiological timescale as well as in an evolutionary context (Carroll et al., 2007; Hoffmann and Willi, 2008; Pfrender, 2012; for a review see Hodgins-Davis and Townsend, 2009). Cladocerans lend themselves to such research, as their ecology is well studied (at least for the planktonic groups) and the genomic resources required are being rapidly developed; "by examining genome structure and the functional responses of genes to environmental conditions within species with traceable ecologies, we further our understanding of gene-environment interactions in an evolutionary context" (Colbourne et al., 2011).

Current genomic research seeks to address the following key questions in cladoceran physiology:

1. Are physiological responses reflected by responses in the functional genome?
2. What are the genomic elements or regulatory programs that allow plastic responses (i.e. by which mechanisms is a change in the environment sensed, integrated, and transformed into distinct physiological responses)?
3. How are physiological systems integrated, and how do genes interact to describe the regulatory networks underlying the biological processes and pathways involved?

In this chapter, we explore the current knowledge on the genetic basis of cladoceran physiology, drawing largely from our growing understanding of *Daphnia*, which has been the most extensively studied genus within the group.

18.2 A LONG HISTORY OF RESEARCH: THE PRE-GENOMICS ERA

The genus *Daphnia* has been the focus of biological and ecological research for over a century, making it one of the oldest study systems in ecology, physiology, and evolutionary biology among freshwater invertebrates. When it comes to genetics, it is the most well-known cladoceran. Today's in-depth knowledge of these organisms results from their prominent ecological role as keystone species in many freshwater ecosystems. It is worth noting that while the genetic investigation of physiological traits in cladocerans has only recently been a major focus of investigation, the framework for detailed inquiry is well in place. Extensive genetic investigation at the population level and the formation of explicit phylogenetic hypotheses for several groups of cladocerans form the backdrop for understanding the evolution of physiological traits. In the pre-genomics era of the past century, the genetic architecture of natural populations of *Daphnia* was investigated through population genetic, quantitative genetic, and phylogenetic approaches. Among the early studies were examinations of the characteristic alterations in the phenotype in response to predator chemical cues (Woltereck, 1909) and regulation of sexuality and diapause (Banta and Brown, 1939). With the advent of readily available molecular genetic techniques (i.e. protein gel electrophoresis) in the 1970s, genetic studies revealed populations and species with high levels of genetic variation indicative of local adaptation. In the following decades, research on cladoceran genetics shifted with the techniques available, from allozyme studies (1970–1990) to the amplification and the sequencing of nuclear and mitochondrial genes (Hebert and Taylor, 1997). Whereas allozyme analysis provided a useful overview of population structure, and is still often used as a diagnostic tool to identify naturally occurring hybrids (Hebert et al., 1989; Cerny and Hebert, 1999), DNA sequencing became the most widespread technique to understand speciation and phylogeny in the group. Cladoceran phylogenies have been variously constructed with mitochondrial genes (*12S*, *16S*, and *COI*) (e.g. Crease and Little, 1997; Colbourne et al., 1998, Crease, 1999), nuclear genes (*18S*, *28S*, and *EF-1*) (e.g. Crease and Taylor, 1998), or combinations of both (e.g. Adamowicz et al., 2009). Within cladoceran families, or at the level of genera, DNA markers have proven to be very useful, as repeatedly shown for *Daphnia* in a genus-wide phylogenetic context (e.g. Lehman et al., 1995; Adamowicz et al., 2009), or in the exploration of the genetic diversity and biogeography of selected species groups (Taylor et al., 1996; Hebert et al., 2003; Kotov and Taylor, 2010; Crease et al., 2012).

Somewhat surprisingly, research that examines phylogenetic relationships in Cladocera as a whole, or in the higher taxonomic level of Anomopoda, is currently limited. Molecular phylogenies at these phylogenetic levels often fail to resolve relationships among the deeper cladoceran branches (de Waard et al., 2006; Stenderup et al., 2006; Van Damme et al., 2007). The phylogenetic relationships within many cladoceran families have not been well investigated, and for some groups the first molecular phylogenetic studies emerged only recently, (e.g. Bosminidae; Kotov et al., 2009; Chydoridae; Sacherová and Hebert, 2003; Belyaeva and Taylor, 2009; Leptodoridae Xu L. et al., 2011; Polyphemidae; Xu S. et al., 2009). As a result, no other cladoceran genus has been genetically characterized as well as *Daphnia*. The rapid development of a genetic characterization of *Daphnia* prompted Hebert (1987) to predict that "the future may well also see the broader use of cladocerans as model systems to examine problems of more general theoretical importance." The extensive understanding of phylogenetic diversity and ecology in *Daphnia* has made this genus the model group of cladocerans. Expansion of research into other relatively unexplored cladoceran lineages will probably reveal a vast amount of ecological, genetic, and physiological variation.

Parallel to the development of population genetic and phylogenetic studies using molecular markers, quantitative genetic studies leveraging the clonal reproductive mode of *Daphnia*, investigated the extent to which phenotypic variation in life-history traits and fitness has a genetic basis. These studies provided links between phenotypic variation, mutation, and heritable genetic variation (e.g. Lynch, 1985; Spitze, 1993; Pfrender and Lynch, 2000). Building on this base of population- and quantitative genetic studies, novel approaches linking quantitative genetic variation with environmental context were advanced in *Daphnia*. These studies make a statistical association between the patterns of population subdivision in quantitative traits and molecular genetic markers. Excess levels of population subdivision in quantitative traits and molecular genetic markers provide evidence of natural selection and local adaptation (Spitze, 1993; Lynch et al., 1999; Morgan et al., 2001). This approach has recently been extended to examine patterns of adaptation through time by leveraging the historical legacy genotypes in diapausing embryos trapped in lake-bottom sediments (see Section 18.7) (Cousyn et al., 2001; Brendonck and De Meester, 2003; Orsini et al., 2012).

While these studies have been essential in establishing the framework for the detailed investigation of physiology and physiological mechanisms, they are best suited to describing patterns of genetic variation within and among species, and provide limited insight into mechanisms. Linking the extensive phenotypic variation in physiology and the wide range of habitats occupied by cladocerans requires new approaches that leverage modern high-throughput DNA sequencing technologies and draw on the rapidly expanding resource of genome sequences, functional genomic assays of gene regulation, and proteomic and metabolomics data. This post-genomic investigation into the tremendous phenotypic and physiological variation in cladocerans will provide insights into physiological mechanisms that were unattainable in the past.

18.3 *DAPHNIA* AS A MODEL SYSTEM FOR ECOLOGICAL AND EVOLUTIONARY PHYSIOLOGY

The field of cladoceran genomics is in the early stages of development, yet advances rapidly. A decade ago, in 2002, the first genome project for a cladoceran species (i.e. *Daphnia pulex*) was launched (see Section 18.4). This initiative aimed at creating a leading model system for ecological, ecotoxicological, and evolutionary genomics (e.g. Colbourne et al., 2011; Lampert, 2011; Miner et al., 2012). With the first draft of the *D. pulex* genome made available in 2006, a recent publication of the first annotated crustacean genome (Colbourne et al., 2011) marks the beginning of a new and promising research era, which seeks to bridge "the gap between genotype and phenotype" in an ecologically relevant framework (Dow, 2007; Gregersen, 2009). But why is *Daphnia* a useful model?

One of the most intriguing features in *Daphnia* biology is its extraordinary ability to cope with changes in the environment. In response to environmental challenges, cladocerans can produce extremely divergent phenotypes, which include altered morphologies (Tollrian and Dodson, 1999; Laforsch et al., 2004), changes in their physiology (Kobayashi and Yamagata, 2000; Paul et al., 2004) and behavior (Wiggins and Frappell, 2002; Zeis et al., 2004; Zeis et al., 2005), and switches in the reproductive mode from parthenogenetic to sexual reproduction cycles (Zaffagnini, 1987; Olmstead and LeBlanc, 2003; Decaestecker et al., 2009). Modern genomic approaches have transformed cladoceran research. Investigators working with this group of organisms are poised to tackle one of the main objectives in modern biology to identify the mechanistic basis of plastic responses to environmental change. At present, researchers explore how the developmental and physiological "program" of an organism can be modified in response to distinct environmental stimuli. However, the latter requires "first and foremost the understanding of its genomic make-up" (Aubin-Horth and Renn, 2009). Even though numerous investigations in diverse taxa have reported the existence of environmentally induced phenotypic plasticity, there has been relatively little success in identifying the empirical patterns of cellular control mechanisms that govern the plastic traits in question. This lack is mainly due a paucity of study organisms with well-developed genomic resources that are also empirical models for investigating phenotypic plasticity. Currently, knowledge on genes, genomes and their regulation is predominantly based on traditional model systems such as *Caenorhabditis elegans*, *Drosophila melanogaster*, zebrafish, and mice. As all these species are of limited relevance in their natural habitats and ecosystems, and consequently lack a significant context outside the laboratory, there is limited data available on how environmental factors may contribute to genetic and phenotypic evolution. Moreover, many genes that respond to ecological conditions have unknown functions, and information from laboratory model species may be insufficient. Consequently, ecological genomic approaches require empirical annotations of new genome sequences from a broader range of species tested under a variety of natural and ecologically relevant conditions (see Section 18.6).

As a model for ecological and evolutionary research (e.g. Lynch and Spitze, 1994; Lampert, 2011; Miner et al., 2012), *Daphnia* now bridges this gap, as it allows us to increase our understanding of the linkages between phenotypic responses and underlying genotypic and environmental effects. There are several compelling reasons why *Daphnia* is particularly well suited to this research agenda:

1. A complete genome sequence of one species is now available.
2. The data resources are publicly accessible (including genomic and cDNA libraries, microarrays, tiling arrays, genetic linkage maps, web-based bioinformatics portals, and annotation and gene expression databases).
3. The phylogenetic position of *Daphnia* (i.e. branchiopods, which are basal crustaceans) is excellent for comparative genomics.
4. Their maintenance in the laboratory is comparably cheap and easy.
5. Their large brood sizes and fast reproductive cycle are ideal for experimental genetics (with the generation time in the laboratory rivaling that of nearly all model eukaryotic systems).
6. The clonal nature of this organism provides an outstanding opportunity to study the genetic responses to environmental stimuli in a defined and constant genetic background with unlimited replication.
7. *Daphnia* are transparent throughout life, allowing studies of tissue-specific gene expression at any life stage (e.g. to changes in oxygen).
8. The extensive phenotypic diversity of this species provides ample raw material to study gene function and genome by environment interactions.
9. *Daphnia*-specific genes have been shown to respond rapidly to environmental disturbances.
10. Research is supported by a large, and expanding, community of scientists, i.e. the *Daphnia* Genomics Consortium.

At present, our understanding of the cellular processes by which plastic responses to environmental stimuli are triggered is still very limited (Aubin-Horth and Renn, 2009). Genome-wide association studies that aim at linking genes to phenotypes strongly suggest that complex traits are governed by many loci (Flint and Mackay, 2009). In *Daphnia*, most of the alternative phenotypes can be generated from the same genetic background through the perception of cues from the environment. These cues lead to modifications of gene regulation that alter the physiological state and developmental trajectory of individuals (e.g. the formation of males). The clonal character of daphniids facilitates dissection of the genetic and environmental components of plastic responses (Shaw et al., 2008, Simon et al., 2012). The remarkable ability of these organisms to cope with environmental change is associated in part with its unique gene inventory and genomic structure (see Section 18.4).

To decipher the relationship between phenotypic plasticity, and adaptive evolution, it is important to understand the genetic basis of interactions between genotype and environment (Miner et al., 2005; Pigliucci, 2005; West-Eberhard, 2005). The regulation, expression, and function of many genes are highly context dependent and are only manifested under particular environmental conditions. It is presently unclear to what extent such genetic mechanisms are species-specific. Ecological genomics aim to understand how natural populations adapt to their local environments. Daphniids are known to have colonized radically different environments on multiple occasions, with a characteristic pattern of convergence of adaptive traits linked to specific habitats (e.g. Colbourne et al., 1997). Consequently, the genetic variation found within natural populations of *Daphnia* provides an exceptional opportunity to evaluate the nature of genomic and phenotypic adaptations (e.g. De Meester et al., 2011). Comparing different *Daphnia* lineages that show habitat-specific reaction norms allows us to decipher whether similar environments can and do induce similar genetic outcomes, and whether different *Daphnia* lineages evolve the same way to meet these environmental challenges (see e.g. Colbourne et al., 1997; Pfrender et al., 2000; Scoville and Pfrender, 2010).

The recent publication of the *Daphnia* genome (see Section 18.4), coupled with a wide range of phenotypic diversity, make this organism an outstanding model system to explore the genetic basis of complex phenotypic traits, including physiology (e.g. Miner et al., 2012). It allows unprecedented insight into the responses to environmental perturbations by tackling physiological acclimation and evolutionary adaptation processes in response to the environment. Whereas earlier investigations were often constrained to hunting for single genes associated with particular phenotypes, employing functional genomic methodologies at genome-wide scales offers a more comprehensive strategy. These tools will facilitate a genuine understanding of how the environment interacts with (and shapes) the genome and how genome regulation and variation are linked to phenotypic plasticity and phenotypic evolution.

18.4 *DAPHNIA'S* ECORESPONSIVE GENOME

Modern physiological investigations will increasingly draw on a genomic perspective to discover underlying mechanisms and to understand the patterns of variation in natural populations occupying diverse environments and confronted with varied physiological challenges. Developing a genomic perspective requires a high level of genomic infrastructure, including genome sequences and functional genomic tools. Of all Cladocera (in fact, of all crustaceans), the widespread *D. pulex* (e.g. Crease et al., 2012) is the best-developed species in this area. It is the first cladoceran species to have the complete genome sequenced and annotated (Colbourne et al., 2011). A second species, *D. magna*, already a well-studied ecotoxicological model, is currently undergoing similar development. What makes the first cladoceran genome so special? Even though the *D. pulex* genome is small compared to that of other organisms, measuring only circa 200 Mb, it contains almost twice as many genes as any other arthropod yet sequenced, and has a greater number of genes than in most other eukaryotic genomes, including the human genome. So far, 30,907 protein-coding genes are predicted on the 12 chromosomes. Sequencing and assembling the waterflea's genome revealed a significant number of non-protein-coding genes including microRNAs, ribosomal RNAs, transfer RNAs, and several families of transposable elements. For a detailed discussion on the *D. pulex* genome and its potential as a genetic model, we refer to Colbourne et al. (2011), on which the current section is based (unless stated otherwise). In addition to the latter work, the authors selected 50 thematic publications (the "companion papers," by the *Daphnia* Genomics Consortium, published between 2005 and 2011), which constitute the initial exploration of the genome (see http://www.biomedcentral.com/series/*Daphnia*). The genome itself can be explored through the publicly available, interactive genome database, wFleaBase (Colbourne et al., 2005), which also includes data on the expression patterns of a large number of protein-coding genes under different experimental conditions.

Comparisons of the *D. pulex* genome with that of several insects revealed that intron sizes appear to be rather reduced in the *Daphnia* genome, with an average intron length of only 170 bp. Moreover, the intron density in the cladoceran seems similar to that observed in *Apis mellifera*, but there are at least twice as many introns per gene than in *Drosophila*, for example. As outlined by Colbourne et al. (2011), the majority (i.e. 78%) of the intron gains observed in the *D. pulex* genome appear to be unique to the *Daphnia* lineage (or perhaps to the branchiopods). That is to say, that the *Daphnia* genome is more compact with respect to the size of its introns, but not in the number of introns acquired.

Exploring the *Daphnia* gene inventory revealed a higher propensity for gene duplication and retention when compared to other organisms. Gene duplication events account for approximately half of the genes in the genome. In fact, the rate of gene duplication in *D. pulex* seems three times as high as that observed for *Drosophila* and *Caenorhabditis*, and about 30% greater than that of the human genome. As Colbourne et al. (2011) showed, the elevated number of genes in the *D. pulex* genome results not only from acquiring but also from retaining a large number of genes. The retention of duplicate genes and the neofunctionalization of these paralogues may be directly linked with the plasticity of cladocerans and their ability to respond to a large range of environmental challenges (Colbourne et al., 2011; Simon et al., 2012).

As gene duplication events are considered to play a key role in shaping genomes, they represent one of the most significant sources of evolutionary novelty. If, after duplication, selection maintains both copies of a gene, then one gene version commonly retains its original function while the other may functionally diverge; the latter is called a *paralogue* (for reviews see e.g. Innan and Kondrashov, 2010; Proulx, 2012). Most of *Daphnia*'s paralogues have evolved context-dependent functional specifications and the expression patterns diverge with the age of the gene. Whereas more recent duplicates, which have different gene sequences, showed rather indistinguishable gene expression patterns under several tested environmental conditions, more ancient duplicates revealed divergent expression patterns over time. These duplication and divergence events in various gene families represent an impressive evolutionary mechanism that enables these animals to express a selective suite of functionally distinct molecules depending on the environmental condition. The genome can thus mediate diverse cellular processes to maintain or restore cellular integrity in response to the environmental challenges faced. However, the *D. pulex* genome was also shown to contain a number of recent duplicates that have nearly identical nucleotide sequences but show rather differential gene expression patterns in response to different environmental parameters. These expression patterns may arise by integrating novel genes into a new genomic location (chromosome) or dissociating them from their previous regulatory framework (e.g. pathways).

Due to their high degree of homology, duplicated genes (and especially tandemly clustered genes) are prone to undergo genetic exchanges via unequal crossing over and unidirectional gene conversion (GC) (Hoffmann et al., 2008). In the *Daphnia* genome, 47% of all genes were revealed to contain tracts of GC; in comparison, in five different *Drosophila* species, only 12–18% of genes were shown to be affected by such genetic homogenization events. However, GC events were shown to be less common among more recent duplicates and within gene families containing only few paralogues. An exceptional example of widespread GC events in the *Daphnia* genome comprises the di-domain hemoglobin (Hb) gene family (see Section 18.5).

Even though the high number of genes in the *Daphnia* genome may indicate that these genes arose by whole-genome duplication, the number of synonymous substitutions among all pairs of duplicated genes in *D. pulex* rather suggests high and steady rates of tandem duplication events in the lineage's evolutionary history. Similar to other genomes in which new duplicates are found in clusters, the *D. pulex* genome has approximately 20% of its genes in tightly arranged groupings. Each cluster can contain up to 80 paralogues. Two gene families that have gained and retained an exceptional number of duplicated genes in *D. pulex* include the photoresponsive genes of the opsin family (Rivera et al., 2010) and di-domain Hbs (Colbourne et al., 2011), which play critical roles in coping with complex light regimes or allowing adequate responses to oxygen deprivation in aquatic environments, respectively.

The expansion of gene families in *D. pulex* was shown to be also associated with distinct metabolic processes. By analyzing the functional role of paralogues, Colbourne et al. (2011) could show the coexpansion of genes in distinct metabolic pathways (e.g. androgen-estrogen metabolism and sphingolipid biosynthesis metabolism), which indicates that the duplicated genes may be interdependent. In fact, half of the expanded metabolic genes were found to belong to a subset of seven distinct pathways, and the expression patterns of these genes codiverged according to their pathway and not according to their evolutionary history, which suggests the functional divergence of expanding genes within pathways.

Finally, a large number of genes (36%) in the *D. pulex* genome lack any detectable homology with any other species yet sequenced. Phylogenetic analyses that account for the expansion and contraction of all gene families within pancrustacean and several deuterostome genomes suggest a net increase in the number of paralogues within the lineage leading to *Daphnia*. That is to say, the vast number of lineage-specific genes appears to result from disproportionate expansions of particular gene families distinctive to this crustacean lineage and the fast divergence of most of these genes. Using whole-genome tiling arrays or EST (i.e. expressed sequence tag) sequencing, it was shown that *Daphnia*-specific genes in particular (as far as it is known) appear to play important roles in the organism's ecology. These "orphan genes" may particularly account for the plastic responses of daphniids, as they were shown to have distinct gene expression patterns that are context specific (and may thus only be expressed under particular ecological conditions). Similarly, another set of *D. pulex* transcripts showed condition-dependent gene expression patterns in response to diverse environmental challenges. These transcriptionally active regions (TARs), which have predictable exon-intron intervals, are not yet assigned to any known gene annotation models, and thus need to be functionally and structurally characterized.

18.5 THE GENETIC BASIS OF PHYSIOLOGICAL PLASTICITY: A CASE STUDY

18.5.1 *Daphnia*'s Hemoglobin Genes

Hb is found in virtually all kingdoms of living organisms and is arguably among the best-studied proteins tied to natural conditions (Weber and Vinogradov, 2001). First characterized by Hünefeld in 1840, its central role in maintaining cellular oxygen homeostasis has garnered markedly detailed molecular biological investigations since the late 1990s, including the comparative study of DNA and protein sequence, and of the molecular assemblage and functional properties of this respiratory protein across diverse lineages of animals (for reviews see Weber and Vinogradov, 2001; Hardison, 2012; Storz et al., 2013). Although Hbs are remarkably diverse in structure and function, their sequence similarity and conserved gene structures indicate that all globins described to date are descended from a shared ancestral gene (Goodman et al., 1987; Goodman et al., 1988; Hardison, 1998; Vinogradov et al., 2005). Expansion and diversification of the Hb gene family, both within and across species lineages, is frequently shown to correlate with the physiological oxygen demands that organisms face in their natural habitats (Weber et al., 2002; Storz et al., 2007; Storz and Moriyama, 2008; Storz et al., 2010; Tufts et al., 2013). Because an animal's ability to cope with environmental change is predominantly determined by its capacity to maintain performance and oxidative metabolism (e.g. Pörtner and Knust, 2007), variations in the protein structure and differential expression of distinct *hb* genes consequently represent central mechanisms that enable proper system functioning and survival. Due to the extensive background knowledge on the physiological role of Hb and its structure-function relationships, critical insights into the genetic and mechanistic bases of physiological acclimation and evolutionary adaptation processes are gained by investigating the structural variation and regulation of this protein in Cladocera.

Numerous investigations on the functional processes involved in oxygen regulation by Hb have been carried out in *Daphnia*, with initial studies dating back to the late 1940s and early 1950s (e.g. Fox et al., 1949; Fox, 1950; Green, 1956b). These aquatic poikilotherms are exposed to a wide range of environmental conditions (Lampert, 2004; Lampert and Sommer, 2007; Brede et al., 2009; Weider et al., 2010; Paul et al., 2012); both diurnal and seasonal fluctuations in oxygen and temperature are among the major abiotic challenges that *Daphnia* face. Their diurnal vertical migration (DVM) between the epilimnion and hypolimnion in deeper lakes demands a high plasticity in their molecular responses to cope with different oxygen conditions in nature within short time intervals. Both environmental hypoxia and temperature-induced mismatches between oxygen supply and demand provoke cellular oxygen deficiency in aquatic organisms (e.g. Ekau et al., 2010; Pörtner, 2010). Daphniids in particular, but branchiopods in general, show an extraordinary plasticity to cope with such environmental conditions; they have evolved various regulatory mechanisms that perceive change and compensate for substantial variation in ambient conditions (Guadagnoli et al., 2005). So far, the mechanisms by which daphniids sense changes in the amount of oxygen in the ambient environment are not entirely understood. However, some studies (e.g. Gorr et al., 2004; Gorr et al., 2006b) have revealed that cellular responses to environmental challenges in *Daphnia* are mediated via distinct regulatory elements that bind to intergenic regions and control the gene expression level of physiologically important proteins (see below).

In response to short-term changes in the ambient conditions, waterfleas can counteract oxygen deprivation via alterations in their ventilation and perfusion rates, which restore adequate oxygen supply to peripheral tissues and cells (Lamkemeyer et al., 2003; Paul et al., 2004; Pirow and Buchen, 2004). If oxygen shortage persists for a prolonged period of time (i.e. a few hours) (see e.g. Zeis et al., 2004; Becker et al., 2011a), then the microcrustacean's oxygen transport capacity is improved by the induction of Hb and even a modification of the Hb structure (Fox et al., 1951; Kobayashi and Hoshi, 1982; Tokishita et al., 1997; Pirow et al., 2001; Zeis et al., 2003; Gorr et al., 2004; Pirow et al., 2004; Gerke et al., 2011; Zeis et al., 2013). In response to severe hypoxia (approximately 2–3 kPa) (e.g. Zeis et al., 2003; Gerke et al., 2011), *Daphnia* Hb levels drastically rise by a factor of 15–20, thus notably coloring adult animals red within a single molting cycle (Kobayashi and Hoshi, 1982; Pirow et al., 2001; Zeis et al., 2003; Gerke et al., 2011). Other branchiopods, such as *Triops*, are known to show a similar response to hypoxia by modifying their Hb structure or elevating *hb* gene expression (Guadagnoli et al., 2005). Earlier investigations on the *de novo* synthesis of the respiratory protein in *D. magna* provided evidence of highly controlled and localized gene expression, with Hb synthesis sites restricted to the fat cells and the epipodite epithelia cells (Goldmann et al., 1999).

Oxygen acquisition in the hemolymph of *Daphnia* is not only adjusted by changes in the Hb concentration but also arises from an altered oxygen-binding affinity of the respiratory protein. Expression of higher-affinity Hbs in animals facing environmental change is assumed to reduce Hb synthesis costs due to an enhanced oxygen transport efficiency of these molecules in comparison to low-affinity variants (Kobayashi et al., 1994; Pirow et al., 2001). In the long term, adjustments in the cellular Hb repertoire allow daphniids to inhabit oxygen-poor water layers, thereby providing access to their respective food resources, which may ultimately increase the animals' overall fitness (Fox et al., 1951; Sell, 1998; Wiggins and Frappell, 2002; Duffy, 2010).

Several investigations on the Hb protein level have thus far revealed that changes in the oxygen-binding affinity of the multimeric Hb molecule are a direct consequence of modifications in the Hb subunit composition (Kobayashi and Takahashi, 1994; Kimura et al., 1999; Lamkemeyer et al., 2003; Zeis et al., 2003; Gorr et al., 2004; Gerke et al., 2011). Compared to vertebrates, invertebrate Hbs are known to exhibit a much broader variability in their molecular structures (Weber and Vinogradov, 2001), presumably indicating adaptations to the wider ranges of environmental conditions to which these animals are subjected in nature. Studies on the molecular structure of Hb in daphniids provide evidence that the extracellular protein is composed of multiple 31–37 kDa di-domain subunits (Dangott and Terwilliger, 1980; Ilan et al., 1982; Peeters et al., 1990; Kimura et al., 1999; Lamkemeyer et al., 2003; Zeis et al., 2003; Gerke et al., 2011). As previously demonstrated (Tokishita et al., 1997), there is remarkably low similarity between the amino acid sequences of the first and second Hb domains (approximately 24%). This suggests that *Daphnia*'s di-domain Hb subunits derive from an ancient duplication via unequal crossing over of two single-domain globin genes, which has been frequently shown for members of the Hb gene family (see Weber and Vinogradov, 2001). Each subunit consists of two heme-containing globin domains (Dangott and Terwilliger, 1980) and has a characteristic structure of a pre-A segment, eight alpha helices and five interhelical regions (Tokishita et al., 1997; Kato et al., 2001). So far, 12 (*D. pulex*) to 16 (*D. magna*) subunits are known to aggregate, forming functional macromolecular multimers (Dangott and Terwilliger, 1980; Peeters et al., 1990; Lamkemeyer et al., 2006) that are freely dissolved in the hemolymph (Pirow et al., 2001).

Recent proteomic analyses of *D. pulex* that were acclimated at different temperature and oxygen conditions revealed the differential expression of a set of seven different Hb subunits, each encoded by a separate gene locus (Gerke et al., 2011). Similar results were obtained from studies on *D. magna* (Zeis et al., 2003; Lamkemeyer et al., 2006). As demonstrated by the Gerke et al. (2011), oxygen and temperature acclimation mainly caused similar changes in the Hb subunit composition. However, further investigations under different experimental or methodical conditions are likely to facilitate the identification of further Hb subunits, which may be only expressed in a condition-specific mode.

Whereas the presence of multiple genes encoding the different Hb subunits provides the basis for alterations in the protein's structural and functional characteristics, the differential gene expression of specific sets of Hb isoforms is regulated via the binding of distinct transcription factors to the promoter regions (Kimura et al., 1999) of the respective *hb* genes. In response to hypoxia, binding of the transcription factor HIF (hypoxia-inducible factor; a protein) to hypoxia-response elements (HREs) in the upstream regulatory (promoter) regions of genes was shown to have a direct impact on the induction of multiple Hb isoforms in *D. magna* (Gorr et al., 2004). HIF-1 is a heterodimer comprised of two bHLH-PAS factors, named HIF-1α and ARNT (aryl hydrocarbon nuclear translocator), which are members of the family of basic helix loop helix-Per/ARNT/Sim (bHLH-PAS) transcription factors (Wang et al., 1995). Stability of HIF-1α is controlled by ambient oxygen conditions, while ARNT is a partner of various other bHLH-PAS proteins and is stable regardless of ambient oxygen conditions (Salceda and Caro, 1997). Stability of HIF-1α is regulated by a member of 2-oxoglutarate-dependent dioxygenase super family, named HIF prolyl hydroxylase (HIF-PHD). It was reported that the *C. elegans* prolyl hydroxylase homologue, EGL-9, plays an important role in regulation of stability of the HIF-1α homologue in response to hypoxia (Epstein et al., 2001). A proline residue within the conserved LXXLAP motif within the oxygen-dependent degradation domain (ODDD) of HIF-1α is hydroxylated by HIF-PHD in response to increased oxygen concentration in mammals. Hydroxylation of this proline residue results in degradation of HIF-1α mediated by the ubiquitin-proteasome pathway (Jaakkola et al., 2001). HIF-1α and its oxygen-dependent regulation are conserved in various organisms such as nematodes, insects, fishes, amphibians, and mammals. HIF-1α recognizes cis-acting elements called HREs in the regulatory region of target genes; the T/G/CACGTG hexanucleotide is a core sequence of HREs (Wenger et al., 2005). As to crustaceans, cDNAs encoding HIF-1α homologues were isolated more recently from grass shrimp *Palaemonetes pugio*, Dungeness crab *Cancer magister*, and *D. magna* (Li and Brouwer, 2007; Tokishita et al., 2006). These HREs are also found upstream of the *D. pulex hb* genes (Colbourne et al., 2011; Gerke et al., 2011), with analogue roles for differential isoform expression in response to disturbed oxygen homeostasis. By employing heterologous transfection studies, Gorr et al. (2004) provided evidence that the extent of hypoxia-induced differential gene expression is not only governed by the number of HREs in the intergenic regions (Kimura et al., 1999; Nunes et al., 2005) but may also involve interfering effects of other regulating elements. Thus, the role of HIF binding, the position of binding sites, and possible interactions with other transcription factors still need to be addressed.

It is therefore clear that *hb* gene expression is a complex mechanism that includes numerous regulatory components and processes, which contribute to a highly fine-tuned cellular response. Besides HIF-1-induced Hb expression, further transcriptional control elements were shown to regulate the Hb concentration in daphniids. Under normoxic conditions, Gorr et al. (2006) revealed Hb levels in *D. magna* to be predominantly under the regulatory control of methyl farnesoate or related juvenoid hormones, which bind to their respective response elements [i.e. juvenoid response elements (JREs)]. However, a recent study (Zeis et al., 2013) using *D. pulex* discovered that Hb transcript and protein levels only poorly correlate, suggesting that *Daphnia*'s Hbs are also posttranscriptionally regulated. By exploiting the exceptional knowledge available on *Daphnia*'s Hbs, further studies on this multigene family in particular, will allow the identification of major regulatory networks linked to plastic responses.

The remarkable flexibility of *Daphnia* at responding to altered oxygen availabilities and demands, by adjusting both the quantity and the quality of Hb, is assumed to be a direct consequence of their exceptional genomic characteristics. Although a number of investigations in the late 1990s (Tokishita et al., 1997; Hebert et al., 1999; Kimura et al., 1999) identified multiple *hb* genes encoding the different Hb subunits in several species of *Daphnia* (i.e. in *D. magna* and *D. exilis*), our understanding has been significantly improved by genome sequencing of *D. pulex* and *D. magna* (Colbourne et al., 2011). The draft *D. pulex* genome sequence assembly and annotation revealed 11 di-domain *hb* genes (*hb1*–*hb11*); 8 of these (*hb1*–*hb8*) are found arranged within a tandem gene cluster on chromosome 7 (Colbourne et al., 2011). Gene duplication events are considered important for shaping genomes and are significant sources of evolutionary novelty (for reviews, see Innan and Kondrashov, 2010; Proulx, 2012). *Daphnia*'s di-domain *hb* genes thus provide a vivid illustration of how paralogues can evolve context-dependent functional specifications. Exposing *D. pulex* to eight different environmental conditions and investigating the respective gene expression profiles of all di-domain *hb* genes revealed a significant age-related trend in the functional divergence among the paralogues recent duplicates, which are most similar in their gene sequences, generally showed indistinguishable gene expression patterns for the tested conditions, while more ancient duplicates within *Daphnia*'s *hb* gene family were more divergent in their expression patterns (Colbourne et al., 2011, Fig. S29). Consequently, duplication followed by functional divergence in the *hb* gene family of *Daphnia* represents an impressive evolutionary mechanism, which enables these animals to express a selective suite of functionally distinct Hb isoforms. These isoforms mediate adequate oxygen transport capacities to maintain or restore cellular integrity in response to the environmental conditions faced (Tokishita et al., 1997; Zeis et al., 2003; Gerke et al., 2011).

Although many duplicates tend to diverge in sequence and function over time, paralogous genes arranged within tandemly duplicated gene clusters are predisposed to homogenization events by unequal crossing over and unidirectional GC due to their high degree of sequence homology (Hoffmann et al., 2008). In the genus *Daphnia*, GC events are a common feature, and one example of widespread GC is found in the di-domain *hb* genes (Hebert et al., 1999; Sutton and Hebert, 2002; Colbourne et al., 2011). By shuffling nucleotide variation among related genes, these genetic exchanges result in a substantial reduction in divergence among duplicates as their sequence evolution becomes concerted. Multigene families that undergo such a concerted evolution are characterized by a higher than expected sequence similarity among paralogous genes within a species, while there is distinct divergence from the orthologous gene family in other species.

Cladocerans are an attractive target for comparative studies of Hb evolution (Hebert et al., 1999). Several investigations have explored the origin and evolution of duplicated di-domain *hb* genes and the consequences of their structural arrangements in different genera of the Daphniidae (Dewilde et al., 1999; Hebert et al., 1999; Kimura et al., 1999; Sutton and Hebert, 2002; Colbourne et al., 2011), yet virtually nothing is known from other cladoceran families. Comparative analyses of the *hb* gene clusters in two *Daphnia* lineages, *D. pulex* and *D. magna*, revealed nearly identical genomic arrangements within a 23.5-kb interval (Colbourne et al., 2011). However, *D. pulex* possesses an extra *hb* gene (i.e. *hb6*) within the tandem gene cluster, and three additional, non-clustered *hb* genes (i.e. *hb9–hb11*) are found in other genomic regions, with *hb9* being located on the antisense strand (Gerke et al., 2011). All clustered *hb* genes, plus *hb9* in *D. pulex*, are composed of seven exons that are separated by six introns. In contrast, *hb10* and *hb11* in *D. pulex* consist of six exons, with the second intron deleted from the ancestral gene structure (Colbourne et al., 2011). Similar intron losses in di-domain *hb* genes have also been reported (Kato et al., 2001) in another cladoceran species, *Moina macrocopa*. Other than the obvious absence of *hb6* from the *D. magna*, cluster, elements in synteny between the two species are seemingly preserved from a duplication history that predates the split between the *Ctenodaphnia* and *Daphnia* subgenera (Colbourne et al., 2011), including a non-coding RNA gene that interrupts the cluster between *hb4* and *hb5*. Although the *hb* gene clusters in both species are homologues due to ancestral gene duplication events, phylogenetic reconstructions of all di-domain *hb* genes in *D. pulex* and *D. magna* provided evidence that GC tracts have heavily homogenized the protein-coding regions, whereas the intergenic regions of all *hbs* show less divergence between the two *Daphnia* lineages (Colbourne et al., 2011).

In contrast to laboratory studies, investigations into *Daphnia*'s Hb function in natural populations are scarce (but, see Schwerin et al., 2010). However, different clones and populations of *Daphnia* have been shown to vary in their ability to synthesize adequate levels of Hb in response to distinct oxygen concentrations in the ambient environment (Carvalho, 1984; Weider and Lampert, 1985; Duffy, 2010). Since spatial and temporal changes in dissolved oxygen concentration may be an important selective force influencing the clonal (genotypic) composition of natural populations, investigating Hb variations and tracing the variation of individual globin genes within and among species allows the exploration of regulatory mechanisms that enable animals to adapt to different environmental conditions (see e.g. Storz et al., 2007; Storz et al., 2009). As this case study illustrates, organisms can survive and function under a wide range of conditions in nature owing to a distinct set of adaptive responses. The remarkable plasticity, characterized by differential Hb subunit expression in response to altered oxygen conditions, is not only found in cladocerans but also occurs in other branchiopods such as *Triops* (Notostraca) (e.g. Guadagnoli et al., 2005). As these authors show, Hb subunit expression is not reversed in *Triops* when conditions return to normoxia, which contrasts with hitherto studied cladocerans. In combination with the differences in Hb domains of the two *Daphnia* subgenera mentioned above, this shows that genetic expression mechanisms may differ significantly within the Branchiopoda, and even within the Cladocera, depending on the lineage.

The structural and regulatory polymorphisms of cladoceran *hbs* represent a superb model for studying oxygen-responsive genes and the mechanisms involved in their regulation. Hence, they highlight the relationship between environmental stimuli and downstream functional molecular responses. The detailed knowledge available on Hb structure-function relationships and its key role in oxygen homeostasis consequently promote a deep understanding of the cellular mechanisms that control a divergent and fine-tuned expression of context-dependent suites of molecules under environmental constraints, and ultimately provides genuine insight into the nature of acclimation and adaptation processes.

18.6 HUNTING FOR PHYSIOLOGICALLY RELEVANT GENES AND REGULATORY NETWORKS

18.6.1 Linking Gene Expression Profiles to Physiological Traits and Responses

Genome-wide expression profiling, also known as *functional genomics* or *transcriptomics*, is a powerful tool for understanding how organisms develop, function, and respond to environmental factors. Although posttranscriptional processing of mRNA and downstream regulation of protein degradation and translation can modify the effects of gene expression on organism physiology and cell function, gene expression profiling still provides important information about the genetic control of biological processes and how variation in gene expression affects phenotypic variation among individuals and among taxa (Gibson and Muse, 2009).

In the absence of global gene expression studies, the expression of candidate genes known or suspected to be involved in the biological process of interest can be assayed by northern blot or quantitative polymerase chain reaction (qPCR) methods, allowing targeted questions about gene function under different environmental conditions, mutational perturbations, or across different taxa (e.g. Scoville and Pfrender, 2010; Schwarzenberger et al., 2012). Assaying genome-wide patterns of gene expression frees researchers from the requirement of *a priori* genetic knowledge, and allows the discovery of previously unknown components important to physiological functioning. Hierarchical cluster analysis of genome-wide expression patterns reveals the coordination of gene function within pathways and provides insight into the interaction of pathways through regulatory networks. A network perspective makes it possible to infer the functional roles of unannotated genes if they share expression profiles with annotated genes. Principal component analysis of gene expression data can also reveal underlying similarities and differences in physiological responses to different kinds of perturbations (Gibson and Muse 2009). Global gene expression studies are hypothesis generating, identifying genes that can most usefully be subjected to the genetic and biochemical analysis necessary to pinpoint the molecular basis of physiological functioning. Transcriptomics data can be integrated with other genome-wide data sets, such as those generated by comparative phylogenomics, proteomics, epigenomics, and metabolomics, for a systems biology approach that works to understand holistically how (1) organisms develop, function, and respond to their environments and (2) the basis of physiological and phenotypic diversity within and between lineages. These datasets can also yield insights into the emergent properties of gene regulatory networks, thus furthering our understanding of the sorts of gene regulatory network architectures that exist and how different expression patterns have evolved (Scoville and Pfrender, 2010; Latta et al., 2012; Whitehead, 2012).

Currently, work on cladoceran genome-wide gene regulation utilizes microarray platforms developed for *D. pulex* or *D. magna*, particularly cDNA or long oligonucleotide microarrays that permit competitive hybridization of fluorescently labeled, amplified cDNA to a set of probes representing expressed genes, predicted genes, and other transcriptionally active genome features. Hybridization of two samples per array (e.g. control vs. treatment, adult vs. neonate) allows a direct comparison of gene expression; many experimental designs also permit indirect comparisons of samples across arrays, e.g. comparisons among different treatments that have each been directly compared to a common reference. As sequencing costs decrease and better bioinformatics tools are developed, inferring gene expression from direct sequencing of amplified cDNA fragments via next-generation sequencing (NGS) platforms, or RNA-Seq, will become more attractive and has the potential to overcome some of the limitations of oligonucleotide microarray platforms, such as incomplete probe sets and a limited ability to capture differential transcription of splice variants and allele-specific transcription.

An early and crucially important step in genome-wide expression studies is developing the experimental design that will best answer the research question with the limited resources available: which genotypes (e.g. species, populations, mutants, or wild-type individuals); life stages (neonate, juvenile, or reproductive adult); and environments (e.g. chemical exposure, chemical concentration, and the duration of exposure, as well as basic culture decisions about light, temperature, and feeding regimes); and whether to collect time series data as animals undergo physiological changes or to assay organisms in physiological equilibrium. Finally, decisions about replication (technical vs. biological), and, if microarray platforms are used, which samples will be directly compared and which will be compared indirectly, have important implications on the power to detect differential gene expression (Gibson and Muse, 2009). After exposures, sacrifice of animals, RNA extraction, preparation of labeled cDNA, microarray hybridization, and collection of fluorescence intensity data, the data must be analyzed with appropriate statistical methods to identify those genes with differential expression across comparisons.

Finally these data must be further analyzed for biological interpretation and visualization of significant genes within pathways and regulatory networks. Functional analysis is based on homology to annotated genes in databases such as the Gene Ontology (GO) database, which assigns each gene to an inferred biological process, molecular function, and cellular component. Other databases infer membership in gene pathways [e.g. the Kyoto Encyclopedia of Genes and Genomes (KEGG)] (reviewed in Tipney and Hunter, 2010). It is worth noting that about 36% of *Daphnia* genes show no homology to other sequenced organisms (Colbourne et al., 2011); these lineage-specific genes are therefore excluded from homology-based analyzes. After the categorization of significant genes, a variety of methods exist to test whether sets of significant genes are enriched for GO categories (e.g. Blast2GO), KEGG pathways [Database for Annotation, Visualization and Integrated Discovery (DAVID)], or previously identified gene lists [e.g. Gene Set Enrichment Analysis (GSEA)]. These higher order analyses allow the biologist to make sense of and visualize the millions of data points that comprise a microarray experiment (Tipney and Hunter, 2010).

18.6.2 Case Studies in Environmental Functional Genomics

The study of genome-wide responses to environmental conditions is a young but rapidly growing field and many of the early applications involve exposure to environmental stressors and toxicants. Environmental biologists are exploring ways to incorporate the results of genome-wide expression studies in biomonitoring efforts, seeking to identify gene expression patterns specific to conditions, and using functional genomics to illuminate the physiological mechanisms underlying these responses so that better estimates of individual effects and better predictions about the population and community impacts of exposures can be made. Environmental stressors often do not affect just a single gene or gene pathway, but can instead have cascading effects, ranging from a highly specific response (e.g. to lowered oxygen concentrations) to general stress responses (e.g. to altered food quality and quantity), while simultaneously involving many aspects of organismal functioning. Therefore, such stressor exposures can provide a suitable model for exploring the intersections of multiple physiological pathways using a systems biology approach (Eads et al., 2008; Shaw et al., 2008). *Daphnia* is often used as a surrogate species to understand the genomic responses to environmental stressors that are important factors in human health and well-being. So far, tolerance limits and regulatory processes that allow organisms (including humans) to cope with pollutants in the short-term span of an individual's lifetime and in the longer time frame of evolutionary change, are still poorly understood. *Daphnia* constitute an excellent model for physiological functional genomics because a long history of environmental and ecological research has provided a wealth of physiological and toxicological information that can be exploited when designing genomics experiments. Its importance as an ecotoxicological model make it imperative that the research community rapidly and collaboratively use these relatively new genomic approaches to understand *Daphnia*'s baseline physiology and the ways that individuals and populations function in the face of environmental perturbations (Shaw et al., 2008). In recognition of its relevance to understanding basic biological processes and as a sensitive organism for environmental monitoring, just recently *D. pulex* was listed as one of the few selected model organisms for biomedical research at the National Institute of Health (http://www.nih.gov/science/models/).

David et al. (2011) explored the global transcriptional response to genotoxicants, a study that also compared the sensitivity of gene expression profiling to the commonly used comet assay for DNA damage. *D. magna* neonates and 7-day-old adults were exposed to a mixture of benzo(a)pyrene and sodium dichromate. Gene expression was found to be a more sensitive indicator of genotoxicant effect than the comet assay even though the comet assay was negative, significant differential gene expression was detected. Gene expression differed in neonate and adult *Daphnia*, and adults showed a stronger transcriptional response to genotoxicants, with three times as many genes showing differential expression (106 genes in adults vs. 34 genes in neonates). Adults had higher constitutive expression of DNA repair and oxidative stress genes than neonates under unexposed conditions, suggesting that adults are better prepared to deal with DNA damage and, perhaps, that adults activate these repair and protective pathways in response to their higher metabolic rates and the associated production of reactive oxygen species.

Exposed adults also showed induction of DNA repair genes, while neonates did not. Additionally, in adults oxidative stress genes and *hb* genes were induced, but chitinase and other proteases were downregulated upon exposure. Since these catabolic genes are implicated in molting associated with reproduction, suppressed gene expression is consistent with reproductive deficits and population level consequences after genotoxicant exposure. Unexposed neonates showed a greater expression of genes involved in polysaccharide binding, pattern binding, and chitin binding than unexposed adults, consistent with greater cuticle formation in developing neonates. While neonates did not show induction of DNA repair pathways, they did show induction of oxidative stress response genes. This study demonstrated the potential of gene expression-based biomarkers for genotoxicant exposure and showed that gene expression assays can be more sensitive than the comet assay following nonlethal exposure levels that nevertheless can induce reproductive effects. It also showed that gene expression profiles can differ dramatically between adult and neonate *Daphnia*. Adults were more transcriptionally responsive to the compounds tested and may therefore be more appropriate subjects for bioassays than the neonates traditionally used in toxicology testing.

Another functional genomics study explored the effects of ibuprofen on gene expression in *D. magna* (Heckmann et al., 2008). The observed effects of ibuprofen had parallels to its effects on invertebrates, including early disruption of eicosanoid metabolism. In *Daphnia*, there were also effects on the endocrine system, juvenile hormone metabolism, and oogenesis, thus providing strong evidence for a mechanism underlying observed reproductive effects in ibuprofen-exposed daphniids. By comparing the expression profile of ibuprofen exposure to that of exposure to a very different stressor, cadmium (Cd) (Poynton et al., 2007), the authors were also able to identify common transcriptional patterns associated with a general stress response in *D. magna*. These patterns included induction of glycolytic, proteolytic, homeostatic, and heat shock protein genes, and downregulation of genes involved with energy metabolism and translation. The authors suggested that the ibuprofen-specific transcriptional response could be useful as an early indicator of significant population ibuprofen exposure with risk to *Daphnia* reproductive biology.

The *Daphnia* transcriptional response to Cd has also been investigated in *D. pulex*. Shaw et al. (2007) used a first-generation cDNA microarray to identify transcriptional responses in adult animals exposed for 48 h to sublethal concentrations of Cd. The study revealed many Cd-responsive genes to be associated with molting and metal detoxification, including the metallothionein gene. Importantly, this led to the discovery of additional novel metallothionein genes in the genome, with low homology to any previously known metallothionein genes. This paper highlights the utility of microarray studies, not only for determining environmental toxicogenomic signatures and exploring physiological pathways involved in stressor response but also for gene discovery.

A more recent study of the Cd response in *D. magna* is notable for combining microarray-based transcriptional profiling with metabolomics data to investigate the role of nutrient uptake, metabolism, and decreased energy reserves in chronic toxicity (Poynton et al., 2011). Mass spectrometry of hemolymph of *Daphnia* exposed for 24 h to sublethal Cd concentrations revealed disruptions in both amino acids and fatty acids, while differential gene expression analysis showed reduced expression of genes encoding digestive enzymes (consistent with a role for impaired nutrient absorption in Cd toxicity), as well as upregulation of genes for cuticle proteins and oxidative stress response. This study focused on the initial phase Cd response but revealed metabolic changes that could lead to decreased fitness during prolonged exposure. Assaying gene transcription and metabolite composition in parallel has allowed researchers to explore the hypothesized roles of gene expression changes and more fully describe the physiological mechanisms of Cd toxicity.

Another recent paper addressed the effects of the algal toxin microcystin on *D. pulex* (Asselman et al., 2012). Microcystins are produced by the harmful cyanobacteria *Microcystis aeruginosa*, an organism associated with toxic algal blooms. Microarray analysis of gene expression supports a characteristic stress response and a role for energy budget disturbances in microcystin toxicity. *Daphnia* fed a diet contaminated with *Microcystis* for 16 days show differential gene expression of ribosomal genes, oxidative phosphorylation genes, protein export genes, and genes associated with mitochondrial function. Gene expression patterns suggest that microcystin toxicity may involve upregulation of protein synthesis and energy production pathways. *Microcystis* also affected expression of several paralogous gene clusters, and some paralogous genes were variably regulated. Taken together, these results suggest physiologically important roles for duplicated genes and indicate that paralogues can assume different functional roles. The authors caution that statistical methods tailored to analyze the overabundance of duplicated genes in *Daphnia* genomes might be necessary to accurately assess functional gene and pathway enrichment in these species.

18.7 PALEOGENOMICS

An emerging area of research, with a huge potential for the study of cladoceran physiology, is the field of paleogenetics and, more recently, paleogenomics. Diapause in the Cladocera allows the study of populations through time and an assessment of historical changes to fitness and genes. Paleogenomics, i.e. the study of extinct genomes, can be directly applied to extinct populations through the use of cladoceran resting stages. The method can be combined with other techniques that have been used to study adaptability of populations through time. For example, the term *resurrection ecology* is widely used (Kerfoot et al., 1999) for the study of reviving specimens from diapausing or dormant stages and comparing them to present populations in order to examine fitness and gene responses under certain (experimental) conditions. This approach allows an analysis of neutral versus adaptive variation of the genes and an increased understanding of evolutionary processes, particularly those operating on adaptive genetic variation (see Orsini et al., 2013). The ecological significance lies in the fact that microevolutionary changes at the population level can be studied in response to a wide range of human-induced impacts such as climate change, eutrophication, landscape changes, and pollution. As Hairston and De Meester (2008) noted in a recent review "The animals hatched from decades-old sediments are living organisms with genotypes representative of past populations. Evolutionary changes in plastic characters such as physiology and behavior can thus be reconstructed by comparing the performance of genetic lineages obtained from different sediment layers in a common set of environmental conditions. This approach has successfully been applied for a growing range of traits and environmental changes."

As all branchiopods can produce dormant stages, the genetic responses to environmental change over time can be studied for a wide range of taxa in the form of genetic archives in freshwater lake sediments deposited as egg banks (Brendonck and De Meester, 2003). Yet, despite its wide occurrence over a wide phylogenetic range, nearly all the research on cladoceran resurrection ecology and paleogenetics gravitates around *Daphnia*, in which the dormant embryos are encased in ephippia (as for all Anomopoda, but not for all Cladocera). A few studies exist on other genera, e.g. *Ceriodaphnia*, which is still within the daphniid family (Reinikainen and Åhlén, 2012). The extent of the time dimension in which the genes can be studied through resurrection ecology is determined by the viability of the ephippia, which mostly stretches to decades and rarely to centuries (Frisch et al., unpubl. data), yet paleogenomics allows us to examine traits (through genes) even further, to centuries and even millennia (e.g. Frisch et al., subm.; Mergeay et al., 2007; Morton et al., 2012). Besides being ecologically relevant, the importance of such techniques lies in the possibility to examine the genetic adaptivity in plastic responses, the evolution of phenotypic plasticity (physiological and behavioral), and rapid evolutionary changes under strong selection pressures induced by humans or biological interactions (e.g. Decaestecker et al., 2007; Hairston and De Meester, 2008). Hairston and De Meester (2008) reviewed two case studies on human-induced selection pressures that strongly alter aquatic ecosystems (eutrophication and fish introduction) and elicit rapid genetic responses in *Daphnia*. Such studies provide hard evidence for the evolution of cladoceran physiological traits, as well as the range (plasticity) and limits of acclimation and adaptation (evolutionary constraints).

Therefore, resurrection ecology and paleogenomics provide a powerful tool for understanding the genetic basis of phenotypic plasticity (Orsini et al., 2013). Using a combination of these methods, Orsini et al. (2012) examined the genome of *D. magna* in space and time, and correlated the response of genes clusters to three major selection pressures (land-use, fish predation, and parasitism) that are known to elicit a rapid (i.e. microevolutionary) genetic and phenotypic response in cladocerans. Such studies illustrate that the genetic background of cladoceran physiology remains a vast *terra incognita*. "Of the 164 genes identified, only 88 (54%) could be annotated and 86 of these have a putative function based on the homology with other organisms. 109 genes were linked to general stress response; of these, 53 (49%) returned unknown function" (Orsini et al., 2012). The authors found repeated local adaptation by the populations and also identified the mechanisms for rapidly responding to the three selection pressures (through homology) as genes involved in proteolysis, metabolic processes, neuronal development, and transcriptional and translational regulation. This provides hard evidence that even human impact can elicit a strong and rapid genetic response in cladocerans and that genotypes in the industrialized world have gone through this selection and are not representative of the original "wild" genotypes. Similarly, De Meester et al. (2011) have shown that temperature changes can quickly result in genetic adjustments in *D. magna*, even over very short time spans (months to years). Human-induced impacts on freshwater ecosystems might (irreversibly) change cladoceran physiology by driving selection. In a recent study, Frisch et al. (unpubl. data) were able to hatch live *D. pulicaria* from ephippia derived from a core taken in South Central Lake (U.S.) from as far back as 1418 AD, the oldest resurrected cladoceran to date. The authors compared the genetic structure of populations through time in relation to the lake's eutrophication, using a combination of resurrection ecology, paleogenomics, and historical records to investigate the response of extinct versus extant populations in relation to eutrophication levels (throughout the lake's history). As De Meester et al. (2011) illustrated for temperature, the work by Frisch et al. (unpubl. data) suggests that cladocerans can adapt quickly to environmental conditions and that such changes can be irreversible. Loci with a signature of directional selection were related to historic phosphorus levels in the lake and an adaptive shift occurred to higher phosphorus concentrations. Resurrected clones from the fifteenth century grew faster under low phosphorus conditions and showed higher phosphorus retention when compared to the twenty-first century population. Therefore, after agricultural expansion around the lake (around the 1920s), *D. hyalina* lost the physiological ability to retain phosphorus at low levels due to clonal selection and genetic erosion, in contrast to the "extinct" populations from before eutrophication. This shows that over the course of a few hundred years (which is a very short geological time, considering the age of the lineage), a cladoceran species can undergo significant and rapid genetic shifts that alter its tolerance to the environment.

The key to using paleogenomics to its full potential lies in identification of the genes that determine ecologically relevant trait variation (Miner et al., 2012). Identifying the large number of unknown genes in cladocerans (*Daphnia*; Colbourne et al., 2011) is a task for the future, and would help to identify the important genes involved in determining plasticity. Other issues also remain challenging, such as the extraction and sequencing of ancient DNA from ephippia, and the use of homologies to identify unknown genes and gene functions. Furthermore, all resurrection ecology and paleogenomics is currently focused on Daphniidae and therefore on the planktonic filter feeders in the aquatic ecosystem. No studies have yet been carried out on benthic/littoral (e.g. Chydoridae) or predatory cladocerans (e.g. Leptodoridae). Well studied in paleolimnology and better preserved in the sedimentary record (e.g. Frey, 1962), benthic/littoral chydorids have a particular potential for paleogenomics, yet the level of genetic knowledge is nowhere near that of daphniids. In time, our knowledge in all of the aforementioned fields will undoubtedly expand (*vide infra*). For further reading on paleogenomics or resurrection ecology, we refer the reader to recent reviews by Brendonck and De Meester (2003), Hairston and De Meester (2008), and Orsini et al. (2013).

18.8 FUTURE DIRECTIONS: EXPLORING PHYSIOLOGICAL VARIATION WITH FUNCTIONAL GENOMICS

In order to understand physiology in an evolutionary and ecological context, the similarities and differences among individuals within and among populations and among species must be understood and exploited. Physiological responses can vary, but little is known about the mechanisms underlying variation or the evolutionary forces generating, maintaining, or constraining such variation. Comparative approaches to identify apparently (i.e. phenotypically) universal physiologies and those that vary among populations and species will be facilitated by the further development of cladocerans as models for physiological genomic research. Importantly, working in species with well-characterized ecologies will allow us to ask how physiologies and genetic mechanisms vary with environmental factors. Comparative functional genomics can test whether or not similar physiologies have been achieved by regulating the same genetic pathways in the same way or through the independent evolution of substantially different mechanisms, and can help to unravel the basis of qualitative and quantitative differences in physiological responses (Whitehead, 2012).

Ultimately, researchers will want to know the genetic and genomic differences responsible for the physiological variation and associated gene expression variation that we observe. Gaining this knowledge will require the incorporation of evolutionary genetics and genomics [e.g. quantitative trait locus (QTL) mapping, expression QTL (eQTL) mapping, genome-wide association studies, and comparative genomics], molecular genetics (e.g. RNAi), and other molecular and bioinformatics methods that can identify variation in regulatory elements associated with gene regulation (e.g. experimental identification of transcription factor-binding sites and sites of epigenetic modification, and computational searches for conserved regulatory motifs and their variation within and among populations). The integration of physiology and functional genomics with ecological and evolutionary genomic analysis has the potential to enrich many aspects of our understanding of *Daphnia*.

As outlined in this chapter, the field of (physiological) genomics is rapidly advancing and Cladocera are on the frontline. Studies on *Daphnia* genomics facilitate the identification of ecologically relevant genes that enable context-dependent responses. By linking different research disciplines, such as evolutionary ecology, physiology, molecular biology, and genomics, current research on *Daphnia* aims to increase our understanding of organisms as integrated units of biological organization shaped by the context of the ecological and the history of evolution. Studying adaptive traits using such a systems-level approach will ultimately provide a radically new perspective on the nature of genetic constraints and the evolution of adaptation and speciation. Consequently, unraveling key adaptive responses in *Daphnia* will positively trigger a cascade of subsequent studies (in *Daphnia* and other organisms) that may have a great impact on forecasting the fate of keystone species, and therefore ecosystem stability, under growing environmental challenges. But where do we go from here? There can be no doubt that our knowledge of the genetic mechanisms of cladoceran physiology has rapidly increased since the late twentieth century. With the first cladoceran genome available (Colbourne et al., 2011), we are now on the brink of a new era. Because of their ecological and genetic characteristics, cladocerans can play an important role in exploring relevant biological questions. Containing model organisms with well-studied ecologies and a major role in ecosystems, future study of the Cladocera has the potential to "revolutionize genomic research and enable[s] us to focus on a number of outstanding questions" (De Meester et al., 2011). For example, by approaching cladoceran physiology and biology from a functional genomics perspective, we can advance our understanding of gene evolution processes and the genetic basis of phenotypic plasticity. Research on cladoceran genomics is also relevant in the context of biodiversity and conservation because of the presence of human-induced, irreversible genetic changes in natural populations. As cladocerans play a pivotal ecological role, their system functioning and fitness are important indicators for freshwater ecosystem health (e.g. Rapport et al., 1998). Therefore, genetic methods combined with physiology provide a powerful tool for monitoring (and perhaps forecasting) ecosystem stability.

18.8.1 Trends and Suggestions for Future Research

First, new state-of-the-art high-throughput techniques such as NGS and RNA interference (RNAi) have now become more accessible. Promising new methods include advances in bioinformatics, which allow the manipulation of large datasets, such as gene expression profiles. RNAi is a powerful tool to investigate gene function by knocking down distinct target genes. Recently, RNAi has been developed for cladocerans (i.e. the *Dll* gene in *D. magna*) (Kato et al., 2011). But, there is more. Cladocera also appear to be promising model systems for epigenetics (*D. magna*; Vandegehuchte et al., 2010a,b; Harris et al., 2012; Robichaud et al., 2012). Sex determination and sexual reproduction in cladocerans may be epigenetically controlled and DNA methylation patterns are responsive to toxicants and heavy metals (Harris et al., 2012). However, this field is still young, especially with regards to the potential of cladocerans as model systems. Examining protein expression profiles under different environmental conditions falls under the area of *proteomics*, and the *D. pulex* genome provides a major step forward in this area of research as well (Fröhlich et al., 2009; Schwerin et al., 2009; Zeis et al., 2009). The same can be said for *metabolomics*, which examines cladoceran metabolites in response to different environmental conditions or stressors (e.g. toxins; Taylor et al., 2010). For paleogenomics, new techniques are expected to increase the molecular resolution of ancient DNA samples (Orsini et al., 2013).

Second, a major challenge for the future is the characterization and exploration of function (and structure) of the circa 13,000 (forming approximately 36% of the genome) unknown "*Daphnia*-specific" protein-coding genes (Colbourne et al., 2011). With more crustacean genomes expected to be sequenced in the near future (Stillman et al., 2008), the evolution and the function of these genes will become clearer.

At present, these still-to-be-annotated genes are not known to occur in other crustacean or cladoceran lineages. Until now, nearly all the genetic research has focused on Daphniidae (in fact, on *Daphnia*), leaving the majority of Cladocera untouched. Other groups take up equally important, yet different roles in aquatic ecosystems, such as the predatory haplopods and the benthic/littoral chydorids (the most speciose group in the Cladocera) (Forró et al., 2008).

Third, and finally, the majority of physiological studies in cladocerans have hitherto focused on species acclim(atiz)ation and adaptation processes in response to (natural or human-induced) environmental stimuli. In order to understand the mechanistic basis of physiological responses (which include phenotypic plasticity), holistic approaches are needed. This integration can only be achieved by analyzing physiological traits in a functional genomics context, and *Daphnia* is an ideal model to do so.

GLOSSARY TERMS

Acclimation Phenotypic change in an individual organism as a physiological response to its environment.

Adaptation Phenotypic change in a population due to natural selection favoring alleles that confer increased fitness.

Allele Any variant of a gene at a particular locus.

Annotation (of genes/genomes) The process of assigning biological information, such as the identity and function of genes and other genetic elements, to positions in a genome sequence. See also *functional genomics.*

Complementary DNA (cDNA) DNA synthesized from an mRNA template by reverse transcription of RNA to DNA followed by DNA polymerization.

cDNA library The complete mRNA collection of an organism or sample, which is stored in a host (e.g. bacterial plasmid, yeast, and bacterial artificial chromosomes) as cDNA. See also *transcriptomics.*

Enrichment analysis A statistical test of whether a set of genes (e.g. a set of differentially expressed genes) contains more members of a genetic pathway or gene class than would be expected by chance.

Epigenetics Modification to the genome produced through environmental effects (e.g. DNA methylation) that in some cases comprise inheritable factors.

eQTL mapping A quantitative trait loci mapping study for which the continuous variable of interest is the expression level of one or more genes; eQTLs are regions in the genome (loci) that regulate the expression of other areas in the genome. See also *QTL mapping, gene expression.*

Expression profiles See *gene expression.*

Functional genomics The study of genomes through the characterization of the effect of sequences on the phenotype at any level (molecular, cellular, tissue, and whole organism). See also *transcriptomics.*

Gene cluster A set of genes located in the same genomic region. See also *synteny.*

Gene conversion (GC) Replacement of one allele by another through non-homologous recombination.

Gene duplication The duplication of a gene, resulting in the presence of two or more copies in the same genome. It can be caused by errors in replication, by recombination, or by retrotransposons. The *Daphnia* genome is characterised by a high number of gene duplications, resulting in many paralogues of the same gene, as in the *hb* gene cluster for example. See also *paralogues* and *tandemly repeated genes.*

Gene-environment interaction Environmental effects on genetic function that contribute to phenotype.

Gene expression The conversion of the information encoded in genes (as a DNA sequence) into RNA. Examination of gene expression patterns and profiles relates to the study of genes that are differentially regulated and transcribed under certain conditions of interest (e.g. metallothionines under metal pollution stress in *Daphnia*).

Genetic linkage The probability of two genes being inherited together in a non-random fashion due to their physical proximity on a chromosome, or patterns of linkage disequilibrium.

Genome All inheritable genetic material that is present in an organism (i.e. DNA).

Genome assembly The process of uniting individual DNA sequences generated by a sequencing project into long (ideally chromosome-length) DNA sequences.

Genomics The study of the structure and function of genomes.

Genotoxicants DNA-damaging mutagenic agents (e.g. UV rays and chemicals).

Heterologous transfection Insertion of an extraneous DNA fragment into a genome. Often used to study gene expression and the functional consequences of genetic variants in a non-native genetic background.

Hypoxia-inducible factor 1 (HIF-1) Oxygen-responsive hetero dimeric transcription factor that mediates cellular responses to tissue hypoxia by the regulation of distinct gene expression events.

Homologue Identity by descent; homologous genes have the same origin but have evolved independently in different lineages through time. See also *paralogue, orthologue.*

Hypoxia-response element (HRE) Located in the regulatory region (i.e., promoter region) of hypoxia-responsive genes; binding to transcription factors such as HIF-1 induces gene expression of distinct target genes.

Intergenic Regions of the genome that fall outside protein-coding regions.

Isoforms Variants of a protein expressed by the same gene or closely related genes. Can be the result of alternative splicing.

Metabolomics The study of all metabolites in an organism and the chemical reactions they are involved in (structures, pathways, etc.).

Microarray Technique for the quantification of whole-genome transcription levels. All nucleic acids extracted from a sample are labeled with fluorescent dyes and hybridized to a microarray chip. The targets for binding on the chip are short oligonucleotides matching unique regions of the genome. Fluorescence levels are proportional to RNA abundance and gene expression.

Messenger RNA (mRNA) Molecules that convey the genetic sequence of an actively transcribed gene from the nucleus to the ribosome, where it specifies the sequence of amino acids to be synthesized into a protein.

Model organism An organism that is amenable to scientific investigation and of which the biology can provide insights on other species. *Daphnia* can reproduce clonally, is easy to culture, and is a model organism for the study of genotype-environment interactions, epigenetics, phenotypic plasticity, and ecological adaptation.

Next-generation sequencing (NGS) A variety of methods of high-throughput, low-cost sequencing, which works by generating thousands of short sequencing reactions in parallel. These methods produce a high number of short (length depends on the platform, e.g. 50–400 bp) sequences (or "reads").

Orthologue (or orthologous gene) Genes that share a common ancestor and are present in different species (not in the same genome). See also *homologue, paralogue.*

Paleogenomics The study of ancient/extinct genomes. In the case of cladocerans, paleogenomics is carried out on embryos from ephippia that are stored in lake sediments.

Paralogue (or paralogous gene) Genes that share a common ancestor and are present in the same genome. The result of gene duplication events. See also *gene duplication, homologue, orthologue*.

Phenotypic plasticity Condition-dependent development that allows an organism to respond to environmental change by altering its phenotype

Phylogeny A hypothesis about the evolutionary relationships among species or populations

Polygenic trait (quantitative trait) A trait that is affected by several to many genes

Proteomics The study of the complete set of proteins present in a given sample (cell, tissue, organism).

Quantitative polymerase chain reaction (qPCR) Quantitative or real-time PCR is a technique that allows the quantification of the abundance of selected DNA sequences in a sample. Can be used on cDNA to quantify gene expression levels and on genomic DNA to measure gene copy number variations.

QTL mapping Quantitative trait loci mapping, a statistical association of genetic polymorphisms (markers) in a genome and variation of a quantitative phenotypic trait. It allows quantification of the relative genetic contribution of each locus to the total observed variation in the trait. See *eQTL*.

Quantitative genetics The study of the effect of genetic variation on quantitative or continuous traits (e.g. body size). See also *QTL mapping*.

Regulatory elements DNA elements that regulate or modify the expression of distinct genes or gene expression events. Includes transcription factors, promotors, and transcription factor binding sites.

Resurrection ecology The study of the ecology of populations or species revived from the past by breaking dormancy (e.g. resurrection ecology of *Daphnia* by hatching populations from ephippia that are stored in lake sediments). See *paleogenomics*.

RNA interference (RNAi) RNA-directed post-transcriptional mechanism for gene downregulation. Short (approximately 20-nt long) RNA fragments guide a multiprotein complex towards specific transcripts, causing their degradation.

RNA sequencing (RNA-Seq) Technique for the identification and quantification of whole-genome transcription levels. The total RNA extracted from the sample is sequenced using high-throughput methods. The number of times that a sequence is read is proportional to its expression level. Can identify polymorphisms and splice variants with base-pair resolution. See *cDNA, NGS*.

Synteny Conservation of the position of genes along the chromosome in different lineages of organisms. For example, in multiple *Daphnia* species, genes in the *hb* gene cluster are arranged in the same order. See *gene cluster*.

Tandem gene cluster Gene cluster generated by a series of tandem duplications. For example the *hb* genes in *Daphnia*. See *tandemly repeated genes, gene cluster*.

Tandemly repeated genes Duplicated genes located in adjacent positions in the genome. Typically the result of small local duplications. See also *gene duplication*.

Transcriptionally active regions Portions of the genome that show evidence of being transcribed under certain conditions. Can be associated with genes in the genome or with transcription of non-coding sequences (e.g. regions encoding short and long RNAs elements).

Transcriptomics The study of gene expression through the measurement of all RNA species [e.g. mRNA, small interfering RNA (siRNA), and microRNA, i.e. the transcriptome] in a given sample. See also *microarray* and *RNA sequencing*.

Conclusions: Special Traits of Cladoceran Physiology

Some representatives of the Cladocera occur in enormous quantities, but other species are rare and confined to narrow ecological niches, potentially partly due to their physiological adaptations.

Cladocera consume small algae, bacteria, and detritus, thus forming a link between primary production and the predators that eat them. A high intensity of feeding is characteristic of Cladocera: the ingested food stays in the intestine for several minutes up to about half an hour. During this time, the Cladocera extract sufficient material to support their intensive reproduction. They also synthesize abundant chitin.

Particular body constituents of Cladocera are dynamic in relation to the season, the chemical composition of their food and of the environment, and starvation. Data are now available on the quantity of glycogen, chitin, lipids, slimes, and pigments [e.g. carotenoids, melanin, hemoglobin (Hb)] consumed and their metabolism, and on the dynamics of particular elements (including phosphorus, calcium, strontium, sodium, and iron). Cladocera may accumulate physiologically important substances, those of no such importance, and toxicants.

Cladocera consume foods that have low quantities of nitrogen and phosphorus and an excess of carbon. Thus, they have to get rid of excess carbon. Only a very small amount of the consumed lipids are transformed. With excessive food consumption (i.e. *luxury consumption*), the proportion of ingested food that is digested may be low. Cladocera consume a lot of chlorophyll; during digestion, only a little is slightly reduced to pheopigments and it is not used in the construction of Hb. Starvation is accompanied by profound changes in the chemical composition of the body.

Respiration occurs through the body surface of cladocera. Littoral, and especially bottom-living species, exist under oxygen stress but are surrounded by abundant food resources, whereas planktonic species usually enjoy a good oxygen environment, but their food resources are periodically either abundant or scarce.

Littoral and bottom-dwelling species live under conditions of hypoxia and anoxia. Cladocera frequently contain Hb but can manage well without it. Studies on the impact of xenobiotics on respiration have been reviewed.

The heart is myogenic and acetylcholine is the inhibitory transmitter substance (Postmes et al., 1989). Two kinds of blood cells have now been identified (with reference to *Daphnia*), one of which performs phagocytosis. The heart rate is similar for all Cladocera species, except for *Ilyocryptus*, in which the heart rate is 2–3 times lower. The heart rate noticeably increases with various disturbances and drops only before death. Heart arrest occasionally occurs without harming the individual

Physiology of the Cladocera.
DOI: http://dx.doi.org/10.1016/B978-0-12-396953-8.00029-X

and it can resume beating without an obvious reason. Adhesion of blood cells (which normally drift in the blood) to the surface of organs may occur for unknown physiological reasons.

Excretion is principally carried out by paired maxillary glands. The main final product of protein metabolism is ammonia (NH_3), accompanied by smaller amounts of urea. Depending on their structure, xenobiotics are transformed within the body and may be passed into the next generations. Within certain limits, Cladocera can support homeostasis of various processes within a dynamic environment, including homeostatic maintenance of chemical constituents and osmotic pressure.

The principal types of osmotic regulation are hyperosmotic regulation (in fresh water), hypoosmotic regulation (in marine *Penilia*), and amphiosmotic regulation. Xenobiotics affect osmotic regulation.

Cladocera live for a week up to several months, depending on the species and the environmental conditions. Body length increments occur between molts. Mechanical damage is repaired by regeneration, which produces either normal or abnormal structures. The current status of investigations into senescence and mortality is described.

Cladocera are remarkable for their predominantly parthenogenetic reproduction, which is sometimes interrupted by bisexual reproduction. The appearance of males depends on environmental conditions, food composition, and hormonal regulation. With the exception of *Leptodora*, there are no larval stages in the free-living period of their life cycle.

The trajectories of littoral and pelagic species differ greatly but little is known about this. The complete muscle system has only once been described, with reference to *Daphnia*. Therefore, comparative investigations of the muscles of littoral species are urgently needed. Published studies are available on immobilization, fatigue, stress, and the impact of xenobiotics on locomotion.

Cladocera can discern polarized and colored light, can orientate themselves in space, and manifest endogenous rhythms.

Cladocera also exhibit complex behavior, which differs in particular species. This includes migration, swarming, akinesis, and escape behavior. Published data are available on disturbances to their behavior by xenobiotics.

Ecological aspects considered include the physiological background of limits to physiological factors, synergism and antagonism among factors, lipid pathways from algae via Cladocera to fish, and environmental conditioning by Cladocera.

In Cladocera, some functions are facultative, such as the presence and formation of Hb, beating of the heart, and vision by means of their eyes. Nevertheless, without the latter, some cladocerans do quite well. Hb generally contributes to the activity of cladocerans. However, it may be absent or its function may be blocked by carbon monoxide, and such specimens continue successful life activities. It must be remembered that in vertebrates the role of Hb is obligate (with the exception of one group of fish: the Chaenichtyids).

Cladocera cannot be considered to be physiologically primitive.

In Cladocera, the absolute value of particular parameters is not species specific but depends on clonal composition and on previous acclimation.

Seemingly chaotic (although within certain limits), dynamic combinations of dozens of Cladocera species live in an environment containing hundreds of dynamic multidirectional factors. Each specimen simultaneously perceives all of these factors with its sense organs. Due to such a dynamic environment and their physiological individuality, particular species can form resultant population peaks, whose position in time and their absolute value may differ greatly in different years.

Initially, toxic and inhibitory pollutants are diluted and may stimulate various vital processes. Their combination with sex hormone-like pesticides can cause multidirectional effects and lead to disturbances in the natural balance.

The available data on cladoceran physiology mostly concern daphnids living in open water; bottom-living and littoral Cladocera have been little studied. The latter species live under conditions of hypoxia and anoxia with an excess of food (organic debris at all stages of decomposition); thus, their investigation may reveal specific adaptations that differ from those characteristic of open-water species. In addition, some issues of cladoceran physiology are well studied (e.g. some aspects of respiration, osmoregulation, and neurosecretion), while others are still awaiting investigation.

Cladocera exhibit physiological radiation, as seen in their various functions. Planktonic Cladocera are normoxic; bottom-living ones prosper in hypoxic or anoxic conditions. Sometimes, physiological radiation is formed in the background of conservative morphology, for example in species of the genus *Moina*, some of which are freshwater species and some of which live in saline lakes.

Let everything that has breath praise the Lord.
Psalm 150.6.

References

Abe, T., Saito, H., Niikura, Y., Nakano, Y., 2001. Embryonic development assay with *Daphnia magna*: application to toxicity of aniline derivatives. Chemosphere. 45 (4–5), 487–495.

Abrusán, G., 2004. Filamentous cyanobacteria, temperature and *Daphnia* growth: the role of fluid mechanics. Oecologia. 141, 395–401.

Acharya, K., Kyle, M., Elser, J.J., 2004. Biological stoichiometry of *Daphnia* growth, an ecophysiological test of the growth rate hypothesis. Limnol. Oceanogr. 49 (1), 656–665.

Achuthankutty, C.T., Shrivastara, Y., Mahamre, G.G., Goswami, S.C., Madhupratap, M., 2000. Parthenogenetic reproduction of *Diaphanosoma celebensis* (Crustacea: Cladocera): influence of salinity on feeding, survival, growth and neonate production. Marine Biol. 137 (1), 19–22.

Adam, C., Garnier-Laplace, J., Baudin, J.P., 2002. Bioaccumulation of 110MAg, 60Co, 137Cs and 54Mn by the freshwater crustacean *Daphnia magna* from dietary sources *Scenedesmus obliquus* and *Cyclotella meneghiniana*. Water Air Soil Pollut. 136, 125–146.

Adamowicz, S.J., Gregory, T.R., Marinone, M.C., Hebert, P.D.N., 2002. New insights into the distribution of polyploid *Daphnia*: the Holarctic revisited and Argentina explored. Molec. Ecol. 11, 1209–1217.

Adamowicz, S.J., Petrusek, A., Colbourne, J.K., Witt, J.D.S., Hebert, P.D.N., 2009. The scale of divergence: a phylogenetic appraisal of intercontinental allopatric speciation in a passively dispersed freshwater zooplankton genus. Mol. Phylogenet. Evol. 423–436.

Adema, D.M.M., 1978. *Daphnia magna* as a test animal in acute and chronic toxicity tests. Hydrobiologia. 59 (2), 125–134.

Agar, W.E., 1920 (reprinted 1963). The genetics of a *Daphnia hyalina* hybrid during parthenogenesis. J. Genet. 10, 303–330.

Agar, W.E., 1930. A statistical study of regeneration of two species of Crustacea. J. Exp. Biol. VII. 4, 349–369.

Agrawal, M.K., Bagchi, D., Bagchi, S., 2001. Acute inhibition of protease and suppression of growth in zooplankter, *Moina macrocopa*, by *Microcystis* blooms collected in Central India. Hydrobiologia. 464, 37–44.

Agrawal, M.K., Bagchi, D., Bagchi, S., 2005a. Cysteine and serine protease-mediated proteolysis in body homogenate of a zooplankter, *Moina macrocopa*, is inhibited by the toxic cyanobacterium, *Microcystis aeruginosa* PCC7806. Comp. Biochem. Physiol. Part B. 141 (1), 33–41.

Agrawal, M.K., Zitt, A., Bagchi, D., Weckesser, J., Bagchi, S.N., von Elert, E., 2005b. Characterization of proteases in guts of *Daphnia magna* and their inhibition by *Microcystis aeruginosa* PCC 7806. Environ. Toxicol. 20 (3), 314–322.

Ahlgren, G., Lundstedt, L., Brett, M., Fosberg, C., 1990. Lipid composition and food quality of some freshwater phytoplankton for cladoceran zooplankters. J. Plankton Res. 12 (4), 809–818.

Akkanen, J., Kukkonen, J.V.K., 2003. Biotransformation and bioconcentration of pyrene in *Daphnia magna*. Aquat. Toxicol. 64 (1), 53–61.

Aksenova, E.I., Bogucharskova, G.I., Zozulina, M.M., 1969. Role of phyto- and zooplankton in the diet of dominant cladocerans of the Lower Don. Gidrobiologicheskii Zhurnal. 5 (5), 41–48.

Aladin, N.V., 1978. Hypoosmotic regulation in the marine cladoceran *Penilia avirostris*. Zhurnal Evolutsionnoi biokhimii i fiziologii. 14 (2), 599–601.

Aladin, N.V., 1979. Morpho-Physiological Adaptations of Marine Cladocerans. Abstract, Candidate's Dissertation. Zoological Institute of the Russian Academy of Sciences, Leningrad, p. 23.

Aladin, N.V., 1981. Osmoregulatory abilities of Cladocera and prognostication their fauna in water bodies with changing salinity. IV s'ezd Vessoyuznogo gidrobiologcheskogo obshchestva, Kiev, 4, 3–4.

Aladin, N.V., 1982a. Salinity adaptations and osmoregulation abilities of the Cladocera. 1. Forms from open seas and oceans. Zoologicheskii Zhurnal. 61 (3), 341–351.

Aladin, N.V., 1982b. Salinity adaptations and osmoregulation abilities of the Cladocera. 2. Forms from Caspian and Aral seas. Zoologicheskii Zhurnal. 61 (4), 507–514.

Aladin, N.V., 1982c. Salinity adaptations and osmoregulation abilities of the Cladocera. 3. Forms from brackish and fresh water. Zologicheskii Zhurnal. 61 (6), 851–859.

Aladin, N.V., 1983. Amphiosmotic regulation in euryhaline cladoceran *Moina mongolica* Daday. Gidrobiologicheskii Zhurnal. 19 (1), 76–81.

Aladin, N.V., 1991. Salinity tolerance and morphology of the osmoregulation organs of Cladocera with special reference to Cladocera from the Aral Sea. Hydrobiologia. 225, 291–299.

Aladin, N.V., 1994. Adaptations to Salinity in Ostracoda and Branchiopoda. Extended Abstract, Doctor's Dissertation. Zoological Institute of the Russian Academy of Sciences, St. Petersburg, p. 45.

Aladin, N.V., 1996. Salinity adaptations in Ostracoda and Branchiopoda. Proceedings of the Zoological Institute, Russian Academy of Sciences, p. 265. 206 (in Russian).

Aladin, N.V., Plotnikov, I.S., 1985. Microcryoscopic investigation of the fluid from the maxillary gland of *Daphnia magna* Straus at acclimation to water with different salinity. Gidrobiologicheskii Zhurnal. 21 (4), 62–65.

Aladin, N.V., Potts, W.T.W., 1995. Osmoregulatory capacity of the Cladocera. J. Comp. Physiol. B.164671–164683.

Aladin, N.V., Valdivia Vilar, R.S., 1987. Microcryoscopic investigation of fluid from eggs and embryos of Cladocera. Gidrobiologicheskii Zhurnal. 23 (2), 93–97.

Alberdi, J.L., Santamaria, N.O., Hernández, D.A., 1990. Effects of deltamethrin and ethanol on survival, and mechanical response of *Daphnia spinulata*. Environ. Contam. Toxicol. 45 (2), 266–271.

Alekseev, V.R., 2010. Physiological and molecular biological mechanisms underlying diapause in aquatic invertebrates. Ontogenez. 41 (2), 83–93.

Alonzo, F., Gilbin, R., Bourrachot, S., Floriani, M., Morello, M., Garnier-Laplace, J., 2006. Effects of chronic internal alpha irradiation on physiology, growth and reproductive success of *Daphnia magna*. Aquat. Toxicol. 80 (3), 228–236.

Alstad, N.E.W., Skardal, L., Hessen, D.O., 1999. Effect of calcium concentration on the calcification of *Daphnia magna*. Limnol. Oceanogr. 44 (8), 2011–2016.

Altermatt, F., Ebert, D., 2007. Genotype specific competitive ability does not correlate with infection in natural *Daphnia magna* populations. PLoS One. 2, e1280.

Altshuler, I., Demiri, B., Xu, S., Constantin, A., Yan, N.D., Cristescu, M.E., 2011. An integrated multi-disciplinary approach for studying multiple stressors in freshwater ecosystems: *Daphnia* as a model organism. Integr. Comp. Biol. 51 (4), 623–633.

Andersen, T., Hessen, D.O., 1991. Carbon, nitrogen, and phosphorus content of freshwater zooplankton. Limnol. Oceanogr. 36 (4), 807–814.

Andersen, T.H., Tjørnhøj, R., Wollenberger, L., Slothuus, T., Baun, A., 2006. Acute and chronic effects of pulse exposure of *Daphnia magna* to dimethoate and primicarb. Environ. Toxicol. Chem. 25 (5), 1187–1195.

Anderson, B.G., 1932. The number of pre-adult instars, growth, relative growth, and variation in *Daphnia magna*. Biol. Bull. 42 (5), 81–98.

Anderson, B.G., 1933. Regeneration in the carapace of *Daphnia magna*. Biol. Bull. 64 (1), 70–85.

Anderson, B.G., 1944. The toxicity thresholds of various substances found in industrial wastes as determined by the use of *Daphnia magna*. Sewage Work. J. 16 (6), 1156–1165.

Anderson, B.G., Brown, L.A., 1930. A study of chitin secretion in *Daphnia magna*. Physiol. Zool. III (4), 485–493.

Anderson, T.R., Hessen, D.O., 2005. Threshold elemental ratios for carbon versus phosphorus limitation in *Daphnia*. Freshw. Biol. 50, 2063–2075.

Anderson, T.R., Hessen, D.O., Elser, J.J., Urabe, J., 2005. Metabolic stoichiometry and the fate of excess carbon and nutrients in consumers. Am. Nat. 165 (1), 1–15.

Andrew, T.E., Herzig, A., 1984. The respiration rate of the resting eggs of *Leptodora kindtii* (Focke 1844) and *Bythotrephes longimanus* (Leydig 1869) at environmentally encountered temperatures. Oecologia. 64 (2), 241–244.

Andronnikova, I.N., 1996. Structural and Functional Organization of Zooplankton in Lake Ecosystems of Different Trophic Status. Nauka, S.-Peterburg, p. 189. [in Russian].

Andrushaitis, G.P., 1972. Influence of pesticides on the environment. Latv. PSR Zinatn. Akad. Vestis. 4, 44–49.

Angel, M.V., 1967. A histological experimental approach to neurosecretion in *Daphnia magna*. In: Stutinsky, F. (Ed.), Neurosecretion. Springer, Berlin, Heidelberg, pp. 229–237.

Angeler, D.G., Martin, S., Moreno, J.M., 2005. *Daphnia* emergence: a sensitive indicator of a fire-retardant stress to temporary wetlands. Environ. Int. 31, 615–620.

Anisimov, V.N., Bakeeva, L.E., Egormin, P.A., Filenko, O.F., Isakova, E.F., et al., 2008. A derivative of plastoquinone addressed to mitochondria as a means interrupting the senescence program. Biokhimiya. 73 (12), 1655–1670.

Armitage, R.B., Landau, L.M., 1982. The effects of photoperiod and temperature on growth and reproduction of *Daphnia ambigua*. Comp. Biochem. Physiol. 71A, 137–140.

Artemov, N.M., Mitropolitanskaya, R.L., 1938. On presence of acetylcholine-like substances in the nervous tissue and of cholinesterase in hemolymph of crustaceans. Byulleten eksperimentalnoi biologii i meditisyny. 5 (4), 382–385.

Arts, M.T., Sprules, W.G., 1987. Energy reserves of three zooplankton species from two lakes with different metal concentrations. Can. J. Fish. Aquat. Sci. 44 (2), 458–466.

Arts, M.T., Ackman, R.G., Holub, B.J., 2001. "Essential fatty acids" in aquatic ecosystems: a crucial link between diet and human health and evolution. Can. J. Fish. Aquat. Sci. 58, 1320137.

Arts, M.T., Evans, M.S., Robarts, R.D., 1992. Seasonal patterns of total energy reserve lipids of dominant zooplankton crustaceans from a hyper-eutrophic lake. Oecologia. 90, 560–571.

Arts, M.T., Robarts, R.D., Evans, M.S., 1993. Energy reserve lipids of zooplanktonic crustaceans from an oligotrophic saline lake in relation to food resources and temperature. Can. J. Fish. Aquat. Sci. 50 (11), 2404–2420.

Arts, M.T., Robarts, R.D., Evans, M.S., 1997. Seasonal changes in particulate and dissolved lipids in a eutrophic prairie lake. Freshw. Biol. 38 (3), 525–537.

Asselman, J., De Coninck, D.I.M., Glaholt, S., Colbourne, J.K., Janssen, C.R., Shaw, J.R., et al., 2012. Identification of pathways, gene networks, and paralogous gene families in *Daphnia pulex* responding to exposure to the toxic cyanobacterium *Microcystis aeruginosa*. Envir. Sci. Tech. 46, 8448–8457.

Ataev, G.L., Coustau, C., 1999. Cellular response to *Echinostoma caproni* infection in *Biomphalaria glabrata* strains selected for susceptibility/resistance. Dev. Comp. Immunol. 23, 187–198.

Aubin-Horth, N., Renn, S.C., 2009. Genomic reaction norms: Using integrative biology to understand molecular mechanisms of phenotypic plasticity. Mol. Ecol. 18, 3763–3780.

Auld, S.K.J.R., Scholefield, J.A., Little, T.J., 2010. Genetic variation in the cellular response of *Daphnia magna* (Crustacea: Cladocera) to its bacterial parasite. Proc. R. Soc. B. 277, 3291–3297.

Auld, S.K.J.R., Edel, K.H., Little, T.J., 2012a. The cellular immune response of *Daphnia magna* under host-parasite genetic variation and variation in initial dose. Evolution. 66, 3287–3293.

Auld, S.K.J.R., Graham, A.L., Wilson, P.J., Little, T.J., 2012b. Elevated haemocyte number is associated with infection and low fitness potential in wild *Daphnia magna*. Funct. Ecol. 26, 434–440.

Auld, S.K.J.R., Hall, S.R., Duffy, M.A., 2012c. Epidemiology of a *Daphnia*-multiparasite system and its implications for the Red Queen. PLOS One. 7, e39564.

Auld, S.K.J.R., Penczykowski, R.M., J. Housley Ochs, Grippi, D. C., Hall, S.R., Duffy, M.A., submitted. Variation in costs of parasite resistance among natural host populations.

Auld, S.K.J.R., Wilson, P.J., Little, T.J., In press. Rapid change in parasite infection traits over the course of an epidemic in a wild host-parasite population. Oikos. DOI: 10.1111/j.1600-0706.2013.00720.x

Avtsyn, A.P., Petrova, T.P., 1986. Resistance of the alimentary canal to enteropathogenic NAG-vibrios. Byulleten eksperimentalnoi biologii i meditisyny. 102 (9), 342–345.

Ayroles, J.F., Carbone, M.A., Stone, E.A., Jordan, K.W., Lyman, R.F., Magwire, M.M., et al., 2009. Systems genetics of complex traits in *Drosophila melanogaster*. Nat. Genet. 41, 299–307.

Back, R.C., Watras, C.J., 1995. Mercury in zooplankton of northern Wisconsin lakes: taxonomic and site–specific trends. Water Air Soil Pollut. 80 (1–4), 931–938.

Baer, K.N., Owens, K.D., 1999. Evaluation of selected endocrine disrupting compounds on sex determination in *Daphnia magna* using reduced photoperiod and different feeding rates. Bull. Environ. Contam. Toxicol. 62 (2), 214–221.

Baillieul, M., Blust, R., 1999. Analysis of the swimming velocity of cadmium-stressed *Daphnia magna*. Aquat. Toxicol. 44 (4), 245–254.

Baillieul, M., De Wachter, B., Blust, R., 1998. Effect of salinity on the swimming velocity of the water flea *Daphnia magna*. Physiol. Zool. 71 (6), 703–707.

Bainbridge, R., 1961. Migrations, The Physiology of Crustacea, vol. 2. Academic Press, New York-London. 431–453.

Baird, D.J., Barber, I., Bradley, M., Calow, P., Soares, A.M. V.M., 1989. The *Daphnia* bioassay: a critique. Hydrobiologia. 188/189, 403–406.

Balayla, D.J., Moss, B., 2004. Relative importance of grazing on algae by plant-associated open-water microcrustacea (Cladocera). Arch. Hudrobiol. 161 (2), 199–224.

Baldwin, W.C., LeBlanc, G.A., 1994a. Identification of multiple steroid hydroxylases in *Daphnia magna* and their modulation by xenobiotics. Environ. Toxicol. Chem. 13 (7), 1013–1021.

Baldwin, W.C., LeBlanc, G.A., 1994b. In vivo transformation of testosterone by phase I and II detoxification enzymes and their modification by 20-hydroxyedysone in *Daphnia magna*. Aquat. Toxicol. 29, 103–117.

Baldwin, W.C., LeBlanc, G.A., 1996. Expression and induction of an immunochemically related class of glutathione S-transferases in *Daphnia magna*. Comp. Biochem. Physiol. 113B (2), 261–267.

Baldwin, W.S., Bailey, R., Long, K.E., Klaine, S., 2001. Incomplete ecdysis is an indicator of ecdysteroid exposure in *Daphnia magna*. Environ. Toxicol. Chem. 20 (7), 1564–1569.

Baldwin, W.S., Graham, S.E., Shea, D., LeBlanc, G.A., 1997. Metabolic androgenization of female *Daphnia magna* by the xenoestrogen 4-nonylphenol. Environ. Toxicol. Chem. 16 (9), 1905–1911.

Baldwin, W.S., Milan, D.L., LeBlanc, G.A., 1995. Physiological and biochemical perturbations in *Daphnia magna* following exposure to the model environmental estrogen diethylstylbestrol. Environ. Toxicol. Chem. 14 (6), 945–952.

Balseiro, E.G., Modenutti, B.E., Queilmlinos, C.P., 1997. Nutrient recycling and shifts in N:P ratio by different zooplankton structures in a South Andes lake. J. Plankton Res. 19 (7), 805–817.

Banta, A.M., 1921. An eyeless daphnid, with remarks on the possible origin of eyeless cave animals. Science. 53, 462–463.

Banta, A.M., 1925. A thelytokopus race of Cladocera in which pseudo-sexual reproduction occurs.

Zeitschrift für induktive Abstammungs und Vererbungslehre. 40 (1/2), 28–41.

Banta, A.M., Brown, L.A., 1922. More eyeless Cladocera. Am. Nat. 56, 568–571.

Banta, A.M., Brown, L.A., 1924. Rate of metabolism and sex determination in Cladocera. Proc. Soc. Exper. Biol. Med., N. Y. 22, 77–79.

Banta, A.M., Brown, L.A., 1929. Control of sex in Cladocera. I. Crowding the mothers as a means of controlling male production. Physiol. Zool. 2 (1), 80–92.

Banta, A.M., Brown, L.A., 1930. Control of sex in Cladocera. VI. By treatment with drugs. Physiol. Zool. 3 (1), 48–55.

Banta, A.M., Brown, L.A., 1939. Control of male and sexual egg production. In: Banta, A.M. (Ed.), Studies on the Physiology, Genetics and Evolution of Some Cladocera. Carnegie Institute, Washington (DC), pp. 106–129.

Banta, A.M., Wood, T.R., 1937. The accumulation of recessive physiological mutations during long-continued parthenogenesis. Genetics. 22, 183–184.

Barashkov, G.K., 1972. Comparative Biochemistry of Algae. Pishchepromizdat, Moscow, p. 336. (in Russian).

Barata, C., Baird, D.J., Amat, F., Soares, A.M.V.M., 2000. Comparing population response to contaminants between laboratory and field: an approach using *Daphnia magna* ephippial egg banks. Funct. Ecol. 14 (4), 513–523.

Barata, C., Baird, D.J., Markish, S.J., 1998. Influence of genetic and environmental factors on the tolerance of *Daphnia magna* Strauss to essential and non-essential metals. Aquat. Toxicol. 42 (2), 115–137.

Barata, C., Baird, D.J., Mitchell, S.E., Soares, A.M.V.M., 2002a. Among- and within-population variability in tolerance to cadmium stress in natural populations of *Daphnia magna*: implications for ecological risk assessment. Environ. Toxicol. Chem. 21 (5), 1058–1064.

Barata, C., Markich, S.J., Baird, D.J., Soares, A.M.V.M., 2002b. The relative importance of water and food as cadmium sources to *Daphnia magna* Straus. Aquat. Toxicol. 61, 143–154.

Barata, C., Solanyan, A., Porte, C., 2004. Role of B-esterases in assessing toxicity of organophosphorus (chlorpiryfos, malathion) and carbamate (carbofuran) pesticides to *Daphnia magna*. Aquat. Toxicol. 66 (2), 125–139.

Barata, C., Varo, I., Navarro, J.C., Arun, S., Porte, C., 2005. Antioxidant enzyme activities and lipid peroxidation in the freshwater cladoceran *Daphnia magna* exposed to redox cycling compounds. Comp. Biochem. Physiol. C. 140 (2), 175–186.

Barber, I., Baird, D.J., Calow, P., 1994. Effect of cadmium and ration level on oxygen consumption, RNA concentration, and RNA-DNA ratio in two clones of *Daphnia magna* Straus. Aquat. Toxicol. 30, 249–258.

Barros, P., Fidalgo, M.L., Soares, A.M.V.M., 2000. Sub-lethal effects of *Microcystis aeruginosa* on *Daphnia magna* after an acute exposure. Verhandl. Internat. Vereinig. Limnol. 27, 2107–2110.

Barry, M.J., 2002. Progress toward understanding the neurophysiological basis of predator-induced morphology in *Daphnia pulex*. Physiol. Biochem. Zool. 75 (2), 179–186.

Barry, M.J., Meehan, B.J., 2000. The acute and chronic toxicity of lanthanum to *Daphnia carinata*. Chemosphere. 41, 1669–1674.

Baudouin, M.F., Ravera, O., 1972. Weight, size, and chemical composition of some freshwater zooplankters: *Daphnia hyalina* (Leydig). Limnol. Oceanogr. 17 (4), 645–649.

Bäumer, C., Pirow, R., Paul, R.J., 2002. Circulatory oxygen transport in the water flea *Daphnia magna*. J. Comp. Physiol. B. 172 (4), 275–286.

Baylor, E.R., 1942. Cardiac pharmacology of the cladoceran, *Daphnia*. Biol. Bull. 83 (2), 165–172.

Baylor, E.R., Smith, F.E., 1953. The orientation of Cladocera to polarized light. Am. Nat. 87 (833), 97–101.

Baylor, E.R., Smith, F.E., 1957. Diurnal migration of plankton crustaceans. In: Sheer, B.T. (Ed.), Advances in Invertebrate Physiology. Univ. of Oregon, pp. 21–35.

Beaton, M.J., 1988. Genome Size Variation in the Cladocera. M.Sc. Thesis, University of Windsor, Windsor, Ontario, Canada, p. 97.

Beaton, M.J., Hebert, P.D.N., 1989. Miniature genomes and endopolyploidy in cladoceran crustaceans. Genome. 32, 1048–1053.

Beaton, M.J., 1995. Patterns of Endopolyploidy and Genome Size Variation in *Daphnia*. Ph. D. Thesis, University of Guelph, Guelph, Ontario, Canada, p. 156.

Beaton, M.J., Hebert, P.D.N., 1994a. Patterns of DNA synthesis and mitotic activity during the intermoult of *Daphnia*. J. Exp. Zool. 268, 400–409.

Beaton, M.J., Hebert, P.D.N., 1994b. Variation in chromosome numbers of *Daphnia* (Crustacea, Cladocera). Hereditas. 120, 275–279.

Beaton, M.J., Hebert, P.D.N., 1997. The cellular basis of divergent head morphologies in *Daphnia*. Limnol. Oceanogr. 42 (2), 346–356.

Beaton, M.J., Hebert, P.D.N., 1999. Shifts in postembryonic somatic ploidy levels in *Daphnia pulex*. Hydrobiologia. 394, 29–39.

Becker, C., Boersma, M., 2005. Differential effects of phosphorus and fatty acids on *Daphnia magna* growth and reproduction. Limnol. Oceanogr. 50 (1), 388–397.

Becker, C., Boersma, M., 2007. Effects of essential fatty acids on the reproduction of a generalist herbivore. J. Plankton Res. 29 (5), 463–470.

Becker, D., Brinkmann, B.F., Zeis, B., Paul, R.J., 2011a. Acute changes in temperature or oxygen availability induce ROS fluctuations in *Daphnia magna* linked with fluctuations of reduced and oxidized glutathione, catalase activity and gene (haemoglobin) expression. Biol. Cell. 103, 351–363.

Becker, D., Paul, R.J., Zeis, B., 2011b. Cellular stress responses in the microcrustacean *Daphnia* due to acute and chronic environmental changes. Abstract book. IXth International symposium on Cladocera, Verbania, p. 2.

Bednarska, A., Ślusarczyk, M., 2011. Effect of cyanobacteria on eggs abortion in *Daphnia pulicaria* at various thermal conditions. IXth International symposium on Cladocera, Verbania, p. 2.

Behning, A., 1941. Cladocera of the Caucasus. Gruzmedgiz, Tbilisi, p. 384. (in Russian).

Beim, A.M., Lavrentieva, I.F., 1981. Investigation of functional state of the digestive system of Cladocera. Problemy klinicheskoi i eksperimentalnoi enterologii. 1, 65–70.

Beim, A.M., Dorzhieva, A.N., Shistareva, E.V., 1994. Acid-basic homeostasis in the intestinal tube of Cladocera. Kislotno-osnovnoi i temperaturnyi gomeostaz: fisziolohiya, biokhimiya i klinika. Syktyvkar, 42–46.

Beim, A.M., Kodratieva, I.A., Sukhovich, L.K., 1970. Whether the cardiotest of *Daphna magna* is specific? Materialy VI mediko-biologicheskoi konferentsii posvyashchennoi 5—letiyu Karelskoi ASSR. Petrozavodsk, p. 199.

Bekker, E.I., Kotov, A.A., Taylor, D.J., 2012. A revision of the subgenus *Eurycercus* (*Euirycercus*) Bair, 1843 emend. nov. (Cladocera, Eurycercidae) in the Holarctic with description of a new species from Alaska. Zootaxa. 3206, 1–40.

Bekker, J.M., Krijsman, B.J., 1951. Physiological investigations into the heart function of *Daphnia*. J. Physiol. 115 (3), 249–257.

Beklemishev, V.N., 1923a. Preliminary investigations on survival of Entomostraca in solutions of salts I. Izvestiya nauchno-issledovatelskogo instituta i Biologicheskoi stantsii pri Permskom gosudarstvennom universitete 2, 2, 153–163.

Beklemishev, V.N., 1923b. On the process of extinction of Entomostraca in solutions of salts. Trudy pervogo Vserossiyskogo syezda zoologov, anatomov i gistologov, 1922. Petrograd, p. 142.

Beklemishev, V.N., 1924. Preliminary investigations on survival of Entomostraca in solutions of salts II. Izvestiya nauchno-issledovatelskogo institute i Biologicheskoi stantsii pri Permskom gosudarstvennom universitete 3, 6, 205–214.

Belayev, G.M., 1950. Osmoregulatory ability of Entomostraca of continental water bodies. Trudy Vsesoyuznogo gidrobiologicheskogo obshchestva. 2, 194–313.

Bell, A.M., Aubin-Horth, N., 2010. What can whole genome expression data tell us about the ecology and evolution of personality? Philos. Trans. R. Soc., B. 365, 4001–4012.

Belyaeva, M., Taylor, D.J., 2009. Cryptic species within the *Chydorus sphaericus* complex (Crustacea: Cladocera) revealed by molecular markers and sexual stage morphology. Mol. Phylogenet. Evol. 50, 534–546.

Bengtsson, G., Hansson, L.A., Montenegro, K., 2004. Reduced grazing rates in *Daphnia pulex* caused by contaminants: implications for trophic cascades. Environ. Toxicol. Chem. 23 (11), 2641–2648.

Bennett, M.D., 1972. Nuclear DNA content and minimum generation time in herbaceous plants. Proc. R. Soc. Lond. B. 181, 109–135.

Benzie, J.A.H., 2005. Cladocera: The Genus *Daphnia* (including *Daphniopsis*). Guides to the identification of the microinvertebrates of the continental waters of the World (Zooplankton Guides) 21, Kenobi, Gent; Backhuys, Leyden, p. 376.

Berges, J.A., Ballantyne, J.S., 1991. Size scaling of whole-body enzyme activities in aquatic crustaceans. Can. J. Fish. Aquat. Sci. 48 (12), 2385–2394.

Berglind, R., 1985. The effect of cadmium on ALA-D activity, growth and haemoglobin content in the water flea, *Daphnia magna*. Comp. Biochem. Physiol. 80C (2), 407–410.

Berglind, R., 1986. Combined and separate effect of cadmium, lead and zinc on ALA-D activity, growth and hemoglobin content in *Daphnia magna*. Environ. Toxicol. Chem. 5 (11), 989–995.

Bergtrom, G., Laufer, H., Rogers, R., 1976. Fat body: a site of hemoglobin synthesis in *Chironomus thummi* (Diptera). J Cell Biol. 69, 264–274.

Berner, D.B., Nguen, L., Nguy, S., Burton, S., 1991. Photoperiod and temperature as indices of gamogenesis in a dicyclic population of *Scapholeberis armata* Herrick (Crustacea: Cladocera: Daphniidae). Hydrobiologia. 225, 269–280.

Bianchini, A., Wood, C.M., 2003a. Mechanism of acute silver toxicity in *Daphnia magna*. Environ. Toxicol. Chem. 22 (6), 1361–1367.

Bianchini, A., Wood, C.M., 2003b. Physiological effects of chronic silver exposure in *Daphnia magna*. Comp. Biochem. Physiol. C. 133 (1–2), 137–145.

Bianchini, A., Wood, C.M., 2008. Sodium uptake in different life stages of crustaceans: the water flea *Daphnia magna* Strauss. J. Exp. Biol. 211, 539–547.

Bianchini, A., Bowler, K.C., Brauner, C.J., Gorsuch, J.W., Kramer, J.R., Wood, C.M., 2002a. Evaluation of the effect of reactive sulfide on the acute toxicity of silver (I) to *Daphnia magna*. Part 2: Toxicity results. Environ. Toxicol. Chem. 21 (6), 1294–1300.

Bianchini, A., Grossell, M., Gregory, S.M., Wood, C.M., 2002b. Acute silver toxicity in aquatic animals is a function of sodium uptake rate. Environ. Sci. Technol. 36 (8), 1763–1766.

Bianchini, A., Rouleu, C., Wood, C.M., 2005. Silver accumulation in *Daphnia magna* in the presence of reactive sulfide. Aquat. Toxicol. 72 (4), 339–349.

Bidder, G., 1929. Geotropism and antennae. Nature. 123 (3108), 799.

Biesinger, K.E., Cristensen, G.M., 1972. Effects of various metals on survival, growth, reproduction, and metabolism of *Daphnia magna*. J. Fish. Res. Board Can. 29 (12), 1691–1700.

Bilkovic, D.M., Lehman, J.T., 1997. Lipid concentration and size variation of *Bythotrephes* (Cladocera: Cercopagidae). J. Great Lakes Res. 23, 149–159.

Billiard, G., 1925. La daphnie (*Daphnia pulex*) consideree comme reactif biologique de la toxicite des liquides humoraux. Comptes rendus hebdomadaires des la Societe de Biologie, Paris 92, 1352–1354.

Billiard, G., 1926. La daphnie (*Daphnia pulex*) reactif biologique de la toxicite urinaire. Sa courbe de resistance vitale. Comptes rendus hebdomadaires des la Societe de Biologie, Paris, I, 13–14.

Billiard, G., Perrot, J., 1926. Les variations de la toxite urinaire au cours de qualques maladies. Albuminuries (*Daphnia pulex* reactif). Comptes rendus hebdomadaires des la Societe de Biologie, Paris, I, 15–16.

Binder, G., 1929. Das Muskelsystem von *Daphnia*. Lotos. 77, 30–32.

Binder, G., 1931. Das Muskelsystem von *Daphnia*. Intern. Revue Hydrobiol. 26, 54–111.

Biotechnics of *Daphnia* culture at fish farms. 1958. Moscow 42 pp. (in Russian).

Birge, E.A., 1898. Plankton studies on Lake Mendota. II. Trans. Wisc. Acad. Sci. Arts Lett. 11, 274–448.

Birge, E.A., Juday, C., 1922. The inland lakes of Wisconsin. The Plankton I. Wisconsin geological and natural history survey, bulletin 64, scientific series13, 1–222.

Bishop, J.G., Dean, A.M., Mitchell-Olds, T., 2000. Rapid evolution in plant chitinases: Molecular targets of selection in plant-pathogen coevolution. Proc. Natl. Acad. Sci. U.S.A. 97, 5322–5327.

Bittner, K., Ebert, D., Rothhaupt, K.O., 2002. Ecological interactions of the microparasite *Caullerya mesnili* and its host *Daphnia galeata*. Limnol. Oceanogr. 47, 300–305.

Black, A.R., 1993. Predator-induced phenotypic plasticity in *Daphnia pulex*: life history and morphological responses to *Notonecta* and *Chaoborus*. Limnol. Oceanogr. 38 (5), 986–996.

Blaustein, L., 1997. Non-consumptive effects of larval Salamander on crustacean prey: can eggs detect predators? Oecologia. 110 (2), 212–217.

Blažka, P., 1966. The rate of crude protein, glycogen and fat in the individual steps of the production chain. In: Hrbaček J., (Ed.), Hydrobiological Studies Prague, I, 395–409.

Boavida, M.J., Heath, R.T., 1984. Are the phosphatases released by *Daphnia magna* components of its food? Limnol. Oceanogr. 29 (3), 641–645.

Bodar, C.W.M., van der Sluis, I., van Montfort, J.C.P., Voogt, P.A., Zandee, D.I., 1990a. Cadmium resistance in *Daphnia magna*. Aquat. Toxicol. 16, 33–40.

Bodar, C.W.M., van Donselaar, E.G., Herwig, H.J., 1990b. Cytopathological investigations of digestive tract and storage cells in *Daphnia magna* exposed to cadmium and tributyltin. Aquat. Toxicol. 17, 325–338.

Bodar, C.W.M., Voogt, P.A., Zandee, D.I., 1990c. Ecdysteroids in *Daphnia magna*: their role in moulting and reproduction and their levels upon exposure to cadmium. Aquat. Toxicol. 17 (4), 339–350.

Boersma, M., Kreutzer, C., 2002. Life at the edge: is food quality really of minor importance at low quantities. Ecology. 83 (9), 2552–2561.

Boersma, M., Vijverberg, J., 1995. The significance of nonviable eggs for *Daphnia* population dynamics. Limnol. Oceanogr. 40 (7), 1215–1224.

Bogatova, I.B., 1962. Lethal ranges of oxygen content, temperature and pH for some representatives of the family Chydoridae. Zoologicheskii Zhurnal. 41 (1), 58–62.

Bogatova, I.B., 1963. Cultivation of *Daphnia* and *Chydorus sphaericus* at the Vyg Fish Hatchery. Trudy vsesoyuznogo naucho-issled. Instituta prudovogo rybnogo khozyaistva 12, 169–177.

Bogatova, I.B., 1980. Fish-cultural Hydrobiology. Pishchevaya promyshlennost, Moscow, p. 168. (in Russian).

Bogatova, I.B., 1992. Industrial propagation of the Cladocera in the USSR. In: Sovremennye problemy izucheniya vetvistousykh rakoobraznykh. S.-Peterburg: Gidrometeoizdat, 122–129.

Bogatova, I.B., Shcherbina, M.A., Ovinnikova, V.V., Tagirova, N.A., 1971. Chemical composition of some planktonic animals under different conditions of growing. Gidrobiologicheskii Zhurnal. 7 (5), 54–58.

Bond, J.A., Bradley, B.P., 1995. Heat-shock reduces the toxicity of malathion in *Daphnia magna*. Mar. Environ. Res. 39, 209–212.

Bond, J.-A., Gonzalez, C.R.M., Bradley, B.P., 1993. Age-dependent expression of proteins in the cladoceran *Daphnia magna* under normal and heat-stress conditions. Comp. Biochem. Physiol. 106B (4), 913–917.

Borgeraas, J., Hessen, D.O., 2000. UB-B induced mortality and antioxidant enzyme activities in *Daphnia magna* at different oxygen concentrations and temperatures. J. Plankton Res. 22 (6), 1167–1183.

Borgeraas, J., Hessen, D.O., 2002a. Variations of antioxidant enzymes in *Daphnia* species and populations as related to ambient UV exposure. Hydrobiologia. 477 (1), 15–30.

Borgeraas, J., Hessen, D.O., 2002b. Diurnal patterns of antioxidant activity in alpine and arctic *Daphnia* under in situ UV-radiation. Arch. Hydrobiol. 156 (1), 83–95.

Bosch de Aguilar, Ph. van den, 1969. Nouvelles donnees morphologiques et hypotheses sur le role du systeme neurosecreteure chez *Daphnia pulex* (Crustacea: Cladocera). Ann. Soc. R. Zool. Belg. 99 (1–2), 27–44.

Bosch de Aguilar, Ph. van den, 1971. Le systeme neurosecreteur de Podon intermedius (Crustacea: Cladocera). Ann. Soc. R. Zool. Belg. 101 (1–2), 57–63.

Bosch de Aguilar, Ph. van den, 1972. Les characteristiques tinctorials des cellules neurosecretices chez *Daphnia pulex* (Crustacea: Cladocera). General and Comp. Endocrinol. 18 (1), 140–145.

Bossuyt, B.T.A., Janssen, C.R., 2004. Influence of multigeneration acclimation to copper on tolerance, energy reserves, and homeostasis of *Daphnia magna* Straus. Environ. Toxicol. Chem. 23 (8), 2029–2037.

Bossuyt, B.T.A., Janssen, C.R., 2005. Copper toxicity to different field-collected cladoceran species: intra- and inter-species sensitivity. Environ. Pollut. 136, 145–154.

Bossuyt, B.T.A., Escolar, Y.R., Janssen, C.R., 2005. Multigenerational acclimation of *Daphnia magna* Straus to different bioavailable copper concentrations. Ecotoxicol. Environ. Saf. 61, 327–336.

Boudou, A., Ribeyre, F., 1981. Comparative study of the trophic transfer of two mercury compounds—$HgCl_2$ and CH_3HgCl—between *Chlorella vulgaris* and *Daphnia magna*. Influence of temperature. Bull. Environ. Contam. Toxicol. 27, 624–629.

Boudou, A., Ribeyre, F., 1983. Contamination of aquatic biocenoses by mercury compounds: an experimental ecotoxicological approach. In: Nriagu, J.O. (Ed.), Advances in Environmental Science and Technology, vol. 13. Aquatic Toxicology. John Wiley & Sons, NY, etc., pp. 73–116.

Bourdier, G., Bauchart, D., 1986–1987. Compsition en lipides et en acides gras de trois crustacés zooplanktoniques dy Lac d'Aydat (63). Résultats préliminaires. Annales des Sciences Naturelles, Paris, 13e série, 8 (4), 243–248.

Bowles, K.C., Bianchini, A., Brauner, C.J., Kramer, J.R., Wood, C.M., 2002. Evaluation of the effect of reactive sulfide on the acute toxicity of silver (I) to *Daphnia magna*. Part 1: Description of the chemical system. Environ. Toxicol. Chem. 21 (6), 1286–1293.

Bradley, M.C., Baird, D.J., Calow, P., 1991a. Mechanisms of energy allocation to reproduction in the cladoceran *Daphnia magna* Straus. Biol. J. Linn. Soc. Lond. 44 (4), 325–333.

Bradley, M.C., Perrin, N., Calow, P., 1991b. Energy allocation in the cladoceran *Daphnia magna* Straus, under starvation and refeeding. Oecologia. 86 (3), 414–418.

Brancelj, A., Čellar, T., Šiško, M., 1996. Four different head shapes in *Daphnia hyalina* (Leydig) induced by presence of larvae of *Chaoborus flavicans* (Meigen). Hydrobiologia. 339 (1–3), 37–45.

Von Brand, T., 1946. Anaerobiosis in invertebrates. Biodynamica, Normandy.

Brandl, Z., Komárková, J., Macek, M., et al. 2010a. 49th annual report. Institute of Hydrobiology, České Budějovice, p. 19.

Brandl, Z., Macek, M., Malá, A., et al. 2010b. 50th annual report. Institute of Hydrobiology, České Budějovice, p. 24.

Brede, N., Sandrock, C., Straile, D., Spaak, P., Jankowski, T., Streit, B., et al., 2009. The impact of human-made ecological changes on the genetic architecture of *Daphnia* species. Proc. Natl. Acad. Sci. 106, 4758–4763.

Brendonck, L., De Meester, L., 2003. Egg banks in freshwater zooplankton: evolutionary and ecological archives in the sediment. Hydrobiologia. 491, 65–84.

Brett, M.T., Müller-Navarra, D.C., 1997. The role of highly unsaturated fatty acids in aquatic foodweb processes. Freshw. Biol. 38, 483–499.

Brett, M.T., Müller-Navarra, D.C., Sang-Kyu Park, 2000. Empirical analysis of the phosphorus limitation on algal food quality for freshwater zooplankton. Limnol. Oceanogr. 45 (7), 1564–1575.

Brett, M.T., Müller-Navarra, D.C., Ballantyne, A.P., Ravet, J.L., Goldman, C.R., 2006. *Daphnia* fatty acid composition reflects that of their diet. Limnol. Oceanogr. 51 (5), 2428–2437.

Brett, M.T., Kainz, M.J., Taipale, S.J., Seshan, H., 2009. Phytoplankton, not allochthonous carbon, sustains herbivorous plankton production. PNAS. 106 (50), 21187–21201.

Breukelman, J., 1932. Speed of toxic action of mercuric chloride on daphnids. Physiol. Zool. 5 (2), 207–217.

Brewer, M.C., 1998. Mating behaviours of *Daphnia pulicaria*, a cyclic parthenogen: comparisons with copepods. Phil. Trans. Royal Soc. Lond. B. 353, 805–815.

Brites, D., Encinas-Viso, F., Ebert, D., Du Pasquier, L., Haag, C.R., 2011. Population genetics of duplicated alternatively spliced exons of the Dscam gene in *Daphnia* and *Drosophila*. PLOS ONE. 6, e27947.

Brites, D., McTaggart, S.J., Morris, K., Anderson, J., Thomas, K., Colson, I.F., et al., 2008. The Dscam homologue of the crustacean *Daphnia* is diversified by alternative splicing like in insects. Mol. Biol. Evol. 25, 1429–1439.

Brown, L.A., Crozier, W.J., 1927. The rate of killing of cladocerans by higher temperatures. J. Gen. Physiol. XI. 1, 25–36.

Brown, S.C., 1970. Histochemical studies on the hyaline capsule of *Holopedium gibberum* Zaddach (Crustacea, Cladocera). Histochemie. 21 (2), 160–169.

Brucet, S., Boix, D., Lopez-Flores, R., 2005. Ontogenetic changes of amino acid composition in planktonic crustacean species. Marine Biol. 148 (1), 131–139.

Bryukhatova, A., 1928. Change of sign of phototaxis in *Chysorus sphaericus* and *Chydorus ovalis* under influence of acids. Primenenie metodov fizicheskoy khimii k izucheniyu biologii presnykh vod. Moscow 124–138.

Buikema Jr., A.L., 1968. Effects of varying wavelengths, intensities and polarized light on population dynamics of *Daphnia pulex* Leydig, 1860, emend. Richard, 1896 (Cladocera). Crustaceana. 14 (1), 45–55.

Buikema Jr., A.L., 1972. Oxygen consumption of the cladoceran *Daphnia pulex* as a function of body size, light, and light acclimation. Comp. Bioch. Physiol. A. 42, 877–888.

Buikema Jr., A.L., 1975. Some effects of light on the energetics of *Daphnia pulex* and implications for the significance of vertical migration. Hydrobiologia. 47, 43–58.

Burgis, M.J., 1967. A quantitative study of reproduction in some species of *Ceriodaphnia* (Crustacea: Cladocera). J. Anim. Ecol. 36 (1), 61–74.

Burns, C.W., 1968a. Direct observations of mechanisms regulating feeding behavior of *Daphnia*, in lakewater. Internat. Revue ges. Hydrobiol. 53 (1), 83–100.

Burns, C.W., 1968b. The relationship between body size of filter-feeding Cladocera and the maximum size of particle ingested. Limnol. Oceanogr. 13 (4), 675–678.

Burns, C.W., 1969. Particle size and sedimentation in the feeding behavior of two species of *Daphnia*. Limnol. Oceanogr. 14 (3), 392–401.

Burns, C.W., 2000. Crowding-induced changes in growth, reproduction and morphology of *Daphnia*. Freshw. Biol. 43, 19–29.

Butorina, L.G., 1973. On content of organic carbon in the body of *Polyphemus pediculus* (L.). Gidrobiologicheskii Zhurnal IX. 3, 69–73.

Butorina, L.G., 1979. Dependence of metabolic rate in *Polyphemus pediculus* (L.) (Cladocera) on visual perception of environmental conditions. Ekologiya. 5, 66–71.

Butorina, L.G., 1980. Effect of visual perception on respiration rate of *Polyphemus pediculus* (L.) (Cladocera). Trudy Instituta Biologii vnutrennikh vod AN SSSR 44(47), 28–39.

Bychek, E.A., Guschina, I.A., 1999. The age changes of lipid composition in *Daphnia*. Biokhimiya. 64 (5), 652–655.

Bychek, E.A., Guschina, I., 2001. The transfer of fatty acids in a freshwater planktonic foodweb of the Kuibyshevskoe Reservoir (middle reaches of the Volga). Hydrobiologia. 442, 261–268.

Bychek, E.A., Dobson, G.A., Harwood, J.L., Guschina, I.A., 2005. *Daphnia magna* can tolerate short-term starvation without major changes in lipid metabolism. Lipids. 40 (6), 599–608.

Cáceres, C.E., 1998. Interspecific variation in the abundance, production, and emergence of *Daphnia* diapausing eggs. Ecology. 79 (5), 1699–1710.

Cáceres, C.E., Christoff, A.N., Boeing, W.J., 2007. Variation in ephippial buoyancy in *Daphnia pulcaria*. Freshw. Biol. 52, 313–318.

Cáceres, C.E., Hall, S.R., Duffy, M.A., Tessier, A.J., Helmle, C., MacIntyre, S., 2006. Physical structure of lakes constrains epidemics in *Daphnia* populations. Ecology. 87, 1438–1444.

Campbell, A.K., Wann, K.T., Matthews, S.B., 2004. Lactose causes heart arrhythmia in the water flea *Daphnia pulex*. Comp. Biochem. Physiol. Part B. 139, 225–234.

Canesi, L., Gallo, G., Gavioli, M., Pruzzo, C., 2002. Bacteria-hemocyte interactions and phagocytosis in marine bivalves. Microsc. Res. Tech. 57, 469–476.

Canli, M., 2005. The transfer of zinc in two linked trophic levels in fresh water and its effect on the reproduction of *Daphnia magna*. J. Freshwater Ecol. 20 (2), 269–276.

Cannon, H.G., 1922. On the labral glands of a cladoceran (*Simocephalus vetulus*), with a description of its mode of feeding. Quart. J. Microsc. Sci. 66 (262, part II), 213–234.

Cannon, H.G., 1933. On the feeding mechanism of the Branchiopoda. With an appendix on the mouth parts of the Branchiopoda. Phil. Trans. R. Soc. Lond., B. 222, 267–352.

Canton, J.H., Adema, D.M.M., 1978. Reproducibility of short-term and reproduction toxicity experiments with *Daphnia magna* and comparison of the sensitivity of *Daphnia magna* with *Daphnia pulex* and *Daphnia cucullata* in short-term experiments. Hydrobiologia. 59 (2), 135–140.

Carius, H.J., Little, T.J., Ebert, D., 2001. Genetic variation in a host-parasite association: Potential for coevolution and frequency-dependent selection. Evolution. 55, 1136–1145.

Carmona, M.J., Snell, T.W., 1995. Glycoproteins in daphnids: potential signals for mating? Arch. Hydrobiol. 134 (3), 273–279.

Carney, G.C., Shore, P., Chandra, H., 1986. The uptake of cadmium from a dietary and soluble source by the crustacean *Daphnia magna*. Environ. Res. 39 (2), 290–298.

Carroll, L., Tenniel, J., Green, R.L., 1971. Alice's adventures in Wonderland: And, Through the Looking Glass and What Alice Found There. Oxford University Press, London.

Carroll, S., Hendry, A., Reznick, D., Fox, C., 2007. Evolution on ecological time-scales. Funct. Ecol. 21, 387–393.

Carvalho, G.R., 1984. Haemoglobin synthesis in *Daphnia magna* Straus (Crustacea: Cladocera): ecological differentiation between neighbouring populations. Freshwater Biol. 14, 501–506.

Carvalho, G.R., Hughes, R., 1983. The effect of food availability, female culture-density and photoperiod on ephippia production in *Daphnia magna* Straus (Crustacea: Cladocera). Freshw. Biol. 3 (1), 37–46.

Cauchie, H.M., Hoffmann, L., Jaspar-Versali, M.F., Salvia, M., Thomé, J.P., 1995. *Daphnia magna* Straus as a source of chitin: ecological aspects. Belg. J. Zool. 125 (1), 67–78.

Cauchie, H.M, Jaspar-Versali, M.-F., Hoffmann, L., Thomé, J.P., 1999. Analysis of the seasonal variation in biochemical composition of *Daphnia magna* Straus (Crustacea: Branchiopoda: Anomopoda) from an aerated wastewater stabilization pond. Ann. Limnol. 35 (4), 223–231.

Cauchie, H.M., Joachim-Justro, C., Hoffmann, L., Thomé, J.P., Thys, I., 2000. A note on the use of fluorescently labelled algae for the determination of gut passage time. In *Bosmina* and *Daphnia*. Vernandl. Internat. Vereinig. Limnol. 27 (5), 2987–2991.

Cavalier-Smith, T., 1985. The Evolution of Genome Size. John Wiley Publ. 523p.

Čehović, G., 1954a. Effet stimulant des gonadotrophines de mammifères sur la reproduction des Daphnies. Comptes rendus hebdomadaires des séances de l'Académie des sciences. Paris 239, 4, 363–365.

Čehović, G., 1954b. Analyse de l'effet stimulant des gonadotrophines de mammifères sur la reproduction des Daphnies. Comptes rendus hebdomadaires des séances de l'Académie des sciences. Paris 239, 12, 722–724.

Cellier, S., Rehaïlia, M., Berthon, J.-L., Buisson, B., 1998. Le rôle de l'œil dans les rhythmes migratoires de *Daphnia magna* et *Daphnia longispina* (cladocères). Ann. Limnol. 34 (2), 159–164.

Cellier-Michel S., Berthon, J.-L., 2003. Variations structurales de l'oeil, liées aux migrations verticales journalières de *Daphnia longispina*. Académie des Sciences, Paris, Comptes rendus 326, 12, 1157–1162.

Cerny, M., Hebert, P.D.N., 1999. Intercontinental allozyme differentiation among four Holarctic *Daphnia* species. Limnol. Oceanogr. 44, 1381–1387.

Chadwick, W., Little, T.J., 2005. A parasite-mediated life-history shift in *Daphnia magna*. Proc. R. Soc. B., Biol. Sci. 272, 505–509.

Chalikov, B.G., 1951. Methods of raising of juvenile acipenserids and of inconnu. Thudy VNIRO. 19, 4–24.

Chambers, R., 1913. The spermatogenesis of a daphnid, *Simocephalus vetulus*. A preliminary paper. Biol. Bull. 25, 134–140.

Chandler, A., 1954. Causes of variation in the haemoglobin content of *Daphnia* (Crustacea: Cladocera) in nature. Proc. Zoolog. Soc. Lond. 124 (Part 3), 625–630.

Chaparro-Herrera, D. de J., Sarma, N., Sarma, S.S.S., Fernandez, R., 2011. Effect of temperature on the demography of *Latonopsis* cf. *australis* Sars and *Diaphanosoma birgei* Korinek (Cladocera: Sididae). Abstract book IXth International symposium on Cladocera, October 2–8, 2011. Verbania, Italy, p. 6.

Chatton, E., 1920. Les membranes péritrophique des Drosophiles (Diptères) et des Daphnies (Cladocères) leur genese et leur role a l'égard des parasites intestinaux. Bull. Soc. Zool. France. 25 (2), 265–280.

Chernykh, S.I., Panasyuk, T.D., 1964. Distribution of *Daphnia magna* under conditions of temperature, oxygen, and illumination gradients. Zoologicheskii Zhurnal. 43 (11), 1715–1716.

Chieco, P., Derenzini, M., 1999. The Feulgen reaction 75 years on. Histochem. Cell Biol. 111, 34–58.

Chirkova, Z.N., 1972a. Some observations on behavior of *Ilyocryptus sordidus* (Cladocera, Macrothricidae). Biologiya vnutrennikh vod. 13, 23–26.

Chirkova. Z.N., 1972b. On "family" aggregations of *Ilyocryptus* (Cladocera, Macrothricidae). Povedeniye vodnyhkh bespozvonochnykh, 1 Vsesoyuznyi simpozium, Borok, 1972, 146–148.

Chouderi, R.B., Gusso-Chouderi, K.P., Melão, M., da, G.G., Lomnbardi, A.T., Vieira, A.A.H., 2009. The influence of cyanobacterium exudates on copper uptake and toxicity to a tropical freshwater cladoceran. J. Plankton Res. 31 (10), 1225–1233.

Christensen, B.T., Lauridsen, T.L., Ravn, H.W., Bayley, M., 2006. A comparison of feeding efficiency and swimming ability of *Daphnia magna* exposed to cypermethrin. Aquat. Toxicol. 73, 210–220.

Christofersen, K., 1988. Effect of food concentration on gut evacuation in *Daphnia pulicaria* and *Daphnia longispina* measured by the fluorescence technique. Verhandl. Internat. Ver. Limnol. 23 (4), 2050–2056.

Chuah, T.S., Loh, J.Y., Hii, Y.S., 2007. Acute and chronic effects of the insecticide-endosulfan on freshwater cladoceran, *Moina macrocopa* Straus. Bull. Environm Contam. Toxicol. 79 (5), 557–561.

Clarke, G.L., 1930. Change of phototropic and geotropic signs in *Daphnia* induced by changes of light intensity. J. Exp. Biol. VII (2), 109–131.

Clarke, G.L., 1932. Quantitative aspects of the change of phototropic sign in *Daphnia*. J. Exp. Biol. 9 (2), 180–211.

Clarke, G.L., Wolf, E., 1933. The mechanisms of tropistic reactions and the strychnine effect in *Daphnia*. J. Gen. Physiol. 16 (1), 99–195.

Claus, C., 1875a. Die Schalendrüse der Daphnien. Zeitschrift für wissenschaftliche Zoologie. 25 (2), 165–173.

Claus, C., 1875b. Grundzüge de Zoologie. Marburg and Leipzig, 356 pp.

Claus, C., 1876. Zur Kenntniss der Organisation und des feinern Baues der Daphniden und verwandter Cladoceren. Zeitschrift für wissenschaftliche Zoologie. 27 (3), 362–402.

Clifford, M., McGeer, J.C., 2009. Development of a biotic ligand model for the acute toxicity of zinc to *Daphnia pulex* in soft waters. Aquat. Toxicol. 91, 26–32.

Colbourne, J.K., Hebert, P.D.N., 1996. The systematics of North American *Daphnia* (Crustacea: Anomopoda): a molecular phylogeny. Phil. Trans. R. Soc. Lond. B. 351, 349–360.

Colbourne, J.K., Hebert, P.D., Taylor, D.J., 1997. Evolutionary origins of phenotypic diversity in *Daphnia*. In: Givnish, T.J., Sytsma, K.J. (Eds.), Molecular Evolution and Adaptive Radiation. Cambridge University Press, pp. 163–188.

Colbourne, J.K., Crease, T.J., Weider, L.J., Hebert, P.D.N., Dufresne, F., Hobæk, A., 1998. Phylogenetics and evolution of a circumarctic species complex (Cladocera: *Daphnia pulex*). Biol. J. Linn. Soc. 65, 347–365.

Colbourne, J.K., Singan, V.R., Gilbert, D.G., 2005. wFleaBase: the *Daphnia* genome database. BMC Bioinf. 6 (1), 45.

Colbourne, J.K., Pfrender, M.E., Gilbert, D., Thomas, W.K., Tucker, A., Oakley, T.H., et al., 2011. The ecoresponsive genome of *Daphnia pulex*. Science. 331, 555–561.

Colmorgen, M., Paul, R.J., 1995. Imaging of physiological functions in transparent animals (*Agonus cataphractus*, *Daphnia magna*, *Pholcus phalangoides*) by video microscopy and digital image processing. Comp. Biochem. Physiol. 111A (4), 583–595.

Combes, C., 2001. Parasitism: the Ecology and Evolution of Intimate Interactions. University of Chicago Press, Chicago.

Connelly, S.J., Moeller, R.E., Sanchez, G., Mitchell, D.L., 2009. Temperature effects on survival and DNA relair in four freshwater cladoceran *Daphnia* species exposed to UV radiation. Photochem. Photobiol. 85 (1), 144–152.

Connon, R., Dewhurst, R.E., Crane, M., Callaghan, A., 2003. Haem peroxidase activity in *Daphnia magna* for sub-lethal toxicity assessment of kerosene-contaminated groundwater. Ecotoxicology. 12 (5), 387–395.

Consi, T.R., Macagno, E.R., 1985. The spectral sensitivity of eye movements in response to light flashes in *Daphnia magna*. J. Comp. Physiol. A. 156 (1), 135–141.

Consi, T.R., Macagno, E.R., Neeles, N., 1987. The oculomotor system of *Daphnia magna*. The eye muscles and their motor neurons. Cell Tissue Res. (Zeitsch. für Zellforschung und Miskroskopische Anatomie). 247 (3), 515–523.

Consi, T.R., Passani, M.B., Macagno, E.R., 1990. Eye movements in *Daphnia magna*. J. Comp. Physiol. A. 166, 411–420.

Cousyn, C., De Meester, L., Colbourne, J.K., Brendonck, L., Verschuren, D., Volckaert, F., 2001. Rapid, local adaptation of zooplankton behavior to changes in predation pressure in the absence of neutral genetic changes. Proc. Natl. Acad. Sci. U.S.A. 98, 6256–6260.

Cowgill, U.M., Burns, C.W., 1975. Differences in chemical composition between two species of *Daphnia* and some freshwater algae cultured in the laboratory. Limnol. Oceanogr. 20 (6), 1005–1011.

Cowgill, U.M., Williams, D.M., Esquivel, J.B., 1984. Effects of maternal nutrition on fat content and longevity of neonates of *Daphnia magna*. J. Crust. Biol. 4 (2), 173–190.

Crease, T., Little, T., 1997. Partial sequence of the mitochondrial genome of the crustacean *Daphnia pulex*. Curr. Genet. 31, 48–54.

Crease, T., 1999. The complete sequence of the mitochondrial genome of *Daphnia pulex* (Cladocera: Crustacea). Gene. 233, 89–99.

Crease, T.J., Omilian, A.R., Costanzo, K.S., Taylor, D.J., 2012. Transcontinental phylogeography of the *Daphnia pulex* species complex. PloS one. 7, e46620.

Crease, T.J., Taylor, D.J., 1998. The origin and evolution of variable-region helices in V4 and V7 of the small-subunit ribosomal RNA of branchiopod crustaceans. Mol. Biol. Evol. 15, 1430–1446.

Crider, J.Y., Wilhm, J., Harman, H.J., 1982. Effects of naphthalene on the hemoglobin concentration and oxygen uptake of *Daphnia magna*. Bull. Environ. Contamin. Toxicol. 28, 52–57.

Crispim, M.C., Watanabe, T., 2001. What can dry reservoir sediments in a semi-arid region in Brazil tell us about Cladocera? Hydrobiologia. 442, 101–105.

Crosby, D.G, Tucker, R.K., 1971. Accumulation of DDT by *Daphnia magna*. Environ. Sci. Technol. 5 (8), 714–716.

Cuenot, L., 1891. Études sur le sang et les glandes lymphatiques dans la serie animale. Archives de zoologie experimentale, 2 serie, IX, 13–90.

Czarnecki, O., Henning, M., Lippert, I., Welker, M., 2006. Identification of peptide metabolites of *Microcystis* (Cyanobacteria) that inhibit trypsin-like activity in planktonic herbiviorous *Daphnia* (Cladocera). Environ. Microbiol. 8 (1), 77–87.

Czeczuga, B., 1965. Haemoglobin in *Ceriodaphnia quadrangula* (O.F. Müller). Nature. 208 (5013), 907–908.

Czeczuga, B., 1976. Studies on carotenoids in *Hydra vulgaris* Pallas—*Ceriodaphnia reticulata* Jur. trophic system. Bull. de l'Académie Polonaise des sciences, ser. des sci. biol. 24 (1), 53–56.

Czeczuga, B., 1984. Investigations of carotenoprotein complexes in animals. XI. Carotenoprotein and carotenoids in *Daphnia magna* Straus. Polskie archiwum hydrobiologii. 31 (2), 91–97.

D'Abramo, L.R., 1979. Dietary fatty acid and temperature effects on the productivity of the cladoceran, *Moina macrocopa*. Biol. Bull. 137 (2), 231–248.

Dabrowski, K., Rusiecki, M., 1983. Content of total and free amino acids in zooplanktonic food of fish larvae. Aquaculture. 30 (1–4), 31–42.

Dangott, L.J., Terwilliger, R.C., 1980. The subunit structure of *Daphnia pulex* hemoglobin. Comp. Biochem. Physiol. B. 67, 301–306.

Danielli, A., Loukeris, T.G., Lageux, M., Muller, H.M., Richman, A., Kafatos, F.C., 2000. A modular chitin-binding protease associated with hemocytes and hemolymph in the mosquito *Anopheles gambiae*. Proc. Natl. Acad. Sci. U.S.A. 97, 7136–7141.

Darchambeau, F., Thys, I., 2005. In situ filtration responses of *Daphnia galeata* to changes in food quality. J. Plankton Res. 27 (3), 227–236.

Darchambeau, F., Faervig, P.J., Hessen, D.O., 2003. How *Daphnia* copes with excess carbon in its food. Oecologia. 136, 336–346.

Dave, G., 1984. Effects of waterborne iron on growth, reproduction, survival and haemoglobin in *Daphnia magna*. Comp. Biochem. Physiol. 78C (2), 433–438.

David, R.M., Dakic, V., Williams, T.D., Winter, M.J., Chipman, J.K., 2011. Transcriptional responses in neonate and adult *Daphnia magna* in relation to relative susceptibility to genotoxicants. Aquat. Toxicol. (Amsterdam). 104, 192–204.

Davison, J., 1969. Activation of the ephippial egg of *Daphnia pulex*. J. Gen. Physiol. 53 (5), 562–575.

Day, K., Kaushik, N.K., 1987. Short-term exposure of zooplankton to the synthetic pyrethroid, fenvalerate, and its effects on rates of filtration and assimilation of the alga. Arch. Environ. Contam. Toxicol. 16 (4), 423–432.

Day, K.E., Scott, I.M., 1990. Use of acetylcholinesterase activity to detect sublethal toxicity in stream invertebrates exposed to low concentrations of organophosphate insecticides. Aquat. Toxicol. 18 (2), 101–114.

De Coen, W., Janssen, C.R., 1997. The use of biomarkers in *Daphnia magna* toxicity testing. II. Digestive enzyme activity in *Daphnia magna* exposed to sublethal concentrations of cadmium, chromium and mercury. Chemosphere. 35 (5), 1053–1067.

De Coen, W., Vangheluwe, M.L., Jansen, C.R., 1998. The use of biomarkers in *Daphnia magna* toxicity testing. III. Rapid toxicity testing of pure chemicals and sediment pore waters usng ingestion and digestive enzyme activity. Chemosphere. 37 (13), 2677–2694.

De Lange, H.J., Van Donk, Hessen, D.O., 2000. In situ effects of UV radiation on four species of phytoplankton and two morphs of *Daphnia longispina* in an alpine lake (Finse, Norway). Verhandl. Internat. Verenig. Limnol. 27 (4), 2008–2013.

De Meester, L., Dumont, H.J., 1988. The genetics of phototaxis in *Daphnia magna*: existence of three phenotypes for vertical migration among parthenogenetic females. Hydrobiologia. 162, 47–55.

De Meester, L., Dumont, H.J., 1989. Phototaxis in *Daphnia*: interaction of hunger and genotypes. Limnol. Oceanogr. 14 (7), 1323–1325.

De Meester, L., Van Doorslaer, W., Geerts, A., Orsini, L., Stoks, R., 2011. Thermal genetic adaptation in the water flea *Daphnia* and its impact: an evolving metacommunity approach. Integr. Comp. Biol. 51, 703–718.

De Schamphelaere, K.A.C., Bossuyt, B.T.A., Janssen, C.R., 2007. Variability of the protective effect of sodium on the acute toxicity of copper to freshwater Cladocerans. Environ. Toxicol. Chem. 26 (3), 535–542.

De Schamphelaere, K.A.C., Vasconcelos, F.M., Tack, F.M.G., Allen, H.E., Janssen, C.R., 2004. Effect of dissolved organic matter source on acute copper toxicity to *Daphnia magna*. Environ. Toxicol. Chem. 23 (5), 1248–1255.

de Waard, J.R., Sacherova, V., Cristescu, M.E.A., Remigio, E.A., Crease, T.J., Hebert, P.D.N., 2006. Probing the relationships of the branchiopod crustaceans. Mol. Phylogenet. Evol. 69, 491–502.

de Waard, J.R., Sacherova, V., Cristescu, M.E.A., Remigio, E.A., Crease, T.J., Hebert, P.D.N., 2006. Probing the origins of the branchiopod crustaceans. Mol. Phylog. Evol. 39, 491–502.

Dearborn, G.U.N., 1903. Some physiological observations on a crustacean heart. Med. News. 82 (544–550), 596–601.

Deardorff, A.D., Stark, J.D., 2009. Acute toxicity and hazard assessment of spinosad and R-11 to three cladoceran species and coho salmon. Bull. Environm. Toxicol. 82 (5), 549–553.

de Boer, J.P., Creasey, A.A., Chang, A., Abbink, J.J., Roem, D., Eerenberg, A.J., et al., 1993. Alpha-2-macroglobulin functions as an inhibitor of fibrinolytic, clotting, and neutrophilic proteinases in sepsis: studies using a baboon model. Infect. Immun. 61 (12), 5035–5043.

Decaestecker, E., De Meester, L., Ebert, D., 2002. In deep trouble. Habitat selection constrained by multiple enemies in zooplankton. Proc. Natl. Acad. Sci. U.S.A. 99, 5481–5485.

Decaestecker E., De Meester L., Mergeay J., 2009. Cyclical parthenogenesis in *Daphnia*: Sexual versus asexual reproduction. In: Van Dijk P., Maertens K., Schoen I. (eds.) Lost sex: the evolutionary biology of parthenogenesis. Springer Netherlands (Dordrecht), pp. 295–316.

Decaestecker, E., Gaba, S., Raeymaekers, J.A.M., Stoks, R., Van Kerckoven, L., Ebert, D., et al., 2007. Host-parasite 'Red Queen' dynamics archived in pond sediment. Nature. 450, 870–873.

Decaestecker, E., Labbé, P., Ellegaard, K., Allen, J.E., Little, T.J., 2011. Candidate innate immune system gene expression in the ecological model *Daphnia*. Dev. Comp. Immunol. 35, 1066–1075.

Decaestecker, E., Lefever, C., De Meester, L., Ebert, D., 2004. Haunted by the past: Evidence for dormant stage banks of microparasites and epibionts of *Daphnia*. Limnol. Oceanogr. 49, 1355–1364.

Decaestecker, E., Vergote, A., Ebert, D., De Meester, L., 2003. Evidence for strong host clone-parasite species interactions in the *Daphnia* microparasite system. Evolution. 57, 784–792.

Dediu, I.I., 1989. Ecological encyclopedic dictionary. Kishinev (in Russian).

Dediu, I.I., Garbalau, A.P., Patalashco, E.P., 1995. The evaluation of toxic influence of heavy metals combinations on *Daphnia magna* Strauss. Bulletenul Academici de ştiinţe a Respublicii Moldova, biol. şi chim. 4, 58–64.

Deines, P., Fink, P., 2011. The potential of methanotrophic bacteria to compensate for food quantity or food quality limitations in *Daphnia*. Aquat. Microb. Ecol. 65, 197–206.

Dejdar, E., 1930. Vitale Elektivefärbungen der rudimentäre Antennendrüse von Cladoceren. Zeitschrift für wissenschaftliche Biologie—Abteilung A: Zeitschrift Morphol. und Oekol. der Tiere. 17, 768–777.

Dekker, T., Greve, G.D., Ter Laak, T.L., Boivin, M.E., Veuger, B., Gortzak, G., et al., 2006. Development and application of a sediment toxicity test using the benthic cladoceran *Chydorus sphaericus*. Environ. Pollut. 140 (2), 231–238.

Deleebeeck, N.M.E., De Schamphelaere, K.A.C., Heijerick, D.G., Bossuyt, B.T.A., Janssen, C.R., 2008. The acute toxicity of nickel to *Daphnia magna*: predictive capacity of bioavailability models in artificial and natural waters. Ecotoxicol. Environ. Saf. 70, 67–78.

DeMott, W.R., 1982. Feeding selectivities and relative ingestion rates of *Daphnia* and *Bosmina*. Limnol. Oceanogr. 27 (3), 518–527.

DeMott, W.R., Müller-Navarra, D.C., 1997. The importance of highly unsaturated fatty acids in zooplankton nutrition: evidence from experiments with *Daphnia*, a cyanobacterium and liopid emulsions. Freshw. Biol. 38 (3), 649–664.

DeMott, W.R., Gulati, R.D., Siewertsen, K., 1998. Effects of phosphorus deficiency on the carbon and phosphorus balance of *Daphnia magna*. Limnol. Oceanogr. 43 (6), 1147–1161.

DeMott, W.R., Pape, B.J., Tessier, A.J., 2004. Patterns and sources of variation in *Daphnia* phosphorus content in nature. Aquat. Ecol. 38 (3), 433–440.

Desvilettes, C., Bourder, G., Amblard, C., Barth, B., 1997. Use of fatty acids for the assessment of zooplankton grazing on bacteria, protozoans and microalgae. Freshw. Biol. 38, 629–637.

Desvilettes, C., Bourder, G., Breton, J.C., Combrouze, P., 1994. Fatty acids as organic markers for the study of trophic relationships in littoral cladoceran communities of a pond. J. Plankton Res. 16 (6), 643–659.

Dewey, E.S., Parker, B.L., 1964. Mass rearing of *Daphnia magna* for insecticide bioassay. J. Econ. Ent. 57 (6), 821–825.

Dewilde, S., van Hauwaert, M.L., Peeters, K., Vanfleteren, J., Moens, L., 1999. *Daphnia pulex* didomain hemoglobin: structure and evolution of polymeric hemoglobins and their coding genes. Mol. Biol. Evol. 16, 1208–1218.

di Delupis, G.D., Macri, A., Civitareale, C., Migliore, L., 1992. Antibiotics of zootechnical use: effects of acute high and low dose contamination on *Daphnia magna* Straus. Aquat. Toxicol. 22, 53–60.

Diamantino, T.C., Almeida, E., Soares, A.V.M., Guilhermino, L., 2001. Lactate dehydrogenase activity as an effect criterion in toxivcity test with *Daphnia magna* Straus. Chemosphere. 45 (4–5), 553–560.

Diamantino, T.C., Almeida, E., Soares, A.V.M., Guilhermino, L., 2003. Characterization of cholinesterases from *Daphnia magna* Straus and their inhibition by zinc. Bull. Environm. Contam. Toxicol. 71 (2), 219–225.

Diamantino, T.C., Guilhermino, L., Almeida, E., Soares, M. V.M., 2000. Toxicity of sodium molybdate and sodium dichromate to *Daphnia magna* Straus evaluated in acute, chronic, and acetylcholinesterase inhibition tests. Ecotoxicol. Environ. Saf. 45, 253–259.

Dierickx, P.J., 1987a. Soluble glutathione transferase isoenzymes in *Daphnia magna* Straus and their interaction with 2,4-dichlorophenoxyacetic acid and 1,4-benzoquinone. Insect Biochem. 17 (1), 1–6.

Dierickx, P.J., 1987b. Glutathione transferases and detoxification function of glutathione in the waterflea *Daphnia magna* Straus. Société Belge de Biochimie, Louvain-La-Neuve, p. B121.

Dodson, S.I., 1989. The ecological role of chemical stimuli for the zooplankton: predator-induced morphology in *Daphnia*. Oecologia. 78, 361–367.

Dodson, S.I., 2005. Introduction to Limnology. McGraw-Hill, Boston, etc. 400 pp.

Dodson, S.I., Frey, D.G., 1991. Cladocera and other Branchiopoda. In: Thorp, J.H, Corviett, A.P. (Eds.), Ecology and Classification of North American Freshwater Invertebrates. Academic Press, San Diego, California, pp. 723–776.

Dodson, S.I., Ramcharan, C., 1991. Size-specific swimming behavior of *Daphnia pulex*. J. Plankton Res. 13 (6), 1367–1379.

Dodson, S.I., Hanazato, T., Gorski, P.R., 1995. Behavioral responses of *Daphnia pulex* exposed to carbaryl and *Chaoborus* kairomone. Environ. Toxicol. Chem. 14 (1), 43–50.

Dodson, S.I., Merritt, C.M., Shannahan, J.-P., Shults, C.M., 1999. Low exposure concentrations of atrazine increase male production in *Daphnia pulicaria*. Environ. Toxicol. Chem. 18 (7), 1568–1573.

Dodson, S.I., Ryan, S., Tollrian, R., Lampert, W., 1997. Individual swimming behavior of *Daphnia*: effects of food, light and container size in four clones. J. Plankton Res. 19 (10), 1537–1552.

Domal-Kwiatkowska, Sosak-Świderska, B., Mazurek, U., Tyrawska, D., 1994. The effect of cadmium on the survival and filtering rate of *Daphnia magna*, Straus 1820. Polskie archiwum hydrobiologii. 41 (4), 465–474.

Dortland, R.J., 1978. Aliesterase (ALI-E) activity in *Daphnia magna* Straus as a parameter for exposure to parathion. Hydrobiologia. 59 (2), 141–144.

Dow, J.A., 2007. Integrative physiology, functional genomics and the phenotype gap: a guide for comparative physiologists. J. Exp. Biol. 210, 1632–1640.

Downing, A.C., 1974. The hydraulic suspension of the *Daphnia* eye—a new kind of universal joint? Vision Res. 14 (8), 647–652.

Drenner, R.W., Strickler, J.R., O'Brien, W.J., 1978. Capture probability: the role of zooplankter escape in the selective feeding of planktivorous fish. J. Fish. Res. Board Can. 35, 1370–1373.

Dresel, E.I.B., 1948. Passage of haemoglobin from blood into eggs of *Daphnia*. Nature. 162 (4123), 736–737.

Du Pasquier, L., 2005. Insects diversify one molecule to serve two systems. Science. 309, 1826–1827.

Dubovskaya, O.P., Gladyshev, M.I., Gubanov, V.G., Makhutova, O.N., 2003. Study of non-consumptive mortality of crustacean plankton in a Siberian reservoir using staining for live/dead sorting and sediment traps. Hydrobiologia. 504, 223–227.

Duchet, C., Inafuku, M.M., Caquet, T., Larroque, M., Franquet, E., Lagneau, C., et al., 2011. Chitobiase activity as an indicator of altered survival, growth and reproduction in *Daphnia pulex* and *Daphnia magna* (Crustacea: Cladocera) exposed to spinosad and diflubenzuron. Ecotoxicol. Environ. Saf. 74, 800–810.

Dudycha, J.L., 2003. A multi-environment comparison of senescence between sister species of *Daphnia*. Oecologia. 135, 555–563.

Duffy, M.A., 2010. Ecological consequences of intraspecific variation in lake *Daphnia*. Freshwat. Biol. 55, 995–1004.

Duffy, M.A., Sivars-Becker, L., 2007. Rapid evolution and ecological host-parasite dynamics. Ecol. Lett. 10, 44–53.

Duffy, M.A., Brassil, C.E., Hall, S.R., Tessier, A.J., Cáceres, C.E., Conner, J.K., 2008. Parasite-mediated disruptive selection in a natural *Daphnia* population. BMC. Evol. Biol. 8, 80.

Duffy, M.A., Cáceres, C.E., Hall, S.R., Tessier, A.J., Ives, A.R., 2010. Temporal, spatial and between-host comparisons of patterns of parasitism in lake zooplankton. Ecology. 91, 3322–3331.

Duffy, M.A., Housley Ochs, J., Penczykowski, R.M., Civitello, D.J., Klausmeier, C.A., Hall, S.R., 2012. Ecological context influences epidemic size and parasite-mediated selection. Science. 335, 1636–1638.

Dufresne, F., Hebert, P.D.N., 1995. Polyploid and clonal diversity in an arctic cladoceran. Heredity. 75, 45–53.

Dumont, H.J., 1967. A five day study of patchiness in *Bosmina coregoni* Baird in a shallow eutrophic lake. Mem. Ist. Ital. Idrobiol. 22, 81–103.

Dumont, H.J., 1987. Groundwater Cladocera: a synopsis. Hydrobiologia. 145, 169–173.

Dumont, H.J., 1995. The evolution of groundwater Cladocera. Hydrobiologia. 307, 69–74.

Dumont, H.J., Pensaert, J., 1983. A revision of the Scapholeberinae. Hydrobiologia. 100, 3–45.

Dumont, H.J., Silva-Briano, M., 1997. Sensory and glandular equipment of the trunk limbs of the Chydoridae and Macrothricidae (Crustacea: Anomopoda). Hydrobiologia. 360, 33–46.

Dumont, H.J., Van de Velde, I., 1976. Some types of head-pores in the Cladocera as seen by scanning electron microscopy and their possible functions. Biol. Jb. Dodonea. 44, 135–142.

Dumont, H.J., Guisez, Y., Carels, I., Verheye, H.M., 1985. Experimental isolation of positively and negatively phototactic phenotypes from natural population of *Daphnia magna* Strauss: a contribution to the genetics of vertical migration. Hydrobiologia. 126, 121–127.

Dumont, H.J., Van de Velde, I., Dumont, S., 1975. The dry weight estimate of biomass in a selection of Cladocera, Copepoda and Rotifera from the plankton, periphyton and benthos of continental waters. Oecologia. 19, 75–97.

Duncan, A.B., Little, T.J., 2007. Parasite-driven genetic change in a natural population of *Daphnia*. Evolution. 61, 796–803.

Duncan, A.B., Hall, S., Little, T.J., 2009. Parasitism and environmental sex determination in *Daphnia*. Evol. Ecol. Res. 11, 965–973.

Duncan, A.B., Mitchell, S.E., Little, T.J., 2006. Parasite-mediated selection and the role of sex and diapause in *Daphnia*. J. Evol. Biol. 19, 1183–1189.

Duneau, D., Luijckx, P., Ben-Ami, F., Laforsch, C., Ebert, D., 2011. Resolving the infection process reveals striking differences in the contribution of environment, genetics and phylogeny to host-parasite interactions. BMC Biol. 9, 11.

Dzialowski, E.M., Turner, P.K., Brooks, B.W., 2006. Physiological and reproductive effects of beta adrenergic receptor antagonists in *Daphnia magna*. Arch. Environ. Contam. Toxicol. 50 (4), 503–510.

Eads, B.D., Andrews, J., Colbourne, J.K., 2008. Ecological genomics in *Daphnia*: stress responses and environmental sex determination. Heredity. 100, 184–190.

Eagle, S.H.C., Crease, T.J., 2012. Copy number variation of ribosomal DNA and Pokey transposons in natural populations of *Daphnia*. Mobile DNA. 3, 4. 10.1186/1759-8753-3-4.

Ebert, D., 2005. Ecology, Epidemiology and Evolution of Parasitism in *Daphnia*. National Center for Biotechnology Information (US), Bethesda (MD). Available from: <http://www.ncbi.nlm.nih.gov/books/NBK2036/>.

Ebert, D., 2008. Host-parasite coevolution: Insights from the *Daphnia*-parasite model system. Curr. Opin. Microbiol. 11, 290–301.

Ebert, D., Carius, H.J., Little, T., Decaestecker, E., 2004. The evolution of virulence when parasites cause host castration and gigantism. Am. Nat. 164, S19–S32.

Ebert, D., Rainey, P., Embley, T.M., Scholz, D., 1996. Development, life cycle, intrastructure and phylogenetic position of Pasteuria ramosa Metchnikoff 1888: rediscovery of an obligate endoparasite of *Daphnia magna*. Phil. Trans. R. Soc. Lond. B. 351 (1348), 1689–1701.

Ebert, D., Yampolsky Yu, L., 1992. Family planning in *Daphnia*: when is clutch size determined? Russian J. Aquat. Ecol. 1 (2), 143–147.

Egloff, D.A., Fofonoff, P.W., Onbé, T., 1997. Reproductive biology of marine cladocerans. Adv. Mar. Biol. 31, 79–167.

Ehrenström, F., Berglind, R., 1988. Determination of biogenic amines in the water flea, *Daphnia magna* (Cladocera, Crustacea) and their diurnal variations using ion-pair reversed phase HPLC with electrochemical detection. Comparat. Biochem. Physiol. 90C (1), 123–132.

Ejsmont-Karabin, J., 1984. Phosphorus and nitrogen excretion rate by lake zooplankton. Ekologia polska. 32 (1), 3–42.

Ekau, W., Auel, H., Pörtner, H.O., Gilbert, D., 2010. Impacts of hypoxia on the structure and processes in pelagic communities (zooplankton, macro-invertebrates and fish). Biogeosciences. 7 (5), 1669–1699.

Elendt, B.P., 1989. Effects of starvation on growth, reproduction, survival and biochemical composition of *Daphnia magna*. Ach. Hydrobiol. 116 (4), 415–433.

Elendt, B.P., 1990. Selenium deficiency in Crustacea. Protoplasma. 154, 25–33.

Elendt, B.P., Storch, V., 1990. Starvation-induced alterations of the ultrastructure of the midgut of *Daphnia magna* Straus, 1829 (Cladocera). J. Crust. Biol. 10 (1), 79–86.

Elliott, W.H., Elliott, D.C., 1997. Biochemistry and Molecular Biology. Oxford University, Oxford, UK.

Elrod-Erickson, M., Mishra, M., Schneider, D., 2000. Interactions between the cellular and humoral immune responses in *Drosophila*. Curr. Biol. 10, 781–784.

Elser, J.J., Hayakawa, J., Urabe, J., 2001. Nutrient limitation reduces food quality for zooplankton: *Daphnia* response to seston phosphorus enrichment. Ecology. 92 (3), 898–903.

Epstein, A.C., Gleadle, J.M., McNeill, L.A., Hewitson, K.S., O'Rourke, J., Mole, D.R., et al., 2001. *C. elegans* EGL-9 and mammalian homologs define a family of dioxygenases that regulate HIF by prolyl hydroxylation. Cell. 107, 43–54.

Eremova, N.G., 1991. Metabolic adaptations to temperature in the period of embryonal development of *Daphnias*. Vestnik Belorusskogo gos. Universiteta, ser. 2 (1), 38–41.

Ermakov, N.V., 1927. Regeneration in freshwater Cladocera and the notion of organizational centers (First communication). Russkii gidrobiologicheskii Zhurnal. 6 (11–12), 213–224.

Ermakov, N.V., 1929. Regeneration in freshwater Cladocera and the notion of organizational centers (Second communication). Russkii gidrobiologicheskii Zhurnal. 8 (1–3), 54–62.

Ermakov, M.V., 1936. Nervous system and heart activity in invertebrates. Observations on the heart of Cladocera. Medichnyi Zhurnal AN Ukr. SSR. 6 (3), 755–782.

Ermakov, M.V., 1937. Nervous system and heart activity in invertebrates. II. On innervation of the heart in Cladocera. Medichnyi Zhurnal AN Ukr. SSR. 7 (1), 263–275.

Esipova, M.A., 1969. Growth and reproduction of *Daphnia magna* (Straus) and *D. longispina* (O.F. Müller) when feeding organic detrite. Gidrobiologichskii Zhurnal V. 5, 30–36.

Esipova, M.A., 1971. On the role of organic detritus in nutrition of Cladocera. Trudy Vsesoyuznogo nauchno-issledovartelskogo instituta. prudovogo rybnogo khozyaistva 18, 69–75.

Espie, P.J., Roff, J.C., 1995. Characterization of chitobiase from *Daphnia magna* and its relation to chitin flux. Physiol. Zool. 68 (5), 727–748.

Evens, R., De Schamphelaere, K.A.C., Janssen, C.R., 2009. The effects of dietary nickel exposure on growth and reproduction of *Daphnia magna*. Aqauat. Toxicol. 94, 138–144.

Ewald, W.F., 1914. Versuche zur Analyze der Licht- und Farbenreactonen eines Wirbellosen (*Daphnia pulex*). Zeitschrift für Psychologie und Physiologie der Sinnesorgane. II. Abteilung. Zeitschrift für Sinnesphysiologie. 48 B, 285–324.

Farkas, T., 1958. Néhány édesvizi rák karotinoidatalmárol. Magy. Tud. Akad., Tihanyi biol. kutatóinter. évk. 25, 187–195.

Farkas, T., 1970. Fats in freshwater crustaceans. I. Fatty acid composition of lipids from *Eudiaptomus gracilis* G.O. Sars (Copepoda) and *Daphnia cucullata* G.O. Sars (Cladocera). Acta Biol. Acad. Sci. Hung. 21 (2), 225–233.

Farkas, T., Kariko, K., Csengeri, I., 1981. Incorporation of [1-14C] acetate into fatty acids of the crustacean

Daphnia magna and *Cyclops strenus* in relation to temperature. Lipids. 6 (6), 418–422.

Farkas, T., Nemecz, G.Y., Csengeri, I., 1984. Differential response of lipid metabolism and membrane physical state by an actively and passively overwintering planktonic crustacean. Lipids. 19 (6), 436–442.

Fedorov, V.K., Sorokin Yu, I., 1967. Determination of the degree of algae, yeast and bacteria assimilation by certain representatives of Cladocera. Doklady Akad. Nauk. SSSR. 174 (4), 969–970.

Feldmanová, M., Hikscherová, K., Maršálek, B., Bláha, L., 2006. Effects of N-heterocyclic polyaromatic hydrocarbons on survival, reproduction, and biochemical parameters in *Daphnia magna*. Environ. Toxicol. 21 (4), 425–431.

Fernández-Casalderrey, A., Ferrando, M.D., Andreu-Moliner, E., 1993. Effect of the insecticide methylparathion on filtration and ingestion rates of *Brachyonus calyciflorus* and *Daphnia magna*. Sci. Total Environ. (Suppl. 2), 867–876.

Feroz, M., Podowski, A.A., Khan, M.A., 1990. Oxidative dehydrochlorination of heptachlor by *Daphnia magna*. Pestic. Biochem. Physiol. 36, 101–105.

Ferrando, M.D., Sancho, E., Andreu-Moliner, E., 1996. Accumulation of tetradifon in an algae (*Nannochloris oculata*) and the cladoceran, *Daphnia magna*. Bull. Environ. Contam. Toxicol. 57 (1), 139–145.

Ferrandon, D., Imler, J.L., Hoffmann, J.A., 2003. Sensing infection in *Drosophila*: Toll and beyond. Semin. Immunol. 16, 43–53.

Ferrão-Filho, A., da, S., Tessier, A.J., DeMott, W.R., 2007. Sensitivity of herbivorous zooplankton to phosphorus-deficient diets: testing stoichiometric theory and the growth rate hypothesis. Limnol. Oceanogr. 52 (1), 407–415.

Filenko, O.F., Lazareva, V.V., 1989. Influence of toxic agents on general biological and cytogenetic parameters in *Daphnia*. Gidrobiologiheskii Zhurnal. 25 (3), 56–59.

Filipe, S.R., Tomasz, A., Ligoxygakis, P., 2005. Requirements of peptidoglycan structure that allow detection by the *Drosophila* Toll pathway. EMBO Rep. 6, 327–333.

Filippov, A.A., Aminov, A.I., Golovanova, I.L., 2010. Influence of herbicide roundup on activity of carbohydrases of aquatic invertebrates. In: Ekologiya vodnykh bespozvonochnykh, Borok, 315–317.

Fischel, A., 1908. Untersuchungen über vitale Färbung an Süsswassertieren, insbesondere bei Cladoceren. Werner Klinkhardt, Leipzig. 69S. (same in: Internationale Revue der gesamt. Hydrobiol. und Hydrogr. 1, 73–141).

Fischer, J.M., Fields, P.A., Pryzbylkowski, P.G., Nicolai, J.L., Neale, P.J., 2006. Sublethal exposure to UV radiation affects respiration rates of the freshwater cladocera *Daphnia catawba*. Photochem. Photobiol. 82 (2), 547–550.

Fish, W., Morel, F.M.M., 1983. Characterization of organic copper-complexing agents released by *Daphnia magna*. Can. J. Fish. Aquat. Sci. 40 (8), 1270–1277.

Flaherty, C.M., Dodson, S.I., 2005. Effects of pharmaceuticals on *Daphnia* survival, growth, and reproduction. Chemosphere. 61, 200–207.

Flerov, B.A., Zhmur, N.S., Ochirova, M.N., Chalova, I.V., 1988. Method of biotesting of natural and effluent waters using *Ceriodaphnia dubia*. In: Metody biotestirovaniya vod. Chernogolovka Institute of Chemical Physics of AN SSSR p. 111–114.

Flint, J., Mackay, T.F., 2009. Genetic architecture of quantitative traits in mice, flies, and humans. Genome Res. 19, 723–733.

Flössner, D., 2000. Die Haplopoda aund Cladocera (ohne Bosminidae) Mitteleuropas. Backhyus, Leiden, p. 428.

Forró, L., Korovchinsky, N., Kotov, A., Petrusek, A., 2008. Global diversity of Cladocerans (Cladocera; Crustacea) in freshwater. Hydrobiologia. 595, 177–184.

Fox, H.M., Hardcastle, S.M., Dresel, E.I.B., 1949. Fluctuations in the haemoglobin content of *Daphnia*. Proc. R. Soc. Lond. B. 136, 388–399.

Flückiger, E., 1951. A case of disturbed metabolism in *Daphnia magna*. Nature (London). 168 (4266), 211.

Flückiger, E., 1952. Ueber die Wirkung von Sympathomimetica und Dihydroergotamin bei Daphnien. I. Wirkung auf Muskelfunktionen. Acta Physiol. Scand. 27 (2–3), 206–216.

Flückiger, E., 1953. Ueber die Wirkung von Sympathomimetica und Dihyroergotamin bei Daphnien. II. Wirkung auf Stoffwechsel. Acta Physiol. Scand. 30 (1), 33–44.

Flückiger, E., Flück, H., 1952. Die Wirkung vn Vitamin B1 am Herz von Daphnien. Experientia VIII. 6, 223–224.

Fogg, G.E., 1953. The Metabolism of Algae. London: Methuen, New York: Wiley.

Folt, C., 2005. *Daphnia* as a keystone conduit for metal trophic transfer. 7th International; Symposium on Cladocera, p. 12.

Fonviller, P.A., Itkin, S.I., 1938. On determination of permeability in *Daphnia magna*. Bulleten experimentalnoy biologii i meditsiny VI. 3, 271–274.

Forasacco, E., Fontvieille, D., 2008. Evaluation of respiration parameters in a few generations and clones *Simocephalus vetulus* (Crustacea, Cladocera) under controlled conditions. VIIIth International Symposium on Cladocera. Aguascalientes, Mexico. 21–25 October 2008, p. 32.

Forsyth, D.J., Haney, J.F., James, M.R., 1992. Direct observation of toxic effects of cyanobacterial extracellular products on *Daphnia*. Hydrobiologia. 228, 151–155.

Foss, P., Partali, V., Olsen, Y., Borch, G., Liaaen-Jensen, S., 1986. Animal carotenoids 29. New (2R)-2-hydroxy-4-keto-β-type carotenoids from *Daphnia magna* (Crustacea). Acta Chemica Scandnavica B. 40, 157–162.

Fox, H.M., 1945. The oxygen affinities of certain invertebrate haemoglobins. J. Exp. Biol. 21 (3–4), 161–165.

Fox, H.M., 1948. The haemoglobin of *Daphnia*. Proc. R. Soc., B. 135, 195–212.

Fox, H.M., 1950. Hemoglobin production in *Daphnia*. Nature. 166, 609–610.

Fox, H.M., 1952. Anal and oral intake of water by Crustacea. J. Exp. Biol. 29 (4), 583–599.

Fox, H.M., 1955. The effect of oxygen on the concentration of haem in invertebrates. Proc. R. Soc. B. 143 (911), 203–214.

Fox, H.M., Mitchell, Y., 1953. Relation of the rate of antennal movement in *Daphnia* to the number of eggs carried in the brood pouch. J. Exp. Biol. 30, 238–242.

Fox, H.M., hear, E.A., 1953. Factors influencing haemoglobin synthesis by *Daphnia*. Proc. R. Soc. B. 141, 179–189.

Fox, H.M., Hardcastle, S.M., Dresel, E.I.B., 1949. Fluctuations in the haemoglobin content of *Daphnia*. Proc. R. Soc. B. 143, 136, 884, 388-399.

Fox, H.M., Gilchrist, B.M., Phear, E.A., 1951. Functions of haemoglobin in *Daphnia*. Proc. R. Soc. B. 138, 514–528.

Fraysse, B., Geffard, O., Berthet, B., Biaganti-Risbourg, S., Geffard, A., 2006. Importance of metallothioneins in the cadmium detoxification in *Daphnia magna*. Comp. Biochem. Physiol., C. 144 (3), 286–293.

Frear, D.E.H., Boyd, J.E., 1967. Use of *Daphnia magna* for the microbioassay of pesticides. I. Development of standardized technique for rearing *Daphnia* and preparation of dosage-mortality curves for pesticides. J. Econ. Entomol. 60 (5), 1228–1236.

Freund, J.A., Schimansky-Geier, L., Besner, B., Neiman, A., Russell, D.F., Yakusheva, T., et al., 2002. Behavioral stochastic resonance: how the noise from a *Daphnia* swarm enhances individual prey capture by juvenile paddlefish. J. Theor. Boil. 214, 71–83.

Frey, D.G., 1962. Cladocera from the Eemian interglacial of Denmark. J. Paleontol. 36, 1133–1154.

Frey, D.G., 1965. Gynandromorphism in the chydoid Cladocera. Limnol. Oceanogr. 10 (Supplement), R103–R114.

Frey, D.G., 1980. The non-swimming chydorid cladocera of wet forests, with descriptions of a new genus and two new species. Intern. Revue ges. Hydrobiol. 65, 613–641.

Frey, D.G., 1982. Contrasting strategies of gamogenesis in northern and southern populations of Cladocera. Ecology. 63 (1), 223–241.

Frey, D.G., 1993. The penetration of cladocerans into saline lakes. Hydrobiologia. 267, 233–248.

Frey, D.G., Hann, B.J., 1985. Growth in Cladocera. Crustacean Issues. 2, 315–335.

Friberg-Jensen, U., Nachman, G., 2010. Early signs of lethal effects in *Daphnia magna* (Branchiopoda, Cladocera) exposed to the insecticide cypermethrin and the fungicide azosistrobin. Environ. Toxicol. Chem. 29 (10), 2371–2378.

Fries, G., 1964. Ueber die Einwirkung der Tagesperiodik und der Temperatur auf den Generationswechsel, die Weiblichengroesse und die Eier von *Daphnia magna* Straus. Zeitschr. Morphol. und Oekol. Tiere. 53 (5), 475–516.

Frisch, D., Morton, P.K., Chowdhury, P.R., Culver, B.W., Colbourne, J.K., Weider, L.J., et al. Accentuated selection signatures in response to environmental change in a millennial-scale *Daphnia* eggbank. Submitted.

Fritsch, R.H., 1953. Die Lebensdauer von *Daphnia* spec. bei verschiedner Ernährung bei Zugabe von Pantothensäure. Zeitschr. Wissensch. Zool. 157 (1), 35–56.

Fritsche, H., 1917. Studien über Schwankungen des osmotischen Druckes der Koerperflussigkeit bei *Daphnia magna*. Int. Rev. ges. Hydrobiol. und Hydrogr. 8 (1), 22–80, 124–203.

Fröhlich, T., Arnold, G., Fritsch, R., Mayr, T., Laforsch, C., 2009. LC-MS/MS-based proteome profiling in *Daphnia pulex* and *Daphnia longicephala*: the *Daphnia pulex* genome database as a key for high throughput proteomics in *Daphnia*. BMC Genomics. 10, 171.

Frost, B.J., 1975. Eye movement in *Daphnia pulex* (De Geer). J. Experim. Biol. 62 (1), 175–187.

Fryer, G., 1953. An eyless *Daphnia* and a note on its progeny. Naturalist, 118.

Fryer, G., 1962. Secretions of the labral and trunk limb glands in the cladoceran *Eurycercus lamellatus*. Nature. 195 (4836), 97.

Fryer, G., 1963. The functional morphology and feeding mechanism of the chydorid cladoceran *Eurycercus lamellatus* (O.F. Müller). Trans. R. Soc. Edinb. 65 (14), 335–381.

Fryer, G., 1968. Evolution and adaptive radiation in the Chydoridae (Crustacea: Cladocera): a study in comparative functional morphology and ecology. Philos. Trans. R. Soc. Lond., B. Biol. 254, 221–385.

Fryer, G., 1969. Tubular and glandular organs in the Cladocera, Chydoridae. Zool. J. Linn. Soc. 48, 1–8.

Fryer, G., 1970. Defecation in some macrothricid and chydorid cladocerans and some problems of water intake and digestion in the Anomopoda. Zool. J. Linn. Soc. 49 (4), 255–270.

Fryer, G., 1972. Observations on the ephippia of certain macrothricid cladocerans. Zool. J. Linn. Soc. 51, 79–96.

Fryer, G., 1974. Evolution and adaptive radiation in the Macrothricidae (Crustacea: Cladocera): a study in comparative functional morphology and ecology. Philos. Trans. R. Soc. Lond., B. Biol. 269 (898), 137–274.

Fryer, G., 1991. Functional morphology and adaptive radiation of the Daphniidae (Branchiopoda: Anomopoda). Philos. Trans. R. Soc. Lond., B. Biol. 331 (1259), 1–99.

Fryer, G., 1993. The Freshwater Crustacea of Yorkshire. Yorkshire Naturalists' Union & Leeds Philosophical and Literary Society, Kendal, p. 312.

Fryer, G., 1994. Natural history and the exploration of nature. Naturalist. 119, 3–22.

Fryer, G., 1996. Diapause, a potent force in the evolution of freshwater crustaceans. Hydrobiologia. 320, 1–14.

Fryer, G., 2004. *Daphnia* shoals: their behaviour and coordination. A problem for the field naturalist. Naturalist. 129, 41–60.

Fryer, G., 2006. The brine shrimp's tale: a topsy turvy evolutionary fable. J. Linn. Soc. 88, 377–382.

Fuchikawa, Y.-I., Fukunoti, N., Doida, Y., 1995. Effects of x-ray and ultrasound on formation of compound eye and normal evelopment of water flea. III. J. Radiat. Res. 36 (4), 313.

Fundel, B., Stich, H.-B., Schmid, H., Maier, G., 1998. Can phaeopigments be used as markers for *Daphnia* grazing in Lake Constance. J. Plankton Res. 20 (8), 1449–1462.

Gainutdinov, M.K., Yargunov, V.G., Krasnoperova, I.A., Plekahnova, V.A., Loboda, V.I., Kuznetsova, T.V., et al., 1997a. On a possibility of existence of the program of death of *Daphnia magna*. Doklady Akademii nauk. 355 (5), 708–711.

Gainutdinov, M.Kh., Yargunov, V.G., Krasnoperova, I.A., Kalinnikova, T.B., et al., 1997b. Mono- and disaccarides as signal molecules in the control system of hydrobios associations. Doklady Akademii nauk. 357 (2), 278–280.

Gainutdinov, M.Kh., Yargunov, V.G., Krasnoperova, I.A., Varlamov, V.E., 2000. Coding of chemosensory information by the nervous system of *Daphnia magna* and *Enchytraeus albidus* under high environmental temperature. Zhurnal evolutsionnoi biokhimii i fiziologii. 36 (3), 214–218.

Gajewskaja, N.S., 1938. On some new methods in studying the nutrition of water animals. II. The getting of bacteria free cultures of Cladocera, Ostracoda, Copepoda, and Rotatoria. Zoolocheskii Zhurnal. 17 (6), 1003–1017.

Gajewskaja, N.S., 1940. On methods of raising of living feed for fish. Trudy Moskovskogo tekhnicheskogo instituta rybnoi promychelnnosti i khozyaistva. 3, 3–16.

Gajewskaja, N.S., 1945. Attempt at determination of feeding ratio of algal feed for *Daphnia magna* under field conditions. Zoologichskii Zhurnal. 24 (2), 79–89.

Gajewskaja, N.S., 1948. Trophological trend in hydrobiology, its object, some principal problems and tasks. In: Pamyati akademika S.A. Zernova, Moscow-Leningrad: AN SSSR, 27–47.

Gajewskaja, N.S., 1949. On feeding selectivity in filter-feeders. Trudy Vsesoyuznogo Gidrobiologochskogo Obshchestva. 1, 150–174.

Galkovskaya, G.A., 1970. Food utilization by some species of pond zooplankton. Gidrobiologicheskii Zhurnal VI. 2, 57–62.

Gapochka, L.D., Gapochka, M.D., Dushkina, D.S., Isakova, E.F., Pavlova, A.S., Shavyrina, O.B., 2012. Irritation effects with *Daphnia magna* culture at the different stage of development by the low-intensity electromagnetic field of the millimetric range. Vestnik MGU.(2), 43–48, ser. 16.

Garcia, G.G., Nandini, S., Sarma, S.S.S., 2004. Effect of cadmium on the population dynamics of *Moina macrocopa* and *Macrothrix triserialis* (Cladocera). Bull. Environ. Contam. Toxicol. 72 (4), 717–724.

Gardner, W.S., Miller III, W.H., 1981. Intracellular composition and net release rates of free amino acids in *Daphnia magna*. Can. J. Fish. Aquatic sci. 38 (2), 157–162.

Gardner, W.S., Scavia, D., 1981. Kinetic examination of nitrogen release by zooplankters. Limnol. Oceanogr. 26 (5), 801–810.

Geller, W., 1975. Food ingestion of *Daphnia pulex* as a function of food concentration, temperature, animal's body length and hunger. Archiv. für Hydrobiol. Suppl. 48 (1), 47–107.

Gerke, P., Bording, C., Zeis, B., Paul, R.J., 2011. Adaptive haemoglobin gene control in *Daphnia pulex* at different oxygen and temperature conditions. Comp. Biochem. Physiol. A Mol. Integr. Physiol. 159, 56–65.

Gerrish, G.A., Cáceres, C.E., 2003. Genetic versus environmental infuence on pigment variation in the ephippia of *Daphnia pulicaria*. Freshw. Biol. 48, 1971–1982.

Gerritsen, J., 1982. Behavioral response of *Daphnia* to rate of temperature change; possible enhancement of vertical migration. Limnol. Oceanogr. 27 (2), 254–261.

Gerritsen, J., Porter, K.G., 1982. The role of surface chemistry in filter feeding by zooplankton. Science. 216 (4551), 1225–1227.

Gerschler, W., 1910. Über ein Extremitätienorgan zur Regelung der Blutcircilation bei *Leptodora kindtii* (Focke). Zool. Anzeig. 36B (18/19), 329–336.

Gershkovich, D.M., Filenko, O.F., Isakova, E.F., 2012. Potential toxicants as stimulants of vital function of hydrobionts. Materialy 4 vserossiyskoy konferentsii po vodnoi ekotoksikologii. Borok. 1, 80–83.

Gibble, R., Baer, K.N., 2003. Effects of 4-nonylphenol on sexual maturation in *Daphnia*. Bull. Environ. Contam. Toxicol. 70 (2), 315–321.

Gibson, G., Muse, S.V., 2009. A Primier of Genome Science. third ed. Sinauer Associates, Sunderland, Massachusetts.

Gicklhorn, J., 1925. Beobachtungen ueber die Kalkinkrustation der Schale der Cladoceren. Lotos (Prag). 73, 157–166.

Gicklhorn, J., 1931a. Elektive Vitalfärbungen im Dienste der Anatomie und Physiologie der Excretionsorgane vin Wirbellosen (Cladoceren als Beispiel). Protoplasma 13, 3/4 (Sonderheft), 701–724.

Gicklhorn, J., 1931b. Beobachtungen an den lateralen Frontalorganen von *Daphnia magna* M. nach elektiver Vital Färbung. Protoplasma 13, 3/4 (Sonderheft), 725–739.

Gicklhorn, J., 1931c. Elektive Vitalfärbungen. Ergebnisse Biol. 7, 549–685.

Gicklhorn, J., Keller, R., 1926. Neue Methoden der elektiven Vitalfärbung zwecks organspezifiescher Differenzierung bei Wirbellosen (Üeber den Bau, die Innervierung und Funktion der Riechstäbe von *Daphnia magna*). Zeitschrift wissensch. Zool. 127, 244–296.

Gillis, P.L., Chow-Frazer, P., Ranville, C.F., Ross, P.E., Wood, C.M., 2005. *Daphnia* need to be gut-cleared too: the effect of exposure to and ingestion of metal-contaminated sediment on the gut-clearance patterns of *D. magna*. Aquat. Toxicol. 71, 143–154.

Ginatullina, E.N., Kamaya, M., 2012. Application of filtering ability of *Daphnia magna* in elaboration of acute toxicity test by means of fluorescing microbeads. Materialy 4 vserossiyskoy konferentsii po vodnoi Ekotoksikologii. Borok. 1, 83–87.

Gingrich, D., Shaw II, C.F., Seidman, L., Remsen, C., 1984. A reductively labile cadmium-binding protein in *Daphnia pulicaria*. Mar. Environ. Res. 14 (1–4), 454–455.

Gladyshev, M.I., Dubovskaya, O.P., Gubanov, V.G., Makhotova, O.N., 2003. Evaluation of non-predatory mortality of two *Daphnia species* in a Siberian reservoir. J. Plankton Res. 25 (8), 999–1003.

Gladyshev, M.I., Sushchik, N.N., Kravchuk, E.S., Ivanova, E.A., Ageev, A.V., Kovaleva, G.S., 2005. Seasonal dynamics of standing stock of essential polyunsaturated fatty acids in biomass of phyto- and zoobenthos at a littoral station in the Eniei River. Doklady Akademii nauk. 403 (2), 277–278.

Glazier, D.S., 1991. Separating the respiration rates of embryos and brooding females of *Daphnia magna*: implications for the cost of brooding and the allometry of metabolic rate. Limnol. Oceanogr. 36 (2), 354–362.

Gliwicz, Z.M., 1968. The use of anaesthetizing substance in studies on the food habits of zooplankton communities. Ekologia polska, seria A. XVI (13), 279–295.

Gliwicz, Z.M., 1986. A lunar cycle in zooplankton. Ecology. 67 (4), 883–897.

Gliwicz, Z.M., Guisande, C., 1992. Family planning in *Daphnia*: resistance to starvation in offspring born to mothers grown at different food levels. Oecologia. 91, 463–467.

Gliwicz, Z.M., Noavida, M.J., 1996. Clutch size at first reproduction in *Daphnia pulicaria* at different levels of food and predation. J. Plankton Res. 18 (6), 863–880.

Glover, C.N., Wood, C.W., 2005. Physiological characterization of a pH- and calcium-dependent sodium uptake mechanism in a freshwater crustacean, *Daphnia magna*. J. Exp. Biol. 208 (5), 951–959.

Glover, C.N., Pane, E.F., Wood, C.M., 2005a. Humic substances influence sodium metabolism in the freshwater crustacean *Daphnia magna*. Physiol. Biochem. Zool. 78 (3), 405–416.

Glover, C.N., Playle, R.C., Wood, C.W., 2005b. Heterogeneity of natural organic matter amelioration of silver toxicity to *Daphnia magna*: effect of source and equilibration time. Environ. Toxicol. Chem. 24 (11), 2934–2940.

Gobert, V., Gottar, M., Matskevich, A.A., Rutschmann, S., Royet, J., Belvin, M., et al., 2003. Dual activation of the *Drosophila* Toll pathway by two pattern recognition receptors. Science. 302, 2126–2130.

Goksen, J.E., McNaught, D.C., 1995. A behavioral bioassay employing *Daphnia* for detection of sublethal effects: response to polarized light. Mar. Fresh. Behav. Physiol. 26, 267–272.

Goldmann, T., Becher, B., Wiedorn, K.H., Pirow, R., Deutschbein, M.E., Vollmer, E., et al., 1999. Epipodite and fat cells as sites of hemoglobin synthesis in the branchiopod crustacean *Daphnia magna*. Histochem. Cell Biol. 112, 335–339.

Gomes, R.L., Deacon, H.E., Lai, K.M., Birkett, J.W., Scrimshaw, M.D., Lester, J.N., 2004. An assessment of the bioaccumulation of estrone in *Daphnia magna*. Environ. Toxicol. Chem. 23 (1), 105–108.

Goodman, M., Czelusniak, J., Koop, B., Tagle, D., Slightom, J., 1987. Globins: a case study in molecular phylogeny, Cold Spring Harbor Symposia on Quantitative Biology, vol. 52. Cold Spring Harbor Laboratory Press, pp. 875–890.

Goodman, M., Pedwaydon, J., Czelusniak, J., Suzuki, T., Gotoh, T., Moens, L., et al., 1988. An evolutionary tree for invertebrate globin sequences. J. Mol. Evol. 27, 236–249.

Goodrich, M.S., Lech, J.J., 1990. A behavioral screening assay for *Daphnia magna*: a method to assess the effects of xenobiotics on spatial orientation. Environ. Contam. Toxicol. 9 (1), 21–30.

Goodwin, T.W., 1955. Carotenoids. Ann. Rev. Biochem. 24, 497–522.

Goodwin, T.W., 1960. Biochemistry of pigments. In: Waterman, T.H. (Ed.), The Physiology of Crustacea, vol. 1. Academic Press, New York and London, pp. 101–140.

Gophen, M., 1976. Temperature dependence of food intake, ammonia excretion in *Ceriodaphina reticulata* (Jurine) (Lake Kinneret, Israel). Freshw. Biol. 6 (5), 451–455.

Gophen, M., Geller, W., 1984. Filter mesh size and food particle uptake by *Daphnia*. Oecologia. 64 (3), 408–412.

Gophen, M., Gold, B., 1981. The use of inorganic substances to stimulate gut evacuation in *Daphnia magna*. Hydrobiologia. 80, 43–45.

Gorbunov, A.K., Gorbunova, A.V., Kosova, A.A., Semenova, N.N., Ivanov, V.M., 1995. Disease agents of invertebrates in the Volga Delta. The Caspian Sea—present and future. Abstracts, International Conference, Astrakhan, November 16–17, 1995, 73–74.

Gorbunova, A.V., 1993. Impact of water turbidity on plankton filter-feeding crustaceans. Diss. Abstract. Moscow State University. p. 23.

Goren, L., Ben-Ami, F., 2013. Ecological correlates between cladocerans and their endoparasites from permanent and rainy pools: patterns in community composition and diversity. Hydrobiologia. 701, 13–23.

Gorokhova, E., Kyle, M., 2002. Analysis of nucleic acids in *Daphnia*: development of methods and ontogenetic variations in RNA-DNA content. J. Plankton Res. 24 (5), 511–522.

Gorr, T.A., 2004. *Daphnia* and *Drosophila*: two invertebrate models for O2 responsive and HIF-mediated regulation of genes and genomes. Int. Congr. Series. 1275, 55–62.

Gorr, T.A., Cahn, J,S., Yamagata, H., Bunn, H.F., 2004. Hypoxia-induced synthesis of hemoglobin in the crustacean *Daphnia magna* is hypoxia-inducible factor-dependent. J. Biol. Chem. 279 (34), 36038–36047.

Gorr, T.A., Gassmann, M., Wappner, P., 2006a. Sensing and responding to hypoxia via HIF in model invertebrates. J. Insect. Physiol. 52, 349–364.

Gorr, T.A., Rider, C.V., Wang, H.Y., Olmstead, A.W., LaBlanc, G.A., 2006b. A candidate juvenoid hormone receptor *cis*-element in the *Daphnia magna* hb2 hemoglo bin gene promoter. Mol. Cell. Endocrinol. 247 (1–2), 91–102.

Gorski, P.R., Dodson, S.I., 1996. Free-swimming *Daphnia pulex* can avoid following Stokes' law. Limnol. Oceanogr. 41 (8), 1815–1821.

Goss, L.B., Bunting, D.L., 1976. Thermal tolerance of zooplankton. Water Res. 10 (5), 361–365.

Goto, T., Hiromi, J., 2003. Toxicity of 17α-ethynylestradiol and norethindrone, constituents of an oral contraceptive pill to the swimming and reproduction of cladoceran *Daphnia magna*, with special reference to their synergic effect. Mar. Pollut. Bull. 47 (1–6), 139–142.

Goulden, C.E., 1966. Co-occurence of moinid Cladocera and possible isolating mechanisms. Verhandl. Internat. Verein. Limnol. 16 (3), 1669–1672.

Goulden, C.E., 1968. The systematics and evolution of the Moinidae. Trans. Am. Philo. Soc. New series. 58 (Part 6), 1–101.

Goulden, C.E, Hornig, L.L., 1980. Population oscillations and energy reserves in planktonic cladocera and their consequences to competition. Proc. Nat. Acad. Sci. 77 (3), 1716–1720.

Goulden, C.E., Place, A.R., 1993. Lipid accumulation and allocation in daphniid Cladocera. Bull. Marine Sci. 53 (1), 106–114.

Goulden, C.E., Comotto, R.M., Hendrickson, J.A. Jr., Hornig, L.L., Johnson, K.L., 1982. Procedures and recommendations for the culture and use of *Daphnia* in bioassay studies. In: Aquatic toxicology and hazard assessment, 5th conference, ASTM. Pearson J.G., Foster R.B., abd Bishop W.E., (Eds.), 139–160.

Goulden, C.E., Henry, L., Berrigan, D., 1987. Egg size, post-embryonic yolk, and survival ability. Oecologia. 72 (1), 28–31.

Gourlay, C., Tusseau-Vuillemin, M.-H., Garric, J., Mouchel, J.-M., 2003. Effect of dissolved organic matter of various origins and biodegradabilities on the bioaccumulation of polycyclic aromatic hydrocarbons in *Daphnia magna*. Environ. Toxicol. Chem. 22 (6), 1288–1294.

Graham, A.L., Shuker, D.M., Pollitt, L.C., Auld, S.K.J.R, Wilson, A., Little, T.J., 2011. Fitness consequences of immune responses: strengthening the empirical framework for ecoimmunology. Funct. Ecol. 25, 5–17.

Grant, J.W.G., Bayly, I.A.E., 1981. Predator induction of crests in morphs of the *Daphnia carinata* King complex. Limnol. Oceanogr. 26 (2), 201–218.

Green, J., 1954. Size and reproduction in *Daphnia magna* (Crustacea: Cladocera). Proc. Zool. Soc. Lond. 124 (3), 535–545.

Green, J., 1955. Haemoglobin in the fat cells of *Daphnia*. Quart. J. Microsc. Sci. 96 (2), 173–176.

Green, J., 1956a. Growth, size and reproduction in *Daphnia* (Crustacea: Cladocera). Proc. Zool. Soc. Lond. 126 (2), 173–204.

Green, J., 1956b. Variation in the haemoglobin content of *Daphnia*. Proc. R. Soc. B. 919 (145), 214–232.

Green, J., 1956c. The structure and function of the post-abdomen of Camptocercus (Crustacea, Cladocera). Proc. Zool. Soc. Lond. 126, 283–290.

Green, J., 1957. Carotenoids in *Daphnia*. Proc. R. Soc. B. 928 (147), 392–401.

Green, J., 1961a. Biliverdin in the eyes of *Polyphemus pediculus* (L.) (Crustacea, Cladocera). Nature. 4760, 227–228.

Green, J., 1961b. A Biology of Crustacea. Quadranglebooks, Chicago, p. 180.

Green, J., 1966a. Seasonal variation in eggs production by Cladocera. J. Anim. Ecol. 35 (1), 77–104.

Green, J., 1966b. Variation in carotenoid pigmentation of *Simocephalus vetulus* (Crustacea: Cladocera). J. Zool. 149 (2), 174–187.

Green, J., 1971. Pigments in Cladocera. Trans. Amer. Microsc. Soc. 80 (1), 115.

Green, J., 1974. Parasites and epibionts of Cladocera. Trans. Zool. Soc. Lon. 32, 417–515.

Gregersen, P.K., 2009. Closing the gap between genotype and phenotype. Nat. Genet. 41, 958–959.

Gregory, T.R., 2002. Genome size and developmental complexity. Genetica. 115, 131–146.

Gregory, T.R., 2005. Genome size evolution in animals. In: Gregory, T.R. (Ed.), The Evolution of the Genome. Elsevier, San Diego, pp. 3–87.

Gregory, T.R., 2013. Animal genome size database. <http://www.genomesize.com>. (accessed 14.06.2013.).

Gremiachikh, V.A., Tomilina, I.I., 2010. Peculiarities of mercury compounds accumulation by planktonic crustacean *Ceriodaphnia affinis*. Gidrobioliogicheskii Zhurnal. 46 (4), 63–74.

Grey, J, Jones, R.I., 2007. Seasonal changes in the importance of the source of organic matter to the diet of zooplankton in Loch Ness, as indicated by stable isotope analysis. Limnol. Oceanogr. 46 (3), 505–513.

Gries, T., Gude, H., 1999. Estimates of nonconsumptive mortality of mesoplankton by measurements of sedimentation losses. Limnol. Oceanogr. 44 (2), 459–465.

Griffiths, P.R.E., 1980. Morphological and ultrastructural effects of sublethal cadmium poisoning on *Daphnia*. Environ. Res. 22 (2), 227–284.

Grobben, C., 1879. Die Entwicklungsgeschichte der *Moina rectirostris*. Arb. Zool. Inst. Wien. 2, 203.

Gromov, A.E., 1992. Influence of light with different wave length on phototaxis of *Daphnia*. Gidrobiologicheskii Zhurnal. 28 (4), 63–67.

Grosser, B.I., Baylor, E.R., Smith, F.E., 1953. Analysis of geotactic responses in *Daphnia magna*. Ecology. 34 (4), 804–805.

Gruithuisen, F.V.P., 1828. Über die Daphnia sima und ihrer Blutkreislauf. Verhandlungen der Keiserlichen Leopoldinisch-Carolinischen Akademie der Naturforscher 6 (Nova acta physico-medica Acadeniae Cesareae Leopoldino-Carolinae naturae curiosorum 14, 2) 399–406.

Guadagnoli, J., Braun, A., Roberts, S., Reiber, C., 2005. Environmental hypoxia influences hemoglobin subunit composition in the branchiopod crustacean *Triops longicaudatus*. J. Exp. Biol. 208, 3543–3551.

Guan, R., Wang, W.-X., 2004a. Dietary assimilation and elimination of Cd, Se, and Zn by *Daphnia magna* at different metal concentrations. Environ. Toxicol. Chem. 23 (11), 2689–2698.

Guan, R., Wang, W.-X., 2004b. Cd and Zn uptake kinetics in *Daphnia magna* in relation to Cd exposure history. Environ. Sci. Technol. 38, 6051–6058.

Guan, R., Wang, W.-X., 2006. Multigenerational cadmium acclimation and biokinetics in *Daphnia magna*. Environ. Pollut. 141 (2), 343–352.

Gudkov, D.I., Kipnis, L.S., 1995. Chronic effect of tritium water on biological parameters of *Daphnia magna* Straus. Gidrobiologicheskii Zhurnal. 31 (4), 96–103.

Gueldner, F.H., Wolff, J., 1970. Ueber die Ultrastruktur des Komplexauges des Wasserflohs *Daphnia pulex*. Zeitschr. Zellforsch. 104 (2), 259–274.

Guenzl, H., 1991. The ultrastructure of the posterior gut and caecum in *Alona affinis* (Crustacea, Cladocera). Zoomorphology. 110, 139–144.

Guides to the Identification of the Macroinvertebrates of the Continental Waters of the World, H.J. F. Dumont, Ed., vol. 1 (Macrothricidae), 1992, p. 143.; 3 (Ctenopoda), 1992, p. 82.; 11 (Chydorinae and Sayciinae), 1996, p. 197.; 13 (Predatory Cladocera—Rivier); 17 (Simocephalus), 2001, p. 130.; 21 (Daphnia), 2005, p. 376.; 22 (Ilyocryptidae—Kotov and Stifter), 2006, p. 172.

Guilhermino, L., Lacerda, M.N., Nogueira, A.S.A., Soares, A.M.V.M., 2000. *In vitro* and *in vivo* inhibition of *Daphnia magna* acetylcholinesterase by surfactant agents: possible implications for contamination biomonitoring. Sci. Total Environ. 247 (2–3), 137–141.

Guisande, C., Toja, J., Mazualos, N., 1991. The effect of food on protein content in rotifer and cladoceran species; a field correlative study. Freshw. Biol. 26 (3), 433–438.

Gulati, R.D., 1978. The ecology of common planktonic crustacean of the freshwaters in Netherlands. Hydrobiiologia. 59 (2), 101–112.

Hackman, R.Y., 1964. Chemistry of the chitin cuticle. In: Rockstein, R. (Ed.), The Physiology of Insecta, vol. III. Academic Press, New York and London, pp. 471–506. Chapter 8.

Hairston, N.G., De Meester, L., 2008. *Daphnia* paleogenetics and environmental change: deconstructing the evolution of plasticity. Int. Rev. Hydrobiol. 93, 578–592.

Halcrow, K., 1969. Sites of presumed neurosecretory activity in *Daphnia magna* Straus. Can. J. Zool. 47 (4), 575–577.

Halcrow, K., 1982. Some ultrastructural features of the nuchal organ of *Daphnia magna* Straus (Crustacea, Branchiopoda). Can. J. Zool. 60, 1257–1264.

Halcrow, K., 1985. A note on the significance of the neck organ of *Leptodora kindtii* (Focke) (Crustacea, Cladocera). Can. J. Zool. 63, 738–740.

Hall, T.M., 1982. Free ionic nickel accumulation and localization in the freshwater zooplankter, *Daphnia magna*. Limnol. Oceanogr. 27 (4), 718–727.

Hall, S.R., Becker, C.R., Duffy, M.A., Cáceres, C.E., 2010. Variation in resource acquisition and use among hosts can create key epidemiological tradeoffs. Am. Nat. 176, 557–565.

Hallam, T.G., Lassiter, R.R., Li, J., Suarez, L.A., 1990. Modeling individuals employing an integrated energy response: application to *Daphnia*. Ecology. 71 (3), 938–954.

Hamilton, W.D., 1980. Sex versus non-sex versus parasite. Oikos. 35, 282–290.

Hanazato, T., 1990. Induction of helmet development by a Chaoborus factor in *Daphnia ambigua* during juvenile stages. J. Plankton Res. 12 (6), 1287–1294.

Hanazato, T., 1991. Influence of food density on the effects of a Chaoborus-released chemical on *Daphnia ambigua*. Freshw. Biol. 25, 477–483.

Hanazato, T., 1995. Combined effect of the insecticide carbaryl and the *Chaoborus* kairomone on helmet development in *Daphnia ambigua*. Hydrobiologia. 310, 95–100.

Hanazato, T., 1996. Combined effects of food shortage and oxygen deficiency on life history characteristics and filter screens of *Daphnia*. J. Plankton Res. 18 (5), 757–765.

Hanazato, T., Dodson, S.I., 1992. Complex effects of a kairomone of Chaoborus and an insecticide on *Daphnia pulex*. J. Plankton Res. 14 (12), 1743–1755.

Hanazato, T., Ooi, T., 1992. Morphological responses of *Daphnia ambigua* to different concentrations of a chemical extract from Chaoborus. Freshw. Biol. 27 (3), 379–385.

Hancy, J.F., Hall, D.J., 1975. Diel vertical migration and filter-feeding activities of *Daphnia*. Archiv. für Hydrobiol. 75 (4), 413–441.

Haney, J.F., 1985. Regulation of cladoceran feeding rates in nature by body size, food concentration, and diel feeding patterns. Limnol. Oceanogr. 10 (2), 397–411.

Haney, J.F., Brauer, M., Nurnberg, G., 1986. Feeding and egestion rates of individual zooplankton using Cherenkov counting. Hydrobiologia. 141 (3), 165–174.

Hann, B., 1980. Population differentiation in Eurycercus (Bullatifrons) species complex (Cladocera, Chysoridae) in Eastern North America. Ph. D. thesis. Indiana University.

Hann, B.J., 1987. Naturally occurring interspecific hybridization in *Simocephalus* (Cladocera, Daphniidae): its potential significance. Hydrobiologia. 145, 219–224.

Hansen, P.D., 1980. Uptake and transfer of the chlorinated hydrocarbon lindane (γ-BHC) in a laboratory freshwater food chain. Environ. Pollut. 21 (2), 97–108.

Hansen, U., Peters, W., 1997. Structure and permeability of the peritrophic membranes of some small crustaceans. Zoolog. Anzeiger. 236 (2–3), 103–108.

Hardie, D.C., Gregory, T.R., Hebert, P.D.N., 2002. From pixels to pictograms: a beginners' guide to genome quantification by Feulgen image analysis. J. Histochem. Cytochem. 50, 735–749.

Hardie, D.C., Hebert, P.D.N., 2004. Genome-size evolution in fishes. Can. J. Fish. Aq. Sci. 61, 1636–1646.

Hardison, R., 1998. Hemoglobins from bacteria to man: evolution of different patterns of gene expression. J. Exp. Biol. 201, 1099–1117.

Hardison, R.C., 2012. Evolution of hemoglobin and its genes. Cold Spring Harbor Perspect. Med. 2, a011627.

Harris, K.D., Bartlett, N.J., Lloyd, V.K., 2012. *Daphnia* as an emerging epigenetic model organism. Genet. Res. Int. 2012, 147892.

Hardy, W.B., 1892. The blood corpuscles of the Crustacea, together with a suggestion as to the origin of the crustacean fibrin-ferment. J. Physiol. (Lond.). 13, 165–190.

Hardy, W.B., McDougall, W., 1894. On the structure and functions of the alimentary canal of *Daphnia*. Proc. Camb. Philos. Soc. VIII (II), 41–50.

Harnisch, O., 1949. Vergleichende Beobachtungen zum Nahrungserwerb von *Daphnia* und *Ceriodaphnia*. Zool. Jahrb., Abt. Syst. Oekol. Tiere. 78 (2), 173–192.

Harris, J.E., 1953. Physical factors involved in the vertical migration of plankton. Quart. J. Microsc. Sci. 94 (4), 537–550.

Harris, J.E., 1963. The role of endogenous rhythms in vertical migration. J. Marine Biol. Assoc. U. K. 43 (1), 153–166.

Harris, J.E., Mason, P., 1956. Vertical migration in eyeless *Daphnia*. Proc. R. Soc., B. 145 (919), 280–290.

Harvey, A., 1972. A semi-continuous culture technique for *Daphnia pulex*. J. Appl. Ecol. 9 (3), 831–834.

Harwood, J., Jones, A.L., 1989. Lipoid metabolism in algae. Adv. Bot. Res. 16, 1–53.

Hasler, A.D., 1937. The physiology of digestion in plankton Crustacea. II. Further studies on the digestive enzymes of (A) *Daphnia* and *Polyphemus*; (B) *Diaptomus* and *Calanus*. Biol. Bull. 72 (3), 290–298.

Hatch, A.C., Burton Jr., G.A., 1999. Phototoxicity of fluoranthene to two freshwater crustaceans, *Hyalella azteca* and *Daphnia magna*: measures of feeding inhibition as a toxicological endpoint. Hydrobiologia. 400, 243–248.

Havas, M., Likens, E., 1985. Toxicity of aluminum and hydrogen ions to *Daphnia catawba, Holopedium gibberum, Chaoborus punctipennis, Chironomus anthracinus* from Mirror Lake, New Hampshire. Can. J. Zool. 63 (5), 1114–1119.

Havel, J.E., 1985. Cyclomorphosis of *Daphnia pulex* spined morphs. Limnol. Oceanogr. 30 (4), 853–861.

Havens, K.E., 1992. Acid and aluminum effects on sodium homeostasis and survival of acid-sensitive and acid-tolerant Cladocera. Can. J. Fish. Aquat. Sci. 49 (11), 2392–2398.

Hazen, W.E., Baylor, E.R., 1962. Behavior of *Daphnia* in polarized light. Biol. Bull. 123 (2), 243–252.

He, W., Megharaj, M., Naidu, R., 2009. Toxicity of tri- and pentavalent arsenic, alone and in combination, to the cladoceran *Daphnia carinata*: the influence of microbial transformation in natural waters. Environ. Geochem. Health. 31 (Suppl. 1), 133–141.

He, X., Wang, W.-S., 2006a. Releases of ingested phytoplankton carbon by *Daphnia magna*. Freshw. Biol. 51, 649–665.

He, X., Wang, W.-S., 2006b. Relative importance of inefficient feeding and consumer excretion to organic carbon flux from *Daphnia*. Freshw. Biol. 51, 649–665.

Heberdey, R.F., Kupka, E., 1942. Hellichkeitsunter scheidungvermoegen von *Daphnia pulex*. Zeitschrift für vergleichende Physiologie. 29 (4), 541–582.

Hebert, P.D.H., 1973. Phenotypic variability of lactate dehydrogenase in *Daphnia magna*. J. Exp. Zool. 186 (1), 33–38.

Hebert, P.D.N., 1978a. The population biology of *Daphnia* (Crustacea, Daphnidae). Biol. Rev. 53, 387–426.

Hebert, P.D.N., 1978b. The adaptive significance of cyclomorphosis in *Daphnia*: more possibilities. Freshw. Biol. 8, 313–320.

Hebert, P.D.N., 1987. Genotypic characteristics of the Cladocera. Hydrobiologia. 145, 183–193.

Hebert, P.D.N., Emery, C.J., 1990. The adaptive significance of cuticular pigmentation in *Daphnia*. Funct. Ecol. 4, 703–710.

Hebert, P.D.N., Finston, T.L., 1996. A taxonomic reevaluation of North American *Daphnia* (Crustacea: Cladocera). II. New species in the *Daphnia pulex* group from the south-central United States and Mexico. Can. J. Zool. 74 (4), 632–653.

Hebert, P.D.N., Finston, T.L., 1997. A taxonomic reevaluation of North American *Daphnia* (Crustacea: Cladocera). III. The *D. catawba* complex. Can. J. Zool. 75 (8), 1254–1997.

Hebert, P.D.N., Grewe, P.M., 1985. Chaoborus-induced shifts in the morphology of *Daphnia ambigua*. Limnol. Oceanogr. 30 (6), 1291–1297.

Hebert, P.D.N., McWalter, D.B., 1983. Cuticular pigmentation in arctic *Daphnia*: adaptive diversification of asexual lineages. Am. Nat. 122 (2), 286–291.

Hebert, P.D.N., Moran, C., 1980. Enzyme variability in natural populations of *Daphnia carinata* King. Heredity. 45 (3), 313–321.

Hebert, P.D.N., Taylor, D.J., 1997. The future cladoceran genetics: methodologies and targets. Hydrobiologia. 360, 295–299.

Hebert, P.D.N., Wilson, C.C., 2000. Diversity of the genus *Daphniopsis* in the saline waters of Australia. Can. J. Zool. 78 (5), 794–808.

Hebert, P.D.N., Beaton, M.J., Schwartz, S.S., Stanton, D.J., 1989. Polyphyletic origins of asexuality in *Daphnia pulex*. I. Breeding-system variation and levels of clonal diversity. Evolution. 43, 1004–1015.

Hebert, P.D.N., Schwartz, S.S., Hrbacek, J., 1989. Patterns of genotypic diversity in Czechoslovakian *Daphnia*. Heredity. 62, 207–216.

Hebert, P.D., Um, Y.M., Prokopowich, C.D., Taylor, D.J., 1999. Gene conversion and evolution of daphniid hemoglobins (Crustacea, Cladocera). J. Mol. Evol. 49, 769–779.

Hebert, P.D.N., Witt, J.D.S., Adamowicz, S.J., 2003. Phylogeographic patterning in *Daphnia ambigua*: regional divergence and intercontinental cohesion. Limnol. Oceanogr. 48, 261–268.

Heckmann, L.H., Sibly, R.M., Connon, R., Hooper, H.L., Hutchinson, T.H., Maund, S.J., et al., 2008. Systems biology meets stress ecology: linking molecular and organismal stress responses in *Daphnia magna*. Genome Biol. 9, R40.

Heintz, E., 1964. Le comportement de *Daphnia* dans les solutions de coumarine et de sulfate de cuivre. Influence de la concentration. Com. Rendues Acad. Sci. 258 (13), 3572–3574.

Heisey, D., Porter, K.G., 1977. Th effect of ambient oxygen concentration on filtering and respiration rate of *Daphnia galeata mendotae* and *Daphnia magna*. Limnol. Oceanogr. 22 (5), 839–845.

Heisig-Gunkel, Gunkel, G., 1982. Distribution of a herbicide (atrazin, s-triazine) in *Daphnia pulicaria*: a new approach to determination. Archiv. für Hydrobiol., Suppl. 59 (4), 359–376.

Herbert, M.R., 1954. The tolerance of oxygen deficiency in the water by certain Cladocera. Mem. Istituto Italiano di Idrobiologia. 8, 97–107.

Herbes, S.E., Risi, G.F., 1978. Metabolic alteration and excretion of anthracene by *Daphnia pulex*. Bull. Environ. Contam. Toxicol. 19 (2), 147–155.

Herodek, S., Farkas, T., 1967. Gas chromatographic studies on the fatty acid composition of some fresh-water crustaceans. Annal. Biol. Tihany. 34, 147–152.

Hérouard, E., 1905. La circulation chez les daphnies. Memoires de la Soc. Zool. de Paris. 18, 214–232.

Herrick, C.L., 1884. A final report on the Crustacea of Minnesota included in the orders Cladocera and Copepoda. The Geological and Natural History Survey of Minnesota. Minneapolis. p. 192.

Herring, P.J., 1968. The carotenoid pigments of *Daphnia magna* Straus–I. The pigments of animals fed *Chlorella pyrenoidosa* and pure carotenoids. II. Aspects of pigmentary metabolism. Comp. Biochem. Physiol. 24 (1), 187–221.

Hessen, D.O., 1985. Filtering structures and particle size selection in coexisting Cladocera. Oecologia (Berlin). 66 (3), 368–372.

Hessen, D.O., 1990. Carbon, nitrogen and phosphorus status in *Daphnia* at various food conditions. J. Plankton Res. 12 (6), 1239–1249.

Hessen, D.O., 1999. UV-B susceptibility and photoprotection of Arctic *Daphnia* morphotypes. Polar Res. 18, 345–352.

Hessen, D.O., Lyche, A., 1991. Inter- and intraspecific variation in zooplankton element composition. Arch. Hydrobiol. 121 (3), 343–353.

Hessen, D.O., Rukke, N.A., 2000a. The costs of moulting in *Daphnia*; mineral regulation of carbon budgets. Freshw. Biol. 45, 169–178.

Hessen, D.O., Rukke, N.A., 2000b. UV radiation and low calcium as mutual stressors for *Daphnia*. Limnol. Oceanogr. 45 (8), 1834–1838.

Hessen, D.O., Alstad, N.E.W., Skardal, L., 2000a. Calcium limitation in *Daphnia magna*. J. Plankton Res. 22 (3), 553–568.

Hessen, D.O., Borgeraas, J., Ørbæk, J.B., 2002. Responses in pigmentation and anti-oxidant expression in Arctic *Daphnia* along gradients of DOC and UV exposure. J. Plankton Res. 24 (10), 1009–1017.

Heugens, E.H.W., Jager, T., Creyghton, R., Kraak, M.H.S., van Straelen, N.M., Admiraal, W., 2003. Temperature-dependent effect of cadmium on *Daphnia magna*: accumulation versus sensitivity. Environ. Sci. Technol. 37 (10), 2145–2151.

Hildemann, W.H., Keighley, G., 1955. Techniques for studies of hemoglobin synthesis in *Daphnia*. Am. Nat. 89 (846), 169–174.

Hillbricht-Ilkowska, A., Karabin, A., 1979. An attempt to estimate consumption, respiration, and production of *Leptofora kindtii* (Focke) in field and laboratory conditions. Polske archwum hydrobiologii. 17 (30), 1/2, 81–86.

Hirosaki, Y., 1953. Studies on life-history of some Cladocerans. Miscellaneous reports of the Research Institute for Natural Resources, Tokyo, 32, 84–91.

Hiruta, C., Tochinai, S., 2012. Spindle assembly and spatial distribution of γ-tubulin during abortive meiosis and cleavage division in the parthenogenetic water flea *Daphnia pulex*. Zool. Sci. 29, 733–737.

Hiruta, C., Nishida, C., Tochinai, S., 2010. Abortive meiosis in the oogenesis of parthenogenetic *Daphnia pulex*. Chromosome Res. 18, 833–840.

Hobæck, A., Wolf, H.G., 1991. Ecological genetics of Norwegian *Daphnia*. II. Distribution of *Daphnia longispina* genotypes in relation to shortwave radiation and water colour. Hydrobiologia. 225, 229–243.

Hobæk, A., Skage, M., Schwenk, K., 2004. *Daphnia galeata* × D. longispina hybrids in western Norway. Hydrobiologia. 526, 55–62.

Hodgins-Davis, A., Townsend, J.P., 2009. Evolving gene expression: from G to E to G × E. Trends Ecol. Evol. 24, 649–658.

Hoffmann, A.A., Willi, Y., 2008. Detecting genetic responses to environmental change. Nat. Rev. Genet. 9, 421–432.

Hoffmann, F.G., Opazo, J.C., Storz, J.F., 2008. Rapid rates of lineage-specific gene duplication and deletion in the α-globin gene family. Mol. Biol. Evol. 25, 591–602.

Hofmann, G.E., Todgham, A.E., 2010. Living in the now: physiological mechanisms to tolerate a rapidly changing environment. Annu. Rev. Physiol. 72, 127–145.

Hoenicke, R., Goldman, C.R., 1987. Resource dynamics and seasonal changes in competitive interactions among three Cladocera species. J. Plankt. Res. 9 (3), 397–417.

Hohnke, L., Scheer, B.T., 1970. Carbohydrate metabolism in Crustacea. Chemical Zoology, vol. V. Arthropoda, Part A. Academic Press, New York and London, p. 147–197.

Holm, N.P., Shapiro, J., 1984. An examination of lipid reserves and the nutritional status of *Daphnia pulex* fed *Aphanozomenin flos-aquae*. Limnol. Oceanogr. 29 (5), 1137–1140.

Holm, N.P., Ganf, G.G., Shapiro, J., 1983. Feeding and assimilation rates of *Daphnia pulex* fed *Aphanizomenon flos-aquae*. Limnol. Oceanogr. 28 (4), 677–687.

Holm-Jensen, I., 1948. Osmotic regulation in *Daphnia magna* under physiological conditions and in the presence of heavy metals. Det Kgl. Danske Videnskabernes Selskab. Biologiske Meddelelser. 11, 1–64.

Holst, L.L., Giesy, J.P., 1989. Chronic effect of the photoinduced toxicity of anthracene on *Daphnia magna* reproduction. Environm. Toxicol. Chem. 8 (10), 933–942.

Homer, D.H., Waller, W., 1983. Chronic effect of reduced dissolved oxygen on *Daphnia magna*. Water Air Soil Pollut. 20 (1), 23–28.

Hoshi, T., 1950a. Studies on physiology and ecology of plankton. III. Changes in respiratory quotient during embryonic development of a *Daphnia*, *Simocephalus vetulus* (O.F. Müller). Sci. Rep. Tohoku Univ., 4th ser. (Biol.). 18 (3), 316–323.

Hoshi, T., 1950b. Studies on physiology and ecology of plankton. V. Fatty substance in development of *Simocephalus vetulus* with reference to behavior of yolk granule. Sci. Rep. Tohoku Univ., 4th ser. (Biol.). 18 (4), 464–466.

Hoshi, T., 1951a. Studies on physiology and ecology of plankton. VI. Glycogen in embryonic life of *Simocephalus vetulus*, with some notes on the energy source of development. Sci. Rep. Tohoku Univ., 4th ser. (Biol.). 19 (1), 123–132.

Hoshi, T., 1951b. Studies on physiology and ecology of plankton. VII. Hatchability of the summer egg of a *Daphnia*, *Simocephalus vetulus*. Sci. Rep. Tohoku Univ., 4th ser. (Biol.). 19 (2), 145–151.

Hoshi, T., 1953. Studies on physiology and ecology of plankton. IX. Changes in the glycogen content during development of the daphnid, *Simocephalus vetulus*, under aerobic and anaerobic conditions. Sci. Rep. Tohoku Univ., 4th ser. (Biol.). 20 (1), 6–10.

Hoshi, T., 1957. Studies on physiology and ecology of plankton. XIII. Haemoglobin and its role in the respiration of the daphnid, *Simocephalus vetulus*. Sci. Rep. Tohoku Univ., 4th ser. (Biol.). 23 (1), 35–58.

Hoshi, T., 1963a. Studies on physiology and ecology of plankton. XV. Crystal of pyridine-haemochromogen from the blood haemoglobin of the daphnid, *Simocephalus vetulus*. J. Fac. Sci. Niigata Univ., ser II. 4, 87–92.

Hoshi, T., 1963b. Studies on physiology and ecology of plankton. XVI. Crystal of haemin from the blood haemoglobin of the daphnid, *Simocephalus vetulus*. J. Fac. Sci. Niigata Univ., ser. II. 4, 93–95.

Hoshi, T., Akuzawa, K., 1979. Studies on physiology and ecology of plankton. XXXII. Physiological notes on O2-behavior in *Simocephalus* and *Daphnia* from O2-response of haemoglobin and cytochrome. Sci. Rep. Niigata Univ., ser. D (Biol.). 16, 13–18.

Hoshi, T., Inada, Y.-Y., 1973. Studies on physiology and ecology of plankton. XVI. O2-consuption, thoracic limb movement and O2 dissociation from haemoglobin of *Daphnia magna in vivo*. Sci. Rep. Niigata Univ., ser. D (Biol.). 10, 79–86.

Hoshi, T., Kobayashi, K., 1971. Studies on physiology and ecology of plankton. XXV. Iron-content and multi-molar extinction coefficient of the *Daphnia*-haemoglobin. Sci. Rep. Niigata Univ., ser. D (Biol.). 8, 65–68.

Hoshi, T., Nagumo, S., 1964. Studies on physiology and ecology of plankton. XVII. Notes on the blood haemoglobin of a water flea, *Moina macrocopa*. Sci. Rep. Niigata Univ. (formerly: J. Fac. Sci. Niigata Univ.). 1, 85–95.

Hoshi, T., Sato, M, 1974. Studies on physiology and ecology of plankton. XXVIII. Uptake of radioactive iron in the blood of *Daphnia magna* with reference to haemoglobin synthesis. Sci. Rep. Niigata Univ. (formerly: J. Fac. Sci. Niigata Univ.). 11, 11–19.

Hoshi, T., Kobayashi, K., Sugano, H., 1967. Studies on physiology and ecology of plankton. XXI. Iron-content and subunit of the haemoglobin from the fresh-water Cladocera, *Moina macrocopa*. Sci. Rep. Niigata Univ., ser. D (Biol.). 4, 1–8.

Hoshi, T., Kobayashi, K., Sugano, H., 1968. Studies on physiology and ecology of plankton. XXII. Oxygen dissociation curve and spectral properties of haemoglobin of fresh-water Cladocera, *Moina macrocopa*. Sci. Rep. Niigata Univ., ser. D (Biol.). 5, 87–98.

Hoshi, T., Yahagi, N., Watanabe, T., 1974. Studies on physiology and ecology of plankton. XXXI. Further studies on O2-response to haemoglobin of *Daphnia magna in vitro*. Sci. Rep. Niigata Univ. 14, 7–13.

Hrbácek, J., 1959. Circulation of water as a main factor influencing the development of helmets in *Daphnia cucullata* Sars. Hydrobiologia. 13 (1–2), 170–185.

Hrbácek, J., Hrbácková, M. 1980. Planktonic species of Cladocera (Crustacea) as biological indicators of eutrophication. Proceedings of the IIId Intern. Conf. bioindicat., etc. 1977. Praha, 1980, 49–54.

Huang, X.-Q, Wang, W., Cheng, Y.-X., 2001. Influence of different feeding conditions and nutritional environment on total lipid and fatty acid composition of *Daphnia magna*. J. Shanghai Fish. Univ. 10 (1), 49–54.

Hubareva, E.S., 2000a. Effect of oxygen concentration on metabolism and locomotion of planktonic crustaceans *Moina micrura* (Hellich) (Cladocera) and *Calanus exinus Hulsemann* (Copepoda). Extended abstract of candidate's dissertation. Sevastopol.

Hubareva, E.S., 2000b. The mechanisms of physiological adaptation to short- and long-term hypoxia in *Moina micrura* (Cladocera). Biologiya morya. 53, 31–35.

Hubareva, E.S., Svetlichnyi, L.S., 1998. Modification of metabolism in the cladoceran *Moina micrura* at adaptation to hypoxia. Dopovidi Natsionalnoi Akademii nauk Ukrainy. 9, 195–199.

Huchzermeyer, E.W., 1963. Herzfrequenz und Lebensablauf von Macrothrix hirsuticornis. Zeitschrift für wissenschaftliche Zoologie. 168 (1/2), 119–132.

Hultmark, D., 2003. *Drosophila* immunity: paths and patterns. Curr. Opin. Immunol. 15, 12–19.

Hünefeld, F.L., 1840. Die Chemismus in der thierischen Organization. Brockhaus, Leipzig, pp. 158–163.

Hunter, K., Pyle, G., 2004. Morphological responses of *Daphnia pulex* to *Chaoborus americanus* kairomone in the presence and absence of metals. Environ. Toxicol. Chem. 23 (5), 1311–1316.

Huovinen, P.S., Soimasuo, M.R., Oikari, A.O.J., 2001. Photoinduced toxicity of retene to *Daphnia magna* under enhanced UV-B radiation. Chemosphere. 45 (4–5), 683–691.

Hurtubise, R.D., Havel, J.E., Little, E.E., 1998. The effects of ultraviolet-B radiation on freshwater invertebrates: experiments with a solar simulator. Limnol. Oceanogr. 43 (6), 1082–1088.

Hutchinson, G.E., 1953. Turbulence as random stimulation of sense organs. Cybernetics. Transactions of the Ninth Conference, March 20–21, 1952, New York, 155–158.

Hutchinson, G.E., 1957. A Treatise on Limnology, vol. 1. John Wiley and sons, New York, p. 1015.

Hykes, O.-V., 1926. Mouvements du coeur chez les Daphnies sous l'influence de quelques substances endocrines. Comptes rendues des seances de la Societe Biologie. 95 (20), 58–60.

Ikenaka, Y., Fun, H., Ishizaka, M., Myanara, Y., 2006. Metabolism of pyrene by aquatuc crustacean, *Daphnia magna*. Aquat. Toxicol. 80 (2), 158–162.

Ilan, E., Weisselberg, E., Daniel, E., 1982. Erythrocruorin from the water-flea *Daphnia magna*. Quaternary structure and arrangement of subunits. Biochem. J. 207, 297.

Infante, A., 1973. Untersuchungen über die Ausnutzbarkeit verschiedener Algen durch das Zooplankton. Arch. Hydrobiol. Suppl. 42 (3/4), 340–405.

Infante, A., 1978. A method for the study of foods of herbivorous zooplankton. Trans. Amer. Microsc. Soc. 97 (2), 256–258.

Ingle, L., 1933. Effects of environmental conditions on longevity. Science. 78 (2031), 511–513.

Ingle, L., Wood, T.R., Banta, A.M., 1937. A study of longevity, growth and heart rate in *Daphnia longispina* as influenced by limitations in quantity of food. J. Exp. Zool. 76 (2), 325–352.

Innan, H., Kondrashov, F., 2010. The evolution of gene duplications: classifying and distinguishing between models. Nat. Rev. Genet. 11, 97–108.

Innes, D.J., 1997. Sexual reproduction of *Daphnia pulex* in a temporary habitat. Oecologia. 111 (1), 53–60.

Irving, P., Troxler, L., Heuer, T.S., Belvin, M., Kopczynski, C., Reichhart, J.M., et al., 2001. A genome-wide analysis of immune responses in *Drosophila*. Proc. Natl. Acad. Sci. U.S.A. 98, 15119–15124.

Isakova, E.F., Kolosova, L.V., 1988. Method of biotesting using *Daphnia*. In: Metody biotestirovaniya vod. Chernogolovka 50–57.

Ivanova, M.B., Klekowski, R.Z., 1972. Respiratory and filtration rates in *Simocephalus vetulus* (O.F. Müller) (Cladocera). Polskie Archwum Hydrobiologii. 19 (3), 303–318.

Ivanova, R.P., Moskaleva, N.V., Samylin, A.F., 1989. Impact of 3,4-dichloroaniline on development and respiration of some hydrobionts. Biochemstry of ecto- and endothermal organisms. Petrozavodsk, 139–144.

Ivleva, I.V., 1969. Biological Foundations and Methods of Mass Cultivation of Food Invertebrates. Nauka, Moscow, p. 171. (in Russian).

Ivnitskii, Y., Sofronov, G.A., Nosov, A.V., 1998. Effect of modifiers of energy metabolism on toxicity of cyclophosphane for *Daphnia magna*. Bulleten eksperimentalnoi biologii i meditsiny. 125 (3), 307–309.

Jaakkola, P., Mole, D.R., Tian, Y., Wilson, M.I., Gielbert, J., Gaskell, S.J., et al., 2001. Targeting of HIF-alpha to the von Hippel-Lindau ubiquitylation complex by O2-regulated prolyl hydroxylation. Sci. Signal. 292, 468.

Jacobs, J., 1961. Cyclomorphosis in *Daphnia galeata mendotae* Birge, a case of environmentally controlled allomentry. Arch. Hydrobiol. 58 (1), 7–71.

Jacobs, J., 1964. Hat die hohe Sommerhelm zyclomorpher Daphnien einen Anpassungswert? Verhandl. Internat. Verein. Theoret. und angew. Limnol. 15 (2), 676–683.

Jacobs, J., 1980. Environmental control of cladoceran cyclomorphosis via target-specific growth factors in the animal. In: Kerfoot, W.C., (Ed.), Evolution and Ecology of Zooplankton Communities. Special Symposium, vol. 3. American Society of Limnology and Oceanography. University press of New England, Hanover, etc., 429–437.

Jacobsen, P.J, Jonsen, G.H., 1988. The influence of food limitation on swarming behavior in the waterflea *Bosmina longispina*. Anim. Behav. 36 (4), 991–995.

Jacobsen, T.R., Comita, G.W., 1976. Ammonia-nitrogen excretion in *Daphnia pulex*. Hydrobiology. 51 (3), 195–200.

Jaeger, G., 1935. Über den Fettkoerper von *Daphnia magna*. Zeitschrift für Zellforschung und Mikroskopische Anatomie. 22 (2), 98–181.

Jander, R., 1959. Spontaneous changes in the angular orientation of animals. Anat. Rec. 134 (3), 587.

Jander, R., 1966. Die Phylogenie von Orientierungsmechanismen der Arthropoden. Zoologischer Anzeiger. 29, 266–306, Supplementband.

Jander, R., 1975. Interaction of light and gravity orientation in *Daphnia pulex*. Fortschitte der Zoologie. 23 (1), 174–184.

Janeway Jr., C.A., Medzhitov, R., 2002. Innate immune recognition. Annu. Rev. Immunol. 20, 197–216.

Janeway, C.A., Travers, P., Walport, M., Schlomchik, M., 2001. Immunobiology. fifth ed. Garland Publication, New York.

Jansen, B., Geldof, S., De Meester, L., Orsini, L., 2011. Isolation and characterization of microsatellite markers in the waterflea *Daphnia magna*. Mol. Ecol Resour. 11, 418–421.

Jensen, T., Hessen, D.O., 2007. Does excess carbon affect respiration of *Daphnia*. Oecologia. 152 (2), 191–200.

Jiang, T.-J, Wang, D.-Z., Niu, T., Xu, Y.-X., 2007. Trophic transfer of paralytic shellfish toxins from the Cladoceran (*Moina mongolica*) to larvae of the fish (*Sciaenops ocellata*). Toxicon. 50 (5), 639–645.

Jin, L., Sai, J., Wang, L, Wei, Z., 1998. Quantitative structure-toxicity relationships for benzaldehydes to *Daphnia magna*. Chemosphere. 37 (1), 79–85.

Joaquim-Justo, C., Thomé, J.P., 1998. Quantification of the trophic pathway in the PCB contamination of two zooplanktonic species: *Daphnia magna* Straus and *Brachionus calyciflorus* Pallas. Verhandl. Internat. Vereinig. Limnol. 26 (4), 1957–1961.

Johnson, D.S., Chua, T.E., 1973. Remarkable schooling behaviour of a water flea, *Moina* (Cladocera). Crustaceana. 24 (3), 332–333.

Johnson, P.A., 1989. Morphological changes in *Daphnia magna* (Straus) exposed to inorganic selenium as sodium selenate. Aquat. Toxicol. 14, 95–108.

Johnson, P.T.J., Longcore, J.E., Stanton, D.E., Carnegie, R.B., Shields, J.D., Preu, E.R., 2006. Chytrid infections of *Daphnia pulicaria*: development, ecology, pathology and phylogeny of *Polycaryum laeve*. Freshw. Biol. 51, 634–648.

Jorgensen, O.M., 1933. On the marine Cladocera from the Northumbrian Plankton. J. Mar. Biol. Assoc. U.K. 19, 177–226.

Kainz, M., Arts, M.T., Mazunder, A., 2004. Essential fatty acids in the planktonic food web and their ecological role for higher trophic levels. Limnol. Oceanogr. 49 (5), 1784–1793.

Kalacheva, G.S., Zadereev, E.S., Gubanov, V.G., Gribovskaya, I.V., 2000. Chemical composition of aggregations of *Moina macrocopa* (Crustacea: Cladocera). Ekologiya. 2, 130–134.

Kamjunke, N., Benndorf, A., Wilbert, C., Opitz, M., Kranich, J., Bollenbach, M., et al., 1999. Bacteria injestion by *Daphnia galeata* in a biomanipulated reservoir: a mechanism stabilizing biomanipulation? Hydrobiologia. 403 (1), 109–121.

Kankaala, P., Taipale, S., Grey, J., Sonninen, E., Arvola, L., Jones, R.I., 2006. Experimental δ13C evidence for a contribution of methane to pelagic food webs in lakes. Limnol. Oceanogr. 51 (6), 2821–2827.

Karnauckhov, V.N., Del Rio De Valdivia, M.L., Yashin, V.A., 1986. Comparative spectral studies of pigmented organs and tissues of two cladocera species. Comp. Biochem. Physiol. 85B (1), 279–284.

Karzinkin, G.S., 1951. On norms of feeding of juvenile acipenserids and of inconnu. Trudy VNIRO. 19, 25–38.

Kashian, D.R., 2004. Toxaphene detoxification and acclimation in *Daphnia magna*: do cytochrome P-450 enzymes play a role? Comp. Biochem. Physiol. Part C. 137, 53–63.

Kast-Hutcheson, K., Rider, C.V., Leblanc, G.A., 2001. The fungicide propioconazole interferes with embryonic development of the crustacean *Daphnia magna*. Environ. Toxicol. Chem. 20 (3), 502–509.

Kates, M., Volcani, B.F., 1966. Lipid components of diatoms. Biochemica and biophysica acta. 116 (2), 265–276.

Kato, K., Tokishita, S., Mandokoro, Y., Kimura, S., Ohta, T., Kobayashi, M., et al., 2001. Two-domain hemoglobin gene of the water flea *Moina macrocopa*: duplication in the ancestral cladocera, diversification, and loss of a bridge intron. Gene. 273, 41–50.

Kato, Y., Kobayashi, K., Watanabe, H., Iguchi, T., 2011. Environmental sex determination in the branchiopod crustacean *Daphnia magna*: deep conservation of a doublesex gene in the sex-determining pathway. PLoS Genet. 7, e1001345.

Kato, Y., Matsuura, T., Watanabe, H., 2012. Genomic Integration and Germline Transmission of Plasmid Injected into Crustacean *Daphnia magna* Eggs. Plos One. 7, e45318.

Kawabata, K., Urabe, J., 1998. Length:weight relationships of eight freshwater planktonic crustacean species in Japan. Freshw. Biol. 39, 199–205.

Kawasaki, T., Sakata, M., Namiki, H., 2004a. Elemental characterization of *Daphnia* resting eggs by x-ray analytical microscopy. Zool. Sci. 21 (10), 1019–1023.

Kawasaki, T., Yoshimura, H., Shibue, T., Ikeuchi, Y., Sakata, M., Igarashi, K., et al., 2004b. Crystalline calcium phosphate and magnetic mineral content of *Daphnia* resting eggs. Zool. Sci. 21 (1), 63–67.

Keating, K.I., 1985. The influence of vitamin B12 deficiency on the reproduction of *Daphnia pulex* (Cladocera). J. Crust. Biol. 5 (1), 130–136.

Keating, K.I., Caffrey, P.B., 1989. Selenium deficiency caused by zinc deprivation in a crustacean. Proc. Natl. Acad. Sci. U.S.A. 86 (16), 6436–6440.

Keating, K.I., Dagbusan, B.C., 1984. Effect of selenium deficiency on cuticle integrity in the Cladocera (Crustacea). Proc. Natl. Acad. Sci. U.S.A. 81 (11), 3263–3606.

Keen, R.E., 1971. Dynamics of natural populations of two species of Chydoridae (Cladocera, Crustacea). Abstracts. Limnologorum conventus, Leningrad, 45–46.

Keller, T.A., Tomba, A.M., Moore, P.A., 2001. Orientation in complex chemical landscapes: spatial arrangement of chemical sources influences crayfish food-finding efficiency in artificial streams. Limnol. Oceanogr. 46 (2), 238–247.

Kerfoot, W.C., 1974. Egg-cycle of a cladoceran. Ecology. 55, 1259–1270.

Kerfoot, W.C., 1975. The divergence of adjacent populations. Ecology. 56 (6), 1298–1313.

Kerfoot, W.C., 1978. Combat between predatory copepods and their prey: *Cyclops*, *Epischura*, and *Bosmina*. Limnol. Oceanogr. 23 (6), 1089–1102.

Kerfoot, W.C., Kirk, K., 1991. Degree of taste discrimination among suspension-feeding cladocerans and copepods: implications for detrivory and herbivory. Limnol. Oceanogr. 36 (6), 1107–1123.

Kerfoot, W.C., Peterson, C., 1980. Predatory copepods and *Bosmina*: replacement cycles and further influences of predation upon prey reproduction. Ecology. 6 (2), 417–431.

Kerfoot, W.C., Kellog, D.L., Jr., Strickler, J.R., 1980. Visual observations of live zooplankters: evasion, escape, and chemical defences. Evolution and ecology of zooplankton communities. Special symposium, vol. 3.

Ed. by W.C. Kerfoot. The University of New England, Hanover, 10–27.

Kerfoot, W.C., Robbins, J.A., Weider, L.J., 1999. A new approach to historical reconstruction: combining descriptive and experimental paleolimnology. Limnol. Oceanogr. 44, 1232–1247.

Kersting, K., 1978. Some features of feeding, respiration and energy conversion of *Daphnia magna*. Hydrobiologia. 59 (2), 113–120.

Kersting, K., van der Leew-Leegwater, C., 1976. Effect of food concentration on the respiration of *Daphnia magna*. Hydrobiologia. 49 (2), 137–142.

Kessel, R.G., 1968. Mechanisms of protein yolk synthesis in crustacean oocytes. Zeitschrift für Zellforschung. 89, 17–38.

Khangarot, B.S., Das, S., 2009. Toxicity of mercury on in vitro development of parthenogenetic eggs of a freshwater cladoceran, *Daphnia carinata*. J. Hazard. Mater. 161 (1), 68–73.

Khlebovich, V.V., 1974. Critical Salinity of Biological Processes. Nauka, Leningrad, p. 235. [in Russian].

Khlebovich, V.V., Degtyarev, A.V., 2005. Mechanism of formation of defensive morphs in *Daphnia pulex* (Cladocera, Daphniidae). Doklady AN. 404 (1), 139–141.

Kien, N.B., Michels, E., De Meester, L., 2001. Phototactic behavior of *Daphnia* and the continuous monitoring of water quality: interference of fish kairomones and food quality. Environ. Toxicol. Chem. 20 (5), 1098–1103.

Kiknadze, G.S., Esakov, B.P., Kuzminykh, S.B., Komarov, V.M., 1983. Attempt at estimation of pollution level of aquatic environment by changes in the period of heart rate of *Daphnia*. Akademiya nauk SSSR, preprint, Poushchino p. 14. (in Russian).

Kikuchi, S., 1982. A unique cell membrane with a lining of repeating subunits on the cytoplasmic side of presumably ion-transporting cells in the gill epithelium of *Daphnia magna*. J. Submicrosc. Cytol. 14 (4), 711–715.

Kikuchi, S., 1983. The fine structure of the gill epithelium of freshwater-flea, *Daphnia magna* (Crustacea: Phyllopoda) and changes associated with acclimation to various salinities. Cell Tiss. Res. 229, 253–268.

Kilham, S.S., Kreeger, D.A., Goulden, C.E., Lynn, S.G., 1997. Effects of algal food quality on fecundity and population growth rates of *Daphnia*. Freshw. Biol. 38, 639–647.

Kim, K., Kotov, A.A., Taylor, D.J., 2006. Hormonal induction of undescribed males resolves cryptic species of cladocerans. Proc. R. Soc. Biol. 273, 141–147.

Kimura, S., Tokishita, S., Ohta, T., Kobayashi, M., Yamagata, H., 1999. Heterogeneity and differential expression under hypoxia of two-domain hemoglobin chains in the water flea, *Daphnia magna*. J. Biol. Chem. 274, 10649–10653.

Kirk, K.L., 1991. Suspended clay reduces *Daphnia* feeding rate: behavioural mechanisms. Freshw. Biol. 25, 357–365.

Kirsch, R., Richter, S., 2007. The nervous system of *Leptodora kindtii* (Branchiopoda, Cladocera) surveyed with confocal scanning microscopy (CLSM) including general remarks on the branchiopod neuromorphological ground pattern. Arthropod Struct. Dev. 36 (2), 143–156.

Kitaev, S.P., 2007. Basic General Limnology for Hydrobiologists and Ichthyologists. Russian Academy of Sciences, Petrozavodsk, p. 394. [in Russian].

Klein, B., 2000. Age as a factor influencing results in the acute daphnid test with *Daphnia magna* Straus. Wat. Res. 34 (5), 1419–1424.

Kleiven, O.T., Larsson, P., Hobaek, A., 1992. Sexual reproduction in *Daphnia magna* requires three stimuli. Oikos. 65, 197–206.

Klekowski, R.Z., Zvirgzds, J., 1971. The influence of herbicide 2,4-D-Na on respiration and survival of *Simocephalus vetulus* (O.F. Müller) (Cladocera). Polskie Archiwum hydrobiologii. 18 (4), 393–400.

Knoechel, R., Holtby, B., 1986. Construction and validation of a body-length-based model for the prediciton of cladoceran community filtering rates. Limnol. Oceanogr. 31 (1), 1–16.

Knowles, C.O., Mckee, M.J., Palawski, D.U., 1987. Chronic effect of di-2-ethylhexyl phthalate on biochemical composition, survival and reproduction of *Daphnia magna*. Environ. Toxicol. Chem. 6 (3), 201–208.

Knyazeva, N.I., 1994. The intensity of gas metabolism in *Daphnia* in water containing various copper salts. Zhurnal evolyutsionnoy biokhimii i fiziologii. 30 (3), 322–337.

Kobayashi, M., 1970. Variation in haemoglobin concentration of the fresh water Cladocera, *Moina macrocopa* and *Daphnia magna*. Sci. Rep. Niigata Univ., ser. D (Biol.). 7, 117–120.

Kobayashi, M., 1981a. Influences of haemoglobin concentration and temperature on oxygen uptake of *Daphnia magna* under low oxygen concentration. Comp. Biochem. Physiol. A. 69A (4), 679–682.

Kobayashi, M., 1981b. Respiratory function of haemoglobin in *Moina macrocopa*. Comp. Biochem. Physiol. 70A (3), 381–385.

Kobayashi, M., 1983a. Haemoglobin in *Ceriodaphnia quadrangula*. Comp. Biochem. Physiol. 74A (4), 919–922.

Kobayashi, M., 1983b. Estimation of the haemolymph volume in *Daphnia magna* by haemoglobin determination. Comp. Biochem. Physiol. 76A (4), 803–805.

Kobayashi, M., Hoshi, T., 1980. Effect of carbon monoxide on oxygen-uptake of *Daphnia magna*. Sci. Rep. Niigata Univ., ser. D (Biol.). 17, 1–6.

Kobayashi, M., Hoshi, T., 1982. Relationship between the haemoglobin concentration of *Daphnia magna* and the ambient oxygen concentration. Comp. Biochem. Physiol. 72A (1), 247–249.

Kobayashi, M., Hoshi, T., 1984. Analysis of respiratory role of haemoglobin in *Daphnia magna*. Zool. Sci. 1 (4), 522–532.

Kobayashi, M., Ichikawa, H., 1987. Effect of oxygen pressure on swimming activity of *Ceriodaphnia quadrangula*. J. Crust. Biol. 7 (3), 507–512.

Kobayashi, M., Nezu, T., 1986. Variation of hemoglobin content in *Daphnia magna*. Physiol. Zool. 59 (1), 35–42.

Kobayashi, M., Takahashi, Y., 1993. In vivo oxygenation of hemoglobin in early embryos of *Daphnia magna*. Comp. Biochem. Physiol., Part A. 107 (1), 127–131.

Kobayashi, M., Takahashi, Y., 1994. In vivo oxygenation of hemoglobin in early embryos of *Daphnia magna*. Comp. Biochem. Physiol., Part A: Physiol. 107, 127–131.

Kobayashi, M., Yamagata, H., 2000. Structure and function of hemoglobin in *Daphnia magna*. Trends Comp. Biochem. Physiol. 6, 163–174.

Kobayashi, M., Yoshida, H., 1986. Effect of hemoglobin concentration on swimming activity of *Ceriodaphnia quadrangula* in low oxygen concentration. J. Crust. Biol. 6 (2), 207–213.

Kobayashi, M., Hayakawa, F., Ninomiya, M., 1987. Hatchability and hemoglobin of *Daphnia magna* embryo. Physiol. Zool. 60 (4), 507–512.

Kobayashi, M., Ishigaki, K., Kobayashi, M., Igarashi, Y., Imai, K., 1994. Oxygen transport efficiency of multiple-component hemoglobin in *Daphnia magna*. Can. J. Zool. 72, 2169–2171.

Koch, H.J., 1934. Essai d'interprétation de la goi-disant 'réduction vitale' de sets d'argent par certains organes d'arthropodes. Ann. Soc. Sci. méd. nat. Brux. B. 54, 346.

Kokova, V.E., 1982. Continuous Culture of Invertebrates. Nauka, Novosibirsk, 168 pp. (in Russian).

Kolkwitz, R., Marsson, M., 1909. Ökologie der tierischen Saprobien. Internationale Revue der gesamte Hydrobiologie. 2, 126–152.

Kolmakov, V.I., Levin, L.A., Dubovskskaya, O.P., Gladyshev, M.I., 2002. A circadiam rhythm of grazing of microalgae by a laboratory culture of *Ceriodaphnia quadrangula*. Biofizika. 47 (3), 673–676.

Kolmakova, A.A., 2010. Amino acids of seston of small eutrophic lakes. Extended abstract of candidate's dissertation. Krasnoyarsk, p. 23.

Kolts, J.M., Boese, C.J., Meyer, J.S., 2009. Effects of dietborne copper and silver on reproduction by *Ceriodaphnia dubia*. Environ. Toxicol. Chem. 28 (1), 71–85.

Kolupaev, B.I., 1988. Method of biotesting by changes of respiration and heart activity in *Daphnia*. In: Methody biotestirovaliya vod. Chernogolovka 103–104.

Kolupaev, B.I., 1989. Normal parameters of functioning of the oxygen supply system in *Daphnia*. Biologiya vnutrennikh vod. Informatsionnyi bulleten. 81, 50–51.

Komjarova, I, Blust, R., 2009. Effect of Na, Ca and pH on simultaneous uptake of Cd, Cu, Ni, Pb, and Zn in the water flea *Daphnia magna* measured stable isotopes. Aquat. Toxicol. 94, 81–86.

Konovalov, P.M., Konovalova, Z.A., 1955. State of daphnias as indicator of changes in conditions of a water body. Voprosy ikhtiologii. 3, 135–139.

Konwick, B.J., Fisk, A.T., Garrison, A.W., Avants, J.K., Black, M.C., 2005. Acute enantioselective toxicity of fipronil and its desulfinyl photoproduct to *Ceriodaphnia dubia*. Environ. Toxicol. Chem. 24 (9), 2350–2355.

Kořínek, V, Hebert, P.D.N., 1996. A new species complex of *Daphnia* (Crustacea, Cladocera) from the Pacific Northwest of the United States. Can. J. Zool. 74 (8), 1379–1393.

Korovchinsky, N.M., 1992. Sididae and Holopedidae (Crustacea: Daphniiformes). Guides to the Identification of the Microinvertebrates of the Continental Waters of the World 3. SPB Academic Publishing, The Hague, 82 pp.

Korovchinsky, N.M., 2004. Cladocera of the Order Ctenopoda of the World Fauna. KMK, Moscow, 410 pp. (in Russian).

Korovchinsky, N.M., Boikova, O.S., 2009. The pelagic crustacean zooplanktonof Lake Glubokoe in 1999–2008 and some results of its long-term observations. Proceedings of the Hydrobiological Station "Lake Glubokoe" X, 39–50.

Korovchinsky, N.M., Sergeeva, N., 2008. A new family of the order Ctenopoda (Crustacea: Cladocera). Zootaxa. 57–66.

Korpelainen, H, Ketola, M., Hietala, J., 1997. Somatic polyploidy examined by flow cytometry in *Daphnia*. J. Plank. Res. 19, 2031–2040.

Koshtoyants, Kh.S., 1951. Foundations of Comparative Physiology, 1. USSR Academy of Sciences, Moscow-Leningrad, p. 523. [in Russian].

Kosova, A.A., 1961. Calculated weights of some zooplankton forms of the lower volga delta. Trudy Astrakhanskogo zapovednika. 5, 151–159.

Kotov, A.A., 1997. A special moult after the release of the embryo from the brood pouch of Anomopoda. Hydrobiologia. 354, 83–88.

Kotov, A.A., 1998. Structure of molar plates and analysis of work of mandibles of *Eurycercus lamellatus* (Anomoipoda, Crustacea). Zoologicheskii Zhurnal. 77 (3), 548–557.

Kotov, A.A., 2000. Mixed multispecific schools of littoral Anomopoda (Crustacea: Cladocera). Hydrobiologia. 434, 211–212.

Kotov, A.A., 2009. A revision of Leydigia Kurz, 1875 (Anomopoda, Cladocera, Branchopoda) and subgeneric differentiation within the genus. Zootaxa. 2082, 1–84.

Kotov, A.A., Štifter, P., 2006. Cladocera: Family Ilyocryptidae (Branchiopoda: Cladocera: Anomopoda). Guides to the Identification of the Macroinvertebrates of the Continental Waters of the World. 22. Ghent: Kenobi, Leiden: Backhuys. p. 172.

Kotov, A.A., Taylor, D.J., 2010. A new African lineage of the *Daphnia obtusa* group (Cladocera: Daphniidae) disrupts continental variance patterns. J. Plankton Res. 6, 937–949.

Kotov, A.A., Ishida, S., Taylor, D.J., 2006. A new species in the *Daphnia curvirostris* (Crustacea: Cladocera) complex from the eastern Palearctic with molecular phylogenetic evidence for the independent origin of neckteeth. J. Plank. Res. 28, 1067–1079.

Kotov, A.A., Ishida, S., Taylor, D.J., 2009. Revision of the genus *Bosmina* (Cladocera: Bosminidae), based on evidence from male morphological characters and molecular phylogenies. J. Linn. Soc. Lond. Zool. 156, 1–51.

Kowalczewska-Madura, K., Gołdyn, R, Szyper, H., 2007. Zooplankton phosphorus excretion in Swarzędzkie Lake (Western Poland) and its influence on phytoplankton. Oceanol. Hydrobiol. Studies. 36 (1), 3–16.

Kraaijeveld, A.R., Godfray, H.C.J., 1997. Trade-off between parasitoid resistance and larval competitive ability in *Drosophila melanogaster*. Nature. 389, 278–280.

Kraaijeveld, A.R., Wertheim, B., 2009. Costs and genomic aspects of *Drosophila* immunity to parasites and pathogens. In: Rolff, J., Reynolds, S. (Eds.), Insect Infection and Immunity. Oxford University Press, Oxford.

Kraaijeveld, A.R., Limentani, E.C., Godfray, H.C.J., 2001. Basis of the trade-off between parasitoid resistance and larval competitive ability in *Drosophila melanogaster*. Proc. R. Soc. Lond., B, Biol. Sci. 268, 259–261.

Kretovich, V.L., 1986. Biochemistry of Plants. Vysshaya shkola, Moscow, p. 503 (in Russian).

Krogh, A., 1939. Osmotic regulation in Aquatic Animals. The University Press, Cambridge, p. 242.

Krueger, D.A., Dodson, S.I., 1981. Embryological induction and predation ecology in *Daphnia pulex*. Limnol. Oceanogr. 26 (2), 219–223.

Krylov, V.V., 2008. Impact of alternating electromagnetic field of ultralow and low frequencies upon survival, development and production parameters in *Daphnia magna* Straus (Crustacea, Cladocera). Inland Water Biol. 1 (2), 33–39.

Krylov, V., 2010. Impact of low-frequency magnetic fields and of a typical magnetic storm on parameters of progeny of *Daphnia magna* (comparison of experimental data). In: Biologiya vnutrennikh vod. Materialy XIV shkoly conferentsii molodykh uchenykh (Borok, 26–30 oktyabrya 2010) 64–70.

Krylov, V., 2011. Method for evaluating of biological effects of various influences using the developmental parthenogenetic eggs of *Daphnia magna* Straus in a test-system. Abstract book. IXth International symposium on Cladocera, October 2–8, 2011. Verbania, Italy, p. 6.

Kryuchkova, N.M., 1969. Some regularities of feeding and growth of *Eurycercus lamellatus* (O.F. Muller). Nauchnye doklady vyssjei shkoly, boil. 8, 35–38.

Kryuchkova, N.M., 1989. Trophic Relationships of Zoo- and Phytoplankton. Nauka, Moscow, p. 124. (in Russian).

Kubitz, J.A., Lewek, E.C., Besser, J.M., Drake III, J.B., Giesy, J.P., 1995. Effects of copper-contaminated sediments on *Hyalella azteca*, *Daphnia magna*, and *Ceriodaphnia dubia*: survival, growth, and enzyme inhibition. Arch. Environ. Contam. Toxicol. 29, 97–103.

Kucherenko, N.E., Vasilyev, A.N., 1985. Lipids. Kiev, Vyshcha shkola, p. 247. (in Russian).

Kudikin, N.P., 2001. Special traits of response of crustacean organism to exogenous endocrine compounds. Abstracts. 8s'ezd Gidrobiologicheskoho obshchestva RAN, Kaliningrad, 1, 243–244.

Kukkonen, J.K., 1991. Effects of pH and natural humic substances on the accumulation of organic pollutants on two freshwater invertebrates. Lecture Notes in Earth Sci. 33, 413–422.

Kukkonen, J., Oikari, A., 1988. Sulphate conjugation is the main route of pentachlorophenol metabolism in *Daphnia magna*. Comp. Biochem. Physiol. 91 C (2), 465–468.

Kumar, R., Rao, T.R., Prasad, P., Hwang, J.-S., 2005a. Can Cladocera reproduce for generations under unfed conditions? 7th International Symposium on Cladocera, 23–24.

Kumar, S., Srivastava, A., Charkrabarti, R., 2005b. Study of digestive proteinases and proteinase inhibitors of *Daphnia carinata*. Aquaculture. 243 (1–4), 367–372.

Künne, C., 1926. Schwarmbildung bei *Bosmina longirostris*. Arch. Hydrobiol. 16 (3), 508.

Kuo-cheng Shan, R., 1970. Influence of light on hatching resting eggs of chydorids (Cladocera). Internat. Revue Ges. Hydrobiol. 55 (2), 295–302.

Kurashov, E.A., 2007. Methods and approaches for quantitative investigation of fresh-water meiobenthos. In: Aktualnye voprosy izucheniya mikro-, meiozoobentosa i fauny zaroslei etc. Borok. Nizhnii Novgorod: Vektor, 5–35.

Kusk, K.O., 1996. Bioavailability and effect of pirimicarb on *Daphnia magna* in a laboratory freshwater/sediment system. Arch. Environ. Contam. Toxicol. 31, 252–255.

Kuzmina, V.V., Shekhovtsova, N.V., Bolobonina, V.E., 2010. Activity dynamics of proteinases and glycosidases of fish chymus with exposure in fresh and brackish water. Izvestiya RAN, ser. Biol. 6, 704–710.

Kuznetsov, N.Ya., 1948. Principles of Insect Physiology. vol. I. AN SSSR, Moscow, Leningrad, 380 pp. (in Russian).

Kwassnik, G.M., Vetter, R.J., Atchinson, G.J., 1978. The uptake of manganese-54 by green algae (*Protococcoidal Chlorella*), *Daphnia magna*, and fathead minnows (*Pimephales promelas*). Hydrobiologia. 59 (3), 181–185.

Kyle, M., Acharya, K., Weider, L.J., Looper, K., Elser, J.J., 2006. Coupling of growth rate and body stoichiometry in *Daphnia*: a role for maintenance processes? Freshw. Biol. 51, 2087–2095.

Labbé, P., McTaggart, S.J., Little, T.J., 2009. An ancient immunity gene duplication in *Daphnia magna*: RNA expression and sequence analysis of two nitric oxide synthase genes. Dev. Comp. Immunol. 33, 1000–1010.

Labbé, P., Vale, P.F., Little, T.J., 2010. Successfully resisting a pathogen is rarely costly in *Daphnia magna*. BMC. Evol. Biol. 10, 355.

LaBerge, S., Hann, B.J., 1990. Acute temperature and oxygen stress among genotypes of *Daphnia pulex* and *Simocephalus vetulus* (Cladocera, Daphniidae). Can. J. Zool. 68 (11), 2257–2263.

Lacayo, M., Cruz, A., Lacayo, J., Fomsgaard, I., 1991. Mercury contamination in Lake Xolotlán (Managua). Hydrobiol. Bull. 25 (2), 173–176.

Laforsch, C., Tollrian, R., 2000. A new preparation technique for scanning electron microscopy using hexanethyldisilazane. Archiv Hydrobiol. 149 (4), 587–596.

Laforsch, C., Tollrian, R., 2004a. Extreme helmet formation in *Daphnia cucullata* induced by small-scale turbulence. J. Plankton Res. 26 (1), 81–87.

Laforsch, C., Tollrian, R., 2004b. Embryological aspects of inducible morphological defences in *Daphnia*. J. Morphol. 262, 701–707.

Laforsch, C., Ngwa, W., Grill, W., Tollrian, R., 2004. An acoustic microscopy technique reveals hidden morphological defences in *Daphnia*. PNAS. 101 (45), 15911–15914.

Lagerspetz, K.Y.H., 2000. Thermal avoidance and preference in *Daphnia magna*. J. Therm. Biol. 25, 405–410.

Lair, N., 1991. Grazing and assimilation rates of natural populations of planktonic cladocerans in a eutrophic lake. Hydrobiologia. 215, 51–61.

Lam, I.K.S., Wang, W.-X., 2006. Transgenerational retention and material transfer of selenium in *Daphnia magna*. Environ. Toxicol. Chem. 25 (9), 2519–2525.

Lambrechts, L., 2010. Dissecting the genetic architecture of host-pathogen specificity. PLOS Pathog. 6, e1001019.

Lamkemeyer, T., Zeis, B., Decker, H., Jaenicke, E., Waschbusch, D., Gebauer, W., et al., 2006. Molecular mass of macromolecules and subunits and the quaternary structure of hemoglobin from the microcrustacean *Daphnia magna*. FEBS J. 273, 3393–3410.

Lamkemeyer, T., Zeis, B., Paul, R.J., 2003. Temperature acclimation to moderate influences temperature-related behaviour as well as oxygen transport physiology and biochemistry in the water flea *Daphnia magna*. Can. J. Zool. 81, 237–241.

Lampert, W., 1975. A laboratory system for the cultivation of large numbers of daphniae under controlled conditions. Archiv. für Hydrobiol. 1 (Suppl. 48), 138–140.

Lampert, W., 1978. Release of dissolved organic carbon by grazing zooplankton. Limnol. Oceanogr. 23 (4), 831–834.

Lampert, W., 1986. Response of the respiratory rate of *Daphnia magna* to changing food conditions. Oecologia (Berlin). 70, 495–501.

Lampert, W., 1987. Feeding and nutrition in *Daphnia*. Memorie dell'Istituto Italiano di Idrobiologia. 45, 143–192.

Lampert, W., 1994. Phenotypic plasticity of the filter screens in *Daphnia*: adaptation to a low-food environment. Limnol. Oceanogr. 39 (5), 997–1006.

Lampert, W., 2004. Evolutionary ecology: natural selection in freshwater systems. In: Moya, A., Font, E. (Eds.), Evolution: From Molecules to Ecosystems. Oxford University Press, Oxford, pp. 109–121.

Lampert, W., 2011. *Daphnia*: Development of a Model Organism in Ecology and Evolution. International Ecology Institute Oedendorf.

Lampert, W., Brendelberger, H., 1996. Strategies of phenotypic low food adaptation in *Daphnia*: filter screens mesh sizes, and appendage beat rates. Limnol. Oceanogr. 41 (2), 216–223.

Lampert, W., Rothhaupt, K.O., von Elert, E., 1994. Chemical induction of colony formation in a green alga (*Scenedesmus acutus*) by grazers (*Daphnia*). Limnol. Oceanogr. 39 (7), 1543–1550.

Lampert, W., Sommer, U., 2007. Limnoecology: The Ecology of Lakes and Streams. Oxford University Press, Oxford.

Landon, M.S., Stasiak, R.H., 1983. *Daphnia* hemoglobin concentration as a function of depth and oxygen availability in Arco Lake, Minnesota. Limnol. Oceanogr. 28 (4), 731–737.

Larink, O., 1992. Labrum und Kopfdrüsen eines Conchostracen (Crustacea, Phyllopoda). Morph. Tiere. 72, 341–348.

Larsson, P., Dodson, S., 1993. Chemical communication in planktonic animals. Arch. Hydrobiol. 129 (2), 129–155.

Larsson, P., Aasen, C., Cieplinska, T., Freitas, R., Fullwood, C.F., 2000. Aggregation and swimming speed of daphniids in a homogeneous environment. Verh. Internat. Verein. Limnol. 27, 1803.

Lastochkin, D.A., 1930. Association of animal population of the coastal area of Pereslavl (Pleshcheevo) Lake. Investiya Ivanovo-Voznesenskogo polyteckhnicheskogo instituta. 17, 3–99.

Latta, L.C., Weider, L.J., Colbourne, J.K., Pfrender, M.E., 2012. The evolution of salinity tolerance in *Daphnia*: a functional genomics approach. Ecol. Lett. 15, 794–802.

Lavrentjeva, I.F., Beim, A.M., 1978. Study of physiological processes in *Daphnia* by means of fluorochromes. Gidrobiolohicheskii Zhurnal. 14 (2), 99–102.

Lavrovskaya, N.F., 1965. Content of chlorophyll, carotenoids, vitamins C and E in some freshwater algae and invertebrates collected in natural conditions. Trudy Vsesoyuznogo nauchno-issledovatelskogo instituta prudovogo rybnogo khozhaystva. 13, 142–149.

Lebedinsky, J., 1891. The development of *Daphnia* from summer ovum. Zool. Anzeiger. 14, 149.

LeBlanc, G.A., McLachlan, J.B., 1999. Molt-independent growth inhibition of *Daphnia magna* by a vertebrate antiandrogen. Environ. Toxicol. Chem. 18 (7), 1450–1455.

LeBlanc, G.A., McLachlan, J.B., 2000. Changes in the metabolic elimination profile of testosterone following exposure of the crustacean *Daphnia magna* to tributyltin. Ecotoxicol. Environ. Saf. 45 (3), 296–303.

LeBlanc, G.A., Hilgenberg, B., Cochrane, B.J., 1988. Relationships between the structures of chlorinated phenols, their toxicity, and their ability to induce glutathione S-transferase activity in *Daphnia magna*. Aquat. Toxicol. 12, 147–156.

LeBlanc, G.A., Mu, X., Rider, C.W., 2000. Embryotoxicity of the alkylphenol degradation product 4-nonylphenol to the crustacean *Daphnia magna*. Environ. Health Perspect. 108 (12), 1133–1138.

Leder, H., 1913. Über den feineren Bau des Nervensystems der Cladoceren. Zoologisher Anzeiger. 43 (6), 279–283.

Leder, H., 1915. Über Penilia schmackeri Richard in der Adria. Zoologischer Anzeiger. 45 (6), 350–360.

Lee, H.O., Davidson, J.M., Duronio, R.J., 2009. Endoreplication: polyploidy with purpose. Genes Dev. 23, 2461–2477.

Leech, D.M., Williamson, C.E., 2001. *In situ* exposure to ultraviolet radiation alters the depth distribution of *Daphnia*. Limnol. Oceanogr. 46 (2), 416–420.

Leech, D.M., Williamson, C.E., Moeller, R.E., Hargrave, B.R., 2005. Effects of ultraviolet radiation on the seasonal vertical distribution of zooplankton: a database analysis. Archiv. für Hydrobiol. 162 (4), 445–464.

Leeuwangh, P., 1978. Toxicity tests with daphnids: its application in the management of water quality. Hydrobiologia. 59 (2), 145–148.

Lehman, J.T., 1980. Release and cycling of nutrients between planktonic algae and herbivores. Limnol. Oceanogr. 25 (4), 620–632.

Lehman, J.T., 1993. Efficiencies of ingestion and assimilation by an invertebrate predator using C and P dual isotope labeling. Limnol. Oceanogr. 38 (7), 1550–1554.

Lehman, J.T., Naumoski, T., 1985. Content and turnover rates of phosphorus in *Daphnia pulex*: effect of food quality. Hydrobiologia. 128 (2), 119–125.

Lehman, N., Pfrender, M.E., Morin, P.A., Crease, T.J., Lynch, M., 1995. A hierarchical molecular phylogeny within the genus *Daphnia*. Mol. Phyl. Evol. 4, 395–407.

Lemaitre, B., Nicolas, E., Michaut, L., Reichhart, J.M., Hoffmann, J.A., 1996. The dorsoventral regulatory gene cassette spatzle/Toll/cactus controls the potent antifungal response in *Drosophila* adults. Cell. 86, 973–983.

Lemcke, H.W., Lampert, W., 1975. Changes in weight and chemical composition of *Daphnia pulex* during starvation. Archiv. für Hydrobiol. Suppl. 48 (1), 108–137.

Lesnikov, L.A., 1967. Daphniae as test-organisms in assessment of the level and kind of the waste-water impact on fishery water bodies. Extended abstract of candidate's. dissertation. Leningrad. p. 21.

Leu, E., Fæerøvig, P.J., Hessen, D.O., 2006. UV effects on stoichiometry and PUFAs of *Selenastrum capricornutum* and their consequences for the grazer *Daphnia magna*. Freshw. Biol. 51, 2296–2308.

Levashina, E.A., Moita, L.F., Blandin, S., Vriend, G., Lageux, M., Kafatos, F.C., 2001. Conserved role of a complement-like protein in phagocytosis revealed by dsRNA knockout in cultured cells of the mosquito *Anopheles gambiae*. Cell. 104, 709–718.

Lévy, R., 1927a. Mode de préparation de la daphnie permettant de battre son coeur en milieu artificiel. Comptes rendus Société de Biologie, Paris. 97, 1600–1601.

Lévy, R., 1927b. Sensibilité aux actions chimiques du coeur de la Daphnie préparée, bettant en milieu artificiel. Comptes rendus Société de Biologie, Paris. 97, 1601–1605.

Leydig, F., 1860. Naturgeschichte der Daphniden. Tuebingen, p. 252.

Li, J.L., Li, H.W., 1979. Species-specific factors affecting predator-prey interactions of the copepod *Acanthocylops vernalis* with its natural prey. Limnol. Oceanogr. 24 (4), 613–626.

Li, T., Brouwer, M., 2007. Hypoxia-inducible factor, gsHIF, of the grass shrimp *Palaemonetes pugio*: Molecular characterization and response to hypoxia. Comp. Biochem. Physiol. B Biochem. Mol. Biol. 147 (1), 11–19.

Lidvanov, V.V., L'vova, T.G., 2003. Organization of the nervous system of *Daphnia magna* (Cladocera). In: Teotericheskie i prikladnye aspekty bioekologii. Kaliningrad, 70–77.

Lin, W., Liu, X., 1985. Elementary composition of some dominant zooplankters in Lake Donghu, Wuhan. Acta hydrobiologica sinica. 9 (3), 258–263.

Lincoln, R.J., 1970. A laboratory investigation into the effects of hydrostatic pressure on the vertical migration of planktonic Crustacea. Marine Biol. 6 (1), 5–11.

Little, T.J., Ebert, D., 1999. Associations between parasitism and host genotype in natural populations of *Daphnia* (Crustacea : Cladocera). J. Anim. Ecol. 68, 134–149.

Little, T.J., Killick, S.C., 2007. Evidence for a cost of immunity when the crustacean *Daphnia magna* is exposed to the bacterial pathogen *Pasteuria ramosa*. J. Anim. Ecol. 76, 1202–1207.

Little, T.J., Carius, H.J., Sakwinska, O., Ebert, D., 2002. Competitiveness and life-history characteristics of *Daphnia* with respect to susceptibility to a bacterial pathogen. J. Evol. Biol. 15, 796–802.

Little, T.J., Colbourne, J.K., Crease, T.J., 2004. Molecular evolution of *daphnia* immunity genes: polymorphism in a gram-negative binding protein gene and an alpha-2-macroglobulin gene. J. Mol. Evol. 59, 498–506.

Little, T.J., O'Connor, B., Colegrave, N., Watt, K., Read, A.F., 2003. Maternal transfer of strain-specific immunity in an invertebrate. Curr. Biol. 13, 489–492.

Liu, W., Gan, J., Schlenk, D., Jury, W.A., 2005. Enantioselectivity in environmental safety of current chiral insecticides. Proc. Natl. Acad. Sci. U.S.A. 102 (3), 517–556.

Livanov, N.A., 1955. Pathways of Evolution of the Animal World. Sovetskaya Nauka, Moscow, 400 pp. (in Russian).

Lizenko, E.I., Bushman, L.G., Nefedova, Z.A., 1977. Content of lipids in plankton of some Karelian lakes. Gidrobiologicheskii Zhurnal. 13 (3), 74–80.

Lobashev, M.E., Ivanova, P.G., 1947. Some regularities of ontogenetic adaptation. Dependence of photoreaction in *Daphnia magna* on adaptation to temperature. Doklady AN SSSR LVIII, 1, 127–130.

Lochhead, J., 1961. Locomotion. In: The Physiology of Crustacea, 2. Acad. Press, New York-London. 313–328.

Loeb, J., 1924. Forced Movements, Tropisms, and Behavior of Animals. Moscow. 174 pp. (Russian translation).

Loose, C.J., von Ellert, E., Dawidowicz, P., 1993. Chemically induced diel vertical migration in *Daphnia*: a new bioassay for kairomones exuded by fish. Archiv. für Hydrobiol. 126 (3), 329–337.

Lopatina, T.S., Zadereev, E.S., 2007. Role of interpopulational communications in induction of change of reproduction mode in Cladocera. In: Vetvistousye rakoobraznye: sistematika i biologiya, Borok. Nishnii Novgorod: Vektor, 287–292.

Lopes, I., Baird, D.J., Ribeiro, R., 2005. Genetically determined resistance to lethal levels of copper by *Daphnia longispina* association with sublethal response and multiple resistance. Environ. Toxicol. Chem. 24 (6), 1414–1419.

Lu, F.-Y., He, Z.-H., Qu, Y.-H., 2000. Research on hatching of ephippial eggs. J. Dalian Fish. Univ. 15 (2), 92–97.

Luecke, C., O'Brien, W.J., 1983. Photoereceptive pigments in a pond morph of *Daphnia middendorffiana*. Arctic. 36 (4), 385–388.

Luferova, L.A., 1962. Device for investigation of the effect of current on aquatic invertebrates. Bulletin Instituta biologii vodokhranilishch. 12, 62–64.

Luferova, L.A., Flerov, B.A., 1971a. Investigation of intoxication of *Daphnia* with phenol. Biologiya vnutrennikh vod. 10, 42–46.

Luferova, L.A., Flerov, B.A., 1971b. Comparative resistance of some Phyllopoda to phenol. Biologiya vnutrennikh vod. 11, 43–46.

Luijckx, P., Ben-Ami, F., Mouton, L., Du Pasquier, L., Ebert, D., 2011a. Cloning of the unculturable parasite *Pasteuria ramosa* and its *Daphnia* host reveals extreme genotype-genotype interactions. Ecol. Lett. 14, 125–131.

Luijckx, P., Fienberg, H., Duneau, D., 2011b. Resistance to a bacterial parasite in the crustacean *Daphnia magna* shows Mendelian segregation with dominance. Heredity. 108 (5), 547–551.

Lumer, H., 1932. The reactions of certain Cladocera to colored lights of equal intensity. Ohio J. Sci. 32 (3), 218–231.

Lundqvist, G., 1927. Bodenablagerungen und Entwicklungstypen der Seen. Die Binnengewässer, Band II, Schweizerbart, Stuttgart, p. 124.

Luzgin, V.K., 1982a. Morphological disruption of hepatic cecums of *Daphnia magna* Straus exposed to cobalt chloride. Vestnik Leningradskogo Universiteta 9, biol., 2, 8–13.

Luzgin, V.K., 1982b. The change of succinate dehydrogenase activity of *Daphnia magna* Straus under the influence of cupric chloride. Vestnik Leningradskogo Universiteta 21, biol., 2, 98–10.

Lynch, M., 1989. The life history consequences of resource depression in *Daphnia pulex*. Ecology. 70 (1), 246–256.

Lynch, M., Weider, L.J., Lampert, W., 1986. Measurement of the carbon balance in *Daphnia*. Limnol. Oceanogr. 31 (1), 17–33.

Lynch, M., 1985. Spontaneous mutations for life history characters in an obligate parthenogen. Evolution. 39, 804–818.

Lynch, M., Pfrender, M.E., Spitze, K., Lehman, N., Allen, D., Hicks, J., et al., 1999. The quantitative and molecular genetic architecture of a subdivided species. Evolution. 53, 100–110.

Lynch, M., Spitze, K., 1994. Evolutionary genetics of *Daphnia*. In: Real, L. (Ed.), Ecological Genetics. Princeton University Press, Princeton, NJ, pp. 109–128.

Macan, T.N., 1963. Freshwater Ecology. Longmans, London, 338 pp.

MacArthur, J.W., Baillie, W.H.T., 1929. Metabolic activity and duration of life in *Daphnia magna*. I, II.. J. Exp. Zool. 53 (2), 221–268.

Macfadyen, E., Williamson, C.E., Grad, G., Lowery, M., Jeffrey, W.H., Mitchell, D.L., 2004. Molecular response to climate change: temperature dependence of UV-induced DNA damage and repair in the freshwater crustacean *Daphnia pulicaria*. Global Change Biol. 10, 408–416.

Maeda, S., Ohki, A., Kusadome, K., Kuroiwa, T., Yoshifuku, I., Naka, K., 1992. Bioaccumulation of arsenic and its fate in a freshwater food chain. Appl. Organometallic Chem. 6 (2), 213–219.

Mäemets, A., 1961. Eesti vesikibuliste (Cladocera) okoloogiast ja fenologiast. Hydrobiol. Issledovaniya. 2, 108–158.

Maia-Barbosa, P.M., Bozelli, R.L., 2005. Length-weight relationships for five cladocerans species in an Amazonian lake. Brazilian Arch. Biol. Technol. 48 (2), 303–308.

Main, T.M., Dobberfuhl, D.R., Elser, J.J., 1997. N:P stoichiometry and ontogeny of crustacean zooplankton: a test of the growth rate hypothesis. Limnol. Oceanogr. 42 (6), 1474–1478.

Maity, A.K., Saxena, J., 1979. Color preference of *Daphnia magna* King. Geobios. 6, 193–195.

Makhutova, O.N., Dubovskaya, O.P., Sushchik, N.N., Gladyshev, M.I., 2009. Effect of the content in seston of mineral and biochemical components on the growth rate of *Daphnia* (longispina group). X s'ezd Gidrobiologicheskogo obshchestva pri RAN. Abstracts. Vladivostok, 263–264.

Makino, S., 1951. An Atlas of the Chromosome Numbers in Animals. Iowa State College Press, Ames.

Makrushin, A.V., 1966. Observations on oogenesis of some Cladocera. Institute of Inland water biology USSR. Ac. Sci., Trudy. 12 (15), 175–182.

Makrushin, A.V., 1968. Adaptations related to reproduction of Cladocera. Gidrobiologicheskii Zhurnal. 5, 72–76.

Makrushin, A.V., 1969. A study of oogenesis in Cladocera. Zoologicheskii Zhurnal. 48 (12), 1788–1792.

Makrushin, A.V., 1970. Changes in the organism of females of some species of Cladocera under their transition to gamogenesis. Zoologicheskii Zhurnal. 49 (10), 1573–1574.

Makrushin, A.V., 1972. Protoephippial glands in Cladocera (Crustacea). Zoologicheskii Zhurnal. 51 (11), 1736–1738.

Makrushin, A.V., 1978. Anhydrobiosis and yolk structure of the eggs in Cladocera. Zoologicheskii Zhurnal. 57 (3), 364–374.

Makrushin, A.V., 1980. Anhydrobiosis and evolution of yolk structure in latent eggs of Cladocera (Crustacea). Zoologicheskii Zhurnal. 59 (3), 368–373.

Makrushin, A.V., 2011. Senescence of *Moina macrocopa* (Cladocera, Crustacea). Adv. Gerontol. 24 (1), 24–25.

Makrushin, A.V., Perevoznikov, M.A., Lysak, A.P., 1990. The method of gathering of latent eggs and ephippia of some species of Cladocera and breaking their diapause. Biologiya vnutrennikh vod. 89, 83–87.

Maksimova, L.P. 1969. Metodichesklie ukazaniya po razvedeniyu melkogo planktonnogo rachka Moina macrocopa Straus (Instructions on cultivation of *Moina macrocopa* Straus). Leningrad. p. 22. [in Russian].

Malikova, E.M., 1953. Chemical composition of some food invertebrates. Trudy Latviiskogo otdeleniya VNIRO. 1, 213–224.

Malikova, E.M., 1956. The nutritional value of certain invertebrates as food for fish. Biokhimiya. v. 21 (2).

Mallin, M.A., Partin, W.E., 1989. Thermal tolerances of common Cladocera. J. Freshw. Ecol. 5 (1), 45–51.

Manuilova, E.F. 1951. The first year of investigations in productivity of waterbodies in Novgorod oblast. Trudy problemnykh i tematicheskikh soveshchanii Zoologicheskogo Instituta AN SSSR, I, Problemy gidrobiologii vnutrennikh vod, I, Moscow, Leningrad, 56–60.

Manuilova, E.F., 1955. On conditions of mass development of cladocera. Trudy biologicheskoi stantsii "Borok". 2, 89–107.

Manuilova, E.F., 1956. On relation of variation of pelagic Cladocera with development of bacteria in a water body. Doklady AN SSSR. 103 (6), 1411–1414.

Manuilova, E.F. 1959. On conditions of mass development of Cladocera. Trudy Biologicheskoi stantsii Borok 2, XX–XX.

Marcolini, B.M., 1953. Alcuni esperimenti per l'alimentatione di cladoceri e copepodi d'aqua dolce con fosforo radioactivo (P32). Boll. Pesca, piscisole. Idrobiol. 8 (2), 181–200.

Mardarevich, M.G., Gudkov, D.I., Kipnis, L.C., 2001. Comparative characteristics of sensitivity of *Daphnia magna* Str. and *Hydra attenuata* Pall. to the action of ions of lead and cadmium. Rep. Natl. Acad. Sci. Ukraine. 8, 142–145.

Markowska, M., Pietrzak, B., Bednarska, A., Slusarczyk, M., 2011. Physiological state of *Daphnia pulex* species living in a fish-free high mountain lake measured by RNA:DNA ratio. Abstract book, IXth International symposium on Cladocera, October 2–8, 2011. Verbania, Italy, p. 23.

Marques, C.R., Abrantes, N., Gonçalves, F., 2004. Life-history traits of standard and autochthonous cladocerans: I. Acute and chronic effects of acetylsalicylic acid. Environ. Toxicol. 19 (5), 518–526.

Marshall, S.M., Beeton, A.M., Chandler, D.C., 1964. Role of zooplankton in the freshwater strontium cycle and influence of dissolved salts. Verhandl. Internation. Verein. Limnol. XV (2), 665–672.

Martin-Creuzburg, D., Heilke F., 2011. Food quality of heterotrophic bacteria for *Daphnia*: the role of essential lipids. Abstract book, IXth International symposium on Cladocera, October 2–8, 2011. Verbania, Italy, p. 23.

Martin-Creuzburg, D., Westelund, S.A., Hofflann, K.H., 2007. Ecdysteroid levels in *Daphnia magna* during a molt cycle: determination by radioimmunoassay (RIA) and liquid chromatography-mass spectrometry (LC-MS). Gen. Comp. Endocrinol. 151 (1), 66–71.

Martinez-Tabche, L., Mora, B.R., German Faz, C., Pardo, D., 1998. Acetylcholinesterase activity of the freshwater Cladocera, *Moina macrocopa* used as an *in vitro* screening tool for water quality. Aquatic Ecosyst. Health Manage. 1 (3), 317–322.

Martinez-Tabche, L., Mora, B.R., German Faz, C., Castelan, I.G., Ortiz, M.M., Gonzalez, V.U., et al., 1997. Toxic effect of sodium dodecylbenzenesulfonate, lead, petroleum, and their mixtures on the activity of acetylcholinesterase of *Moina macrocopa in vitro*. Environ. Toxicol. Eater Qual. 12 (3), 211–215.

Martins, N., Lopes, I., Brehm, A., Ribero, R., 2005. Cytochrome B gene partial sequence and RAPD analysis of two *Daphnia longispina* lineages differing in their resistance to copper. Bull. Environ. Contam. Toxicol. 74 (4), 755–760.

Martynova, M.V., 2010.Bottom deposits as a component of limnic systems. Moscow, Nauka. p. 243. [in Russian].

Matthias, U., Puzicha, H., 1990. Erfahrungen mit dem dynamischen Daphnientest—Einfluss von Pestiziden auf das Schwimmverhalten von *Daphnia magna* unter Labor- und Praxisbedingungen. Zeitschrift für Wasser- und Abwasser-Forschung. 23 (5), 193–198.

Maynard, D.M., 1960. Circulation and heart function. In: Waterman, T.H. (Ed.), The Physiology of Crustacea, vol. 1. Academic Press, New York and London, pp. 161–226.

McCallum, J.B., 1905. The action of purgatives in a crustacean (*Sida crystallina*). Univ. Calif. Publ. Physiol. 2 (6), 65–70.

McCarthy, J.F., 1983. Role of particulate organic matter in decreasing accumulation of polynuclear aromatic hydrocarbons by *Daphnia magna*. Arch. Environ. Contam. Toxicol. 12 (5), 559–568.

McCauley, E., Murdoch, W.W., Nisbet, R.M., Gurney, W.S. C., 1990. The physiological ecology of Daphnia: development of a model of growth and reproduction. Ecology. 71 (2), 703–715.

McCoole, M.D., Anderson, N.J., Graham, D.I., Grasserr, E.B., et al., 2012. Genomic analyses of aminergic signaling systems (dopamine, octopamine and serotonin) in *Daphnia pulex*. Comp. Biochem. Physiol. Part D. 7, 35–58.

McCoole, M.D., Baer, K,N., Christie, A.E., 2011. Histaminergic signaling in the nervous system of *Daphnia* and a role for it in the control of phototactic behavior. J. Exp. Biol. 214, 1773–1782.

McKee, M.J., Knowles, C.O., 1986. Protein, nucleic acid and adenylate levels in *Daphnia magna* during chronic exposure to chlordecone. Environ. Poll. Series A.(42), 335–351.

McKee, M.J., Knowles, C.O., 1987. Levels of protein, RNA, DNA, glycogen and lipid during growth and development of *Daphnia magna* Straus (Crustacea: Cladocera). Freshw. Biol. 18 (2), 341–351.

McKenzie, J.D., Calow, P., Nimmo, W.S., 1992. Effects of inhalational general anaesthetics on intact *Daphnia magna* (Cladocera: Crustacea). Comp. Biochem. Physiol. 101C (1), 9–13.

McMahon, J.W., 1968. Environmental factors influencing the feeding behaviour of *Daphnia magna* Straus. Canad. J. Zool. 56 (4), 756–762.

McMahon, J.W., Rigler, F.H., 1963. Mechanisms regulating the feeding rate of *Daphnia magna* Straus. Canad. J. Zool. 46, 321–332.

McMahon, J.W., Rigler, F.H., 1965. Feeding rate of *Daphnia magna* Straus in different foods labeled with radioactive phosphorus. Limnol. Oceanogr. 10 (1), 105–113.

McNaught, D.C., 1966. Depth control by planktonic cladocerans in lake Michigan, great lakes research division. Univ. Michigan. 15, 98–108.

McNaught, D.C., 1971. Plasticity of cladoceran visual systems to environmental changes. Trans. Amer. Micrscop. Soc. 90 (1), 113–114.

McNaught, D.C., Hasler, A.D., 1964. Rate of movement of populations of *Daphnia* in relation to changes in light intensity. J. Fish. Res. Board Can. 21 (2), 291–318.

McNaught, D.C., Hasler, A.D., 1966. Photoenvironments of planktonic Crustacea in Lake Michigan. Verhandl. Internat. Verein. Theoret. und angew. Limnol. 16, 194–203.

McTaggart, S.J., Conlon, C., Colbourne, J.K., Blaxter, M.L., Little, T.J., 2009. The components of the *Daphnia pulex* immune system as revealed by complete genome sequencing. BMC Genom. 10, 175.

McTaggart, S.J., Obbard, D.J., Conlon, C., Little, T.J., 2012a. Immune genes undergo more adaptive evolution than non-immune system genes in *Daphnia pulex*. BMC Evol. Biol. 12, 63.

McTaggart, S.J., Wilson, P.J., Little, T.J., 2012b. *Daphnia magna* shows reduced infection upon secondary exposure to a pathogen. Biol. Lett. 23;8(6):972–5.

Meijering, M.P.D., 1958a. Herzfrequenz und Lebensdauer von *Daphnia magna* Straus. Zeitschrift für wissenschaftliche Zoologie. 161 (3/4), 240–265.

Meijering, M.P.D., 1958b. Herzfrequenz von *Daphnia magna* Straus innerhalb einzelner Häutungstadien. Die Naturwissenschaften. 45, 14.

Meijering, M.P.D., 1960. Herzfrequenz und Herzschlagezahlen zwischen Häuting und Eiablage bei Cladoceren. Zeitschrift für wissenschaftliche Zoologie. 164 (1–2), 127–142.

Meijering, M.P.D., 1962. Längewachtum und Geschlechtsreife bei Männchen von *Daphnia magna* Straus. Zeitschrift für wissenschaftliche Zoologie. 167 (1–2), 116–121.

Meijering, M.P.D., 1975. Life cycles and metabolic activity of *Daphnia*-populations in Arctic and Central-European biotopes. Verh. Internal. Verein Limniol. 19 (4), 2868–2873.

Meijering, M.P.D., 2003. The long-lasting resistance of diapausing eggs from Arctic Cladocera frozen at −18°C. Polish Polar Res. 24 (2), 167–172.

Meijering, M.P.D., Redfern, J.B., 1962. Heart-beat as a measure of biotic time in the life of Daphniidae. Am. Nat. 96 (886), 61–64.

Meijering, M.P.D., von Reden, H.K.J., 1962. Vergleichende Untersuchungen an Zeitplänen dreier Cladocerenarten. Zeitschrift für wissenschaftliche Zoologie. 173 (1/2), 174–190.

Menzikova, O.V., 1988. Characterization of cholinesterases from *Daphnia magna*. Zhurnal evolyutsionnoi biokhimii. i fiziologii. 4, 596–598.

Mergeay, J., Vanoverbeke, J., Verschuren, D., De Meester, L., 2007. Extinction, recolonization, and dispersal through time in a planktonic crustacean. Ecology. 88, 3032–3043.

Miner, B.E., De Meester, L., Pfrender, M.E., Lampert, W., Hairston, N.G., 2012. Linking genes to communities and ecosystems: *Daphnia* as an ecogenomic model. Proc. Biol. Sci. 279, 1873–1882.

Miner, B.G., Sultan, S.E., Morgan, S.G., Padilla, D.K., Relyea, R.A., 2005. Ecological consequences of phenotypic plasticity. Trends Ecol. Evol. 20, 685–692.

Metchnikoff, E., 1884. Über eine Sprosspulzkrankheit der Daphnien. Beitrag zur Lehre über den Kampf der Phagocyten gegen Krankheitserreger. Archiv für pathologische Anatomie und Physiologie und für klinische Medicin. 96 (2), 177–193.

Metchnikoff, I., 1884. A disease of *Daphnia* caused by a yeast. A contribution to the theory of phagocytes as agents for attack on disease-causing organisms. In: Brock, T. (Ed.), Milestones Microbiol. American Society for Microbiology, Washington D.C, pp. 132–138.

Metchnikoff, E. (Mechnikov I.I.), (1885 [1954]). On ability of lower organisms to resist infections. Academic collected works, vol. 5, Moscow, 39–42.

Metchnikoff, E., 1888. Pasteuria ramosa, un representant des bactéries a division longitudinale. Annales de l'Intitut Pasteur. 2 (4), 165–170.

Metchnikoff, E., 1892. Lectures in comparative pathology of inflammation. Rikker: S.-Peterburg. p. 162. [also subsequent editions, the last one of 1947; also in Academic collected works, vol. 5, Moscow [in Russian].

von Metz, H., 1973. Kohlenstoffinhalt, Aufnahme und Abgabe des Kohlenstoffes bei *Daphnia pulex*. Arch. für Hydrobiol. 2 (Suppl. 42), 232–252.

Meurice, J.C., Goffinet, G., 1983. Ultrastructural evidence of the ion-transporting role of the adult and larval neck organ of the marine gymnomeran Cladocera (Crustacea, Branchiopoda). Cell Tiss. Res. 234, 351–363.

Meyers, D.G., 1980. Diurnal vertical migration in aquatic microcrustaceans: light and oxygen responses of littoral zooplankton. In: Evolution and ecology of zooplankton communities. Kerfoot W.Ch., Ed., 80–90.

Meyers, D.G., 1985. Morphological evidence of mechanoreceptive gravity reception in a water flea, *Daphnia magna*. Physiologist. 28 (6), Suppl. S149–S150.

Michael, R.G., Frey, D.G., 1984. Separation of *Disparalona leei* (Chien, 1970) in North America from *D. rostrata* (Koch, 1841) in Europe (Cladocera, Chydoridae). Hydrobiologia. 114, 81–108.

Michel, T., Reichhart, J.M., Hoffmann, J.A., Royet, J., 2001. *Drosophila* Toll is activated by Gram positive bacteria through a circulating peptidoglycan recognition protein. Nature. 414, 756–759.

Midzuno, Hata, Kono, 1960. Seasonality in generation change in *Daphnia pulex* and analysis of environmental factors I. J. Ecol. 10 (1).

Mikhailova, L., Anisimova, S., Strugova, A., 1986. Accumulation of oil hydrocarbons in different links of water reservoir's food chain and their connection with biosubstrate. Exp. Water Toxicol. 11, 27–28, Riga: Zinatne.

Mims, S.D., Webster, C.D., Tidwell, J.H., Yancey, D.H., 1991. Fatty acid composition of *Daphnia pulex* cultured by two different methods. J. World Aquacul. Soc. 22 (2), 153–156.

Mirsky, A.E., Ris, H., 1951. The deoxyribonucleic acid content of animal cells and its evolutionary significance. J. Gen. Physiol. 34, 451–462.

Mirza, R.S., Pyle, G.G., 2009. Waterborne metals impair inducible defences in *Daphnia pulex*: morphology, life-history traits and encounters with predators. Freshw. Biol. 54 (5), 1016–1027.

Mitchell, J., Halves, J., Lampert, W., 2004. Coexistence of similar genotypes of *Daphnia magna* in intermittent populations: response to thermal stress. Oikos. 106, 469–478.

Mitchell, S.E., Read, A.F., Little, T.J., 2004. The effect of a pathogen epidemic on the genetic structure and reproductive strategy of the crustacean *Daphnia magna*. Ecol. Lett. 7, 848–858.

Mityanina, I.F., 1980. Changes in the *Daphnia* and rotifer caloricity under acclimation to constant temperatures. Vesci Akademii navuk BSSR, ser. biol. 4, 88–90.

Miyakawa, H., Imai, M., Sugimoto, N., Ishikawa, Y., Ishikawa, A., Ishigaki, H., et al., 2010. Gene upregulation in response to predator kairomones in the water flea, *Daphnia pulex*. BMC Develop. Biol. 10, 45. 10.1186/1471-213X-10-45.

Moenickes, S., Richter, O., Pirow, R., 2010. Approaching the evolutionary advantage of ancillary types of haemoglobin in *Daphnia magna* by simulation of oxygen supply. J. Exp. Biol. 213 (3), 408–413.

Monakov, A.V., 1998. Feeding of Freshwater Invertebrates. Moscow. 321 pp. (in Russian).

Monakov, A.V., Sorokin, Yu.I., 1960. Experimental investigation of feeding of *Daphnia* by means of C14. Doklady AN SSSR. 135 (6), 161–185.

Mooney, R., Obreshkove, V., 1948. Action of prostigmine, carbaminoyl-choline (doryl) and acetyl-B-methylcholine (mecholyl) on intestine of a cladoceran. Proc. Soc. Exp. Biol. Med. 68 (1), 42–46.

Moore, R., Toro, E., Stanton, M., Khan, M.A.Q., 1977. Absorption and elimination of 14C-alpha- and gamma-chlordane by a freshwater alga, daphnid, and goldfish. Arch. Environ. Contam. Toxicol. 6 (4), 411–420.

Morales Ventura, J., Nandini, S., Sarma, S.S.S., 2011. Gut filling and evacuation time of the cladoceran *Moina macrocopa* on a diet of a probiotic and the green alga *Scenedesmus acutus*. Abstract book, IXth International symposium on Cladocera, October 2–8, 2011. Verbania, Italy, p. 25.

Mordukhai-Boltovskoi, F.D., 1954. Materials on mean weight of aquatic invertebrates of the Don basin. Trudy probemnykh i tematicheskikh soveshchanii, ZIN AN SSSR, 2. 223–241.

Mordukhai-Boltovskoi, F.D., Chirkova, Z.N., 1973. General review of structure and ecology of Ilyocryptus species (Cladocera, Macrothricidae). Biologiya vnutrennikh vod. 18, 33–37.

Morgado, J.M., Soares, A.M.V.M., 1995. Activity of pyruvate kinase and malate dehydrogenase in *Daphnia magna* under 3,4-dichloroaniline stress. Arch. Environ. Contam. Toxicol. 29 (1), 94–96.

Morgan, K.K, Hicks, J., Spitze, K., Latta, L., Pfrender, M., Weaver, C., et al., 2001. Patterns of genetic architecture for life-history traits and molecular markers in a subdivided species. Evolution. 55, 1753–1761.

Mortimer, C.H., 1936. Experimentelle und cytologische Untersuchungen über den Generationswechsel der Cladocera. Zoologische Jahrbücher, Abt. für allgeneine Zoologie und Physiologie der Tiere. 56 (3), 323–388.

Morton, P.K., Frisch, D., Weider, L.J., 2012. Ecological genomics in a time capsule—integrating RAD sequencing and resurrection ecology in the water flea, *Daphnia pulicaria*. Ecological Genomics 10th Anniversary Symposium Abstract Book, pp. 20 (abstract 47).

Mu, X., LeBlanc, G.A., 2002. Environmental antiecdysteroids alter embryo development in the crustacean *Daphnia magna*. J. Exp. Zool. 292, 287–292.

Mu, X., LeBlanc, G.A., 2004. Cross communication between signaling pathways: juvenoid hormones modulate ecdysteroid activity in a crustacean. J. Exp. Zool. 301A, 793–801.

Mucklow, P.T., 2003. Physiology of immunity in the water flea *Daphnia magna*: phenoloxidase activity. Physiol. Biochem. Zool. 76 (6), 831–942.

Mucklow, P.T., Ebert, D., 2003. Physiology of immunity in the water flea *Daphnia magna*: Environmental and genetic aspects of phenoloxidase activity. Physiol. Biochem. Zool. 76, 836–842.

Mucklow, P.T., Vizoso, D.B., Jensen, K.H., Redfardt, D., Ebert, D., 2004. Variation in phenoloxidase activity and its relation to parasite resistance within and between populations of *Daphnia magna*. Proc. R. Soc. Lond., B, Biol. Sci. 271, 1175–1183.

Mudrezowa-Wyss, K.A., 1933. The Effect of Carbon Dioxide, Sulphureted Hydrogen, Methane, and Oxygen Absence on Aquatic Organisms. Gosstroiizdat, Moscow-Leningrad, p. 48. [in Russian].

Müller, O.F., 1785. Entomostraca seu insecta testacea, quae in aquis Daniae et Norvegiae reperit, decripsit et iconibus illustravit Otho Fridericus Müller. 67–121.

Müller-Navarra, D.C., 1995a. Evidence that a highly unsaturated fatty acid limits *Daphnia* growth in nature. Arhiv. für Hydrobiol. 132 (3), 297–307.

Müller-Navarra, D.C., 1995b. Biochemical versus mineral limitation in *Daphnia*. Limnol. Oceanogr. 40 (7), 1209–1214.

Müller-Navarra, D.C., 2006. The nutritional importance of polyunsaturated fatty acids and their use as trophic markers for herbivorous zooplankton: Does it contradict? Arch. Hydrobiol. 167 (1–4), 501–513.

Müller-Navarra, D.C., Brett, M.T., Liston, A.M., Goldman, C.R., 2000. A highly unsaturated fatty acid predicts carbon transfer between primary producers and consumers. Nature. 403, 74–77.

Müller-Navarra, D.C., Brett, M.T., Park, S., Chandra, S., Ballantyne, A.P., Zorita, E., et al., 2004. Unsaturated fatty acids content in seston and tropho-dynamic coupling in lakes. Nature. 427, 69–72.

Munger, C., Hare, L., Craig, A., Charest, P.-M., 1999. Influence of exposure time on the distribution of cadmium within the cladoceran *Ceriodaphnia dubia*. Aquat. Toxicol. 44 (3), 195–200.

Munier, A.I., Medzhitov, R., Janeway Jr., C.A., Doucet, D., Capovilla, M., Lagueux, M., 2004. Graal: a *Drosophila* gene coding for several mosaic serine proteases. Insect Biochem. Mol. Biol. 34, 1025–1035.

Munn, M.D., Gillion, R.J., Moran, P.W., Nowell, L.H., 2006. Pesticide toxicity index for freshwater aquatic organisms. second ed. U.S. Department of the Interior, U.S. Geological Survey. p. 81.

Muños, M.J., Ramos, C., Tarazona, J.V., 1996. Bioaccumulation and toxicity of hexachlorobenzene in *Chlorella vulgaris* and *Daphnia magna*. Aquat. Toxicol. 35 (3–4), 211–220.

Munro, I.G., White, R.W.G., 1975. Comparison of the influence of temperature on the egg development and growth of *Daphnia longispina* (O.F. Müller) (Crustacea: Cladocera) from two habitats in Southern England. Oecologia (Berlin). 20 (2), 157–165.

Murakami, Y., 1961. Studies on the winter eggs of the water flea, *Moina macrocopa* Straus. J. Fac. Fish Anim. Husbandry Hiroshima Univ. 3 (2), 323–346.

Murtaugh, P.A., 1981. Inferring properties of mysid predation from injuries to *Daphnia*. Limnol. Oceanogr. 26 (5), 811–821.

Murtaugh, P.A., 1985. The influence of food concentration and feeding rate on the gut residence time of *Daphnia*. J. Plankton Res. 7 (3), 415–420.

Muyssen, B.T.A., Janssen, C.R., 2002. Acclimation and regulation of zinc in *Daphnia magna*: links with homeostasis and toxicity. Arch. Environ. Contam. Toxicol. 43 (4), 492–496.

Muyssen, B.T.A., Janssen, C.R., Bossuyt, B.T.A., 2002. Tolerance and acclimation to zinc of field-collected *Daphnia magna* populations. Aquat. Toxicol. 56, 69–79.

Myrand, B., de la Noüe, J., 1983. Ingestion and assimilation rates of Oocystis sp. by *Daphnia magna* in treated wastewaters. Environ. Poll., series A. 31 (2), 77–95.

Naberezhnyi, A.I., Gorbatenkiy, G.G., 1973. The effect of different saponin concentrations on the life cycle of *Daphnia magna*. Herald hydrobiol. 5–7.

Naddy, R.B., Klaine, S.J., 2001. Effect of pulse frequency and interval on the toxicity of chlorpyrifos to *Daphnia magna*. Chemosphere. 45 (4–5), 497–506.

Naddy, R.B., Johnson, K.A., Klaine, S.J., 2000. Response of *Daphnia magna* to pulsed exposures of chlorpyrifos. Environ. Toxicol. Chem. 19 (2), 423–431.

Nagl, W., 1978. Endopolyploidy and Polyteny in Differentiation and Evolution. North-Holland Publ., Amsterdam.

Nandini, S., Sarma, S.S.S., Pedro Ramírez-García, 2005. Length-weight relationships of three cladoceran species from a tropical reservoir in Mexico. J. Freshwater Ecol. 20 (2), 405–406.

Nappi, A.J., Ottaviani, E., 2000. Cytotoxicity and cytotoxic molecules in invertebrates. BioEssays. 22, 469–480.

Naumann, E., 1918. Über die natürliche Nahrung des limnischen Zooplanktons. Lunds Universitets Arsskrift, N.F., Avd. 2, Bd 4, Nr. 31, 1–48.

Naumann, E., 1929. Grundlinien der experimentellen Planktonforschung. Die Binnengewässer, vol. 6, Stuttgart, p. 100.

Nikitina, S.M., Polunina, Y., 2000. Quantitative content of glucocorticoids in the body of *Daphnia magna* Str., Hydrobiological studies in the Atlantic Ocean. Trudy AtlantNIRO. 1, 109–114, 141, 147–148.

Nikitinsky, J., Mudrezowa-Wyss, K., 1930. Über die Wirkung der Kolensaure, des Schwefelkwasserstoffs, des Methans und Abwesebheit des Sauerstoffs auf Wasserorganismen. Zentralblatt fur Bakteriologie, Parasitenkunde und Infektionskrankheiten 2 Abt, 81, 8/14, 167–198.

Nikonov, K.P., Kamaeva, A.A., Belyanin, V.A., Chulkov, E. A., 1970. Some data on the effect of ultrasound on heart activities of *Daphnia*. Trudy IX conf. po problemam fiziologii, Kutaisi, 69–71.

Nilov, V.I., 1980. Accumulation of 65Zn and 203Hg by certain species of fresh-water invertebrates. Gidrobiologicheskii Zhurnal. 16 (5), 96–103.

Nilov, V.I., 1983. Concentrating of 137Cs by some species of freshwater invertebrates. Gidrobiologicheskii Zhurnal. 19 (4), 41–46.

Nimmo, D.W., 1966. The role of inorganic ions in maintaining osmotic balance in *Simocephalus*. J. Colorado-Wyoming Acad. Sci. 5 (7), 22.

Nunes, F., Spiering, D., Wolf, M., Wendler, A., Pirow, R., Paul, R.J., 2005. Sequencing of hemoglobin gene 4 (dmhb4) and southern blot analysis provide evidence of more than four members of the *Daphnia magna* globin family. Biosci. Biotechnol. Biochem. 69, 1193–1197.

O'Connor, R.J., 1950. Effect of D-tubocurarine chloride and carbaminoylcholine on the respiration of *Daphnia*. Nature. 166, 4219,441.

O'Keefe, T.C., Brewer, M.C., Dodson, S., 1998. Swimming behavior of *Daphnia*: its role in determining predation risk. J. Plankton Res. 20 (5), 973–984.

Oberdörster, E., Ritteschof, D., Le Blanc, G.A., 1998. Alteration of [14C]-testosterone metabolism after chronic exposure of *Daphnia magna* to tributyltin. Arch. Environ. Chem. Toxicol. 34 (1), 21–25.

Obreshkove, V., 1941a. The action of acetylcholine, atropine, and physostigmine on the intestine of *Daphnia magna*. Biol. Bull. 81 (1), 105–113.

Obreshkove, V., 1941b. Cardiac inhibition of a cladoceran and the action of acetylcholine and physostigmine. Proc. Soc. Exp. Biol. Med. 49 (3), 427–431.

Obreshkove, V., 1942. Cardiac inhibition of a cladocera and the action of acetylcholine and physostigmine. Proc. Soc. Exp. Biol. Med. 49, 427–431.

Obreshkove, V., Banta, A.M., 1930. A study of the rate of oxygen consumption of different cladocera clones derived originally from a single mother. Physiol. Zool. 3 (1), 1–8.

Obreshkove, V., Fraser, A.W., 1940a. Growth and differentiation of *Daphnia magna* eggs *in vitro*. Proc. Soc. Exp. Biol. Med. 423 (3), 543–544.

Obreshkove, V., Fraser, A.W., 1940b. Growth and differentiation of *Daphnia magna* eggs in vitro. Biol. Bull. 78 (3), 428–436.

Obreshkove, V., King, A.L., 1932a. The effect of X-rays on the gaseous metabolism of a cladoceran. Physiol. Zool. 5 (3), 457–471.

Obreshkove, V., King, A.L., 1932b. A change in the morphology of a cladoceran effected by x-ray treatment. Physiol. Zool. 5 (3), 472–478.

Oda, S., Tatarazako, N., Watanabe, H., Morita, M., Iguchi, T., 2005a. Production of male neonates in four cladoceran species exposed to juvenile hormone analog fenoxycarb. Chemosphere. 60 (1), 74–78.

Oda, S., Tatarazako, N., Watanabe, H., Morita, M., Iguchi, T., 2005b. Production of male neonates in *Daphnia magna* (Cladocera, Crustacea) exposed to juvenile hormones and their analogs. Chemosphere. 60 (1), 74–78.

Ogle, R.S., Knight, A.W., 1996. Selenium bioaccumulation in aquatic ecosystems: I. Effects of sulfate on the uptake and toxicity of elenate in *Daphnia magna*. Arch. Environ. Contam. Toxicol. 30 (2), 274–279.

Oikari, A., Kukkonen, J., 1990. Bioavailability of benzo(a) pyrene and dehydroabietic acid from a few lake waters containing varying dissolved organic carbon concentrations to *Daphnia magna*. Bull. Environ. Contam. Toxicol. 45, 54–61.

Ojima, Y., 1954. Some cytological observations on parthenogenesis in *Daphnia pulex*. J. Fac. Sci. Hokkaido Univ. 12, 230–235.

Olesen, J., 1996. External morphology and phylogenetic significance of the dorsal/neck organ in the Conchostraca and the head pores of the cladoceran family Chydoridae (Crustacea, Branchiopoda). Hydrobiologia. 330, 213–226.

Olesen, J., Richter, S., Scholtz, G., 2003. On the ontogeny of *Leptodora kindtii* (Crustacea, Brachiopoda, Cladocera), with notes on the phylogeny of the Cladocera. J. Morphol. 256, 235–259.

Olmstead, A.W., LeBlanc, G.A., 2000. Effects of endocrine-active chemicals on the development of sex characteristics of *Daphnia magna*. Environ. Toxicol. Chem. 19 (8), 2107–2113.

Olmstead, A.W., LeBlanc, G.A., 2001b. Temporal and quantitative changes in sexual reproductive cycling of the cladoceran *Daphnia magna* by a juvenile hormone analog. J. Exp. Zool. 290 (2), 148–155.

Olmstead, A.W., LeBlanc, G.A., 2002. Juvenoid hormone methyl farnesoate is a sex determinant in the Crustacean *Daphnia magna*. J. Exp. Zool. 293 (7), 736–739.

Olmstead, A.W., LeBlanc, G.A., 2003. Insecticidal juvenile hormone analogs stimulate the production of male offspring in the crustacean *Daphnia magna*. Environ. Health Perspect. 111 (7), 919–924.

Olmstead, A.W., LeBlanc, G.A., 2007. The environmental-endocrine basis of gynadromorphism (intersex) in a crustacean. Int. J. Biol. Sci. 3 (2), 77–84.

Olsen, K.Y., Jensen, A., Reinertsen, H., 1986. Dependence of the rate of release of phosphorus by zooplankton on the P:C ratio in the food supply, as indicated by a recycling model. Limnol. Oceanogr. 31 (1), 34–44.

Olsen, Y, Varum, K.M., Jensen, A., 1986. Some characteristics of the carbon compounds released by *Daphnia*. J. Plankt. Res. 8 (3), 505–517.

Olsen, Y., Østgaard, K., 1985. Estimating release rates of phosphorus from zooplankton: model and experimental verification. Limnol. Oceanogr. 30 (4), 844–852.

Olvera-Hernández, E., Martinez-Tabche, L., Martinez-Jerónimo, F., 2004. Bioavailability and effects of malathion in artificial sediments on *Simocephalus vetulus* (Cladocera: Daphniidae). Bull. Environm. Contam. Toxicol. 73, 197–204.

Onbé, T., 2002. Nocturnal release of parthenogenetic neonates of the marine cladocerans, *Pseudevadne tergestina* and *Evadne nordmanni* (Branchiopoda: Onychopoda), in the Inland Sea of Japan with observations by infrared-light video microscopy. Plankton Biol. Ecol. 49 (2), 49–57.

Opute, F.I., 1974. Lipid and fatty acid composition of diatoms. J. Exp. Bot. 25 (97), 823–835.

Orlov, A.V., Cherepanov, O.A., 1986. Metabolic products influence reproduction mode change in *Moina macrocopa*. Ekologia. 5, 86–87.

Orlova-Bienkowskaja, M.J., 2001. Daphniidae: genus *Simocephalus*. Guides to the Identification of the Macroinvertebrates of the Continental Waters of the World 17. Leiden: Backhuys. p. 130.

Orsini, L., Spanier, K.I., De Meester, L., 2012. Genomic signature of natural and anthropogenic stress in wild populations of the waterflea *Daphnia magna*: validation in space, time and experimental evolution. Mol. Ecol. 21, 2160–2175.

Orsini, L., Schwenk, K., De Meester, L., Colbourne, J.K., Pfrender, M.E., Weider, L.J., 2013. The evolutionary time machine: forecasting how populations can adapt to changing environments using dormant propagules. Trends Ecol. Evol. 28, 274–282.

Osipova E.A., Krylov, V.V., Yusupov, V.I., Simonova, N.B., 2010. Effect of low-frequency laser and ultraviolet radiation on early ontogenesis of *Daphnia magna*. In: Biologiya vnutrennikh vod. Materialy XIV shkoly conferentsii molodykh uchenykh (Borok, 26–30 oktyabrya 2010) 96–101.

Ostapenya, A.P., Pavlyutin, A.P., Babitsky, V.A., Inkina, G. A., 1968. Transformation of the food energy by some species of planktonic Cladocera. Zhurnal obshchei biologii. 29 (3), 334–342.

Pacaud, A., 1939. Contribution a l'ecologie des Cladoceres. Supplements au Bull. biol. de France et de Belgique. Suppl. XXV, 1–260.

Padhye, S., Kotov, A.A., 2010. Cladocera (Crustacea: Branchiopoda) from Indian hot water springs. Invertebrate Zool. 7 (2), 155–158.

Pancella, J.R., Stross, R.G., 1963. Light induced hatching of *Daphnia* resting eggs. Chesapeak Sci. 4 (3), 135–140.

Pandolfini, E., Leporcq, B., Descy, J.-P., 2000. Grazing experiments with two freshwater zooplankters: fate of chlorophyll and carotenoid pigments. J. Plankton Res. 22 (2), 305–319.

Pane, E.F., McGeer, J.C., Wood, C.M., 2004. Effects of chronic nickel exposure of two successive generations of *Daphnia magna*. Environ. Toxicol. Chem. 23 (4), 1051–2004.

Pane, E.F., Smith, C., McGeer, J.C., Wood, C.M., 2003. Mechanisms of acute and chronic waterborne nickel toxicity in the freshwater cladoceran, *Daphnia magna*. Environ. Sci. Technol. 37, 4382–4389.

Papanicolau, G., 1910. Experimentelle Untersuchungen uber Fortpflanzungverhaeltnisse der Daphniden (Simocephalus vetulus und Moina rectirostris var. lilljeborgi). Biol. Cenralbl. 30, 689–692, 737–750, 753–774, 785–802.

Papchenkova, G.A., 2012. Chronic effect of the insecticide tanrek on *Daphnia magna* Straus (Cladocera). Materialy 4 vserossiyskoy konferentsii po vodnoi ekotoksikologii. Borok. 1, 184–188.

Papchenkova, G.A., Golovanova, I.L., Ushakova, N.V., 2009. Reproductive parameters, size, and hydrolases activity in successive generations of *Daphnia magna* Straus under effect of "Roundup" herbicide. Inland Water Biol. 3, 105–110.

Pardi, L., Papi, F., 1961. Kinetic and tactic responses, The Physiology of Crustacea, vol. 2. Academic Press, New York-London. 365–399.

Park, S., Brett, M.T., Müller-Navarra, D.C., Goldman, C.R., 2002. Essential fatty acid content and phosphorus to carbon ratio in cultivated algae as indicators of food quality for *Daphnia*. Freshw. Biol. 47, 1377–1390.

Park, S., Brett, M.T., Oshel, E.T., Goldman, C.R., 2003. Seston food quality and *Daphnia* production efficiencies in an oligo-mesotrophic subalpine lake. Aquat. Ecol. 37, 123–136.

Parker, B.L., Dewey, J.E., 1969. Further improvements on the mass rearing of *Daphnia magna*. J. Econ. Entomol. 62 (3), 725–726.

Parker, R.A., 1966. The influence of photoperiod on reproduction and molting of *Daphnia schødleri* Sars. Physiol. Zool. 39, 266–279.

Parker, R.A., Olson, M.I., 1966. The uptake of inorganic phosphate by *Daphnia schødleri* Sars. Physiol. Zool. 39 (1), 53–65.

Parks, L.G., Le Blanc, G.A., 1996. Reductions in steroid hormone biotransformation/elimination as a biomarker of pentachlorophenol chronic toxicity. Aquat. Toxicol. 34 (4), 291–303.

Parry, G., 1960. Excretion. The Physiology of Crustacea, vol. 1. Academic Press, New York and London. 341–366.

Pashkova, I.M., Andronikov, V.B., Korotneva, N.V., 1998. Changes in *Daphnia magna* cardiac muscle thermoresistance in copper sulfate contaminated medium. Tsitologiya. 40 (4), 275–281.

Patt, D.I., 1947. Some Cytological Observations on the Nährboden of *Polyphemus pediculus* Linn. Trans. Amer. Micr. Soc. 66, 344–353.

Paul, R., Colmorgen, M., Hüller, S., Tyroller, F., Zinkler, D., 1997. Circulation and respiratory control in millimeter–sized animals (*Daphnia magna* and *Folsomia candida*) studied by optical methods. J. Comp. Physiol. B. 167 (6), 399–408.

Paul, R., Colmorgen, M., Pirow, R., Chen, Y.-H., Tsai, M.-C., 1998. Systemic and metabolic responses in *Daphnia magna* to anoxia. Comp. Biochem. Physiol. Part A. 120 (3), 519–530.

Paul, R.J., Mertenskötter, A., Pinkhaus, O., Pirow, R., Gigengack, U., Buchen, I., et al., 2012. Seasonal and interannual changes in water temperature affect the genetic structure of a *Daphnia* assemblage (*D. longispina* complex) through genotype-specific thermal tolerances. Limnol. Oceanogr. 57, 619–633.

Paul, R.J., Zeis, B., Lamkemeyer, T., Seidl, M., Pirow, R., 2004. Control of oxygen transport in the microcrustacean *Daphnia*: regulation of haemoglobin expression as central mechanism of adaptation to different oxygen and temperature conditions. Acta Physiol. Scand. 182, 259–275.

Pavey, S.A., Bernatchez, L., Aubin-Horth, N., Landry, C.R., 2012. What is needed for next-generation ecological and evolutionary genomics? Trends Ecol. Evol. 27, 673–678.

Pauwels, K., De Meester, L., Decaestecker, E., Stoks, R., 2011. Phenoloxidase but not lytic activity reflects resistance against *Pasteuria ramosa* in *Daphnia magna*. Biol. Lett. 7, 156–159.

Pauwels, K., Stoks, R., De Meester, L., 2005. A molecular antipredator trait: induction of heat shock proteins in *Daphnia magna*. 7th International Symposium on Cladocera, 34–35.

Pauwels, K., Stoks, R., Verbiest, A., 2007. Biochemical adaptation for dormancy in subitaneous and dormant eggs of *Daphnia magna*. Hydrobiologia. 594 (1), 91–96.

Pavlutin, A.P., 1983. Duration of food passage through bowels of *Daphnia magna*. Biologiya vnutrennikh vod. 61, 27–30.

Pearse, A.G.E., 1969. Histochemistry. Churchill, London [Russian translation, Inostrannaya Literatura, Moscow, 1962].

Peeters, K., Brendonck, L., Moens, L., 1994. The occurrence of artemocyanin in Branchiopoda (Crustacea). Comp. Biochem. Physiol. 109A (3), 773–779.

Peeters, K., Mertens, J., Herbert, B., Moens, L., 1990. The globin composition of *Daphnia pulex* hemoglobin and the comparison of the amino acid composition of invertebrate hemoglobin. Comp. Biochem. Physiol. B. 97, 369–381.

Pfrender, M.E., Lynch, M., 2000. Quantitative genetic variation in *Daphnia*: temporal changes in genetic architecture. Evolution. 54, 1502–1509.

Peñalva-Arana, D.C., Manca, M., 2007. An SEM study of the nuchal organ in *Daphnia himalaya* (nov. sp.) embryos and neonates collected from Khumbu region (Nepalese Himalayas). J. Limnol. 66, 153–159.

Peñalva-Arana, D.C., Moore, P.A., Feinberg, B.A., DeWall, J., Strickler, J.R., 2007. Studying *Daphnia* feeding behavior as a black box: a novel electrochemical approach. Hydrobiologia. 594, 153–163.

Pérez Gutierrez, R.M., Lule, R.P., 2005. Cytotoxic constituents from *Daphnia pulex*. Pharm. Biol. 43 (4), 313–316.

Pérez-Martínez, C., Gulati, R.D., 1998/1999. Species-specific N and P release rates in *Daphnia*. Hydrobiologia. 391, 147–155.

Pérez-Martínez, C., van Tongeren, Gulati, R.D, 1995. Estimation of phosphorus metabolic rates of *Daphnia galeata* using a three-compartment model. J. Plankton Res. 17 (8), 1605–1619.

Persson, J., Vrede, T., 2006. Polyunsaturated fatty acids in zooplankton: variation due to taxonomy and trophic position. Freshw. Biol. 51, 887–900.

Peters, E., 1927. Vergleichende Untersuchungen ueber ein Lichsinn einheimischer Cladocerenarten. Zool. Jahrb., Abt. Allg. Zool. und Physiol. der Tiere. 43 (1), 1–40.

Peters, R.H., 1975. Phosphorus regeneration by natural populations of limnetic zooplankton. Verh. Internat. Verein. Limnol. 19, 273–279.

Peters, R.H., 1987. Metabolism in *Daphnia*. Memorie dell'Istituto Italiano di Idrobiologia. 45, 193–243.

Peters, R.H., Rigler, F.H., 1973. Phosphorus release by *Daphnia*. Limnol. Oceanogr. 18 (6), 821–839.

Peterson, J.K., Kashian, D.R., Dodson, S.I., 2001. Methoprene and 20-OH-ecdysone affect male production in *Daphnia pulex*. Environ. Toxicol. Chem. 20 (3), 582–588.

Pfrender, M.E., 2012. Triangulating the genetic basis of adaptation to multifarious selection. Mol. Ecol. 21, 2051–2053.

Pfrender, M.E., Spitze, K., Lehman, N., 2000. Multi-locus evidence for rapid ecologically-based speciation in *Daphnia*. Mol. Ecol. 9, 1717–1735.

Philippova, T.G., Postnov, A.L., 1988. Relationship between metabolism and feeding in Cladocera (Crustacea). Zhurnal obshchei biologii. 49 (4), 540–551.

Pickering, J.W., 1894. On the action of certain substances on the hearts of daphniae. J. Physiol. 17, 357–363.

Pieters, B.J., 2007. Daphnid population responses to pesticides. Academisch proefschritt. Universiteit van Amsterdam.

Pieters, B.J., Bosman-Mejerman, D., Steenbergen, E., van den Brandhof, E.-J., van Beelen, P., van der Grinten, E., et al., 2008. Ecological quality assessment of Dutch surface waters using a new bioassay with the cladoceran *Chydorus sphaericus*. Proc. Neth. Entonol. Soc. Meeting. 61, 157–164.

Pigliucci, M., 2005. Evolution of phenotypic plasticity: where are we going now? Trends Ecol. Evol. 20, 481–486.

Pijanowska, J., Kowalczewski, A., 1997. Predators can induce swarming behavior and locomotory responses in *Daphnia*. Freshw. Biol. 37 (3), 649–656.

Pilati, A., Wurtsbaugh, W.A., Brindza, N.R., 2004. Evidence of coprophagy in freshwater zooplankton. Freshw. Ecol. 49, 913–918.

Pirow, R., Buchen, I., 2004. The dichotomous oxyregulatory behaviour of the planktonic crustacean *Daphnia magna*. J. Exp. Biol. 207, 683–696.

Pirow, R., Bäumer, C., Paul, R.J., 2001. Benefits of haemoglobin in the cladoceran crustacean *Daphnia magna*. J. Exp. Biol. 204, 3425–3441.

Pirow, R., Bäumer, C., Paul, R., 2004. Crater landscape: two-dimensional oxygen gradients in the circulatory system of the microcrustacean *Daphnia magna*. J. Exp. Biol. 207, 4393–4405.

Pirow, R., Wollinger, F., Paul, R.J., 1999a. The importance of the feeding current for oxygen uptake in the water flea *Daphnia magna*. J. Exp. Biol. 202 (5), 553–562.

Pirow, R., Wollinger, F., Paul, R.J., 1999b. The sites of respiratory gas exchange in the planktonic crustacean *Daphnia magna*: an *in vivo* study employing blood haemoglobin as an internal probe. J. Exp. Biol. 202 (22), 3089–3099.

Plath, K., 1998. Adaptive feeding behavior of *Daphnia magna* in response to short-term starvation. Limnol. Oceanogr. 43 (4), 593–599.

Podosinovikova, N.P., Kosmachev, A.V., Tonkopil, V.D., Zagrebin, A.O., Evdokimova, E.A., Malov, A.M., 2001. New approaches to the analysis of interrelationships between cholinergic and dopaminergic system. Eksperimentalnaya i klinicheskaya farmakologiya. 64 (6), 20–22.

Polunina Yu, Y., 1999. Effect of exogenous oxytocin and estradiol on *Daphnia magna*. In: Teoreticheskie i prikladnye aspekty biologii. Kaliningrad, 70–75.

Porcella, D.B., Rixford, C.E., Slater, J.V., 1967. Factors influencing radiostrontium accumulation in *Daphnia*. Health Phys. 13, 391–399.

Porcella, D.B., Rixford, C.E., Slater, J.V., 1969. Molting and calcification in *Daphnia magna*. Physiol. Zool. 42 (2), 148–159.

Porter, K.G., 1975. Viable gut passage of gelatinous green algae ingested by *Daphnia*. Verh. Internal. Vereinig. Limnol. 19 (4), 2840–2850.

Porter, K.G., Orcutt, J.D., 1980. Nutritional adequacy, manageability, and toxicity as factors that determine the food quality of green and blue- green algae for *Daphnia*. The evolution and ecology of zooplankton communities. Hanover, New Hampshire, 268–281.

Porter, K.G., Gerritsen, J., Orcutt Jr., J.D., 1982. The effect of food concentration on swimming patterns, feeding behavior, ingestion, assimilation, and respiration by *Daphnia*. Limnol. Oceanogr. 27 (5), 935–949.

Pörtner, H.O., 2010. Oxygen-and capacity-limitation of thermal tolerance: a matrix for integrating climate-related stressor effects in marine ecosystems. J. Exp. Biol. 213, 881–893.

Pörtner, H.O., Knust, R., 2007. Climate change affects marine fishes through the oxygen limitation of thermal tolerance. Science. 315, 95–97.

Postmer, T.J., Nacken, G., Nelissen, R.G., 1974. An electronic method for measuring heart frequency of the waterfleas: *Daphnia pulex*. Experientia.1478–1480.

Postmes, T.J., Prick, R., Brorens, I., 1989. The deceleration of the heart frequency in the waterflea *Daphnia magna* by adrenoceptor agonists and antagonists. Hydrobiologia. 171, 141–148.

Postnov, A.L., Philippova, T.G., 1988. The method of cladoceran basal metabolism experimental estimation. Vestnik Moskovskogo universiteta, ser. 16, Biol. 1, 42–47.

Poston, T.M., Klopfer, D.C., Simmonds, M.A., 1990. Short-term bioconcentration studies of Np in freshwater biota. Health Phys. 59 (6), 869–877.

Pott, E., 1982. Experimentelle Untersuchung zur Wirking von Herbiziden auf die Futterausnahme von *Daphnia pulicaria*. Arch. Hydrobiol., Supplementband. 59 (4), 330–358.

Potts, W.T.W., Durning, C.T., 1980. Physiological evolution in the Branchiopods. Comp. Biochem. Physiol. 67B, 475–484.

Potts, W.T.W., Fryer, G., 1979. The effects of pH and salt content on sodium balance in *Daphnia magna* and *Acantholeberis curvirostris* (Crustacea: Cladocera). J. Comp. Physiol. 129, 289–294.

Poupa, O., 1948. Effects of antihistamines on *Daphnia*. Nature. 161 (4085), 235–236.

Poynton, H.C., Taylor, N.S., Hicks, J., Colson, K., Chan, S., Clark, C., et al., 2011. Metabolomics of microliter hemolymph samples enables an improved understanding of the combined metabolic and transcriptional responses of *Daphnia magna* to cadmium. Env. Sci. Tech. 45, 3710–3717.

Poynton, H.C., Varshavsky, J.R., Chang, B., Cavigiolio, G., Chan, S., Holman, P.S., et al., 2007. *Daphnia magna* ecotoxicogenomics provides mechanistic insights into metal toxicity. Env. Sci. Tech. 41, 1044–1050.

Pravda, O., 1950. Fluorescence analysis of some Crustacea. Věst. Čs. Společ. Zool. 42 (2), 148–159.

Présig, M, Véro, M., 1983. A new method for determining the heart beat rate of *Daphnia magna*. Water Res. 17 (10), 1245–1248.

Printes, L.B., Callaham, A., 2003. Intraclonal variability in *Daphnia* acetylcholinesterase activity: the implications for its applicability as a biomarker. Environ. Toxicol. Chem. 22 (9), 2042–2047.

Proctor, V., 1964. Viability of crustacean eggs recovered from ducks. Ecology. 45 (3), 656–658.

Proksova, M., 1950. O srdecni cinnosti perloocki *Daphnia magna* Strauss. Prace Moravskoslezske Akademie ves prirodnich, 22, 2, F 227, 33–56.

Prosser, C.L., Brown, F.A., Jr., 1967. Comparative animal physiology. Moscow: Mir (translation of: Saunders: Philadelphia, London, 1962).

Proulx, S.R., 2012. Multiple routes to subfunctionalization and gene duplicate specialization. Genetics. 190, 737–751.

Putman, A., Panis, B., De Meester, L., 2011. Differences between sexual and asexual eggs of *Daphnia magna*. Abstract book, IXth International symposium on Cladocera, October 2–8, 2011. Verbania, Italy, p. 31.

Pyatakov, M.L., 1956. On seasonal changes of fecundity in cladocera. Zoologicheskii Zhurnal. 35 (12), 1814–1819.

Quaglia, A., Sabelli, B., Villani, L., 1976. Studies on the intestine of Daphniidae (Crustacea, Cladocera). J. Morphol. 150 (3), 711–726.

Rabinovich, E.I., 1945. Photosynthesis and Related Processes, vol. 1. Interscience, New York, p. 599.

Rabinovich, E., 1951. Photosynthesis. Inostrannaya Literatura, Moscow, 648 pp.

Rammner, W., 1929, Über periodische Erscheinungen am Cladoceren-Individuum (Häutung, Fortpflanzung, Wachtum). Intern. Revue gesamt. Hydrobiol. Und hydrogr. 21, 402–420.

Rammner, W., 1933. Wird der Cladoceren Embryo vom Muttertier ernährt. Arch. Hydrobiol. 25 (4), 692–698.

Ramult, M.F., 1926. Development and resisting power of Cladocera embryos in the solutions of certain inorganic salts. Bull. internat. de l'Académie Polonaise des sciences et des lettres. Class des sciences mathemat. et naturelles. Serie B, 135–194.

Rankin, G.P., 1929. The nutritional physiology of cladocera. Contrib. Canad. Biol. 4, 107–113.

Rapport, D., Costanza, R., McMichael, A., 1998. Assessing ecosystem health. Trends Ecol. Evol. 13, 397–402.

Rasch, E.M., 1985. DNA "standards" and the range of accurate DNA estimates by Feulgen absorption microspectrophotometry. Adv. Microsc. 196, 137–160.

Rautenberg, W., 1953. Jahreszeitliche Schwanckungen des Daphnientestes. Pharmazie. 8 (6), 507–508.

Rautio, M., Korhola, A., 2002. UV-induced pigmentation in subarctic *Daphnia*. Limnol. Oceanogr. 47 (1), 295–299.

Ravet, J.L., Brett, M.T., Müller-Navarra, D., 2003. A test of the role of polyunsaturated fatty acids in phytoplankton food quality for *Daphnia* using liposome supplementation. Limnol. Oceanogr. 48 (5), 1938–1947.

Ray, D., David, J.P., Cuany, A., Amichot, M., Heyran, J.C., 2000. Differential sensitivity to vegetable tannins in planktonic crustacea from alpine mosquito breeding sites. Pesticide Biochem. Physiol. 67 (2), 103–113.

Razlutskii, V.I., 1992. Effect of trophic conditions on the rate of biological processes in *Moina macrocopa* Straus and *M. rectirostris* Leydig. Gidrobiologicheskii Zhurnal. 28 (1), 53–59.

Redfield, G.W., 1981. Nutrition and the degeneration of eggs in a limnetic daphnid. Verhandl. Internat. Verein. Limnol. 21, 1550–1554.

Reinikainen, M., Meriluoto, J.A.O., Spoof, L., Harada, K.-I., 2001. The toxicities of a polyunsaturated fatty acid and a microcystin to *Daphnia magna*. Environ. Toxicol. 16 (5), 444–448.

Rellstab, C., Spaak, P., 2007. Starvation with a full gut? Effect of a suspended particles on the fitness of *Daphnia hyalina*. Hydrobiologia. 594, 131–139.

Renn, S.C., Siemens, D.H., 2010. Ecological genomics - changing perspectives on Darwin's basic concerns. Mol. Ecol. 19, 3025–3030.

Reinikainen, M., Åhlén, E., 2012. Resurrected *Ceriodaphnia quadrangula* highlight differences between phenotypic and genotypic expressions. Ecol. Evol. 2, 2989–2998.

Repka, S., Pihlajamaa, K., 1996. Predation-induced phenotypic plasticity in *Daphnia pulex*: uncoupling morphological defenses and life history shifts. Hydrobiologia. 339 (1–3), 67–71.

Rhode, S.C., Rawlowski, M., Tollrian, R., 2001. The impact of ultraviolet radiation on the vertical distribution of zooplankton of the genus *Daphnia*. Nature. 412 (6842), 69–72.

Riccardi, N., Mangoni, M., 1999. Consideration on the biochemical composition of some freshwater zooplankton species. J. Limnol. 58 (1), 58–65.

Richman, S., 1958. The transformation of energy by *Daphnia pulex*. Ecol. Monogr. 28 (3), 273–291.

Rider, C.V., Gorr, T.A., Olmstead, A.W., Wasilak, B.A., LeBlanc, G.A., 2005. Stress signaling: coregulation of hemoglobin and male sex determination through a terpenoid signaling pathway in a crustacean. J. Environ. Biol. 208, 15–23.

Rieder, N., 1987. The ultrastructure of the so-called olfactory setae on the antennula of *Daphnia magna*. Hydrobiologia. 145, 175–181.

Riemann, B., Jørgensen, N.O.G., Lampert, W., Fuhrman, J.A., 1986. Zooplankton induced changes in dissolved free amino acids and in production rates of freshwater bacteria. Micro. Ecol. 12, 247–258.

Riessen, H.P., 1984. The other side of cyclomorphosis: why *Daphnia* lose their helmets. Limnol. Oceanogr. 29 (5), 1123–1127.

Riessen, H., Linley, R.D., Altshuler, J., van Yan, N.D., 2011. Calcium, kairomones, and growth of *Daphnia*. Abstract book, IXth International symposium on Cladocera, October 2–8, 2011. Verbania, Italy, p. 32.

Rigler, F.H., 1961a. The uptake and release of inorganic phosphorus by *Daphnia magna*. Limnol. Oceanogr. 6 (2), 165–174.

Rigler, F.H., 1961b. The relation between concentration of food and feeding rate of *Daphnia magna* Straus. Can. J. Zool. 39, 857–868.

Rimet, M., 1961. Quelques aspects nouveaux du phototropisme de *Daphnia pulex*. Bull. Soc. Zool. France. 86 (2–3), 342–357.

Rimet, M., 1965. Influence des substances pharmacodynamiques sur la photocinèse de *Daphnia pulex* De Geer. Bull. Soc. Zool. France. 90 (1), 143–154.

Ringelberg, J., 1963. The behaviour of *Daphnia* in diffuse light. Naturwissenschaften. 50 (7), 313–314.

Ringelberg, J., 1964. The positively phototactic reaction of *Daphnia magna* Straus. Netherlands J. Sea Res. 2 (3), 319–406.

Ringelberg, J., 1987. Light induced behaviour in *Daphnia*. Memorie dell'Istituto Italiano di Idrobiologia. 45, 285–323.

Ringelberg, J., Royackers, K., 1985. Food uptake by hungry cladocerans. Archiv. für Hydrobiol. Beih., Ergebnisse Limnologie. 21, 199–207.

Ringelberg, J., Servaas, H., 1971. A circadian rhythm in *Daphnia magna*. Oecologia. 6 (3), 289–292.

Rivera, A.S., Pankey, M.S., Plachetzki, D.C., Villacorta, C., Syme, A.E., Serb, J.M., et al., 2010. Gene duplication and the origins of morphological complexity in pancrustacean eyes, a genomic approach. BMC Evol. Biol. 10, 123.

Rivero, A., 2006. Nitric oxide: an antiparasitic molecule of invertebrates. Trends Parasitol. 5, 219–225.

Rivier, I.K., 1974. Disturbances and special traits of embryogenesis in some Cladocera. Biologiya vnutrennikh vod, Informatsionnyi bulleten. 22, 29–34.

Rivier, I.K., 1986. Composition and Ecology of Winter Zooplankton Communities. Nauka, Moscow, 160 pp.

Rivier, I.K., 1992. Ecology of the Cladocera in winter water bodies. In: Sovremennye problemy izucheniya vetvistiusykh rakoobtaznykh. St. Petersburg: Gidrometeoizdat, 65–80.

Rivier, I.K., 1998. The predatory Cladocera of the World. Guides to the Identification of the Macroinvertebrates of the Continental Waters of the World 13. Leiden: Backhuys, 213 pp.

Rivier, I.K., 2005. Advances of investigation of the fam. Polyphemoidae (Cladocera, Onychopoda) in the Laboratory of Ecology of Aquatic Invertebrates. In: Biologicheskie resursy presnykh vod: bespozvonochnye. Rybinsk, 291–310.

Rivier, I.K., 2012. Cold-water zooplankton of lakes of the Upper Volga basin. B;tdcr. 381 pp. [In Russian].

Robertson, A.L., 1990. The population dynamics of Chydoridae and Macrothricidae (Cladocera: Crustacea) from the River Thames. U.K. Freshw. Biol. 24, 375–389.

Robichaud, N.F., Sassine, J., Beaton, M.J., Lloyd, V.K., 2012. The epigenetic repertoire of *Daphnia magna* includes modified histones. Genet. Res. Int.174860.

Robinson, K.A., Baird, D.J., Wrona, F.J., 2003. Surface metal adsorption on zooplankton carapaces: implication for exposure and effects in consumer organisms. Environ. Poll. 122, 159–167.

Rodgers Jr., J.H., Deaver, E., Suedel, B.C., Rodgers, P.L., 1997. Comparative aqueous toxicity of silver compounds: laboratory studies with freshwater species. Bull. Environ. Contam. Toxicol. 58 (6), 851–858.

Rodina, A.G., 1946. The experiments on the feeding habits of *Daphnia magna*. Zoologicheskii Zhurnal. 25 (3), 237–244.

Rodina, A.G., 1948. Dissolved organic substances in feeding of Cladocera. Zoologicheskii Zhurnal. 27 (5), 403–410.

Rodina, A.G., 1950. Experimental investigation of feeding of daphnias. Trudy Vsesoyuznogo gidrobiologicheskogo obshchestva. 2, 169–193.

Rodina, A.G., 1963. Microbiology of detritus of lakes. Limnol. Oceanogr. 8, 388–393.

Røen, U.I., 1962. Studies on freshwater Entomostrataca in Greenland. II.. Medd. Grønland. 170 (2), 1–249.

Rohrlack, T., Christoffersen, K., Kaebernik, M., Neilan, B. A., 2004. Cyanobacterial protease inhibitor microviridin J causes a lethal molting disruption in *Daphnia* pulicaria. Appl. Environ. Microbiol. 70 (8), 5047–5050.

Rohrlack, T., Dittman, E., Börner, T., Christoffersen, K., 2001. Effects of cell-bound microcystins on survival and feeding of *Daphnia* spp. Appl. Environ. Microbiol. 67 (8), 3523–3529.

Rolff, J., Siva-Jothy, M.T., 2003. Invertebrate ecological immunology. Science. 301, 472–475.

Romanenko, V.D., Goncharova, M.T., Konovits, M.T., Kipnis, L.S., Krot, Yu.G., 2011. Method of complex estimation of bottom sediments toxicity using benthic and planktonic organisms. Hydrobiol. J. (Kiev). 47 (6), 32–62.

Romanenko, V.D., Solomatina, V.D., Krot, Yu.G., Lekontseva, T.I., Mogilevich, N.A., 2004. Physio-Biochemical assessment of the state of culture of *Moina macrocopa* Straus at enrichment culturing. Hydrobiol. J. (Kiev). 40 (1), 74–80.

Romanova, E.P., Bondarenko, L.F., 1984. Caloric content of crustaceans of the Kuybyshev Reservoir. Biologiya vnutrennikh vod. Informatsionnyi byulleten. 63, 37–42.

Romanovskiy, A.V., Loganova, E.A., 2010. Effect of a single stepwise change of temperature on growth and fecundity of *Ceriodaphnia quadrangula*. In: Biologita vnutrennikh vod. Materualy XIV shkoly-konferentsii mododykh uchenykh, Borok, 26–30 oktyabrya 2010, 115–120.

Rossi, F., 1980. Comparative observations on the female reproductive system and parthenogenetic oogenesis in Cladocera. Boll. di Zool. 47, 21–38.

Roth, O., Ebert, D., Vizoro, D.B., Bieger, A., Lass, S., 2008. Male-biased sex ratio distortion caused by Octosporea bateri, a vertically and horizontally transmitted parasite of *Daphnia magna*. Int. J. Parasitol. 1–11.

Rudiger, P., Zeis, B., 2011. Seasonal temperature changes affect both the genetic structure of a *Daphnia* assemblage and the hemoglobin concentration of *Daphnia*. Abstract book, IXth International symposium on Cladocera, October 2–8, 2011. Verbania, Italy, p. 27.

Rukke, N.A., 2002. Tolerance to low ambient calcium shows inter-population differences in *Daphnia galeata*. J. Plankton Res. 24 (5), 527–531.

Rylov, V.M., 1940. On the negative effect of mineral seston in the diet of some planktonic Entomostraca. Doklady AN SSSR. 29 (7), 522–524.

Saalfeld, E.V., 1936. Untersuchungen über den Blutkreislauf bei Leptodora hyalina. Zeitschr. Vergleich. Physiol. 24 (1), 58–70.

Sabeter, F., 1987. On the interstitial Cladocera in the River Ter (Catalonia, NE Spain), with a description of the male of *Alona phreatica*. Hydrobiologia. 144, 51–62.

Sacherová, V., Hebert, P.D.N., 2003. The evolutionary history of the Chydoridae (Crustacea: Cladocera). Biol. J. Linn. Soc. 79, 629–643.

Sadd, B., Schmid-Hempel, P., 2009. Principles of ecological immunology. Evol. Appl. 2, 113–121.

Sadykhov, D.A., Bogatova, I.B., Filatov, V.I., 1975. Amino acid composition of certain fresh-water zooplankton representatives. Gidrobiologicheskii Zhurnal. 11 (6), 53–57.

Sakamoto, M., Hanazato, T., Tanaka, Y., 2009. Impact of an insecticide on persistence of inherent antipredator morphology of a small cladoceran, *Bosmina*. Arch. Environ. Contam. Toxicol. 57 (1), 68–76.

Salceda, S., Caro, J., 1997. Hypoxia-inducible factor 1a (HIF-1a) protein is rapidly degraded by the ubiquitin-proteasome system under normoxic conditions. Its stabilization by hypoxia depends on redox-induced changes. J. Biol. Chem. 272, 22642–22647.

Salo, Z.T., 1960. Interrelationship of respiratory and cordial function in lower crustaceans. Nauchnye Doklady Vysshey Shkoly, Biol. 1, 48–50.

Samassa, P., 1891. Untersuchungen über das centrale Nervensystem der Cladoceren. Arch. Mikroscopische Anatomie. 38, 100–141.

Samchyshyna, L.V., 2002. About the viability of eggs of some Crustacea from the golden carp feces. Biologiya vnutrennikh vod. 2, 103–105.

Sang Don Kim, Huizhong Ma, Allen, H.E., Cha, D.K., 1999. Influence of dissolved organic matter on the toxicity of copper to *Ceriodaphnia dubia*: effect of complexation kinetics. Environ. Toxicol. Chem. 18 (11), 2433–2437.

Saprykina, E.A., 1996. Reaction of *Daphnia* in several generations to action of pyrethroid insecticide phastak. Vavilovskie chteniya. Yoshkar-Ola.346–348.

Sargent, J., Bell, G., McEvoy, L., Tocher, D., Estevez, A., 1999. Recent developments in the essential fatty acid nutrition of fish. Aquaculture. 177, 191–199.

Sars, G.O., 1862. On the Freshwater Crustaceans Occurring in the Vicinity of Christiania. University of Bergen Press, Bergen, 197 pp.

Sars, G.O., 1886. On some Australian Cladocera raised from dry mud. Forhandlinger i Videnskabs-Selskabet i Christiania. 8, 1–46.

Sars, G.O., 1888. Additional notes on Australian Cladocera raised from dry mud. Forhandlinger i Videnskabs-Selskabet i Christiania.1–74.

Sarviro, V.S., 1985. Dependence on temperature of development duration and of specific growth rate of immature females of *Daphnia longispina* O.F. Müller (Crustacea, Cladocera) in cage experiments. Gidrobiologocheskiy Zhurnal. 3, 28–33.

Scavia, D., Garner, W.S., 1982. Kinetics of nitrogen and phosphorus release in varying food supplies by *Daphnia magna*. Hydrobiologia. 96 (2), 105–111.

Scavia, D., McFarland, M.J., 1982. Phosphorus release patterns and the effects of reproductive stage and ecdysis in *Daphnia magna*. Can. J. Fish. Aquatic sci. 39 (9), 1310–1314.

Schechter, V., 1950. Effect of heparin and vitamin K on the life span of *Daphnia magna*. Proc. Soc. Exp. Biol. Med. 74 (4), 747–748.

Scheffer, D., Robert, P., Médioni, J., 1958. Reactions oculomotrices de la daphnie (*Daphnia pulex* DeGeer) en response a des lumieres monohromatiques d'egale energie. Comptes Rend. Soc. Biol. 152 (6), 1000–1003.

Schindler, D.W., 1968. Feeding, assimilation and respiration rates of *Daphnia magna* unmder various environmental conditions and their relation to production estimates. J. Anim. Ecol. 37, 369–385.

Schindler, J.E., 1971. Food quality and zooplankton nutrition. J. Anim. Ecol. 49 (3), 589–595.

Schlechtriem, C., Arts, M.T., Zellmer, I.D., 2006. Effect of temperature on the fatty acid composition and temporal trajectories of fatty acids in fasting *Daphnia pulex* (Crustacea, Cladocera). Lipids. 41 (4), 397–400.

Schlotz, N., Martin-Creuzburg, D., 2011. Food quality meditated resistance of *Daphnia magna* to pathogenic bacteria: the role of polyunsaturated fatty acids. Abstract book, IXth International symposium on Cladocera, October 2–8, 2011. Verbania, Italy, p. 34.

Schmid-Hempel, P., 2003. Variation in immune defence as a question of evolutionary ecology. Proc. Biol. Sci. 270, 357–366.

Schmid-Hempel, P., 2011. Evolutionary Parasitology: The Integrated Study of Infections, Immunology, Ecology, and Genetics. Oxford University Press, Oxford, UK.

Schmid-Hempel, P., Ebert, D., 2003. On the evolutionary ecology of specific immune defence. Trends Ecol. Evol. 18, 27–32.

Schmidt, G.W., 1968. Zur Ausnutzung des Nahrungsstickstoffs ducrh *Daphnia magna* Straus. Arch. Hydrobiol. 65 (2), 142–186.

Schmidt, K., Pflugmacher, S., Staaks, G.B.O., Steinberg, C.E.W., 2006. The influence of tributyltin chloride and polychlorinated biphenyls on swimming behavior, body growth, reproduction, and activity of biotransformation enzymes in *Daphnia magna*. J. Freshw. Ecol. 21 (1), 109–120.

Schmidt, K., Steinberg, C.E.W., Staaks, G.B.O., Pflugmacher, S., 2005. Influence of a xenobiotic mixture (PCB and TBT) compared to single substances on swimming behavior or reproduction of *Daphnia magna*. Acta hydrochimica et hydrobiologica. 33 (4), 287–300.

Schmidt, W.J., 1943. Bemerkungen zur Kalkeinlagerung im Cladocerenpanzern. Biol. Generalis (Wien). 16, 445–449.

Schmoker, C., Henández-León, S., 2003. The effect of food on the respiration rates of *Daphnia magna* using a flow-through system. Scientia Marina. 67 (3), 361–365.

Schoenberg, S.A., Maccubbin, A.F., Hodson, R.E., 1984. Cellulose digestion by freshwater microcrustacea. Limnol. Oceanogr. 29 (5), 1132–1136.

Schrader, F., 1925. The cytology of pseudo-sexual eggs in a species of *Daphnia*. Zeitschrift für induktive Abstammungs- und Vererbungslehre XL, 1/2. 1–27.

Schultz, T.W., Freeman, S.R., Dumont, J.N., 1980. Uptake, depuration, and distribution of selenium in *Daphnia* and its effect on survival and ultrastructure. Arch. Environ. Contam. Toxicol. 9 (1), 23–40.

Schultz, T.W., Kennedy, J.R., 1976a. The fine structure of the digestive system of *Daphnia pulex* (Crustacea, Cladocera). Tissue Cell. 8 (3), 479–490.

Schultz, T.W., Kennedy, J.R., 1976b. Cytotoxic effects of the herbicide 3-amino-1,2,4,-triazole on *Daphnia pulex* (Crustacea, Cladocera). Biol. Bull. 151 (2), 370–385.

Schultz, T.W., Kennedy, J.R., 1977. Analysis of the integument and muscle attachment in *Daphnia pulex* (Cladocera, Crustacea). J. Submicr. Cytol. 9 (1), 37–51.

Schulz, H., 1928. Ueber die Bedeutung des Lichtes im Leben niederer Krebse (nach Versuchen an Daphniden). Zeitschrift für vergleichende Physiologie. 7 (3), 488–552.

Schulz, K.L., Sterner, R.W., 1999. Phytoplankton phosphorus limitation and food quality for *Bosmina*. Limnol. Oceanogr. 44 (6), 1549–1556.

Schulz, K.L., Sterner, R.W., 2000. Zooplankton food quality in large lakes—growth of *Daphnia* on high P content seston from Lake Superior. Verhandl. Internat. Vereinig. Limnol. 27 (4), 1855–1860.

Schulze-Robbecke, G., 1951. Untersuchungen über Lebensdauer, Altern und Tod bei Arthropoden (*Daphnia magna* Straus, *Dixippus* (*Carausius*) *morosus* Brunner V. Wattenwyl und *Melolontha melolontha* L.). Zool. Jahrb. Abt. Allgemeine Zoologie und Physiol. Tiere. 62 (4), 366–394.

Schwartz, S.S., Hebert, P.D.N., 1984. Subgeneric distinction in the genus *Daphnia*: a new diagnostic trait. Trans. Am. Microsc. Soc. 103, 341–346.

Schwarzenberger, A., Kuster, C.J., Elert, E., 2012. Molecular mechanisms of tolerance to cyanobacterial protease inhibitors revealed by clonal differences in *Daphnia magna*. Mol. Ecol. 21, 4898–4911.

Schwerin, S., Zeis, B., Horn, W., Horn, H., Paul, R.J., 2010. Hemoglobin concentration in *Daphnia* (*D. galeata-hyalina*) from the epilimnion is related to the state of nutrition and the degree of protein homeostasis. Limnol. Oceanogr. 55, 639.

Schwerin, S., Zeiss, B., Lamkemeyer, T., Paul, R.D., Koch, M., Madlung, J., et al., 2009. Acclimatory response of the *Daphnia pulex* proteome to environmental changes. II. Chronic response to different temperatures (10 and 20 degrees C) mainly affects protein metabolism. BMC Physiol. 21 (9), 8.

Scourfield, D.J., 1894. Entomostraca and the surface-film of water. Linn. Soc. J. Zool. 25, 1–19.

Scourfield, D.J., 1896. The olfactory setae of the Cladocera. J. Quekett Microscop. Club, ser. 2.(6), 280–288.

Scourfield, D.J., 1900a. Note on *Scapholeberis mucronata* and the surface film of water. J. Quekett Microscop. Club, ser. 2.(7), 309–312.

Scourfield, D.J., 1900b. The swimming peculiarities of *Daphnia* and its allies, with an account of a new method of examining living Entomostraca and similar organisms. J. Quekett Microscop. Club, ser. 2.(7), 217–244.

Scourfield, D.J., 1905. Locomotion of microscopic aquatic animals. J. City Lond. Coll. Sci. Soc. X. 7, 63–70.

Scourfield, D.J., 1923. Physical factors in freshwater biological problems. J. Quekett Microscop. Club, ser. 2.(89), 1–18.

Scourfield, D.J., 1929. Water-surface plants and animals: with special reference to the surface-tension factor in their environment. Essex Nat. 22, 217–236.

Scoville, A.G., Pfrender, M.E., 2010. Phenotypic plasticity facilitates recurrent rapid adaptation to introduced predators. Proc. Natl. Acad. Sci. U.S.A. 107, 4260–4263.

Sedova, T.V., 1977. Principles of Cytology of Algae. Nauka, Leningrad, p. 172. [in Russian].

Seely, C.J., Lutnesky, M.M.F., 1998. Odour-induced antipredator behavior of the water flea, *Ceriodaphnia reticulata*, in various predator and prey densities. Freshw. Biol. 40 (1), 17–24.

Seidl, M.D., Paul, R.J., Pirow, R., 2005. Effects of hypoxia acclimation on morpho-physiological traits over three generations of *Daphnia magna*. J. Exp. Biol. 208 (1), 2165–2175.

Seidl, M.D., Pirow, R., Paul, R.J., 2002. Water fleas (*Daphnia magna*) provide a separate ventilatory mechanism for their brood. Zoology. 105, 15–23.

Seidman, L.A., Larsen Jr., J.H., 1979. Ultrastructure of the envelopes of resistant and nonresistant *Daphnia* eggs. Can. J. Zool. 57 (9), 1773–1777.

Sell, A.F., 1998. Adaptation to oxygen deficiency: contrasting patterns of haemoglobin synthesis in two coexisting *Daphnia* species. Comp. Biochem. Physiol. A. 120, 119–125.

Selvaraj, C., 1979. On the hatching of the sexual eggs of a cladoceran, *Daphnia similis* Claus. J. Inland Fish. Soc. India. 11 (1), 34–39.

Semenova, L.M., 1974. On nutrition of *Bosmina coregoni*, Baird (Cladocera). Gidrobiologicheskii Zhurnal. 10 (3), 39–46.

Sent-Ilrt, M. [Saint-Hilaire, M.], 1860. Materials for the Monograph of Daphnid Family. S.-Peterburg, p. 95. [in Russian].

Sergeeva, S.S., 2004. *Pseudopenilia bathyalis* gen. n., sp. n. (Crustacea, Branchiopoda, Ctenopoda)—an inhabitant of the sulphur dioxide zone of the Black Sea. Vestnik zoologii. 38 (3), 37–42.

Seurat, L., Schmitt, F.G., Brewer, M.C., Strickler, R, Soussi, S., 2004. From random walk to multifractal random walk in zooplankton swimming behavior. Zool. Stud. 43 (2), 498–510.

Shan, R.K.-C., 1969. Life cycle of a chydorid cladoceran, *Pleuroxus denticulatus* Birge. Hydrobiologia. 34 (3–5), 513–523.

Shan, R.K.-C., 1970. Influence of light on hatching resting eggs of chydorids (Cladocera). Internat. Revue ges. Hydrobiol. 55 (2), 295–302.

Shan, R.K.-C., 1971. Genetic stability in the Cladocera. Trans. Amer. Microsc. Soc. 90 (1), 116.

Shaw, J.R., Dempsey, T.D., Chen, C.Y., Hamilton, J.W., Folt, C.L., 2006. Comparative toxicity of cadmium, zinc, and mixtures of cadmium and zinc to daphnids. Environ. Toxicol. Chem. 25 (1), 182–189.

Shaw, J.R., Colbourne, J.K., Davey, J.C., Glaholt, S.P., Hampton, T.H., Chen, C.Y., et al., 2007. Gene response profiles for *Daphnia pulex* exposed to the environmental stressor cadmium reveals novel crustacean metallothioneins. BMC Genomics. 8, 477.

Shaw, J.R., Pfrender, M.E., Eads, B.D., Klaper, R., Callaghan, A., Sibly, R.M., et al., 2008. *Daphnia* as an emerging model for toxicological genomics. Adv. Exp. Biol. 2, 165–328.

Shcherbakov, A.P., 1935. On oxygen consumption by some planktonic crustaceans. Arbeiten der Limnologoschen Station zu Kosino. 19, 67–89.

Shcherbakov, A.P., 1967. Lake Glubokoe. Moscow: Nauka. 379 pp. [in Russian].

Shcherban, E.P., 1979. Toxicity of certain surfactants to *Daphnia magna* Straus. Gidrobiolochicheskii Zhurnal. 15 (3), 69–73.

Shcherban, E.P., 1980. Effect of the surfactant (R = C10-C13) on the major bioparameters and productivity of *Daphnia magna* Straus. Gidrobiologcheskii Zhurnal. 16 (2), 102–105.

Shcherban, E., 1986. Toxicity investigation of some substances on Cladocera. Exp. Water Toxi. 11, 137–143, Riga: Zinatne.

Sheldon, B.C., Verhulst, S., 1996. Ecological immunology: costly parasite defences and trade-offs in evolutionary ecology. Trends Ecol. Evol. 11, 319–321.

Sherstyuk, V.V., 1971. Calorific value of forage organisms in Kremenchug Reservoir. Gidrobiologicheskii Zhurnal. 7 (6), 99–103.

Shi, L., Paskewitz, S.M., 2004. Identification and molecular characterization of two immune-responsive chitinase-like proteins from *Anopheles gambiae*. Ins. Mol. Biol. 13, 387–398.

Shimkevich, V., 1925. Biological foundations of zoology. vol. 2. Leningrad, Moscow, Gosudarstvennoe Izdatelstvo. p. 655. [in Russian].

Shimkevich, V., 1909 (estimated publication date). Urodstva i proiskhozhdenie vidov (Abnormalities and the origin of species). St. Petersburg, Moscow: Volf. p. 107. [in Russian].

Shkorbatov, G.L., 1953. Ecologo-physiological traits and conditions of existence of close forms of freshwater animals. Zoologicheskii Zhurnal. 32 (5), 793–803.

Shkorbatov, G.L., 1982. A contribution to the general theory of adaptation. Zhurnal obshchei biologii. 43 (6), 775–787.

Shpet, G.I., 1968. Biological Productivity of Fish and of Other Animals. Urozhai, Kiev, p. 92. [in Russian].

Shuba, T., Costa, R.R., 1972. Development and growth of *Ceriodaphnia reciculata* embryos. Trans. Amer. Microscop. Soc. 91 (3), 429–435.

Shuter, B.J., Thomas, J.E., Taylor, W.D., Zimmerman, A.M., 1983. Phenotypic correlates of genome DNA content in unicellular eukaryotes and other cells. Amer. Nat. 122, 26–44.

Siebeck, O., 1978. Ultraviolet tolerance of planktonic crustaceans. Verhandl. Internat. Vereining. Limnol. 20, 2469–2473.

Simčič, T., Brancelj, A., 1997. Electron transport system (ETS) activity and respiration rate in five *Daphnia* species at different temperatures. Hydrobiologia. 360, 117–125.

Simon, J.-C., Pfrender, M.E., Tollrian, R., Tagus, D., Colbourne, J.K., 2012. Genomics of environmentally induced phenotypes in 2 extremely plastic arthropods. J. Heredity. 102, 512–525.

Sinev Yu, A., La-Orsri Sanoamuang, 2011. Hormonal induction of males as a method for studying tropical cladocerans: description of males of four chydorid species. Zootaxa. 2826, 45–56.

Skadovskiy, S.N., 1939. Physiological analysis of phototaxis in daphnias (*Daphnia pulex* De Geer), Uchenye zapiski Moskovskogo Gosudarstvennogo Universiteta. Gidrobiologiya. 33 (3), 237–246.

Skadovskiy, S.N., 1955. Ecological physiology of aquatic organisms. Moscow, 338 pp. [in Russian].

Sklayrova, T.V., 1938. Influence of phosphate on development of *Daphnia*. Trudy Voronezhskogo universiteta. 10 (3), 115–123.

Sladkova, S.V., Kholodkevich, S.V., 2012. Intensity of oxygen consumption by daphniae as a biomarker of quality of the aquatic environment. Materialy 4 vserossiyskoy konferentsii po vodnoi ekotoksikologii. Borok. 1, 209–212.

Ślusarczyk, M., Dawidowicz, P., Rygelska, E., 2005. Hide, rest or die: a light meditated diapause response in *Daphnia magna* to the threat of fish predation. Freshw. Biol. 50, 141–146.

Smaridge, M.W., 1954. Iron excretion by *Daphnia* during haemoglobin loss. Nature. 173 (4408), 782–783.

Smaridge, M.W., 1956. Distribution of iron in *Daphnia* in relation to haemoglobin synthesis and breakdown. Quart. J. Micro. Sci. 97 (2), 205–214.

Smirnov, N.N., 1962. *Eurycercus lamellatus* (O.F. Mueller): field observations and nutrition. Hydrobiologia. 20 (3), 280–294.

Smirnov, N.N., 1964. *Pleuroxus* (Chydoridae): field observations and growth. Hydrobiologia. 23 (3–4), 305–320.

Smirnov, N.N., 1965a. Life cycle of some Chydoridae. Zoologicheskii Zhurnal. 44 (9), 1409–1411.

Smirnov, N.N., 1965b. Heart rate of cladocerans as a criterion of their ecological traits and condition. Gidrobiologicheskii Zhurnal. 1 (6), 28–34.

Smirnov, N.N., 1966. Littoral Cladocera of the Cherepovets Reservoir. In: Plankton i bentos vnutrennbickh vodoemov. Leningrad: Nauka 145–150.

Smirnov, N.N., 1967. On sensilla on limbs of the Cladocera. Zoologicheskii Zhurnal. 46 (2), 286–288.

Smirnov, N.N., 1969. Morpho-functional grounds of mode of life of Cladocera. II. The functional complex providing digestion. Gidrobiologicheskii Zhurnal. 5 (6), 41–46.

Smirnov, N.N., 1970. Morpho-functional grounds of the mode of life of Cladocera. IV. Respiratory adaptations in chydorids. Zhurnal obshchei biologii. 31 (1), 94–98.

Smirnov, N.N., 1971. Chydoridae. Fauna SSSR, new series, no. 101, Nauka: Leningrad, 531 pp. [in Russian].

Smirnov, N.N., 1974. Chydoridae. Jerusalem. 644 pp. [translation of 1971 edition from Russian].

Smirnov, N.N., 1975. Biology of cladocera. Itogi nauki i tekhniki, zoologiya bespozvonochnykh, vol. 3, p. 116. [in Russian].

Smirnov, N.N., 1977. Morpho-functional grounds of mode of life of Cladocera. VIII. Akinesis in Cladocera. Zoologicheskii Zhurnal. 56 (3), 471–472.

Smirnov, N.N., 1988. Cladocera (Crustacea) from Nicaragua. Hydrobiologia. 160, 63–77.

Smirnov, N.N., 1992. The Macrothricidae of the World. Guides to the Identification of the Microinvertebrates of the Continental Waters of the World 1. SPB Academic Publishing, The Hague, 143 pp.

Smirnov, N.N., 1996. Cladocera: the Chydorinae and Sayciinae (Chydoridae) of the World. Guides to the Identification of the Microinvertebrates of the Continental Waters of the World 1. SPB Academic Publishing, Amsterdam, 197 pp.

Smirnov, N.N., Timms, B.V., 1983. A Revision of Australian Cladocera (Crustacea). Records of the Australian Museum, Sydney, Supplement 1, 132 pp.

Smirnova, L.A., 1960. On the dependence of phototaxis of some crustaceans on the state of sulfhydryl groups of protein bodies. Zoologicheskii Zhurnal. 39 (7), 1012–1017.

Smith, F.E., Baylor, E.R., 1953. Color responses in the Cladocera and their ecological significance. Am. Nat. 83 (832), 49–55.

Smith, K.C., Macagno, E.R., 1990. UV photoreceptors in the compound eye of *Daphnia magna* (Crustacea, Branchiopoda). A fourth spectral class in single ommatidia. J. Comp. Physiol. A. 166, 597–606.

Sobral, O., Chastinet, C., Nogueira, A., Soares, A.M.V.M., Gonçalves, F., Ribeiro, R., 2001. *In vitro* development of parthenogenetic eggs: a fast ecotoxicity test with *Daphnia magna*? Ecotoxicol. Environ. Saf.(50), 174–179.

Soderhall, K., Cerenius, L., 1998. Role of the prophenoloxidase-activating system in invertebrate immunity. Curr. Opin. Immunol. 10, 23–28.

Sofyan, A., Rosita, G., Price, D.J., Birge, J., 2007. Cadmium uptake by *Ceriodaphnia dubia* from different exposures: relevance to body burden and toxicity. Environ. Toxicol. Chem. 26 (3), 470–477.

Sokolova, V.A., 1974. Weight characteristics of crustaceans of Viluisk Reservoir. Byulleten nauchno-tekhnicheskoi informatsii. Biologicheskie problemy Severa. Akademiya nauk SSSR. Yakutskii filial Sibirskogo otdeleniya, 11–12.

Sollman, T., Webb, W., 1941. Pharmacological responses of *Daphnia magna*. J. Exp. Therap. 71 (1), 261–267.

Soma, Y., Itoh, N., Tani, Y., 2005. Sterol composition of steryl chlorine esters (SCEs) formed through grazing of algae by freshwater crustaceans: relevance to the composition of sedimentary SCEs. Limnology (Springer). 6 (1), 45–51.

Somov, A.Y., Malinina, Y.A., 2003. Effect of electromagnetic radiation of industrial frequency 50 Hz on *Daphnia magna* (Straus). Radiatsionnaya biologiya. Radioekologiya. 43 (5), 552–554.

Sørensen, N.A., 1936. Über die Verbreitung des Astacins unter den Evertevbraten. Lipochrome mariner Tiere VII. Det Kgl. Norske Videnskabers Selskabs Skrifter 1, 1–14.

Southward, G.R., Beauchamp, J.J., Schmieder, P.K., 1978. Bioaccumulation potential of polycyclic aromatic hydrocarbons in *Daphnia pulex*. Water Res. 12 (11), 973–977.

Spaak, P., 1995. Sexual reproduction in *Daphnia*: interspecific differences in a hybrid species complex. Oecologia. 104, 501–507.

Spangenberg, F., 1877. Das Centranervensystem von *Daphna magna* und Moina rectirostris. Munchen. p. 37.

Spehar, R.L., Fiandt, J.T., Anderson, R.L., DeFoe, D.L., 1980. Comparative toxicity of arsenic compounds and their accumulation in invertebrates and fish. Arch. Environ. Contam. Toxicol. 9 (1), 53–63.

Sperfeld, E., Wacker, A., 2011. Temperature affects the limitation of *Daphnia magna* by eicosapentaenoic acid and the fatty composition of body tissue and eggs. Freshw. Biol.1–12. 10.1111/j.1365-8427.2011.02719.x

Spitze, K., 1992. Predator meditated plasticity of prey life history and morphology: *Chaoborus americanus* predation on *Daphnia pulex*. Am. Nat. 139 (2), 229–247.

Spitze, K., 1993. Population structure in *Daphnia obtusa*: quantitative genetic and allozymic variation. Genetics. 135, 367–374.

Starkweather, P.L., 1975. Diel patterns of grazing in *Daphnia pulex* Leydig. Verh. Intrnat. Verein. Limnol. 19 (4), 2851–2857.

Starkweather, P.L., 1978. Diel variation in feeding behavior of *Daphnia pulex*. Influences of food density and nutritional history on mandibular activity. Limnol. Oceanogr. 23 (2), 306–317.

Stazi, A.V., Mantovani, A., Fuglieni, F., Dojmi di Delupis, 1994. Observations on fungal infection of the ovary of laboratory-cultured *Daphnia magna*. Bull. Environm. Contam. Toxicol. 53 (5), 699–703.

Stearns, S.C., 1975. Light responses of *Daphnia pulex*. Limnol. Oceanogr. 20 (1), 564–570.

Steinberg, C.E.W., Ouergemmi, N., Herrmann, S., Bouchnak, R., Timofeyev, MA., Mezel, R., 2010. Stress by poor food quality and exposure to humic substances: *Daphnia magna* responds with oxidative stress, lifespan extension, but reduced offspring numbers. Hydrobiologia. 652, 223–236.

Steiner, S., Kasprzak, P., 2000. Cladoceran grazing and the fate of planktonic primary production in a mesotrophic stratified lake. Internat. Vereinig. Limnol. 27 (5), 3005–3008.

Stelzer, C.P., Riss, S., Stadler, P., 2011. Genome size evolution at the speciation level: the cryptic species complex *Brachionus plicatilis* (Rotifera). BMC Evol. Biol. 11, 90. 10.1186/1471-2148-11-90.

Stenderup, J.T., Olesen, J., Glenner, H., 2006. Molecular phylogeny of the Branchiopoda (Crustacea)—multiple approaches suggest a 'diplostracan' ancestry of the Notostraca. Mol. Phylogenet. Evol. 41, 182–194.

Stepanova, G.M., 1967. On age differences of biochemical composition in *Daphnia magna* Claus and *Ceriodaphnia affinis* Lill. Materialy V konferentsii molodykh uchenykh Moldavii (zoologiya etc.). Kishinev. 27–28.

Stepanova, G.M., 1968. Chemical composition and nutritive value of some Daphniidae species. Izvestiya Akademii nauk Moldavskoi SSR. Biol. Chem. 1, 55–57.

Stepanova, G.M., 1974. Chemical composition of Entomostraca from Moldavian water bodies. Biologicheskie Resursy vodoyemov Moldavii. 12, 43–50.

Stepanova, G.M., Borshch, Z.T., 1970. On the content of vitamin B12 in ground and in some freshwater algae and invertebrates. Biologicheskie resursy vodoemov Moldavii. 3, 48–50.

Stepanova, G.M., Naberezhnyi, A.I., 1970. Amino acid composition of *Daphnia magna* Straus and *Daphnia pulex* (De Geer). Biologichekie protsessy v morskikh I kontinentalnykh vodoemakh. Kushinev. 354–355.

Stepanova, G.M., Naberezhnyi, A.I., 1972. Amino acid composition of *Daphnia magna* Straus and *Simocephalus vetulus*. Herald Hydrobiol. p. 1–4.

Stepanova, G.M., Vinogradova, Z.A., 1970. Biochemical characteristics of some components of the hydrofauna of water bodies of Moldavia. Izvestiya Akademii nauk Moldavskoi SSR, Biol. 1, 39–45.

Stepanova, G.M., Naberezhnyi, A.I., Byzgu, S.E., 1971. Dependence of chemical composition of some cladocera on water mineralization and kind of food. Biologicheskie resursy vodoyemov Moldavii. 8, 19–25.

Sterba, G., 1956a. Zytologische Untersuchungen an Grosskernigen Fettzellen von *Daphnia pulex* unter besonderer Beruecksichtigung des Mitochondrien-Formwechsels. Zeitschrift f. Zellforschung. 44, 456–487.

Sterba, G., 1956b. Beobachtungen über die Lebensdauer von *Daphnia pulex* (de Geer) unter natuerlichen Bedingungen. Zoologischer Anzeiger. 157 (9–10), 179–184.

Sterba, G., 1957a. Riesenzellen der Daphnien-Oberlippe. Zeitschr. Zellforsch. 47, 198–213.

Sterba, G., 1957b. Die Neurosekretorischen Zellgruppen einiger Cladoceren (*Daphnia pulex* und *D. magna*, *Simocephalus vetulus*). Zoolog. Jahrb., Abteil. Anat. 76, 303–310.

Sterner, R.W., 1993. *Daphnia* growth on varying quality of Scenedesmus: mineral limitation of zooplankton. Ecology. 74 (8), 2351–2360.

Sterner, R.W., Schwalbach, M.S., 2001. Diel integration of food quality by *Daphnia*: luxury consumption by a freshwater planktonic herbivore. Limnol. Oceanogr. 46 (2), 410–416.

Sterner, R.W., Smith, R.F., 1993. Clearance, ingestion and release of N and P by *Daphnia obtusa* feeding on *Scenedesmus acutus* of varying quality. Bull. Marine Sci. 53 (1), 228–239.

Stobbart, R.H., Keating, J., Earl, R., 1977. A study of sodium uptake by the water flea *Daphnia magna*. Comp. Biochem. Physiol. A. 58 (3A), 299–309.

Sterner, R.W., Hagemeier, D.D., Smith, R.F., Smith, W.L., 1992. Lipid-ovary indices in food limited *Daphnia*. J. Plankton Res. 14 (10), 1449–1460.

Sterner, R.W., Hagemeier, D.D., Smith, R.F., Smith, W.L., 1993. Phytoplankton nutrient limitation and food quality for *Daphnia*. Limnol. Oceanogr. 38 (4), 857–871.

Stillman, J.H., Colbourne, J.K., Lee, C.E., Patel, N.H., Phillips, M.R., Towle, D.W., et al., 2008. Recent advances in crustacean genomics. Integr. Comp. Biol. 48, 852–868.

Stirnadel, H.A., Ebert, D., 1997. Prevalence, host specificity and impact on host fecundity of microparasites and epibionts in three sympatric *Daphnia* species. J. Anim. Ecol. 66, 212–222.

Storch, O., 1929. Analyse der Fangapparate niederer Krebse auf Grund von Micro-Zeitlupeaufnahmen. I. Mitteilung. Der Fangapparat von Sida crystallina O.F. Müller. Biologia Generalis. 5, 1–62.

Storch, O., 1931. Ueber des Mechanik des Herzschlages bei Cladocere. Eine Analyse mit Hilfe der Mikrozeitlupe. Zeitschrift für vergleichende Physiologie. 14 (4), 709–736.

Storz, U.C., Paul, R.J., 1998. Phototaxis in water fleas (*Daphnia magna*) is differently influenced by visible and UV light. J. Comp. Physiol. A. 183 (6), 709–717.

Storz, J.F., Moriyama, H., 2008. Mechanisms of hemoglobin adaptation to high altitude hypoxia. High. Alt. Med. Biol. 9, 148–157.

Storz, J.F., Opazo, J.C., Hoffmann, F.G., 2013. Gene duplication, genome duplication, and the functional diversification of vertebrate globins. Mol. Phylogenet. Evol. 66, 469–478.

Storz, J.F., Runck, A.M., Moriyama, H., Weber, R.E., Fago, A., 2010. Genetic differences in hemoglobin function between highland and lowland deer mice. J. Exp. Biol. 213, 2565–2574.

Storz, J.F., Sabatino, S.J., Hoffmann, F.G., Gering, E.J., Moriyama, H., Ferrand, N., et al., 2007. The molecular basis of high-altitude adaptation in deer mice. PLoS Genet. 3, e45.

Storz, U.C., Paul, R.J., 1998. Phototaxis in water fleas (*Daphnia magna*) is differently influenced by visible and UV light. J. Comp. Physiol. A. 183 (6), 709–717.

Strand, M.R., 2008. The insect cellular immune response. Insect Sci. 15, 1–14.

Straskraba, M., 1964. Preliminary results of a new method for the quantitative sorting of freshwater net plankton into main groups. Limnol. Oceanogr. 9 (2), 268–270.

Stroganov, N.S., 1983. Adaptation and adaptability in the system of interactions of a hydrobiont with a toxicant. In: Reactions of hydrobionts to pollution. Moscow, Nauka, pp. 5–13.

Stroschein-Stevenson, S.L., Foley, E., O'Farrell, P.H., Johnson, A.D., 2006. Identification of *Drosophila* gene products required for phagocytosis of *Candida albicans*. PLoS Biol. 4, 87–99.

Stross, R.G., 1965. Diapause induction in *Daphnia* requires two stimuli. Science. 150 (1702), 1462–1464.

Stross, R.G., 1966. Light and temperature requirements for diapause development and release in *Daphnia*. Ecology. 47 (3/4), 368–374.

Stross, R.G., 1969a. Photoperiod control of diapause in *Daphnia*. I. Induction of winter diapause in the arctic. Biol. Bull. 136 (2), 264–273.

Stross, R.G., 1969b. Photoperiod control of diapause in *Daphnia*. II. Two-stimulus control of long-day, short-day induction. Biol. Bull. 137 (2), 359–374.

Stross, R.G., 1969c. Long-day induction in a short-day arthropod. *Daphnia pulex*. Am. Soc. Zool. 594–595.

Stross, R.G., 1987. Photoperiodism and phased growth in *Daphnia* populations: coactions in perspective. Memorie dell'Istituto Italiano di Idrobiologia. 45, 413–437.

Stross, R.G., Hill, J.C., 1965. Diapause induction in *Daphnia* requires two stimuli. Science. 150 (3702), 1462–1464.

Stross, R.G., Hill, J.C., 1968. Photoperiod control of winter diapause in the freshwater crustacean, *Daphnia*. Biol. Bull. 134 (1), 176–198.

Stuart, C.A., Cooper, H.J., 1932. Food as a sex-determining factor in *Moina macrocopa*. Physiol. Zool. 5 (1), 70–91.

Sturm, A., Hansen, P.-D., 1999. Altered cholinesterase and monooxygenase levels in *Daphnia magna* and *Chironomus riparius* exposed to environmental pollutants. Ecotoxicol. Environ. Saf. 42 (1), 9–15.

Suedel, B.C., Rodgers Jr., J.H., Deaver, E., 1997. Experimental factors that may affect toxicity of cadmium to freshwater organisms. Arch. Environ. Contam. Toxicol. 33 (2), 188–193.

Sundbom, M., Vrede, T., 1997. Effects of fatty acid and phosphorus content of food on the growth, survival and reproduction of *Daphnia*. Freshw. Biol. 38 (3), 665–674.

Suomalainen, P., 1939. The effect of adrenaline and choline on the heart rate in *Daphnia pulex* Geer. Annales Zoologici Societatios Zoologicae-Botanicae Fenjicae Vanamo. 7 (3), 1–6.

Sushchenya, L.M., 1958a. Dependence of filtration rate on in planktonic crustaceans on concentration of food particles. Trudy Biologicheskoi stantsii na ozere Naroch, Belorussk Gos. Univ. 1, 223–240.

Sushchenya, L.M., 1958b. Quantitative data on the fitratory feeding of pelagic Crustacea. Biologicheskie nauki. 1, 16–20.

Sushchenya, L.M., 1972. Respiration intensity in Crustacea. Kiev: Naukova Dumka, p. 196. [in Russian].

Sushchenya, L.M., 1975. Quantitative regularities of feeding of Crustacea. Minsk: Nauka i tekhnika, p. 207. [in Russian].

Sushchik, N.N., 2001. Growth and development of freshwater zooplankton at usual shortage of essential fatty acids in food. Conference of young scientists, December 4–7, 2001, Novosibirsk (Internet).

Sushchik, N.N., 2006. Fatty acids as markers of trophometabolic interactions in freshwater ecosystems. Extended abstract of doctor's dissertation. Krasnoyarsk. p. 50.

Sushchik, N.N., 2008. Role of essential fatty acids in trophometabolic interactions in the freshwater ecosystems (a review). Zhurnal obshchei biologii. 69 (4), 299–316.

Sushchik, N.N., Gladyshev, M.I., Kalacheva, G.S., Dubovskaya, O.P., Kravchuk, Y.S., Ivanova, Y.A., et al., 2002. The seasonal dynamics of zooplankton and contents of essential fatty acids in seston of a small pond. Biologiya vnutrennikh vod. 2, 60–68.

Sushchik, N.N., Gladyshev, M.I., Makhutova, O.N., 2008. Seasonal shift of pool of eicosapentaenoic acid through pelagic food chain. Doklady Akademii nauk. 422 (6), 848–849.

Sutton, R.A., Hebert, P.D., 2002. Patterns of sequence divergence in Daphniid hemoglobin genes. J. Mol. Evol. 55, 375–385.

Svetlichny, L.S., Hubareva, E.S., 2002a. Special traits of metabolic reaction to change of oxygen concentration in *Moina micrura* (Cladocera) grown at hypo- and normoxia. Gidrobiolocheskii Zhurnal. 38 (4), 35–44.

Svetlichny, L.S., Hubareva, E.S., 2002b. Effect of oxygen concentration on metabolism and locomotory activity of *Moina micrura* (Cladocera) cultured at hypo- and normoxia. Marine Biol. 141 (1), 145–151.

Svetlichny, L.S., Hubareva, E.S., 2004. Peculiarities of metabolic response of *Moina micrura* (Cladocera) growth at hypo- and normoxia on changed oxygen concentrations. Gidrobiolocheskii Zhurnal. 40 (3), 199–204.

Swift, H., 1950. The constancy of deoxyribose nucleic acid in plant nuclei. Proc. Natl. Acad. Sci. USA. 36, 643–654.

Szlauer, L., 1962a. On the direction of movements of *Daphnia magna* Straus. Polskie archiwum hydrobiologii. 10 (23), 355–369.

Szlauer, L., 1962b. Wplyw karmienia na zachowanie sie, *Daphnia magna* Straus i *Daphnia pulex* De Geer (Cladocera). Zesz. Nauk ESR Iszt. 13 (1), 169–193.

Szlauer, L., 1963a. Diurnal migrations of minute invertebrates inhabiting the zone of submerged hydrophytes in a lake. Schweiz. Zeitschr. Hydrol. 25 (1), 56–64.

Szlauer, L., 1963b. The speed of the movements of planktonic Crustacea as a factor influencing the vertical stratification of juvenile and adult individuals. Polskie archiwum hydrobiologii. 11 (14), 5–15.

Szlauer, L., 1964. Reaction of *Daphnia pulex* De Geer to the approach of different objects. Polskie archiwum hydrobiologii. 12 (1), 5–16.

Szlauer, L., 1965. The refuge ability of plankton animals before models of plankton-eating animals. Polskie archiwum hydrobiologii XIII (XXVI). 1, 89–95.

Szlauer, L., 1971. Fishing of lake arthropoda into light traps. Polskie archiwum hydrobiologii. 18 (1), 1–91.

Szlauer, L., 1973. Settlement of *Sida crystallina* O.F.M. (Cladocera) on solid objects. Ekol. Pol. 12, 185–192.

Taipale, S., Kankaala, P., Jones, R.I., 2007. Contributions of different organic carbon sources to *Daphnia* in the pelagic foodweb of a small polyhumic lake: results from microcosm DI 13C additions. Ecosystems. 10, 757–772.

Taipale, S.J., Brett, M.T., Pulkkinen, K, Kainz, M.J., 2012. The influence of bacteria-dominated diets on *Daphnia magna* somatic growth, reproduction, and lipid composition. Fed. Europ. Microbiol. Soc. 82 (1), 50–62.

Tasse, J., Camougis, G., 1965. Electrographic studies on cardiac activity of the cladoceran, *Daphnia*. Crustaceana. 8 (2), 197–205.

Tatarazako, N., Oda, S., 2007. The water flea *Daphnia magna* (Crustacea, Cladocera) as a test species for screening and evaluation of chemicals with endocrine disrupting effect on crustaceans. Ecotoxicology. 16 (1), 197–203.

Tatarazako, N., Oda, S., Watanabe, H., Morita, M., Iguchi, T., 2003. Juvenile hormone agonists affect the occurrence of male *Daphnia*. Chemosphere. 53 (8), 827–833.

Taub, F.B., Dollar, A.M., 1968. The nutritional inadequacy of *Chlorella* and *Chlamydomonas* as food for *Daphnia pulex*. Limnol. Oceanogr. 13 (4), 607–617.

Tauson, A., 1924a.The effect of hydrogen ions and antagonistic action of cations on survival of animals. 2. Izvestiya nauchno-issledovatelskogo instituta i Biologicheskoi stantsii pri Permskom gosudarstvennom univerisitete 3, 1, 215–228.

Tauson, A., 1924b.The effect of hydrogen ions and antagonistic action of cations on survival of animals. Izvestiya nauchno-issledovatelskogo instituta i Biologicheskoi stantsii pri Permskom gosudarstvennom univerisitete 3, 6, 53–68.

Tauson, A., 1931. Die Wirkung der aeusseren Bedingungen auf die Veraenderung des Geschlechts und auf die Entwicklung von *Daphnia pulex* De Geer. W. Roux Arch. Entwicklungsmech. Organism. 123B, 80–131.

Taylor, D.J., Hebert, P.D.N., Colbourne, J.K., 1996. Phylogenetics and evolution of the *Daphnia longispina* group (Crustacea) based on 12 S rDNA sequence and allozyme variation. Mol. Phylogenet. Evol. 5, 495–510.

Taylor, N.S., Weber, R.J., White, T.A., Viant, M.R., 2010. Discriminating between different acute chemical toxicities via changes in the daphnid metabolome. Toxicol. Sci. 118, 307–317.

Tazima, I., Hoshi, T., Inoue, S., 1975. Histological distribution of ferruginous compounds in *Daphnia*. Zool. Mag. 84 (1), 67–70.

Teles, L.O., Alpendurada, N.F., Vieira, M.N., 2005. Interaction between cadmium and mercury accumulation by *Daphnia magna* Straus (Crustacea, Branchiopoda). Limnetica. 24 (3–4), 211–224.

Ten Berge, W.F., 1978. Breeding *Daphnia magna*. Hydrobiologia. 59 (2), 121–123.

Terao, A., Tanaka, T., 1928a. Population growth of the water-flea, *Moina macrocopa* Straus. Proc. Imperial Acad. (of Japan). IV, 550–552.

Terao, A., Tanaka, T., 1928b. Influence of temperature upon the rate of reproduction in the water-flea *Moina macrocopa* Straus. Proc. Imperial Acad. (of Japan). IV, 553–555.

Terra, N.R., Feiden, I.R., Lucheta, F., Gonçalves, S.P., Gularte, J.S., 2010. Bioassay using *Daphnia magna* Straus, 1820 to evaluate the sediments of Caí River (Rio Grande do Sul, Brazil). Acta limnologicas brasiliensia. 22 (4), 1–10.

Tessier, A.J., Consolatti, N.L., 1991. Resource quantity and offspring quality in *Daphnia*. Ecology. 72 (2), 468–478.

Tessier, A.J., Goulden, C.E., 1982. Estimating food limitation in cladoceran populations. Limnol. Oceanogr. 27 (4), 707–717.

Tessier, A.J., Henry, L.L., Goulden, C.E., Durand, M.W., 1983. Starvation in *Daphnia*: energy reserves and reproductive allocation. Limnol. Oceanogr. 28 (4), 667–676.

Tezuka, Y., 1971. Feeding of *Daphnia* on planktonic bacteria. Japanese J. Ecol. 21 (3–4), 127–146.

Theopold, U., Schmidt, O., Soderhall, K., Dushay, M.S., 2004. Coagulation in arthropods: defence, wound closure and healing. Trends Immunol. 25, 289–294.

Threlkeld, S.T., 1976. Starvation and the size structure of zooplankton communities. Freshw. Biol. 6 (6), 489–496.

Threlkeld, S.T., 1979. Estimating cladoceran birth rates: The importance of egg mortality and the egg distribution. Limnol. Oceanogr. 24 (4), 601–612.

Tipney, H., Hunter, L., 2010. Knowledge-driven approaches to genome-scale analysis. In: Alterowitz, G., Marco, R. (Eds.), Knowledge-Based Bioinformatics: from Analysis to Interpretation. John Wiley & Sons, New York, pp. 34–65.

Tokishita, S., Kimura, S., Mandokoro, Y., Kato, K., Shiga, Y., Takahashi, Y., et al., 2006. Tissue-specific expression of a bHLH-PAS protein homologous to ARNT during the development of crustacean *Daphnia magna*. Gene. 376, 231–239.

Tokishita, S., Shiga, Y., Kimura, S., Ohta, T., Kobayashi, M., Hanazato, T., et al., 1997. Cloning and analysis of a cDNA encoding a two-domain hemoglobin chain from the water flea *Daphnia magna*. Gene. 189, 73–78.

Tollinger, J., 1909. Der Verdauungstrakt von Lynceus intermedius (G.O. Sars). Ann. Biol. Lacustre. 3 (3), 271–299.

Tollrian, R., 1990. Predator induced helmet formation in *Daphnia cucullata* (Sars). Archiv. für Hydrobiol. 119 (2), 191–196.

Tollrian, R., 1993. Neckteeth formation in *Daphnia pulex* as an example of continuous phenotypic plasticity: morphological effects of *Chaoborus* kairomone concentration and their quantification. J. Plankton Res. 13 (11), 1309–1318.

Tollrian, R., 1994. Fish-kairomone induced morphological changes in *Daphnia lumholtzi* (Sars). Archiv. für Hydrobiol. 130 (1), 69–75.

Tollrian, R., 1995. Predator-induced morphological defences: costs, life-history shifts, and maternal effects in *Daphnia pulex*. Ecology. 76 (6), 1691–1705.

Tollrian, R., Dodson, S.I., 1999. Inducible defenses in Cladocera: constraints, costs, and multipredator environments. Ecol. Evol. Inducib. Def. pp. 177–202.

Tomasik, P., Magadza, C.H.D., Mhizha, S., Chrume, A., 1995. The metal-metal interactions in biological systems. Part III. *Daphnia magna*. Water Air Soil Poll. 82, 695–711.

Tonapi, G.T., Varghese, G., Shinde, S., 1984. Preliminary observations on the cardiophysiology of *Moinodaphnia macleayi* (King). Curr. Sci. 53 (6), 340–341.

Tonkopiy, V.D., Kosmachev, A.B., Zagrebin, A.O., 1994a. Comparative pharmacological analysis of m-cholinoreceptors of *Daphnia* and rats. Byulleten eksperimentalnoi biologii i meditsiny. 5, 497–499.

Tonkopiy, V.D., Kutsenko, S.A., Zagrebin, A.O., Shertsneva, L.A., 1993. Theoretical and experimental substantiation of elaboration of new methods of bioindication of anticholinesterase compounds in aquatic environment. Zhurnal ekologicheskoi khimii. 2, 133–137.

Tonkopiy, V.D., Zagrebin, A.O., Kosmachev, A.B., Kutsenko, S.A., Sherstneva, L.A., 1994b. Pharmacological analysis of cholinergic system in *Daphnia magna*. Zhurnal evolutsionnoi biokhimii i fiziologii. 30 (3), 386–391.

Tonolli, V., 1947. Il ritmo cardiaco della *Daphnia pulex* Degeer. Mem Ist. Ital. idrobiol. 3.

Trentini, M., 1980. Chromosome numbers of nine species of Daphniidae (Crustacea, Cladocera). Genetica. 54, 221–223.

Trofimova, M.G., 1979. Some pathomorphological alternations in *Daphnia magna* under the banlen influence. Trudy GosNIORKh. 144, 102–109.

Trubetskova, I., Lampert, W., 1995. Egg size and eggs mass of *Daphnia magna*: response to food availability. Hydrobiologia. 307, 139–145.

Tsitsilas, A., Barry, M.J., 2002. Optimising hatching conditions for Australian *Daphnia ephippia*. Verhandl. Internat. Vereining. Limnol. 28 (3), 1460–1463.

Tsui, M.T.K., Wang, W.-X., 2004a. Maternal transfer efficiency and transgenerational toxicity of methylmercury in *Daphnia magna*. Environ. Toxicol. Chem. 23 (6), 1504–1511.

Tsui, M.T.K., Wang, W.-X., 2004b. Intake and elimination rates of inorganic mercury and methylmercury in *Daphnia magna*. Environ. Sci. Technol. 38 (3), 808–816.

Tsui, M.T.K., Wang, W.-X., 2005. Multigenerational acclimation of *Daphnia magna* to mercury: relationships between biokinetics and toxicity. Environ. Toxicol. Chem. 24 (11), 2927–2933.

Tsui, M.T.K., Wang, W.-X., Chu, L.M., 2005. Influence of glyphosate and its formulation (Roundup®) on toxicity and bioavailability of metals to *Ceriodaphnia dubia*. Environ. Poll. 138 (1), 59–68.

Tsvetkov, I.L., Popov, A.P., Konichev, A.S., 2012. Activity of acid phosphatase and deoxyribonuclease in hydrobionts under effect of different toxic substances of aquatic medium. Gidrobiologicheskii Zhurnal. 48 (1), 93–108.

Tsvetkov, I.L., Zarubin, S.L., Urvantseva, G.A., Konichev, A.S., Filippovich, Yu.B., 1997. Acid phosphatase of hydrobionts as an enzymatic marker of biochemical adaptation to the action of toxicants. Izvesriya Akademii nauk, Boil. 5, 539–545.

Tufts, D.M., Revsbech, I.G., Cheviron, Z.A., Weber, R.E., Fago, A., Storz, J.F., 2013. Phenotypic plasticity in blood-oxygen transport in highland and lowland deer mice. J. Exp. Biol. 216, 1167–1173.

Tumanov, A.A., Postnov, I.E., Osipova, N.I., Zimin, A.B., 1988. Chemical-biological method of estimation of phenol content in water by means of *Daphnia*. Metody biotestirovaniya vod. Chernogolovka.104–106.

Tumanov, A.A., Postnov, I.E., Osipova, N.I., Zimin, A.B., Filimonova, I.A., 1988. Express method of determination of organophosphate pesticides in water. Metody biotestirovaniya vod.106–111, Chernogolovka.

Untersteiner, H., Kahapka, J., Kaiser, H., 2003. Behavioural response of the cladoceran *Daphnia magna* Straus to sublethal copper stress—validation by image analysis. Aquat. Toxicol. 65 (4), 435–442.

Urabe, J., 1991. Effect of food concentration on the carbon balance of *Bosmina longirostris* (Crustacea: Cladocera). Freshw. Biol. 26, 57–68.

Urabe, J., Watanabe, Y., 1991. Effect of food concentration on the assimilation and production efficiency of *Daphnia galeata* G.O. Sars (Crustacea: Cladocera). Funct. Ecol. 5 (5), 635–641.

Urabe, J., Clasen, J., Sterner, R.W., 1997. Phosphorus limitation of *Daphnia* growth: Is it real? Limnol. Oceanogr. 42 (6), 1436–1443.

Usanov, D.A., Shishkin, G.G., Skripal, A.V., Usanov, A.D., Panasenko, V.I., 2001a. *Daphnias* as bioindicator of electromagnetic influences on the aquatic environment. Peterburgskii Zhurnal elektroniki. 4, 38–42.

Usanov, D.A., Shishkin, G.G., Usanov, A.D., Skripal, A.V., 2001b. Effect of external alternating magnetic fields on heart rate of a freshwater crustacean *Daphnia*. Biomeditsinskie tekhnologii i radioelektronika. 8, 57–61.

Usanov, D.A., Shishkin, G.G., Usanov, A.D., Skripal, A.V., 2003. Effect of alternating magnetic fields of low intensity on heart rate of *Daphnia*. Biomeditsinskie tekhnologii i radioelektronika. 3, 58–62.

Usuki, I., Yamaguchi, K., 1979. Physiological significance of the extracellular haemoglobin on the survival and lactic acid production of *Daphnia magna* Straus in low oxygen conditions. Sci. Rep. of the Niigata Univ., ser. D (Biol.). 16, 5–12.

Uttieri, M., Mazzochi, M.G., Nihongi, A., D'Alcala, R., Strickler, J.R., 2004. Lagranian description of zooplankton swimming trajectories. J. Plankton Res. 26 (1), 99–105.

Uttieri, M., Zanbianchi, E., Strickler, R., Mazzocchi, M.G., 2005. Fractal characterization of three-dimensional zooplankton swimming trajectories. Ecol. Modell. 185 (1), 51–63.

Vadstein, O., Brekke, O., Andersen, T., Olsen, Y., 1995. Estimation of phosphorus release rates from natural zooplankton communities feeding on plankton algae and bacteria. Limnol. Oceanogr. 40 (2), 250–262.

Vaidya, A.R., Gurenlian, L., Brady, L., Romero, N.E., Kohn, R., 2009. Cardioactive effect of diphenhyramine and curcumin in *Daphnia magna*. Impulse: The Premier Journal for Undergraduate Publications in the Neurosciences.1–12.

Vale, P.F., Little, T.J., 2012. Fecundity compensation and tolerance to a sterilizing pathogen in *Daphnia*. J. Evol. Biol. 25, 1888–1896.

Van Damme, K., Dumont, H.J., 2006. Sex in a cyclical parthenogen: mating behavior of *Chydorus sphaericus* (Crustacea; Branchiopoda; Anomopoda. Freshw. Biol. 51, 2334–2346.

Van Damme, K., Dumont, H.J., 2007. Limb morphology of the carnivorous anomopods *Anchistropus emarginatus* Sars, 1862 and *Pseudochydorus globosus* (Baird, 1843) (Crustacea: Branchiopoda: Anomopoda). Ann. Limnol. 43 (4), 271–284.

Van Damme, K., Dumont, H.J., 2010. Cladocera of the Lençóis Maranhenses (NE - Brazil): faunal composition and a reappraisal of Sars' Method. Brazilian J. Biol. 70 (3), 755–779.

Van Damme, K., Shiel, R.J., Dumont, H.J., 2007. *Notothrix halsei* gen.n., sp.n., representative of a new family of freshwater cladocerans (Branchiopoda, Anomopoda) from SW Australia, with a discussion of ancestral traits and a preliminary molecular phylogeny of the order. Zool. Scr. 36, 465–487.

Van de Bund, W.J., 2000. Chironomid faecal pellets as a potential food source for the chydorid cladoceran *Chydorus piger*. In: Late 20th Century on Chironomidae, Shaker: Aschen 455–441.

Van Donk Ellen, Hessen D.O., 1993. Grazing resistance in nutrient-stressed phytoplankton. Oecologia. 93, 508–511.

Vandegehuchte, M.B., De Coninck, D., Vandenbrouck, T., De Coen, W.M., Janssen, C.R., 2010. Gene transcription profiles, global DNA methylation and potential transgenerational epigenetic effects related to Zn exposure history in *Daphnia magna*. Environ. Pollut. 158, 3323–3329.

Vandegehuchte, M.B., Lemiere, F., Vanhaecke, L., Vanden Berghe, W., Janssen, C.R., 2010. Direct and transgenerational impact on *Daphnia magna* of chemicals with a known effect on DNA methylation. Comp. Biochem. Physiol. C Toxicol. Pharmacol. 151, 278–285.

Vandegehuchte, M.B., De Coninck, D., Vandenbrouck, T., De Coen, W.M., Janssen, C.R., 2010a. Gene transcription profiles, global DNA methylation and potential transgenerational epigenetic effects related to Zn exposure history in *Daphnia magna*. Environ. Pollut. 158, 3323–3329.

Vandegehuchte, M.B., Lemiere, F., Vanhaecke, L., Vanden Berghe, W., Janssen, C.R., 2010b. Direct and transgenerational impact on *Daphnia magna* of chemicals with a known effect on DNA methylation. Comp. Biochem. Physiol. C Toxicol. Pharmacol. 151, 278–285.

Vandekerkhove, J., Declerck, S., Brendonk, L., et al., 2005. Hatching of cladoceran resting eggs: temperature and photoperiod. Freshw. Biol. 50, 96–104.

Vandenbrouck, T., Soetaert, A., van der Ven, Blust, R., De Coen, W., 2009. Nickel and binary metal mixture responses in *Daphnia magna*: molecular fingerprints and (sub)organismal effects. Aquat. Toxicol. 92, 18–29.

Vanvlasselaer, E., De Meester, L., 2008. Trapped in diapause? Can we resurrect old dormant eggs of the water flea *Daphnia magna*? VIIIth International Symposium on Cladocera. Aguascalientes, Mexico. 21–15 October 2008, p. 53.

Vareschi, E., Wübben, D., 2001. Vertical migration of *Daphnia pulex* in response to UV radiation. Verhandl. Internat. Verenig. Limnol. 27 (6), 3349–3353.

Vasama, A., Kankaala, P., 1990. Carbon-length regressions of planktonic crustaceans in Lake Ala-Kitka (NE-Finland). Aqua Fennica. 20 (1), 95–102.

Vasilyeva, G.L., Smirnov, N.N., 1975. A contribution to the ecology of Chydoridae (Cladocera) of Lake Baikal. Zoologicheskii Zhurnal. 54 (9), 1293–1305.

Vatovec, S., Timet, D., 1952. About the influence of vertebrates sera on the heart beat of *Daphnia*. Bull. Internation. de l'Academie Jugoslave des Sciences et des Beaux Arts. 7, 93–101.

Vatovec, S., Timet, D., 1955. About the influence of vertebrates sera on the heart beat of *Daphnia*. II. Differences in serum activity of various species. Bull. Internation. de l'Academie Jugoslave des Sciences et des Beaux Arts. 14, 49–58.

Vega, M.P., Pizarro, R.A., 2000. Oxidative stress and defence mechanisms of the freshwater cladoceran *Daphnia longispina* exposed to UV radiation. J. Photochem. Photobiol. B: Biol. 54, 121–125.

Velu, C.S., Czeczuga, B., Munuswamy, N., 2003. Carotenoprotein complexes in entomostracan crustaceans (*Streblocephalus dichotomus* and *Moina micrura*). Comp. Biochem. Physiol. B. 135 (1), 35–42.

Verbitsky, V.B., 1990. Food value of *Bosmina longirostris* for fish larvae. II. Amino acid composition of protein. Biologiya vnutrennikh vod. Informatsionnyi bulleten. 88, 69–73.

Verbitsky, V.B., Verbitskaya, T.I., 2002. The upper thermal tolerances of the *Bosmina longirostris* (O.F. Müller) (Crustacea: Cladocera) and its dependence on environmental temperatures. Biologiya vnutrennikh vod. 2, 55–59.

Verbitsky, V.B., Verbitskaya, T.I., 2009. Influence of arrhythmic step's changes of temperature on dynamics of abundance of *Simocephalus vetulus* (O.F. Müller, 1776) experimental populations. Doklady Akademii Nauk. 429 (6), 838–841.

Verbitsky, V.B., Verbitskaya, T.I., 2011. Final thermal preference of parthenogenetic females of *Daphnia magna* Straus (Crustacea: Cladocera) acclimated to various temperatures. Izvestiya Rossiiskoi Akademii Nauk. Biol. 5, 576–589.

Vereshchagin Yu, G., 1912. Changes of the cyclicity of Cladocera in dependence on the geographical latitude of the locality. Protokoly zasedanii Obshchestva estestvoispytatelei pri Imperatorskom Varshavskom universitete. 23, 241–247.

Vergilino, R., Belzile, C., Dufresne, F., 2009. Genome size evolution and polyploidy in the *Daphnia pulex* complex (Cladocera: Daphniidae). Biol. J. Linn. Soc. 97, 68–79.

Verkhovskaya, I., 1940. Influence of polarized light on phototaxis of some organisms. Bulleten Moskovskpogo Obshchestva Ispytatelei Prirody, Biol. 49. (3–4), 101–113.

Vesel, S., Ondruka, V., Kua, K., Patoka, J., 2006. Freshwater microcrustacean *Daphnia magna* Straus as an early screen model to compare toxicity of acetylcholinesterase inhibitors. J. Appl. Biomed. 4, 105–110.

Vijverberg, J., Frank, T.H., 1976. The chemical composition and energy content of copepods and cladocerans in relation to their size. Freshw. Biol. 6 (4), 333–345.

Vijverberg, J., Kalf, D.F., Boersma, M., 1996. Decrease in *Daphnia* egg viability at elevated pH. Limnol. Oceanogr. 41 (4), 789–794.

Villaroel, M.J., Ferrando, M.D., Andreau, E., 1998. Toxic anorexia as a sensitive endpoint in *Daphnia magna*. J. Environm. Sci. Health, B. 2, 151–160.

Villegas-Navarro, A., Rosas-L, E., Reyes, J.L., 2003. The heart of *Daphnia magna*: effect of four cardioactive drugs. Comp. Biochem. Physiol. Part C136, 127–134.

Vinogradov, S.N., Hoogewijs, D., Bailly, X., Arredondo-Peter, R., Guertin, M., Gough, J., et al., 2005. Three globin lineages belonging to two structural classes in genomes from the three kingdoms of life. Proc. Natl. Acad. Sci. U.S.A. 102, 11385–11389.

Vinokurova, N.V., 1997. Influence of exogenous biologically active compounds on reproduction of *Daphnia magna* in different seasons. In: Teoreticheskie i prikladnye aspekty biologii. Kaliningrad, 69–72.

Volkova, N.S., 1980. The composition of fatty acids in lipids of certain fresh-water diatoms. Gidrobiologicheskii Zhurnal. 16 (5), 79–82.

Vollenweider, R.A., Ravera, O., 1958. Preliminary observations on the oxygen uptake by some freshwater zooplankters. Verhandl. Internat. Vereinig. Limnol. 13, 369–380.

von Dehn, M., 1930. Untersuchungen über die Verdauung bei Daphniden. Zeitschrift vergleich. Physiol. 13 (2), 334–358.

von Dehn, M., 1948. Experimentelle Untersuchungen ueber den Generationswehcsel der Cladoceren. II. Cytologische Untersuchungen bei Moina rectirostris. Chromosoma. 3 (3), 167–194.

von Dehn, M., 1949. Experimentelle untersuchungen über den generationswechsel der Cladoceren. Chromosoma. 3, 167–193.

von Dehn, M., 1950. Der Einfluss der Hefefettstoffe auf die Fortpflanzungart von *Moina rectirostris*. Die Naturwissenschaften. 37 (18), 429–430.

von Dehn, M., 1955. Die Geschlechtsbestimmung der Daphniden. Die Bedeutung der Fettstoffem untersucht an *Moina rectirostris* L. Zool. Jahrb., Abt. 1. 65 (3), 334–356.

von Elert, E., 2000. Fatty acids released by *Daphnia*: do they induce colony formation in *Scenedesmus acutus*? Verhandl. Internat. Vereinig. Limnol. 27 (4), 2128–2131.

von Elert, E., 2004. Food quality constraints in *Daphnia*: interspecific differences in the response to the absence of a long chain polyunsaturated fatty acid in the food source. Hydrobiologia. 526, 187–196.

von Elert, E., Franck, A., 1999. Colony formation in *Scenedesmus*: grazer-mediated release and chemical features of the infochemical. J. Plankton Res. 21 (4), 789–804.

von Elert, E., Agraval, M.K., Gebauer, C., Jaensch, H., Bauer, U., Zitt, A., 2004. Protease activity in gut of *Daphnia magna*: evidence for trypsin and chymotrypsin enzymes. Comp. Biochem. Physiol. B. 137 (3), 287–296.

von Elert, E., Martrin-Creuzburg, D., Le Coz, J.R., 2002. Absence of sterols constrains carbon transfer between cyanobacteria and a freshwater herbivore (*Daphnia galeata*). Proceed. R. Soc. Lond. B. 270, 1209–1214.

Vonk, M.J., 1960. Digestion and metabolism. In: Waterman, T.H. (Ed.), The Physiology of Crustacea, vol. 1. Academic press, New York and London, pp. 291–316.

Voronin, V.N., 1986. Microsporidia of crustaceans. In: Mikrosporidii. Protozoologiya 10, 137–168.

Voronin, V.N., Makrushin, A.V., 2006. The artificial infection of the cladoceran *Moina macrocopa* (Crustacea, Phyllopoda) with the microsporidia *Gurleya* sp. (Microsporidia, Gurleyidae). Parazitologiya. 40 (5), 462–471.

Vorozhun, I.M., 2001. Experimental and model investigations of feeding rate of *Moina macrocopa* Straus (Crustacea, Cladocera) under different temperature and feeding conditions, Ecological Studies, Hazards and Solutions, 5. MAX Press, Moscow, p. 55.

Vrba, J., Machaček, J., 1994. Release of dissolved extracellular β-*N*-acetylglucosaminidase during crustacean molting. Limnol. Oceanogr. 39 (3), 712–716.

Vrede, T., Andersen, T., Hessen, D.O., 1998. Anatomical and biochemical distribution of phosphorus in *Daphnia magna* and *Daphnia galeata*. Verhandl. Internat. Vereinig. Limnologie. 26 (4), 1962.

Vrede, T., Andersen, T., Hessen, D.O., 1999. Phosphorus distribution in three crustacean zooplankton species. Limnol. Oceanogr. 44 (1), 225–229.

Vrede, T., Persson, J., Aronsen, G., 2002. The influence of food quality (P : C ratio) on RNA:DNA ratio and somatic growth rate of *Daphnia*. Limnol. Oceanogr. 47 (2), 487–494.

Wacker, A., von Elert, E., 2001. Polyunsaturated fatty acids: evidence for non-substitutable biochemical resources in *Daphnia galeata*. Ecology. 82 (9), 2507–2520.

Wærvågen, S.B., Rukke, N.A., Hessen, D.O., 2002. Calcium content of crustacean zooplankton and its potential role in species distribution. Freshw. Biol. 47, 1866–1878.

Walker, M., Read, L., Jackson, K., McFarlane, P., 1998. The effects of oestrogenic substances on two components of freshwater ecosystems. J. Biol. Educat. 32 (3), 226–229.

Walseng, B., Schartau, A.-K., 2011. Cladocerans as indicators. Abstract book, IXth International symposium on Cladocera, October 2–8, 2011. Verbania, Italy, p. 40.

Wang, D., Hu, J., Forthaus, B.E, Wang, J., 2011. Synergistic toxic effect of nano-A2O3 and As(V) on *Ceriodaphnia dubia*. Environ. Poll. 159, 3002–3008.

Wang, W.-X., Guan, R., 2010. Subcellular distribution of zinc in *Daphnia magna* and implication for toxicity. Environ. Toxicol. Chem. 29 (8), 1841–1848.

Wang, Z., Yan, C., Hyne, R.V., 2010. Effects of dietary cadmium exposure on saltwater cladoceran *Moina mongolica* Daday: implications in water quality criteria. Environ. Toxicol. Chem. 29 (2), 365–372.

Wang, Z.-S., Kong, H.-N., Wu, D.-Y., 2007. Reproductive toxicity of dietary copper to a saltwater cladoceran, *Moina mongolica* Daday. Environ. Toxicol. Chem. 26 (1), 126–131.

Waterman, T., 1960. Interaction of polarized light and turbidity in the orientation of *Daphnia* and *Mysidium*. Zeitschr. vergleich. Physiol. 43 (2), 149–172.

Waterman, T., 1961. Light sensitivity and vision. In: Waterman, T.H. (Ed.), The Physiol. of Crustacea, vol. 2. Academic Press, New York and London, pp. 1–64.

Waterman, T.H., (Ed.), The Physiology of Crustacea, vol. 1, 1960, p. 670.; vol. 2, 1961, p. 681.

Watson, F.L., Puttmann-Holgado, R., Thomas, F., Lamar, D.L., Hughes, M., Kondo, M., et al., 2005. Extensive diversity of Ig-superfamily proteins in the immune system of insects. Science. 309, 1874–1878.

Weber, A.K., Pirow, R., 2009. Physiological responses of *Daphnia pulex* to acid stress. BMC Physiol. 9 (9), 1–25.

Weber, R.E., Vinogradov, S.N., 2001. Nonvertebrate hemoglobins: functions and molecular adaptations. Physiol. Rev. 81, 569–628.

Weber, R.E., Ostojic, H., Fago, A., Dewilde, S., Van Hauwaert, M.L., Moens, L., et al., 2002. Novel mechanism for high-altitude adaptation in hemoglobin of the Andean frog *Telmatobius peruvianus*. Am. J. Physiol. Regul. Integr. Comp. Physiol. 283, R1052–60.

Wechsler, A.S., St. John, P.A., 1960. The spectral sensitivity of *Daphnia magna* and its variance with age. Anat. Rec. 137 (3), 400.

Weers, P.M.M., Siewertsen, K., Gulati, R.D., 1997. Is the fatty acid composition of *Daphnia galeata* determined by the fatty acid composition of the ingested diet? Freshw. Biol. 38 (3), 731–738.

Weider, L.J., Lampert, W., 1985. Differential response of *Daphnia* genotypes to oxygen stress: respiration rates, hemoglobin content and low-oxygen tolerance. Oecologia. 65, 487–491.

Weider, L.J., Hebert, P.D.N., 1987. Ecological and physiological differentiation among low-arctic clones of *Daphnia pulex*. Ecology. 68, 188–198.

Weider, L.J., Frisch, D., Hebert, P.D., 2010. Long-term changes in metapopulation genetic structure: a quarter-century retrospective study on low-arctic rock pool *Daphnia*. Proc. Biol. Sci. 277, 139–146.

Weismann, A., 1874. Über Bau und Lebenserscheinungen von Leptodora hyalina. Engelmann, p. 70.

Weismann, A., 1877a. Beiträge zur Naturgeschichte der Daphniden. Theil II, III und IV. Zeitschrift Wissenschaftl. Zoologie. 28, 93–254.

Weismann, A., 1877b. Über die Fortpflanzung der Daphnoiden. In: Versammlung Deutscher Naturforscher und Artze in München vom 17–22 September 1877, 172–178.

Weismann, A., 1878. Über die Schmuckfarbungen der Daphnoiden. Zeitschrift Wissenschaftl. Zoologie. 30, 123–178.

Weismann, A., 1880. Beiträge zur Naturgeschichte der Daphnoiden. VI. Zeitschrift wissenschaftl. Zoologie. 33, 55–270.

Weiss, L.C., Tollrian, R., Herbert, Z., Laforsch, C., 2012. Morphology of the *Daphnia* nervous system: a comparative study on *Daphnia pulex*, *Daphnia lumholtzi*, and *Daphnia longicephala*. J. Morphol., 1–14.

Wen, Y.H., Vezina, A., Peters, R.H., 1994. Phosphorus fluxes in limnetic cladocerans: coupling of allometry and compared analysis. Can. J. Fresh. Aquat. Sci. 51, 1055–1064.

Wendel, T., Jüttner, F., 1997. Excretion of heptadecene-1 into lake water by swarms of *Polyphemus pediculus* (Crustacea). Freshw. Biol. 38 (1), 203–207.

Wenger, R.H., Stiehl, D.P., Camenisch, G., 2005. Integration of oxygen signaling at the consensus HRE. Sci. STKE.306, re12.

Wenqing, C., Yuanshao, L., Donghui, G., Liting, Y., 1999. Karyologic observation on *Moina mongolica*, a Cladocera. J. Ocean. Taiwan Str., 01.

West-Eberhard, M.J., 2005. Developmental plasticity and the origin of species differences. Proc. Natl. Acad. Sci. U.S.A. 102 (Suppl. 1), 6543–6549.

Westlake, D.F., 1980. Primary production. In: The functioning of freshwater ecosystems. Cambridge University, Cambridge, etc. International Biological Programme 22, 141–246.

Whitehead, A., 2012. Comparative genomics in ecological physiology: toward a more nuanced understanding of acclimation and adaptation. J. Exp. Biol. 215, 884–891.

Wiggins, P.R., Frappell, P.B., 2000. The influence of haemoglobin on behavioural thermoregulation and oxygen consumption in *Daphnia carinata*. Physiol. Biochem. Zool. 71, 153–160.

Wiggins, P.R., Frappell, P.B., 2002. Behavioural thermoregulation in *Daphnia carinata* from different depths of a natural water body: influence of environmental oxygen levels and temperature. Comp. Biochem. Physiol. A Mol. Integr. Physiol. 133, 771–780.

Williamson, C.E., Neale, P.J., Grad, G., DeLange, H.J., Hargraves, B.R., 2001. Beneficial and detrimental effects of UV on aquatic organisms: implications of spectral variation. Ecol. Appl. 11 (2), 1843–1857.

Wiltshire, K.H., Lampert, W., 1999. Urea excretion by *Daphnia*: a colony-inducing factor in *Scenedesmus*? Limnol. Oceanogr. 44 (8), 894–1903.

Winberg, G.G., 1950. Intensity of metabolism and size of Crustacea. Zhurnal obshchei biologii XI. 5, 367–380.

Winberg, G.G., 1933. On action of ions on biological objects. Biologicheskii Zhurnal. 2 (6), 519–527.

Winder, M., Spaak, P., 2002. Effects of natural UV radiation on the life history of alpine *Daphnia*. Verhandl. Internat. Vereinig. Limnol. 28 (1), 355–359.

Wingstrand, K.G., 1978. Comparative spermatology of the Crustacea Entomostraca. 1. Subclass Branchiopoda. Det Kongelige Danske Videnskabernes Selskab Biologiske Skrifter 22, 1, 1–66, plates 1–20.

Winner, R.W., Gauss, J.D., 1986. Relationship between aquatic toxicity and bioaccumulation of copper, cadmium and zinc as affected by water hardness and humic acid. Aquat. Toxicol. 8 (3), 149–161.

Witschi, E., 1934. On determinative cleavage and yolk formation in the harpactid copepod *Tisbe furcata* (Baird). Biol. Bull. 67, 335–340.

Wojewodzic, M., Kyle, M., Hessen, D., Elser, J.J., Andersen, T., 2011. Joint effect of phosphorus limitation and temperature on alkaline phosphatase activity and somatic growth in *Daphnia magna*. Abstract book, IXth International symposium on Cladocera, October 2–8, 2011. Verbania, Italy, p. 41.

Wojnarovich, E., 1958. A *Daphnia magna* Straus produktiesbiologiai jellem zese. Magy. Tud. Akad. Tihanyi biol. Kutatoint evk. 25, 83–97.

Wojnarovich, E., 1959. Produktionsbiologische Characteresierung der *Daphnia magna* Straus. Acta biol. Acad. Scent. Hung. 9 (3), 229–244.

Wolf, G., Scheunders, P., Selens, M., 1998. Evaluation of the swimming activity of *Daphnia magna* by image analysis after administration of sublethal cadmium concentrations. Comp. Biochem. Physiol. 120 (1), 99–105.

Wolinska, J., Bittner, D., Ebert, D., Spaak, P., 2006. The coexistence of hybrid and parental *Daphnia*: the role of parasites. Proc. R. Soc. Lond., B, Biol. Sci. 273, 1977–1983.

Woltereck, R., 1909. Weitere experimentelle untersuchungen uber art veranderung, speziell uber das wesen quantitativer artunterschiede bei Daphniden. Verh. D Zool. Ges. Leipzig. 19, 110–173.

Wolvecamp, H.P., Waterman, T.H., 1960. Respiration. In: Waterman, T.H. (Ed.), The Physiology of Crustacea, vol. 1. Academic press, New York and London, pp. 35–100.

Wood, T.R., 1932. Resting eggs that fail to rest. Am. Nat. 66 (704), 277–281.

Wood, T.R., Banta, A.M., 1937. Hatchability of *Daphnia* and *Moina* sexual eggs without drying. Internationale Revue der ges. Hydrobiol. und Hydrogr. 35 (1/3), 229–242.

Wright, D., O'Brian, W.J., Vinjard, G.L., 1980. Adaptive value of vertical migration: a simulation model argument for the prediction hypothesis. In: Kerfoot, W.C. (Ed.), Evolution and Ecology of Zooplankton Communities. University press of New England, Hanover, etc, pp. 138–147.

Wu, W.Z., Li, W., Xu, Y., Wang, J.W., 2001. Long-term toxic impact of 2,3,7,8-tetrachlorobenzo-p-dioxin on the reproduction, sex differentiation, and development of different life stages of *Gobiocypris rarus* and *Daphnia magna*. Ecotoxicol. Environ. Saf. 48 (3), 293–300.

Wu, Y., Huang, J., Yuan, L., 2005. Phototactic behavior of *Daphnia carinata* as an indicator of chromium biotoxicity. Chinese J. Appl. Ecol. 16 (1), 171–174.

Wübben, D., Vareschi, E., 2001. UV-radiation sensitivity of freshwater and marine zooplankton tested in a sunshine simulator. Verhandl. Internat. Vereining. Limnol. 27 (6), 3354–3357.

Wübben, D.L., Vareschi, E., Hawes, J., Bodeker, G., 2001. Differential UVB-sensitivities of five New Zealand freshwater zooplankton species. New Zealand J. Marine Freshw. Res. 35 (3), 635 645.

Xie, F., Koziar, S.A., Lampi, M.A., Dixon, D.G., Norwood, W.P., Borgmann, U., et al., 2006. Assessment of the toxicity of mixtures of copper, 9,10-phenanthrenequinone, and phenanthrene to *Daphnia magna*: evidence for a reactive oxygen mechanism. Environ. Toxicol. Chem. 25 (2), 613–622.

Xu, L., Han, B.P., Van Damme, K., Vierstraete, A., Vanfleteren, J.R., Dumont, H.J., 2011. Biogeography and evolution of the Holarctic zooplankton genus *Leptodora* (Crustacea: Branchiopoda: Haplopoda). J. Biogeogr. 38, 359–370.

Xu, S., Hebert, P.D.N., Kotov, A.A., Cristescu, 2009. The noncosmopolitanism paradigm of freshwater zooplankton: insights from the global phyleography of the predatory *Polyphemus pediculus* (Linnaeus, 1761) (Crustacea, Onychopoda). Mol. Ecol. 18, 5161–5179.

Yakovlev, A.M., Zainulgabidinov, E. R, Kuznetsova, T.V., 2000. Experimental estimation of interaction of heavy metals: *Daphnia magna* Strauss. AQUATEK-2000, 4th International congress Water: Ecology and Technology", Moscow, May 30—June 2, p. 205.

Yalynskaya, N.S., 1961. Cultivation of Entomostraca (Cladocera) under northern conditions. Dopovidi ta povidoml. Lvivsk. Univers. 9 (2), 94–96.

Yamamura, M., Hatakeyama, S., Suzuki, K.T., 1983. Cadmium intake and induction of cadmium-binding protein in the waterflea (*Moina macrocopa*). Bull. Environm. Cont. Toxicol. 30 (3), 303–308.

Yan, N.D., Mackie, G.L., Grauds, P., 1990. Control of cadmium levels in *Holopedium gibberum* (Crustacea, Cladocera) in Canadian shield lakes. Environ. Toxicol. Chem. 9, 895–908.

Yasumoto, K., Nishigami, A., Yasumoto, M., Kasai, F., Okada, Y., Kusumi, T., et al., 2005. Aliphatic sulfates released from *Daphnia* induce morphological defence of phytoplankton isolation and synthesis of kairomones. Tetrahedron Lett. 46 (28), 4765–4767.

Yatsenko, A., 1928. Activation of the culture of *Chydorus ovalis* with ethyl alcohol. Trudy Zvenogorodskoi gidrofiziologicheskoi stantsii,112–123.

Yerba, L., Hernández-León, S., 2004. Aminoacyl-tRNA synthetases activity as a growth index in zooplankton. J. Plankton Res. 26 (3), 351–356.

Young, S., 1974. Directional differences in the colour sensitivity of *Daphnia magna*. J. Exp. Biol. 61 (1), 261–267.

Young, S., Downing, A.C., 1976. The receptive fields of *Daphnia ommatidia*. J. Exp. Biol. 64, 185–202.

Young, S., Taylor, V.A., 1990. Swimming tracks in swarms of two cladoceran species. Anim. Behav. 39, 10–16.

Yu, R.-A., Wang, W.-X., 2002. Trace metal assimilation and release budget in *Daphnia magna*. Limnol. Oceanogr. 47 (2), 495–504.

Yukleevskikh Yu, M., 2000. Change of the content of some chemical elements in *Daphnia magna* in laboratory culture. In: Gorizonty fiziko-khimicheskoi biologii. Pushchino, 95–96.

Yurista, P.M., 1999. A model for temperature correction of size-specific respiration in Bythotrephes cederstroemi and *Daphnia middendorffiana*. J. Plankton Res. 21 (4), 721–734.

Zadereev, E.S., Gubanov, V.G., Egorov, I.V., 1998. The influence of food concentration and "water of aggregation" on the reproductive parameters of *Moina macrocopa* (Crustacea: Cladocera) females. Zhurnal obshchei biologii. 59 (1), 45–57.

Zaffagnini, F., 1964. Osservationi comparative sull'accrescimento e la reproduczione in tre specie di Cladoceri. Mem. Ist. Ital. Idrobiol. 17, 103–114.

Zaffagnini, F., (1987) Reproduction in *Daphnia*. In: "*Daphnia*" R.H. Peters and R. deBernardi (eds.), Mem. Ist. Ital. Idrobiol. 45, 245–284.

Zaffagnini, F., Sabelli, B., 1972. Karyologic observations on the maturation of the summer and winter eggs of *Daphnia pulex* and *Daphnia middendorffiana*. Chromosoma. 36, 193–203.

Zaffagnini, F., Trentini, M., 1975. Osservazioni sul corredo chromomico di alcuni Dafnidi (Crostacei, Cladoceri). Atti Acad. Sci. Ist. Bolgna Rc. C.l. Sci. Fish. 13, 63–70.

Zaffagnini, F., Zeni, C., 1986. Considerations on some cytological and ultrastructural observations on fat cells of *Daphnia* (Crustacea, Cladocera). Boll. Zool. 53, 33–39.

Zahid, Z.R., Khalaf, A.N., Al-Haidaby, M.T., 1980. A light microscopical study of the nervous system of *Simocephalus vetulus* Schødler (Cladocera) with special reference to neurosecretion. Crustaceana. 38 (1), 2–18.

Zaret, R.E., 1980. The animal in its viscous environment. In: Evolution and ecology of zooplankton communities. Special symposium, vol. 3. American Soc. of Limnology and Oceanography. Kerfoot W.C., (Ed.), University press of New England, Hanover, etc., 3–9.

Zaret, R.E., Kerfoot, W.C., 1980. The shape and swimming technique of *Bosmina longirostris*. Limnol. Oceanogr. 25 (1), 126–133.

Zarubin, S.L., Tsvetkov, I.L., Urvanteva, G.A., Shlyapnikova, N.A., 1997. Investigation of induced resistance of *Daphnia magna* to prolonged influence of chromium chloride. In: Biologisheskie issledonaniya v Yarolavskom universitete, 70–71.

Zeis, B., Becher, B., Goldmann, T., Clark, R., Vollmer, E., Bolke, B., et al., 2003. Differential haemoglobin gene expression in the crustacean *Daphnia magna* exposed to different oxygen partial pressures. Biol. Chem. 384, 1133–1145.

Zeis, B., Becher, B., Lamkemeyer, T., Rolf, S., Pirow, R., Paul, R.J., 2003. The processes of hypoxic induction of *Daphnia magna* hemoglobin: subunit composition and functional properties. Comp. Biochem. Physiol. B. 134 (2), 243–252.

Zeis, B., D. Becker, P. Gerke, M. Koch, R.J. Paul (2013). Hypoxia-inducible hemoglobins of *Daphnia pulex* and their role in the response to acute and chronic temperature increase. Biochim. Biophys. Acta, Prot. Proteom., in press 10.1016/j.bbapap.2013.01.036

Zeis, B., Lamkemeyer, T., Paul, R.J., 2004. Molecular adaptation of *Daphnia magna* hemoglobin. Micron. 35 (1–2), 47–49.

Zeis, B., Lamkemeyer, T., Paul, R.J., Nunes, F., Schwerin, S., Koch, M., et al., 2009. Acclimatory responses of the *Daphnia pulex* proteome to environmental changes. I. Chronic exposure to hypoxia affects the oxygen transport system and carbohydrate metabolism. BMC Physiol. 9, 7.

Zeis, B., Maurer, J., Pinkhaus, O., Bongartz, E., Paul, R.J., 2004. A swimming activity assay shows that the thermal tolerance of *Daphnia magna* is influenced by temperature acclimation. Can. J. Zool. 82, 1605–1613.

Zeis, B., Pinkhaus, O., Bredebusch, I., Paul, R.J., 2005. Oxygen preference of *Daphnia magna* is influenced by Po2 acclimation and biotic interactions. Physiol. Biochem. Zool. 78, 384–393.

Zeiss, F.P., 1963. Effects of population densities on zooplankton respiration rates. Limnol. Oceanograp. 8 (1), 110–115.

Zellmer, I.D., Arts, M., 2005.Field and laboratory evidence of damage in *Daphnia*'s intestine due to exposure to sub-lethal ultraviolet radiation. 7th Internation Symposium on Cladocera, Abstract book, p. 51.

Zellmer, I.D., Arts, M., Abele, D, Humbeak, K., 2004. Evidence of sublethal damage in *Daphnia* (Cladocera) during exposure to UV radiation in subarctic ponds. Arctic, Antarctic, Alpine Res. 36 (3), 370–377.

Zellmer, I.D., Arts, M., Suster, V., 2006. Food chain effects of sublethal ultraviolet radiation on subarctic *Daphnia pulex*—a field and laboratory study. Archiv für Hydrobiolgie. 167 (1–4), 515–531.

Zeman, F.A., Gilbin, R., Alonso, F., Lecomte-Pradines, C., Garnier-Laplace, J., 2008. Effect of waterborne uranium on survival, growth, reproduction, and physiological processes of the freshwater cladoceran *Daphnia magna*. Aquatic Toxi. 86 (3), 370–378.

Zeni, C., Zaffagnini, F., 1988. Occurrence of innervation of labral glands of *Daphnia obtusa* (Crustacea, Cladocera). J. Morphol. 198 (1), 43–48.

Zhang, P. Zhao, W., Sun, J-X., 2009. The karyotype and evolutionary relationship among 5 species of water fleas *Moina*. J. Dalian Fish. Univ. 02.

Zhang, Q., Keller, R., Dircksen, H., 1997. Crustacean hyperglycaemic hormone in the nervous system of the primitive crustacean species *Daphnia magna* and *Artemia salina* (Crustacea: Branchiopoda). Cell tissue Res. 287 (3), 565–576.

Zhao, C.-M., Fan, W.-H., Wang, W.-X., 2009. Aqueous and dietary copper uptake and elimination in *Daphnia magna* determined by the 67Cu radiotracer. Environ. Toxicol. Chem. 28 (11), 2360–2366.

Zhao, W., Huo, Y.-Z., Gao, J., 2006. Analysis and appraisement of nutrient compositions for *Daphniopsis tibetana* Sars. J. Fish. Sci. China. 13 (3), 446–451.

Zhao, W., Zhang, P., Huo, Y-X., Wang, H-L., 2004. Study of karyotype on *Daphniopsis tibetana* Sars (Cladocera: Daphniidae). J. Dalian Fish. Univ., 03.

Zhao, X.-F., Liu, G.-X., Hu, Z.-Y., 2006. Physico-chemical properties of alkaline phosphatases released by a planktonic crustacean, *Daphnia magna* (Cladocera). Crustaceana. 79 (6), 677–689.

Zhukova N.A. 1955. Cyclomorphosis in daphnias. Leningradskiy gos. Pedagogicheskiy institut, Uchenye zapiski VII, 85–148. Zoological Record, 10 (Crustacea). 1864–2007. Vols. 1–143.

Zou, E., 1997. Effects of sublethal exposure to zinc chloride on the reproduction of the water flea, *Moina irrasa* (Cladocera). Bull. Environm. Contam. Toxicol. 58 (3), 437–441.

Zou, E., Fingerman, M., 1997. Effects of estrogenic xenobiotics on molting of the water flea, *Daphnia magna*. Ecotoxicol. Environ. Saf. 38 (3), 281–285.

Zschokke, F., 1911. Die Tiefseefauna der Seen Mitteleuropas. Leipzig, 246 pp.

Index of Latin Names of Cladocera

Note: Page numbers followed by "*f*", and "*t*" refers to figures and tables respectively.

E

G

H

I

K

L

M

O

P

Index of Chemical Substances

Note: Page numbers followed by "*f*" and "*t*" indicates figure and tables repectively.

Subject Index

Note: Page numbers followed by "f" indicate figure and those followed by "t" indicate table.